UNIT OPERATIONS IN ENVIRONMENTAL ENGINEERING

UNIT OPERATIONS IN ENVIRONMENTAL ENGINEERING

Edited by

Robert Noyes

np

NOYES PUBLICATIONS

Park Ridge, New Jersey, U.S.A.

Library of Congress Catalog Card Number: 94-1324
ISBN: 0-8155-1343-7
Printed in the United States

Published in the United States of America by
Noyes Publications
Mill Road, Park Ridge, New Jersey 07656

10 9 8 7 6 5 4 3 2 1

Library of Congress Cataloging-in-Publications Data

Unit operations in environmental engineering / edited by Robert Noyes
p. cm.
Includes bibliographical references and index.
ISBN 0-8155-1343-7
1. Sanitary engineering. I. Noyes, Robert.
TD145.U46 1994
628--dc20 94-1324
CIP

About the Author

Robert Noyes is a chemical engineer, who after working in industry for a number of years and also serving as a consultant, has been President of Noyes Data Corporation and Noyes Publications, scientific and technical publishers, for many years. Although involved in writing, editing, and publishing in numerous technical disciplines, his primary concern in the past fifteen years has been environmental technology.

He is the author of three previous books: *Handbook of Pollution Control Processes; Handbook of Leak, Spill and Accidental Release Prevention Techniques;* and *Pollution Prevention Technology Handbook.*

Preface

This book discusses the practical aspects of environmental technology organized into eight chapters relating to unit operations as follows:

1. Biological Technology
2. Chemical Technology
3. Containment and Barrier Technology
4. Immobilization Technology
5. Membrane Technology
6. Physical Technology
7. Radiation and Electrical Technology
8. Thermal Destruction Technology

Traditional technologies have been included, as well as those that can be considered innovative, and emerging. The traditional approaches have been the most successful, as contractors are careful about bidding on some of the newer technologies. However, as regulatory requirements increase, markets will open for the innovative and emerging processes. There will be increasing pressure to break down complex waste streams, with each subsequent stream demanding separate treatment. In addition, a number of technologies have been developed by combining processes directly, or in a treatment train, and these developments are expected to assume increasing importance. However, such concerns as uncertainties due to liability, regulatory approval, price competition, and client approval have limited the application of some of these newer technologies.

The purpose of this book is not to describe commercial processes, but a number of proprietary processes are included in order to present additional information. The inclusion or exclusion of any commercial process bears no relationship to its comparative effectiveness in any environmental control situation. Also, keep in mind that various governmental and commercial organizations may use different nomenclature and terminology, for the same technology.

All regulations mentioned in this book are on the Federal level. State regulations could require different treatment standards. Although regulations are mentioned throughout the book, no legal or technical advice is intended, and anyone investigating a hazardous waste problem should obtain appropriate legal and technical guidance.

Condensed Contents

Notice

Contents

1

Biological Technology

Biological treatment can typically be divided into two classifications: aerobic biological treatment and anaerobic biological treatment. Aerobic biological treatment takes place in the presence of oxygen, while anaerobic biological treatment is an oxygen–devoid process.

Aerobic biological treatment is a treatment technology applicable to wastewaters containing biodegradable organic constituents and some nonmetallic inorganic constituents including sulfides and cyanides. Four of the most common aerobic biological treatment processes are (a) activated sludge, (b) aerated lagoon, (c) trickling filter, and (d) rotating biological contactor (RBC). The activated sludge and aerated lagoon processes are suspended–growth processes in which microorganisms are maintained in suspension with the liquid. The trickling filter and the RBC are attached–growth processes in which microorganisms grow on any inert medium such as rocks, slag, or specifically designed ceramic or plastic materials.

Anaerobic digestion is best suited to wastes with a moderate to high pH, nonhalogenated hydrocarbons, moderate to low organic loadings, and low to zero biological oxygen demand. The waste should also be in a semisolid or sludge form. Anaerobic biological treatment typically takes place in an anaerobic digester. There are also anaerobic bioreclamation processes.

Another route is anoxic decomposition in which the microorganisms utilize the nitrate ion; this process is termed denitrification. They can be combined in the design of the aeration basin. Advantages of anoxic selector systems include: (1) control of filamentous organisms, (2) nitrogen removal, and (3) reduced alkalinity consumption.

Biological processes are used in wastewater treatment, and in bioreclamation of contaminated sites. Also biofiltration is used extensively in Europe to treat gases with low concentrations of VOCs, and for odor control.

As regulation of inorganic pollutants in wastewater affluents has developed, so has biological technology. It is now possible to design a single–sludge process that relies on sequential anaerobic, anoxic, and aerobic environments for removal of phosphorus, nitrogen, and BOD.

The sequential exposure of microorganisms to anaerobic (no nitrite or nitrate and no dissolved oxygen), anoxic (nitrite or nitrate present and no dissolved oxygen), and aerobic environments that facilitate biological nitrogen and phosphorus removal shows promise as an effective process for removing a broad range of chemicals.

Biological treatment is a destruction process relying primarily on oxidative or reductive mechanisms. Enzymatic activity can effect lysis, e.g., hydrolysis or dehalogenation. Further, biological activity can result in pH changes in the waste stream which may require adjustment by chemical means. The use of biological treatment processes is directed toward accomplishing (1) destruction of organic contaminants, (2) oxidation of organic chemicals whereby the organic chemicals are broken down into smaller constituents, and (3) dehalogenation of organic chemicals by cleaving a chlorine atom(s) or other halogens from a compound. Biological treatment processes have certain advantages over other common treatment technologies, namely, the organic contaminants to be destroyed are used and transformed by bacteria or other organisms as a source of food. These processes can be employed in soils, slurries, or waters (ponds, groundwater, etc.) to aid in the remediation of a site.

Biological processes can be used on a broad class of biodegradable organic contaminants. Some compounds, called refractiles, are persistent compounds which are not readily biodegradable. It should be noted that very high concentrations as well as very low concentrations of organic contaminants are difficult for biological processes to treat.

The degradation potential for organic compounds is in the following order of increasing difficulty:

1. Straight-chain compounds
2. Aromatic compounds
3. Chlorinated straight-chain compounds
4. Chlorinated aromatic compounds

In treating wastes containing halogenated organic compounds, the effectiveness of the system in removing these compounds is dependent primarily on the microorganisms that are present. Most of these compounds are man-made and, therefore, natural microorganisms did not originally have the ability to degrade these compounds. Through exposure to the compounds, however, some groups of microorganisms have developed enzymatic systems resistant to the toxic compounds and with a capability to degrade them at a slow rate. Treatment systems that are innoculated with these types of microorganisms may have the ability to remove these compounds. There has been specific interest in white rot fungus which is capable of degrading the complex lignin molecule, and therefore has been investigated for degrading other complex molecules.

The two greatest weaknesses of biological systems are seen as:

1. Inability to adapt easily to changes in input, and
2. Their need for operator intervention to control the process.

The two greatest strengths of biological systems are:

1. Low to moderate downtime compared to other technologies, and
2. Cost effectiveness.

Enzymes are simple or combined proteins acting as specific catalysts. The most characteristic property of enzymes is the striking specificity which manifests itself in their catalytic action. A given enzyme catalyzes only the reaction of an individual group of compounds, a single type of compound, or even a certain kind of bond in a given compound. For enzyme treatment of wastes to proceed, the enzymes must first be separated from living cells, segregated into reliably pure and active forms, maintained within the appropriate environment, "immobilized" by attaching them to solid media to avoid their being washed away, and coupled with the specific phase of the chemical reaction they each address and the appropriate cofactors (i.e., energy sources) necessary for their function.

It is doubtful that use of enzyme treatment at a facility receiving varying hazardous waste mixtures would be economically feasible. Enzymes cannot adapt or acclimate to varying substrates. Also, they are inhibited by presence of soluble inorganics and highly sensitive to pH and temperature levels. It is doubtful the enzyme treatment could compare economically with biological treatment, even on relatively pure waste streams, unless resource recovery were the principal objective.

1.1 WASTE CHARACTERISTICS AFFECTING PERFORMANCE (WCAPs)

In determining whether biological treatment will achieve the same level of performance on an untested waste that it achieved on a previously tested waste and whether performance levels can be transferred, EPA examines the following waste characteristics: (a) the ratio of the biological oxygen demand to the total organic carbon content, (b) the concentration of surfactants, and (c) the concentration of toxic constituents and waste characteristics.

Ratio of Biological Oxygen Demand to Total Organic Carbon Content: Because organic constituents in the waste effectively serve as a food supply for the microorganisms, it is necessary that a significant percentage be biodegradable. If they are not, it will be difficult for the microorganisms to successfully acclimate to the waste and achieve effective treatment. The percentage of biodegradable organics can be estimated by the ratio of the biological oxygen demand (BOD) to the total organic carbon (TOC) content. Since the biological oxygen demand is a measure of the amount of oxygen required for complete microbial oxidation of biodegradable organics, the BOD analysis is mostly relevant to aerobic biological treatment. (In anaerobic biological treatment, BOD is one of the main restrictive characteristics in that BOD must be relatively low or zero.) If the ratio of BOD to TOC in an untested waste is significantly lower than that in the tested waste, the system may not achieve the same performance and other, more applicable technologies may need to be considered for treatment of the untested waste.

Concentration of Surfactants: Surfactants can affect biological treatment performance by forming a film on organic constituents, thereby establishing a barrier to effective biodegradation. If the concentration of surfactants in an untested waste is significantly higher than that in the tested waste, the system may not achieve the same performance and other, more applicable technologies may need to be considered for treatment of the untested waste.

Concentration of Toxic Constituents and Waste Characteristics: A number of constituents and waste characteristics have been identified as potentially toxic to microorganisms. Specific toxic concentrations have not been determined for most of these constituents and waste characteristics. The constituents and waste characteristics found to be potentially toxic to microorganisms include metals and oil and grease, ammonia, and phenols. High concentrations of dissolved solids are treated more effectively by anaerobic treatment than by aerobic treatment. If the concentration of toxic constituents and waste characteristics in an untested waste is significantly higher than that in the tested waste, the system may not achieve the same performance and other, more applicable technologies may need to be considered for treatment of the untested waste.

1.2 DESIGN AND OPERATING PARAMETERS

In assessing the effectiveness of the design and operation of a biological treatment system, EPA examines the following parameters: (a) the amount of nutrients, (b) the concentration of dissolved oxygen, (c) the food-to-microorganism ratio, (d) the pH, (e) the biological treatment temperature, (f) the mean cell resistance time, (g) the hydraulic loading rate, (h) the settling time, and (i) the degree of mixing.

For many hazardous organic constituents, analytical methods are not available or the constituent cannot be analyzed in the waste matrix. Therefore, it would normally be impossible to measure the effectiveness of the biological treatment system. In these cases one should identify measurable parameters or constituents that would act as surrogates in order to verify treatment.

For organic constituents, each compound contains a measurable amount of total organic carbon (TOC). Removal of TOC in the biological treatment system indicates removal of organic constituents. Hence, TOC analysis is likely to be an adequate surrogate analysis where the specific organic constituent cannot be measured.

However, TOC analysis may not be able to adequately detect treatment of specific organics in matrices that are heavily organic-laden (i.e., the TOC analysis may not be sensitive enough to detect changes at the milligrams per liter (mg/ℓ) level in matrices where total organic concentrations are hundreds or thousands of milligrams per liter). In these cases other surrogate parameters should be sought. For example, if a specific analyzable constituent is expected to be treated as well as the unanalyzable constituent, the analyzable constituent concentration should be monitored as a surrogate.

Amount of Nutrients: Nutrient addition is important in controlling the growth of microorganisms because an insufficient amount of nutrients results in poor microbial growth with poor biodegradation of organic constituents. The principal inorganic nutrients used are nitrogen and phosphorus. In addition, trace amounts of potassium, calcium, sulfur, magnesium, iron, and manganese are also used for optimum microbial growth. The percent distribution of nitrogen and phosphorus added to microorganisms varies with the age of the organism and the particular environmental conditions. The total amount of nutrients required depends on the net mass of organisms produced.

Concentration of Dissolved Oxygen: A sufficient concentration of dissolved oxygen (DO) is necessary to metabolize and degrade dissolved organic constituents in aerobic treatment. The DO concentration is controlled by adjusting the aeration rate. The aeration rate must be adequate to provide a sufficient DO concentration to satisfy the BOD requirements of the waste, as well as to provide adequate mixing to keep the microbial population in suspension (for activated sludge and aerated lagoon processes). The reverse is true for anaerobic treatment, in that DO must be absent for anaerobic treatment to occur.

Food-to-Microorganism Ratio: The food-to-microorganism (F/M) ratio, which applies only to activated sludge systems, is a measure of the amount of biomass available to metabolize the influent organic loading to the aeration unit. This ratio can be determined by dividing the influent BOD concentration by the concentration of active biomass, also referred to as the mixed liquor volatile suspended solids (MLVSS). The F/M ratio is controlled by adjusting the wastewater feed rate or the sludge recycle rate. If the F/M ratio is too high, too few microorganisms will be available to degrade the organics.

pH: Generally, neutral or slightly alkaline pH favors microorganism growth. The optimum range for most microorganisms used in biological treatment systems is between 6 and 8. Treatment effectiveness is generally insensitive to changes within this range. However, pH values outside the range can lower treatment performance.

Biological Treatment Temperature: Microbial growth can occur under a wide range of temperatures, although the majority of the microbial species used in aerobic biological treatment processes are active between 20° and 35°C (68° to 95°F). For anaerobic systems, the temperature is typically between 30° and 70°C (86° to 158°F). The rate of biochemical reactions in cells increases with temperature up to a maximum above which the rate of activity declines and microorganisms either die off or become less active.

Mean Cell Residence Time: In activated sludge, aerated lagoon, and anaerobic digestion systems, the mean cell residence time (MCRT) or sludge age is the length of time organisms are retained in the unit before being drawn off as waste sludge. By controlling the MCRT, the growth phase of the microbial population can be controlled. The MCRT must be long enough to allow the organisms in the unit to reproduce. The MCRT is determined by dividing the total active microbial mass in the unit (MLVSS) by the total quantity of microbial mass withdrawn daily (wasted).

Hydraulic Loading Rate: The hydraulic loading rate determines the length of time the organic constituents are in contact with the microorganisms and, hence, the extent of biodegradation that occurs. In trickling filters, the hydraulic loading rate also determines the shear velocities on the microbial layer. Excessively high hydraulic loading rates may wash away the microbial layer faster than it can grow back. However, the hydraulic loading rate must be high enough to keep the microbes moist and to remove dead or dying microbes that have lost their ability to cling to the filter media. For all aerobic biological treatment processes, the hydraulic loading rate is controlled by adjusting the wastewater feed rate. In addition, for RBCs, the hydraulic loading rate can be controlled by changing the disk speed or adjusting the submersion depth.

Settling Time: Adequate settling time must be provided to separate the biological solids from the mixed liquor. Activated sludge systems cannot function properly if the solids cannot be effectively separated and a portion returned to the aeration basin.

Degree of Mixing: Mixing provides greater uniformity of the wastewater feed in the equalization basin to reduce variations that may cause process upsets of the microorganisms and diminish treatment efficiency. For activated sludge and aerated lagoon systems, sufficient aeration in the aeration unit provides mixing to ensure adequate contact between the microorganisms and the organic constituents in the wastewater. The quantifiable degree of mixing is a complex assessment that includes, among other factors, the amount of energy supplied, the length of time the material is mixed, and the related turbulence effects of the specific size and shape of the mixing unit. The degree of mixing is beyond the scope of simple measurement.

1.3 AEROBIC PROCESSES

The basic principle of operation for aerobic biological treatment processes is that living, oxygen-requiring microorganisms decompose organic and nonmetallic inorganics constituents into carbon dioxide, water, nitrates, sulfates, simpler low-molecular-weight organic by-products, and cellular biomass. Wastes that can be degraded by a given species or genus of organisms may be very limited. A mixture of organisms may be required to achieve effective treatment, especially for wastes containing mixtures of organic compounds. Nutrients such as nitrogen and phosphorus are also required to aid in the biodegradation process.

Aerobic biological treatment of wastewaters containing organic constituents results in the net accumulation of a biomass of expired microorganisms consisting mainly of cell protein. However, the cellular biomass or sludges may also contain entrained constituents from the wastewater or partially degraded constituents. These sludges must be periodically removed ("wasted") to maintain proper operation of the aerobic biological treatment system.

In aerobic respiration, organic molecules are oxidized to carbon dioxide (CO_2) and water and other end products using molecular oxygen as the terminal electron acceptor. Oxygen may also be incorporated into intermediate products of microbial catabolism through the action of oxidase enzymes, making them more susceptible to further biodegradation. Microorganisms metabolize hydrocarbons by anaerobic respiration in the absence of molecular oxygen using inorganic substrates as terminal electron acceptors. Naturally occurring aerobic bacteria can decompose organic materials of both natural and synthetic origin to harmless or stable forms or both by mineralizing them to CO_2 and water.

Some anthropogenic compounds can appear relatively refractory to biodegradation by naturally occurring microbial populations because of the interactions of environmental influences, lack of solubility, absence of required enzymes, nutrients, or other factors. However, the use of properly selected or engineered microbial populations, maintained under environmental conditions most conducive to their metabolic activity can be an important means of biologically transforming or degrading these otherwise refractory wastes.

All microorganisms require adequate levels of inorganic and organic nutrients, growth factors (vitamins, magnesium, copper, manganese, sulfur, potassium, etc.), water, oxygen, carbon dioxide and sufficient biological space for survival and growth. One or more of these factors are usually in limited supply. In addition, various microbial competitors adversely affect each other through the struggle for these limiting factors.

Other factors which can influence microbial biodegradation rates include microbial inhibition by chemicals in the waste to be treated, the number and physiological state of the organisms as a function of available nutrients, the seasonal state of microbial development, predators, pH and temperature. Interactions between these and other potential factors can cause wide variations in degradation kinetics.

For these and other reasons, aerobic biodegradation is usually carried out in processes in which all or many of the requisite environmental conditions can be controlled. Such processes include conventional activated sludge processes as well as modifications such as sequencing batch reactors, and aerobic-attached growth biological processes such as rotating biological contactors and trickling filters. Recent developments with genetically engineered bacteria have been reported to be effective for biological treatment of specific hazardous wastes which are relatively uniform in composition.

Aerobic processes are used to treat aqueous wastes contaminated with low levels (e.g., BOD less than ≈10,000 mg/ℓ) of nonhalogenated organic and/or certain halogenated organics. The treatment requires consistent, stable operating conditions.

Proper aeration is essential for operation of aerobic biological waste treatment systems. The three types of aerators are diffused air, submerged turbine, and surface aerators.

Another innovation for aerobic treatment is the inclusion of anoxic and anaerobic zones within a basically aerobic system. In a properly designed system, the anoxic zones allow denitrification to occur. Establishment of a preliminary anaerobic zone has been used to enhance phosphorus removal.

Bioreclamation is used to treat contaminated areas through the use of aerobic microbial degradation. It may be accomplished by in–situ treatment using injection/extraction wells or an excavation process. Extracted waters, leachates or wastes are oxygenated, nutrients and bacteria are added and the liquids reinjected in the ground. Bacteria then can degrade wastes still in the soil. The treatment has been successfully applied to biodegradable nonhalogenated organics to reduce the contaminated levels in soils and groundwater.

For in–situ treatment, limitations would include site geology and hydrogeology which could restrict pumping and extraction of hazardous wastes, along with reinjection and recirculation. Ideal soil conditions are those with neutral pH, high permeability and a moisture content of 50 to 75%.

1.4 ANAEROBIC PROCESSES

Anaerobic digestion is a biological treatment process for the degradation of simple organics in an air–free environment. Anaerobic organisms utilize part of the carbon substrate for cell growth, and convert the other part to methane and carbon dioxide gas. Since anaerobic decomposition results in less efficient utilization of organic substrate for cell growth than aerobic decomposition the process has the advantage of low waste solids generation. The reduction of sulfates results in the production of hydrogen sulfide.

Metals concentrations are tolerated in the system as long as they are insoluble. Only soluble metal species are toxic to microbial activity, and generally the heavier the metal ion the greater the inhibition. At the neutral pH levels under which the anaerobic process must operate, most of the metals are precipitated as sulfides.

All anaerobic biological treatment processes achieve the reduction of organic matter, in an oxygen–free environment, to methane and carbon dioxide. This is accomplished by using cultures of bacteria which include facultative and obligate anaerobes. Anaerobic bacterial systems include hydrolytic bacteria (catabolize saccharides, proteins, lipids); hydrogen producing acetogenic bacteria (catabolize the products of hydrolytic bacteria, e.g., fatty acids and neutral end products); homolactic bacteria (catabolize multicarbon compounds to acetic acid); and methanogenic bacteria (metabolize acetic and higher fatty acids to methane and carbon dioxide). The strict anaerobes require totally oxygen–free environments and oxidation reduction potential of less than –0.2 V.

Microorganisms in this group are commonly referred to as methanogenic consortia and are found in anaerobic sediments or sewage sludge digesters. These organisms play an important role in reductive dehalogenation reactions, nitrosamine degradation, reduction of epoxides to olefins, reduction of nitro groups and ring fission of aromatic structures. Available anaerobic treatment concepts are based on such approaches as the classic well–mixed system, the two–stage systems and the fixed bed. In the well–mixed digester system a single vessel is used to contain the wastes being treated and all bacteria must function in that common environment.

Such systems typically require long retention times and the balance between acetogenic and methanogenic populations is easily upset. In the two–stage approach, two vessels are used to maintain separate environments, one optimized for the acetogenic bacteria (pH 5.0), and the other optimized for the methanogenic bacteria (pH 7.0). Retention times are significantly lower and upsets are uncommon in this approach.

The fixed bed approach (for single or two–staged systems) utilizes an inert solid media to which the bacteria attach themselves and low solids wastes are pumped through columns of such bacteria–rich media. Use of such supported cultures allows reduced retention times since bacterial loss through washout is minimized. Organic degradation efficiencies can be quite high. A number of proprietary engineered processes based on these types of systems are actively being marketed, each with distinct features but all utilizing the fundamental anaerobic conversion to methane and carbon dioxide.

This process is used to treat aqueous wastes with low to moderate levels of organics. Anaerobic digestion can handle certain halogenated organics better than aerobic treatment. Stable, consistent operating conditions must be maintained. Anaerobic degradation can take place in native soils but when used as a controlled treatment process, an air–tight reactor is required. Since methane and CO_2 gases are formed, it is common to vent the gases or burn them in flare systems. However, volatile hazardous materials could readily escape via such gas–venting or flare systems. Thus, controlled off–gas burning could be required. Alternatively, depending on the nature of the waste to be treated, the off–gas could be used as a source of energy.

An important point to note is that cost savings in anaerobic systems are found primarily in the costs associated with sludge handling and disposal. This is an area that is becoming particularly restrictive on a regulatory basis, with the resulting rising costs to meet ultimate disposal requirements. The substantial reductions in residual sludge realized by the use of the anaerobic process is a key advantage.

The most recent and significant advances in anaerobic digestion are related to the technology's ability to accommodate relatively high rates of organic loading. Companies are also interested in using anaerobic digestion for the biodestruction of organic materials that are not removed in conventional aerobic treatment. As applications of anaerobic technology to various process streams increase, more successes are inevitable, resulting in industries that are more economically competitive because of their more judicious use of natural resources.

Companies adopting anaerobic digestion fall into the following categories: companies scheduled to expand their production–process capacity and whose treatment facilities are already at capacity; companies that discharge to publicly owned treatment works whose surcharge for treatment has increased substantially; companies that have relatively monotonous, highly concentrated organic waste streams contributing a major portion of total waste load; and companies in areas where extremely high land costs make conventional aerobic digestion too expensive.

Anaerobic processes are being investigated for bioremediation of hazardous waste sites.

The recent success of several research groups investigating the anaerobic transformation of polychlorinated biphenyls (PCBs) has prompted investigation of similar transformations of dioxins and dibenzofurans. In the case of the PCBs, transformation of highly chlorinated biphenyl mixtures such as Arochlor 1260 to lower–chlorite–content biphenyls has been observed. The well–advanced aerobic transformation of PCBs has been judged limited to biphenyls with seven and fewer attached chlorines. The strengths of the aerobic and anaerobic processes are seen as complementary. One direction of potential treatment development has been to treat lower–chlorinated mixtures by aerobic means and to use a sequential treatment of aerobic and anaerobic processing for the higher–chlorinated substances. This same sequence of treatment may be useful for the treatment of dioxins and dibenzofurans

A number of anaerobic digestion plants are in operation, or being built in Europe to handle animal slurry, municipal waste, and industrial waste. The generation of methane–rich biogas is an important factor in these processes.

1.5 BIOLOGICAL WASTE TREATMENT

Application of microbial degradation and removal of undesirable constituents in industrial and municipal wastes is not a new concept. It is a commonly used process for general wastewater treatment activities and has been for many years. As the awareness of chemical contamination of the environment, much research on biological degradation of toxic chemicals has occurred. Among the range of treatment technologies, biological degradation ranks among the most effective. Its management application is enhanced by the potential to apply biological treatment in sequence with other chemical and thermal processes.

Another area for incorporating biological technologies in hazardous waste management activities is the recovery of reusable materials. Metals recovery is a very important area for biological applications. Because these inorganic elements cannot not be destroyed, an important goal is to recover and recycle metals, to the maximum extent possible. Using microbial–based technologies to recover inorganics may become an increasingly important area for further development.

Dilute hazardous wastes can pose a problem for cost–effective management. Many chemical and thermal treatment processes are only cost–effective on concentrated waste constituents. Biological treatment processes that concentrate these mixtures of dilute toxic constituents can be an important component in a sequential management strategy. For example, biological treatment may be used to concentrate organic constituents, followed by thermal treatment of the biological residue.

In the past, the primary function of biological treatment systems has not been to remove toxic organic pollutants, but to remove the conventional, easily biodegradable organic compounds. Recently, however, the biodegradation of toxic compounds has received increasing attention due to its potentially lower cost versus other treatment technologies.

Bioaugmentation is another interesting process in which selected microorganisms are added in order to enhance the microbial efficiency of a treatment facility.

Mixed substrate systems are often encountered in pharmaceutical, food, wastewater processes and chemical manufacturing industries. In wastewater treatment systems, a number of organic compounds are present at the same time. In these cases it is inevitable that the toxic, or inhibitory substrates will be found in mixtures with nontoxic, or conventional wastes. In the presence of alternative carbon sources, a number of possible substrate interactions can occur. Extensive studies on biodegradation of single components have been conducted. However, there is insufficient information on the performance of biological treatment facilities for the removal of a specific chemical from wastewater, consisting of a mixture of organic pollutants. There is a strong need for extensive studies of multisubstrate systems. A broad data base will help to understand the interaction and removal rates of organic compounds in mixtures.

In biological treatment plants, the substrate removal pattern in a multisubstrate system may include simultaneous, preferential, or sequential utilization. The diauxic growth in *Escherichia coli* suggests that the very presence of a particular substrate in a wastewater stream might prevent an organism from acclimatizing to another substrate until the first one has been completely metabolized. The blockage of metabolism of one compound by another may lead to preferential or sequential substrate removal from a multisubstrate environment.

The mechanism of substrate utilization by a bacterial cell can be generally described as a sequence of three complex processes: contact of a cell with the molecule of a substrate; transport of the molecule into the cell; and formation of the substrate intermediate. On the basis of this general mechanism, it is possible to classify various types of substrates into three main groups: (a) single components substrates, which are directly transportable; (b) multicomponent substrates, which are represented by a mixture of several single substrates; (c) complex substrates, which have to be changed externally prior to transportation into the cell.

As a treatment generality, EPA's experience has revealed that biological treatment usually is technically more effective and less costly than physical–chemical treatment for control of organic pollutants in wastewaters, especially those waters with complex mixtures of wastes. In some cases, a combination of biological and physical–chemical treatment may be the optimum treatment combination.

1.5.1 Activated Biofilter

A biofilm first stage is followed by an activated–sludge second stage and a settler. Sludge is recycled to the biofilm stage and to the activated sludge tank. This variation combines biofilm and suspended–growth characteristics.

1.5.2 Activated Sludge

The activated sludge process is a typical type of suspended growth biological treatment system and probably the most widely used biological process for the treatment of organic and industrial waste waters. However, it can only treat aqueous organic wastestreams having less than 1% suspended solid content, and can not tolerate shock loadings of concentrated organics. Therefore, the wastestream entering this process will usually have passed through a pretreatment process which includes a clarifier (primary clarifier) and an equalization basin. The primary clarifier is used for removal of grit, oily and fatty material and gross solid material, while the equalization basin is used to dampen wastewater flow variations and to provide more uniform organic loading to the activated sludge system.

Activated sludge processes are used to treat municipal and industrial wastes since they are versatile, flexible, and can be used to produce an effluent of desired quality by varying process parameters. The process was so-named because it produces an active mass of microorganisms capable of aerobically stabilizing a waste. Many versions of the basic process exist but all are fundamentally similar.

The term activated sludge is applied to both the process and to the biological solids in the treatment unit. The mixed liquor suspended solids or activated sludge contains a variety of heterotrophic microorganisms such as bacteria, protozoa, fungi, and larger microorganisms. The predominance of a particular microbial species depends upon the waste that is treated and the way in which the process is operated.

The activated sludge process is currently the most widely used biological treatment process. This is partly the result of the fact that recirculation of the biomass, which is an integral part of the process, allows microorganisms to adapt to changes in wastewater composition with a relatively short acclimation time and also allows a greater degree of control over the acclimated bacterial population.

An activated sludge system consists of an equalization basin, a settling tank, an aeration basin, a clarifier, and a sludge recycle line. Wastewater is homogenized in an equalization basin to reduce variations in the feed, which may cause process upsets of the microorganisms and diminish treatment efficiency. Settleable solids are then removed in a settling tank.

Next, wastewater enters an aeration basin, where an aerobic bacterial population is maintained in suspension and oxygen, as well as nutrients, are provided. The contents of the basin are referred to as the mixed liquor. Oxygen is supplied to the aeration basin by mechanical or diffused aeration, which also aids in keeping the microbial population in suspension. The mixed liquor is continuously discharged from the aeration basin into a clarifier, where the biomass is separated from the treated wastewater. A portion of the biomass is recycled to the aeration basin to maintain an optimum concentration of acclimated microorganisms in the aeration basin. The remainder of the separated biomass is discharged or "wasted." The biomass may be further dewatered on sludge drying beds or by sludge filtration to disposal. The clarified effluent is discharged.

The recycled biomass is referred to as activated sludge. The term "activated" is used because the biomass contains living and acclimated microorganisms that metabolize and assimilate organic material at a higher rate when returned to the aeration basin. This occurs because of the low food-to-microorganism ratio in the sludge from the clarifier.

For the treatment of industrial wastewater, supplemental nutrient sources are often needed to provide sufficient nitrogen and phosphorus. In most cases, nitrogen is added as ammonia and phosphorus as phosphoric acid. A proper pH range (6 to 8) and a sufficient dissolved oxygen concentration (a minimum of 1 to 2 mg/ℓ) must also be maintained in the aeration basin to support a healthy and active system.

The aeration basin hydraulic retention time (HRT) and sludge residence time (SRT) are important operational factors. HRT is defined as the ratio of the volume of aeration tank to the influent liquid flow rate, and SRT is the total amount of sludge in the system divided by the rate of sludge leaving the system as waste. Sufficient time must be provided to allow the bacteria to assimilate the organic material in the wastewater. The

HRT is usually from 6 to 24 hours and SRT is from 4 to 10 days for the activated sludge process. The optimum operating temperature is in the range of 25° to 32°C.

Although organisms present in activated sludge systems range from viruses to multicellular organisms, the predominant and most active are heterotrophic, and to lesser extent, autotrophic bacteria, which are both aggregated in the sludge flocs and dispersed in the liquid. Heterotrophic bacteria utilize organic material as a source of both carbon and energy, while autotrophic bacteria generally depend on the oxidation of mineral compounds for energy requirements and utilize carbon dioxide as a carbon source. These bacteria are capable of performing hydrolysis and oxidation reactions.

Complex hydrocarbons are oxidized to lower molecular weights by oxygenase enzymes which incorporate oxygen directed into the long chain or cyclic hydrocarbon molecule. Polysaccharides, fats, and proteins are degraded from their polymeric state to monomeric units via hydrolysis. The end–products, i.e., alcohols and acids, from those reactions will enter the microorganism and be metabolized by oxidation reactions catalyzed by endo–enzymes. The oxidation follows the chemical sequence of: alcohols oxidized to aldehydes and then to acids. A portion of the acids are oxidized to carbon dioxide and water to obtain the necessary energy to use remaining acids for cell growth.

Generally, the activated sludge process is readily capable of decomposing alcohols, aldehydes, fatty acids, alkanes, alkenes, cycloalkenes and aromatics. Other compounds such as isoalkanes and halogenated hydrocarbons are more resistant to microbial decomposition. Therefore, the degree of treatment and the rate of decomposition are dependent upon the acclimated biomass in the activated sludge system. However, only dilute aqueous wastes can normally be treated, and most hazardous organic wastes are toxic or inhibitory to the process except at very low concentrations. Therefore, treatment of hazardous wastes by this process is often most practical where the aqueous waste can be mixed with a more readily biodegradable wastewater stream.

Dissolved metal ions and fine metal particles produce an adverse effect on microbial metabolism by binding at the enzyme–active site or causing conformational changes in the enzyme with the activated sludge process. Normally, microorganisms can tolerate only a few milligrams per liter or less of heavy metals. Heavy metals may be kept insoluble by the addition of ferrous sulfate to encourage sulfide precipitation and light metal cations may be detoxified by encouraging formation of carbonates and bicarbonates. In addition to biodegradation, organic materials may be removed by air–stripping, and/or sorption to the sludge.

Pact™ Process: An important variation on the activated sludge process is the Powdered Activated Carbon Treatment (PACT)™ process. This process offers a combined treatment and pretreatment system in which noncompatible and toxic constituents are adsorbed onto activated carbon, while microorganism–compatible waste remains in solution. Powdered activated carbon is added directly to the aeration basin of the activated sludge treatment system. Overall removal efficiency is improved because compounds that are not readily biodegradable or that are toxic to the microorganisms are adsorbed onto the surface of the powdered activated carbon. The carbon is removed from the wastewater in the clarifier along with the biological sludge. Usually, the activated carbon is recovered, regenerated, and recycled. Limitations of the PACT™ system include applicability to dilute liquids and residual sludges. The system is susceptible to clogging when there are high solids or high oil content in the wastewater.

A number of advantages have been reported for this combined physical/biological process. These include the removal of non–biodegradable organics, reduced emission of organics to the air, particularly during the period of acclimation, better settling properties of the biomass/powdered activated carbon sludge, and protection of the microbial population from toxic shocks. In addition, the powdered activated carbon helps reduce effluent concentrations of organics during an acclimation period. Some evidence also exists for the ability of microorganisms to bioregenerate the powdered activated carbon during periods of low organic loading.

The process has been used successfully with several industrial wastes including those from the manufacture of complex organic chemicals and from oil refining. In one full–scale study, greater than 82% removal of the priority pollutants was achieved with the PACT™ process. With a carbon dosage of 100 mg/ℓ, 99.6% removal of benzene and 84% removal of 2,4–dichlorophenol was achieved.

High Biomass Systems: Many current approaches to high biomass systems employ a combination of fixed film and freely suspended biomass in the process. High biomass systems have gained a certain popularity in Europe. During the past few years, a number of investigations undertaken in the Federal Republic of Germany (FRG) have been reported. Among the advantages attributed to such systems have been improvements in nitrification performance, sludge settleability, and effluent quality.

Reasons for selecting high biomass systems over construction of additional aeration tanks and clarifiers (or other secondary treatment processes) include reduced space requirements, increased process stability, and capital/operating cost savings.

High biomass systems call for installation of supplemental equipment over that contained in a conventional activated sludge plant. More installed equipment generally implies more maintenance, and, to some extent, this is true for some of the systems. In addition, the presence of both suspended and fixed biomass forms and higher biomass concentrations may require a certain level of additional operator time to achieve optimum system performance.

The presence of inert support media and higher biomass concentrations in these systems can increase overall power consumption. To achieve desired mixing patterns in retrofitted aeration tanks, power input may have to be increased. Also, the presence of additional biomass increases system oxygen requirements which, in turn, requires additional power input. In addition, high biomass systems generally yield higher levels of nitrification, which also can affect overall power consumption. Such factors should be addressed when analyzing operating costs.

Oxidation Ditches: An oxidation ditch is an activated sludge biological treatment process; commonly operated in the extended aeration mode. Typical oxidation ditch treatment systems consist of single channel or concentric, multichannel configurations.

Some form of preliminary treatment such as bar screens, comminutors, or grit removal normally precede the oxidation ditch. Primary settling prior to an oxidation ditch is sometimes practiced, however, it is not common. Flow to the oxidation ditch is mixed with return sludge from a secondary clarifier and aerated. The aerators may be brush rotors, disc aerators, surface aerators, draft tube aerators, or fine bubble diffusers. The aerators provide mixing and circulation in the ditch, as well as oxygen transfer. A high degree of nitrification occurs in the ditch due to operation in the extended aeration mode. Oxidation ditches are typically designed with a nominal hydraulic detention time at average design flow of greater than 10 hours and a mean cell residence time (sludge age) ranging from 10 to 50 days. Oxidation ditch effluent is usually settled in a separate secondary clarifier, however, intrachannel clarifiers are also used.

Ditches may be constructed of various materials, including concrete, gunite, asphalt, or impervious membranes. Concrete is the most commonly used. The single channel oxidation ditch may be found in a variety of shapes including ovals, horseshoes, or ells, whichever best fits the site. The concentric multichannel ditches may be circular or oval in shape. The addition of an intrachannel clarifier may be incorporated into the ditch design.

An oxidation ditch may be operated with an anoxic zone in the channel to achieve partial denitrification. An anoxic tank upstream of the ditch may be added along with recycle to that tank from the anoxic zone in the channel to achieve higher levels of denitrification. A anaerobic tank may be added prior to the ditch for enhanced biological phosphorus removal.

Oxidation ditches were usually not designed for nitrification or denitrification. Design parameters used, however, often ensured that nitrification occurred. Current concern over nutrient discharges to natural water systems has led to interest in upgrading existing oxidation ditches and modifying the oxidation ditch system design to incorporate biological nutrient removal.

Modifications to the basic oxidation ditch design can be made to achieve nitrogen and phosphorus removal. The key to obtaining nitrogen removal is the proper control of dissolved oxygen levels in different sections of the oxidation ditch, and the maintenance of adequate mass of bacteria under aerobic and anoxic conditions. To meet more stringent total nitrogen effluent limits a separate anoxic channel or basin outside the ditch channels may be added.

Holding mixed liquor under anaerobic conditions is required for enhanced biological phosphorus removal. This can be accomplished in either a nonaerated channel or by adding an anaerobic basin before the aerobic oxidation ditch channel.

A **vertical loop reactor (VLR)** is an aerobic suspended growth activated sludge biological treatment process similar to an oxidation ditch.

Other Variations: A variation to the activated sludge process is the use of high purity oxygen instead of air for aerobic treatment. Oxygen can be supplied from on-site gas generators with liquid oxygen storage as back-up. In addition to oxygen use, the aeration tank is covered which helps to eliminate odors and maintain temperatures in cold-weather periods.

There are many design variations to the conventional activated sludge process besides the use of high purity oxygen. These include: multiple units with series and/or parallel flow patterns; a tapered distribution of air along the tank length; stepwise addition of raw waste; reaeration of the recycled sludge before mixing with the raw influent; and extended aeration, e.g., 24 hours or longer, used for small wastewater flows.

Advantages and Disadvantages: Activated sludge treatment is used extensively in industry. It is probably the most cost-effective manner of destroying organics present in an aqueous waste stream. By using activated sludge modes which ensure complete mixing and high dissolved oxygen levels, high strength organic waste streams can be handled at an industrial waste treatment facility.

Some of the commonly listed disadvantages of the activated sludge process include the high capital investment required, high energy costs, the lengthy start–up time, and sensitivity to toxic and hydraulic shocks. On the other hand, the system can handle high organic loads using relatively short retention times, and can be controlled to achieve various degrees of treatment. Finally, the widespread use of activated–sludge facilities means the process has been well researched and documented.

The treatment requires consistent stable operating conditions. Activated sludge processes are not suitable for removing highly chlorinated organics, aliphatics, amines and aromatic compounds from a waste stream. Some heavy metals and organic chemicals are harmful to the organisms. When utilizing conventional open aeration tanks and clarifiers, this technology can result in the escape of volatile hazardous materials.

The efficiency of this process depends upon the satisfactory functioning of both the biological oxidation and the solids separation processes. Bulking and foaming must be controlled as they inhibit satisfactory separation of sludge solids.

1.5.3 Aerobic Systems

Aerobic biological treatment consists of conventional activated sludge processes as well as modifications of these processes including:

1. Sequential batch reactors.
2. Rotating biological contactors,
3. Trickling filters, and
4. Fixed film reactors.

All of these systems can treat aqueous waste streams contaminated with low levels of non–halogenated organics and/or certain halogenated organics.

Biological reactors require stable operating conditions. Abrupt changes in waste stream characteristics can generate shock loading to the biomass. The maintenance of stable levels is crucial for a number of key environmental parameters in the waste stream, including:

1. Dissolved oxygen (1 to 3 mg/ℓ minimum),
2. pH (6 to 8),
3. Nutrients (phosphorus, nitrogen, carbon),
4. Alkalinity (provides buffering capacity),
5. Minimal levels of suspended solids (particularly for fixed film reactors), and
6. Liquid retention times of 2 to 5 hours.

No process is more fundamental to the successful operation of an aerobic biological treatment system than is the transfer of dissolved oxygen. Unless dissolved oxygen is available where and when the bacterial system requires it, the process will not function. And, if adequate oxygen is available, the process will function almost in spite of all other upsetting conditions. Much research and development has been undertaken to increase the efficiency of oxygen transfer.

Biological aerobic processes include both suspended growth, and fixed film systems.

Suspended Growth Systems: There are many variations of the suspended growth systems currently employed for the removal of organics from municipal wastewaters. The mode of operation can dictate the net amount of organisms produced, the extent of SS degradation, the importance of intracellular materials, the settling characteristics of the organisms and the extent of treatment of priority pollutants. By varying the location of aeration, raw waste input location and organism concentration in the reactor, each of the different suspended growth systems can be made to perform, to a greater or lesser extent, in much the same way. The primary point here is that there is just one biological process and a multitude of physical variations which can be implemented. While each new physical variation is often noted as a separate process, the conversion of organics to cell mass is the primary process involved.

The physical variations of the suspended growth system may be categorized most simply as: (1) no recycle or recycle, (2) completely mixed or plug flow, and (3) single tank or multiple tank. No-recycle systems, e.g., aerated lagoons, containing low mixed liquor suspended solids (MLSS) concentrations, require relatively large quantities of land and are found mostly in rural areas. These systems usually put out high concentrations of SS during warmer periods because of the production and discharge of non-settleable algae. The tank (actually, probably an earthen ditch) contents may or may not be completely mixed. Organisms grown may be allowed to settle either in a quiescent portion of the tank or in a separate quiescent tank. In a recycle system, the organisms grown are returned to the reactor so that the rate of degradation of organics can be increased and the volume of the reactor required decreased.

The generic term for a suspended growth system with recycle is Activated Sludge. The tank may either be completely mixed or channeled such that a general appearance of plug flow is achieved. Plug flow conditions are also simulated by operating several tanks in series. Depending upon the location of aeration, the application points for raw feed, the hydraulic retention time and general physical appearance, the Activated Sludge system may be referred to as Step Feed, Step Aeration, Extended Aeration, Conventional, Completely Mixed, Contact Stabilization, Oxidation Ditch or any one of a variety of other names. Those systems operated such that the organisms are first exposed to high loadings in either the inlet portion of a tank or the first tank of a multiple tank system and then allowed to "burn-off" (i.e., oxidize) the organics in the remainder of the tank or tanks generally produce the highest quality effluent.

Fixed Film Reactors: While approximately two-thirds of Publicly Owned Treatment Works (POTWs) employing some form of biological treatment utilize suspended growth reactors, fixed film systems are the second most common variety of biological treatment. Many of the comments directed at suspended growth systems apply to the fixed film systems. There is just one biological process and many physical plants to house that process. The suspended growth system is based on the premise that the microorganisms selected not only can utilize the organics supplied to the reactor but also can be separated (usually by sedimentation) from the treated wastewaters. Similarly, the fixed film system utilizes the organics, both soluble and insoluble, but selects for organisms which attach to surfaces.

The original fixed film reactors were called Trickling Filters and used rocks for organism attachment. Later systems have employed synthetic plastic media instead of rocks. In the Trickling Filters the medium remains stationary and the organics move past the medium. An alternative form of fixed film system is one in which the film is attached to a drum rotating through the wastewater flow. This system, the Rotating Biological Contactor (RBC), has received considerable attention during the past decade.

Other fixed film systems include those in which the organisms are attached to a medium such as clay, sand or plastic. Attachment may be either on the surface or in the interior of the medium. The individual organism systems thus created may be either mixed (i.e., fluidized) or allowed to remain stationary (fixed). Such reactor configurations (especially the fluidized beds) have features of both fixed film and suspended growth facilities and are likely to have a dramatic impact on the future of biological wastewater treatment.

Efficiency of oxygen transfer and increased concentrations of organics (due to some form of staging) are common features among five new treatment systems:

1. Activated bio-filter
2. Reactor-clarifier
3. Deep shaft
4. Sequencing batch reactor
5. Porous biomass support system

1.5.4 Anaerobic Digestion

The anaerobic digestion process was initially engineered for domestic wastewater sludges, and is widely used in Publicly Owned Treatment Works (POTWs).

Anaerobic biotechnology for industrial waste treatment is steadily expanding in the U.S. and abroad. Applications were initially in the area of food processing wastewaters. Meat packing plant wastewaters received most of the initial attention and rather simple anaerobic ponds were the most common type of unit process. The anaerobic contact process evolved from this effort. Subsequently, the key role of cell immobilization was recognized and the anaerobic upflow filter, upflow anaerobic sludge blanket (UASB) and fluidized bed unit processes evolved. Recently, hybrids of these first two processes have emerged to capitalize on the positive features of each.

Traditionally, anaerobic treatment has not worked well with wastewaters other than those from municipal or food processing (distillery, beverage, vegetable) sources. Industrial wastewaters from the chemical industry are often complex, containing a wide variety of organics unrelated to the carbohydrate structures found in the municipal or feed processing wastes. Though many industrial chemicals are amenable to anaerobic metabolism, process wastes usually do not contain a single chemical component. Complex process intermediates, polymers and toxicants often are encountered which defy any biological treatment. In addition, much of the previous work on evaluation of anaerobic treatment of industrial wastes has not been conducted with knowledge of anaerobic metabolism and has overemphasized yield on methane.

The ever changing regulatory environment demands an ongoing evaluation of existing wastewater treatment schemes for process wastes which have historically been treated using conventional methods. Anaerobic technology offers distinct advantages over its aerobic counterpart: less sludge production, economic operation, low nutrient requirements, high microbial biomass, and potential for energy recovery.

Developments in reactor design and operation have established anaerobic digestion as an accepted process for industrial wastewater treatment. Parallel advances in our understanding of the complex microbiology of the process are providing new insights into microbial interactions and into factors governing the dominance, activity and maintenance of individual species in digester mixed liquors, biofilms and granules. This knowledge will contribute significantly to improved design, start-up, process control and operation of anaerobic digesters, with corresponding effects on the efficiency, stability and cost effectiveness of full-scale anaerobic waste treatment applications.

In anaerobic biological treatment, the influent sludge is settled and equalized, then pumped to the anaerobic digester along with an alkaline adjustment additive. There may or may not be mechanical agitation of the digester. After an adequate residence time to allow for proper digestion, the digester contents are allowed to settle. The supernatant is pumped to an aerobic treatment area (typically to an activated sludge unit), while the sludge is taken to disposal areas or subjected to additional treatment, such as drying or incineration.

Both standard-rate and high-rate systems are utilized. The standard-rate process is a one-tank process that must be large due to long retention times, low loading rates, and thick scum layer formation. Two tanks operating in series are the basis of the high-rate system, and the function of fermentation, and solids/liquid separation are handled separately.

Available anaerobic treatment concepts are based on such approaches as the classic well-mixed system, the two-stage systems and the fixed bed. In the well-mixed digester system a single vessel is used to contain the wastes being treated and all bacteria must function in that common environment. Such systems typically require long retention times and the balance between acetogenic and methanogenic populations is easily upset. In the two-stage approach, two vessels are used to maintain separate environments, one optimized for the acetogenic bacteria (pH 5.0), and the other optimized for the methanogenic bacteria (pH 7.0). Retention times are significantly lower and upsets are uncommon in this approach. The fixed bed approach (for single or two-staged systems) utilizes an inert solid media to which the bacteria attach themselves and low solids wastes are pumped through columns of such bacteria rich media. Use of such supported cultures allows reduced retention times since bacterial loss through washout is minimized. Organic degradation efficiencies can be quite high. A number of proprietary engineered processes based on these types of systems are actively being marketed, each with distinct features but all utilizing the fundamental anaerobic conversion to methane and carbon dioxide.

Suspended growth or fixed film system can be utilized as follows:

1. Suspended-Growth Systems
   - (a) Anaerobic lagoons
   - (b) Anaerobic contact process
   - (c) Anaerobic upflow blanket
2. Fixed-Growth Systems
   - (a) Anaerobic upflow filter
   - (b) Anaerobic downflow filter
   - (c) Anaerobic fluidized bed
3. Combination Suspended/Fixed-Growth Systems

Anaerobic treatment was successfully applied to a textile wastewater which caused uncontrollable foaming in the aeration basin of an aerobic treatment plant, negating successful treatment. By proper adaptation, anaerobic treatment has been applied to organically polluted high salt concentration wastewaters. Another area of application is to use anaerobic treatment of wastewaters containing volatile organic contaminants which would tend to be air stripped in an aeration basin or trickling filter if aerobic treatment was used. A much reduced level of gas stripping occurs in anaerobic treatment and in addition, any stripped volatile organics can be combusted with the methane gas in a boiler or waste gas flare. An example of successful adaptation of anaerobic treatment to biodegradation of an adiponitrile wastewater is quite notable.

General Electric has shown that anaerobic bacteria can remove the very refractory ortho chlorine atom from PCBs, and is researching the use of anaerobic bacteria in removing PCBs from aqueous sediments. A demonstration will be conducted in the Hudson and Housatonic Rivers.

Conventional aerobic processes are often unable to satisfactorily detoxify VOCs due to the extreme volatility of these compounds and because the high aeration rates commonly used in aerobic biological processes result in excessive stripping into the gas phase. Furthermore, while non–chlorinated VOCs seem to be readily biodegraded aerobically, chlorinated VOCs for the most part resist aerobic breakdown and stripping tends to be the dominant mechanism for their removal.

Anaerobic treatment offers two distinct advantages for the treatment of VOCs: first, the effect of stripping is substantially diminished compared to that in aerobic processes. Stripping in an anaerobic process could occur only due to the production of methane gas, and, typically, the amount of gas produced is significantly smaller than the normal aeration rates employed in aerobic processes. Stripping of VOCs will occur to a much greater extent when wastewater is treated aerobically than when it is treated anaerobically.

The second distinct advantage of anaerobic treatment of VOCs over aerobic treatment is that biodegradation of chlorinated compounds under anaerobic conditions occurs by reductive dehalogenation, and, as such, the greater the number of chlorine atoms on a compound the more easily it will be anaerobically degraded. Several recent studies have shown that many of the VOCs appearing on the Resource Conservation and Recovery Act (RCRA) list of compounds are amenable to biodegradation under anaerobic conditions. Thus, anaerobic treatment appears to be a promising technology for the detoxification of many chlorinated VOCs.

Key considerations in investigating the treatability of specific wastestreams are the presence of sufficient nutrients, and the level of sulfate relative to available substrate, and the potential for inhibitory effects from the wastewaters. Application of anaerobic treatment is particularly affected by the level of sulfates, and the need for sulfide control and gas desulfurization.

Anaerobic digestion is a biodegradation process capable of handling high strength aqueous waste streams that would not be efficiently treated by aerobic biodegradation processes. Advantages of anaerobic systems over aerobic systems include:

1. Capability to break down some halogenated organics,
2. Low production of biomass sludges that require further treatment and disposal,
3. Low cost, and
4. Lower energy consumption.

However, anaerobic systems can be less reliable than aerobic systems. Disadvantages of anaerobic systems include:

1. Potential for shock loading of biomass and termination of biodegradation process due to variation in waste stream characteristics,
2. Low throughput due to the slow biodegradation process (two steps),
3. Frequent necessity for further treatment of effluent prior to discharge off–

site or to a municipal treatment system, and
4. Generation of methane gas (a problem if it cannot be readily used on–site for meeting energy requirements).

Careful design and control can often solve these problems. Anaerobic systems are more susceptible to variation in waste stream characteristics and environmental parameters. Fixed anaerobic systems are widely used in industry for treatment of uniform, concentrated biodegradable waste in aqueous waste streams due to the low cost, low residual generation and production of usable methane gas. Anaerobic systems have a good potential as a pretreatment step for an aerobic system that will otherwise be unable to process a high strength waste such as a leachate.

As with aerobic systems, the biodegradation process can be slowed or halted by the following:

1. Abrupt change in waste stream characteristics,
2. Variable environmental conditions (e.g., temperature, pH),
3. Elevated levels of heavy metals or halogenated organics toxic to the biomass,
4. Inadequate nutrient levels.

Environmental impacts include:

1. Methane gas is produced and must be utilized or disposed of,
2. Additional treatment of effluent from the digester will be required,
3. Undesirable odors may be generated, and
4. Disposal of residuals will be required (volume is considerably less than that produced by aerobic systems).

Anaerobic Lagoon: This process is used as a pretreatment by the food industry, prior to discharge. It is a simple process where temperature and other conditions are not closely controlled. Clarifiers and sludge recycle may or may not be used.

Anaerobic Contact Process: This process is basically the anaerobic analog of the activated sludge process, with separate clarification for liquid/solids separation and sludge return.

Solids/liquid separation and sludge return are key to the successful operation of the process. There are various methods used to minimize settling difficulties, including agitation, degasification, flocculating agents, and the addition of inert material. Limitations are evident with each; suggestions include the use of short distances, such as inclined surfaces for settling, with a fill and draw cycle to remove accumulated solids. The solids contact process was commercialized as the Anamet system in Europe; including a mixed anaerobic contact reactor, degasification, clarification and sludge return, followed by aerobic final treatment. It has seen wide application in both Europe and the United States primarily on food wastewaters, including dairy, beet sugar, rum distilling, citric acid and molasses.

The process depends on the symbiotic relation of two classes of microorganisms: acid–forming bacteria and methane–forming bacteria. Facultative and anaerobic acid–forming bacteria first convert complex organic substrates in the wastes to short–chain organic acids (primarily acetic acid, propionic, and lactic acids), alcohols, carbon dioxide and H_2. Then strictly anaerobic methane–forming bacteria convert the volatile acids to methane gas, CO_2 and other trace gases. Methane–forming bacteria are inherently slow growing, with doubling times measured in days. They are also very sensitive to changes in the environment. In contrast, the acid forming bacteria can function over a wide range of environmental conditions and have doubling times measured in hours. When an anaerobic digestion is stressed by sudden changes in organic loads, temperature fluctuations, or an inhibitory material, the activity of methane–forming bacteria begins to lag behind that of acid–forming bacteria. The acids cannot be converted as rapidly as they form and pH drops. The methanogens are further inhibited, and the process eventually fails. Therefore, the overall rates of anaerobic processes are controlled by methane–forming bacteria. When acid builds up, lime or bicarbonate may be added to control the pH. However, the best way is to stop the influent waste, and to allow the methane–forming bacteria to restore balance in the process.

The process is usually operated in a pH range of 6.8 to 7.5 and temperature range of 31° to 35°C. However, some high–rate anaerobic contact digestors are operated at about 37°C to increase the rate of microbial growth. Since the methane–forming bacteria are recognized as the most sensitive microorganisms in anaerobic digestion, inhibition is indicated by the rate of methane gas production. Soluble heavy metals can cause the anaerobic digestor to fail. The light metal cations which come from industrial operations and the addition of alkaline material for pH control also play an important role in anaerobic digestion. They can be either stimulatory or toxic depending on their concentrations in solution.

The soluble heavy metals can be removed by the addition of sulfide compounds. Approximately 0.5 mg/ℓ sulfide is need to precipitate 1 mg/ℓ of heavy metal. However, the soluble sulfides in the solution are toxic to the anaerobic digestion system if the concentration exceeds 200 mg/ℓ. Light metal cations may be detoxified by encouraging formation of carbonates and bicarbonates.

For most industrial wastewaters, nutrients such as nitrogen and phosphorus have to be added. The nitrogen requirement for anaerobic treatment is only a small fraction of that required by the aerobic process. The phosphorus requirement is approximately 15% of the nitrogen requirements. Since the methanogens are unique in the anaerobic digestion process, they also need some unique trace nutrients. Studies have shown that trace amounts of iron, cobalt, nickel, sulfide, molybdenum, tungsten or selenium can stimulate the methane gas production rate.

Anaerobic Filter: The anaerobic filter is generally based about a submerged support medium with the wastewater directed in either an upflow or downflow mode. The media provide the surface area for bacterial attachment. These are available in several varieties; generally they are plastic and installed as fixed packing or as randomly placed units within the reactor. Media selection is an important aspect in the design and operation of the filter and is typically dictated by the type of wastewaters to be treated.

The Bacardi and Celrobic filters are the more notable commercial systems, although the filter in particular is applied as a generic process design. Proprietary claims more typically relate to process operations and specific accessory hardware. The loadings cited are generally higher than those cited for the UASB or suspended growth systems. This is attributed to the higher SRT in the filter, particularly the upflow filter. The systems are designed with or without effluent recycle and external clarification.

The Celrobic™ high–rate anaerobic treatment process was developed by the Celanese Chemical Company, and is currently being used commercially in nine installations. The upflow random packed–bed configuration accounts for the reliability and stability of the process. Its long–term continuous performance relies on the ability to measure and control the quantity of solids that remain in the reactor.

In the Bacardi design for treating rum distilling wastewater, the filter is submerged, wastewater is downflow, with countercurrent recycle flow.

An operational problem generally cited for the filter is the possibility of the media clogging with excess growth. More recent work however suggests that this is a controllable problem with the proper selection of the media and with operational procedures using gas purging, high recycle rates, etc.

Anaerobic Sequencing Batch Reactors: These are anaerobic biological reactors operated in a fill and draw mode. The reactor cycle begins with FILL, a period of time when raw wastewater is pumped into the reactor which contains the biomass from previous cycles. Mixing is provided to promote contact between the biomass and influent organics. At the end of REACT, mixing is discontinued and settling of the biomass is provided, SETTLE. The treated supernatant is removed during DRAW. IDLE is a period between the end of DRAW and the beginning of the next FILL period. The process is being investigated for the biological treatment of coal conversion wastewaters.

Denitrification: Nitrates and nitrites are reduced to nitrogen gas by facultative heterotrophic organisms. A supplemental carbon source, usually methanol, is added. Various systems may be used for denitrification including:

1. Suspended growth
2. Coarse–media attached growth
3. Fine–media attached growth

Fluidized and Expanded Bed Bioreactors: A bed of small particles is fluidized by the upward flow of water. Very high specific surface areas can be achieved without introducing the problem of clogging. Fluidized beds are sometimes called expanded beds.

This process relies on developing attached growth on small inert particles such as sand. The media are kept in a fluidized state by the upflow velocity of the raw and recycle wastewater flow. This fluidized state provides a very large available surface area for growing the biofilm, allowing for very high active sludge inventories. The reactor loadings, on a volumetric basis, are therefore generally significantly higher than possible for the filters or upflow sludge beds (10 to 20 g/ℓ/day). Commercial fluidized bed systems are currently marketed by Ecolotrol, Air Products, Biothane, and Dorr–Oliver.

The fluidized bed is better suited to soluble wastes. Recycle is required to maintain suitable fluidization of the reactor bed. The arrangement is less likely to hold suspended, unattached solids; excessive solids retention may, in fact, interfere with proper maintenance of the reactor bed.

The most obvious advantage of a fluidized–bed biofilm reactor is that it can have a very high specific surface area that is not prone to clogging because the small particles are fluidized. The high specific surface area allows accumulation of a high–volume density of biofilm, which usually has low resistance to external mass transport. This makes it possible to build compact reactors. Detention times measured in minutes are possible, making it possible to process loads many times greater than those treated by conventional aerobic processes.

The Anitron™ system is a highly efficient anaerobic wastewater treatment process which utilizes a fluidized bed reactor. Within the reactor, a fixed–film of microbial growth (supported growth) occurs on the media (usually sand), which are hydraulically supported as a fluidized bed by the incoming wastewater and recycled effluent. Wastewaters with BOD_5 levels of 2,000 mg/ℓ or more, such as found in the food, beverage, and pulp and paper industries, are candidates for treatment with this technology.

The fluidized bed system has also been studied as a two phase anaerobic system. A summary of wastes studied with the fluidized bed included soy bean, dairy whey, wheat starch, corn starch, paper mill whitewater, evaporator condensate, Kraft mill decanter, dairy industry cheese processing, brewery wastes and sludge heat treatment liquor.

The objective of a recent EPA-funded study was to examine the effectiveness of the anaerobic Granulated Activated Carbon (GAC) expanded-bed bioreactor as a pretreatment unit for the detoxification of a simulated high strength industrial wastewater containing several volatile RCRA compounds present in backgrounds consisting of non-RCRA organic compounds. As a pretreatment unit, the goal was not to maximize COD destruction but to reduce the VOC concentrations to acceptable levels. This goal was achieved very satisfactorily. The reactor demonstrated excellent treatment; removals of greater than 97% were achieved for all the VOCs. Chloroform was found to be inhibitory to the system at effluent concentrations of about 100 μg/ℓ. It was found to inhibit the degradation of acetate and acetone, two of the three base flow organic compounds. Chloroform itself, however, was removed to greater than 97%. The only limiting factor in this treatment study was the high effluent COD experienced during the inhibitory phase, which was composed almost entirely of acetate and acetone and as such, should easily be removed by any of several treatment options. The amount of stripping occurring was negligible compared to the amount of stripping anticipated to occur in an aerobic biological process. The anaerobic GAC expanded-bed bioreactor represents an excellent pretreatment unit for the treatment of wastes containing VOCs.

Hybrid Anaerobic Processes: Hybrids of the anaerobic upflow filter, and the upflow anaerobic sludge blanket (UASB), have emerged to capitalize on the positive features of each.

Recent applications suggest combining the filter with a suspended growth system to maximize sludge retention and accomplish possibly higher loadings. Biomass International and Zimpro market a commercial system of this configuration, although generic configurations have also been constructed.

The system can be designed with or without recycle; incorporating recycle has an advantage in keeping the lower sludge blanket zone in suspension. Additionally, recycle of clarifier underflow will help to maximize the sludge retention. Having the lower portion of the upflow reactor designed as a sludge blanket gives it a better capability to handle high raw solids loads. The upper filter zone give it better stability with the fixed film growth; unattached, suspended solids would have a tendency to fall back to the lower bed, or be captured in the sludge bed with recycle.

In order to expand the capacity of the plant, two options existed: either expand the aerobic treatment facility or reduce the biological load by treating the high strength wastes separately. The HYAN hybrid anaerobic process was developed to effectively treat this high strength waste and to generate a continuous and reliable supply of gas energy.

Anaerobic processes are prone to high accumulations of inert solids in their fixed film filter zones when treating high strength wastes. Most designs require frequent shut-down and the use of difficult cleaning procedures to reduce short-circuiting and to maintain performance. The HYAN reactor controls such accumulations by design, keeping its filter media zone free of unwanted solids.

The HYAN concept is also being effectively applied where the upflow sludge blanket process has had difficulty maintaining biomass in the reactor. The HYAN design is proving to be more efficient in those cases by not requiring a separate settling tank to capture and return biomass to the reactor.

The reduced load to the aerobic facility has lowered the overall electrical treatment costs. In addition, the methane produced has replaced most of the natural gas requirements of the Thermal Conditioning Process. The reduced loading on the aerobic system has deferred major capital expenditures for new tankage and aeration systems until increased sewage flows justify additional plant capacity. The HYAN system has also reduced the quantity of solids requiring treatment and disposal, by efficiently converting the organic pollutants to gas.

Sulfate Removal: Inherent to the anaerobic digestion of wastewater containing sulfate or other sulfur–bearing substrates is the generation of hydrogen sulfide along with the biogas. The treatment of biogas for removal of sulfur compounds becomes more and more important as both government regulations restricting sulfur emissions become tighter and the effect of corrosion from biogas usage creates maintenance and operational problems. In most cases, better than 99% sulfur removal is necessary. This means that the hydrogen sulfide level in the effluent gases must often be reduced to 10 ppm or less to compensate for the traces or organic sulfur compounds, carbonyl sulfide, and other sulfurous materials which are more difficult to remove. Assuming an inlet hydrogen sulfide concentration of two volume %, or 20,000 ppm, the reduction down to 10 ppm represents 99.95% removal.

Anaerobic degradation is ideally suited for the pretreatment of high strength industrial effluents. However, many, particularly those from the pulp and paper industry, may contain substantial amounts of sulfur. This will result in sulfide toxicity and inhibit anaerobic degradation.

The presence of sulfates, sulfites, and sulfides has been considered a nuisance. In an anaerobic reactor, the sulfate reducing bacteria (SRB) reduce sulfates to sulfides that can create a toxic environment for the methane forming bacteria (MFB). The sulfides produced end up in the bio–gas formed by the anaerobic reactor. The sulfides present cause corrosion problems and odors.

There has been considerable research and development devoted to sulfate reducing bacteria, as there is an obvious advantage to removing the sulfur compounds before or during anaerobic digestion, to eliminate the hydrogen sulfide and other problems. One example of a recent process is the "Biosulfix" process, wherein sodium bisulfide is recovered.

Upflow Anaerobic Sludge Blanket (UASB): This process configuration relies on the establishment of an active sludge bed which is kept in a suspended, expanded state by the upflow velocity of the liquid and, to some extent, by the gas generated within the bed. The major commercial system is marketed by Biothane, although there are several versions of the process. The key to the successful operation of the upflow anaerobic sludge blanket (UASB) appears to be the development of active, granular, sludge particles, in effect simulating a fixed film fluidized bed process, although with lower upflow velocity requirements. The particles are relatively large (1 to 5 mm). Its primary applications have been to high carbohydrate type wastes such as found in food processing, particularly the brewing industry. The process was developed in the Netherlands.

Essentially this process is an upflow anaerobic sludge contact system, incorporating into the reactor a gas–solids separator. Mechanical mixing is not required and recirculation is minimized to encourage good settling. The system design was based on the development of a good settling, anaerobic sludge.

Important hardware elements are the gas/liquid/solids separation devices used by the equipment manufacturers. These often form the proprietary basis for these types of systems. This generally involves specifically designed baffling devices installed within the reactor. Effluent recycle requirements are dictated by the strength of the incoming wastes and/or upflow velocity criteria. Solids recycle is discouraged, thus external clarification would be required if there is substantial carryover of suspended solids. Sludge recycle is felt to interfere with the development and maintenance of the granular sludge solids bed within the reactor. This same restriction should also apply to the incoming raw solids; wastewaters high in raw solids may be a problem for the UASB configuration because of the problems they pose with the maintenance of an acceptable bed.

Sorption/Anaerobic Stabilization: Many wastewater streams contain dilute concentrations of organic pollutants that are not treated effectively by conventional activated sludge processes. These pollutants, however, can often be treated effectively anaerobically. If the pollutants were treated anaerobically, pass–through of the pollutants to the receiving stream and stripping of volatile compounds during aeration could be minimized. To treat the entire wastewater stream in an anaerobic digester would not be economical. However, if the bulk liquid stream could first be passed through a sorbent bed such as granular activated carbon (GAC) prior to aeration, only the sorbent material, a much smaller volume, would require anaerobic stabilization at elevated temperatures.

A feasibility study, performed at bench–scale with complex real–world wastes, demonstrated that the experimental system was capable of consistently removing 40 to 50% of the influent COD (Chemical Oxygen Demand) for a year–long period. No GAC replacement was necessary during this time. The reduction of COD discharged to the aeration basin would reduce aeration requirements as well as aerobic sludge production in actual application. In addition, the stabilization process produces methane from the removed COD which potentially would be recoverable as fuel for heating the reactor. When hazardous compounds are present in the influent waste, the sorption stage is capable of trapping significant amounts, preventing their pass–through to the aeration basin and subsequent volatilization of the strippable chemicals. The sorption stage also attenuates the effects of shock loads of compounds which may be toxic to the aerobic portion of the plant. In addition, the combined sorption/anaerobic stabilization stage retention time for GAC, and, therefore, biomass and sorbed organics, is extremely high, maximizing the potential for degradation of compounds which are normally recalcitrant at conventional treatment plant retention times.

Methane from Municipal Solid Waste (MSW): The feasibility of methane production from solid waste, with limited additions of sewage sludge; and including an evaluation of gas production as a function of pH, temperature, solids loading, retention time and slurry concentration, and an evaluation of the costs and net economic benefits of the system has been demonstrated. Also, Consolidated Natural Gas Service Company performed laboratory and engineering studies to evaluate biogas production from MSW. These studies reconfirmed the technical feasibility of the anaerobic digestion process to convert organic wastes to pipeline quality fuel gas.

The digestion process utilizes microorganisms to stabilize organic matter and to produce enzymes to catalyze the process. The details of the process are not completely understood because many of the organisms have not yet been isolated. Nonetheless, the biochemistry of the overall process is thought to proceed in three distinct stages. The first stage is fermentation. Faculative bacteria, which can live either in the presence or absence of oxygen, and their enzymes reduce complex molecules (polymeric solids such as cellulose, fats, and proteins) to simple organics (monomers such as sugar, fatty acids, and amino acids). In the second stage, acidogenic bacteria reduce the monomers to acetic acid and hydrogen. In the third stage, methanogenic bacteria use the acetic acid and hydrogen to produce methane and carbon dioxide. The methanogenic bacteria, essential to the success of the system, are strictly anaerobic, and thus must be contained in an airtight reaction vessel. Other essential factors are a neutral pH, proper nutrients (nitrogen, phosphorus, trace metals), absence of toxins, and proper temperature. The microbial population which affects the digestion may be introduced with the organics or may be seeded into the digester when the substrate does not have a large population of its own, as is the case with MSW.

1.5.5 Anoxic Treatment

In the absence of oxygen, certain microorganisms will use nitrate as the terminal electron acceptor. This is termed denitrification. This process produces nitrogen, carbon dioxide, water, and new cell material.

Anoxic zones may be incorporated into the designs of the aeration basins. Anoxic zones aid in controlling filamentous growth and the removal of nitrogen. Mixed liquor is recycled to the head of the aeration basin where under anoxic conditions nitrate is used as the electron acceptor for the uptake of soluble BOD (Biological Oxygen Demand). Metabolization of the BOD occurs in subsequent aerobic zones of the aeration basin. Besides removing nitrogen and controlling filamentous organisms, the anoxic selectors aid in the control of pH, and reduces aeration demands.

Experience at anoxic municipal facilities has shown that:

1. Anoxic selectors effectively control filamentous bulking,
2. Nitrogen removal and alkalinity recovery are functions of BOD loading,
3. Design details facilitate scum handling,
4. Effluent quality is good, and
5. Effluent BOD/TSS varies between facilities.

1.5.6 Aquatic Plant Systems

Aquatic plant systems are engineered and constructed systems that use aquatic plants in the treatment of industrial or domestic wastewater. They are designed to achieve a specific wastewater treatment goal. Aquatic plant systems can be divided into two categories:

1. Systems with floating aquatic plants such as water hyacinth, duckweed, pennywort; and
2. Systems with submerged aquatic plants such as waterweed, water milfoil, and watercress.

Until recently, most of the floating aquatic plant systems have been water hyacinth systems. However, use of water hyacinth has been limited, in geographic location, to warm weather regions because of the sensitivity of water hyacinth to freezing conditions. Water hyacinth systems have been most often used for either removing algae from oxidation pond effluents or for nutrient removal following secondary treatment.

Aquatic treatment systems consist of one or more shallow ponds in which one or more species of water tolerant vascular plants such as water hyacinths or duckweed are grown. The shallower depths and the presence of aquatic macrophytes in place of algae are the major differences between aquatic treatment systems and stabilization ponds. The presence of plants is of great practical significance because the effluent from aquatic systems is of higher quality than the effluent from stabilization pond systems for equivalent or shorter detention times. This is true, particularly when the systems are situated after conventional pond systems which provide greater than primary treatment.

In aquatic systems, wastewater is treated principally by bacterial metabolism and physical sedimentation, as is the case in conventional trickling filter systems. The aquatic plants themselves, bring about very little actual treatment of the wastewater. Their function is to provide components of the aquatic environment that improve the wastewater treatment capability and/or reliability of that environment.

Aquatic plant systems can be designed and operated to accomplish a variety of wastewater treatment tasks, but the designs and the operation are not always simple. Hyacinth systems are susceptible to cold weather and particularly in the southern states, can be affected by biological controls introduced to help control water hyacinths in the natural environment. Concerns of health agencies for mosquitoes can play a very big factor in the design and operation of aquatic plant systems. Finally although water hyacinth systems may be useful in nutrient removal, there are limits to the treatment capacity and dependability of hyacinth systems in terms of meeting very low effluent values.

Scott Cunningham (DuPont) is investigating phytoremediation which uses plants to remove metals. Plants take up the metals in their roots and translocate them to their shoots, which are harvested, burned in a kiln, and the metals recovered and recycled. The challenge is finding or engineering plants that can absorb, translocate, and tolerate heavy metals while producing enough foliage to make the process efficient; ore outcroppings and metal-containing waste sites are good locations for suitable candidates.

1.5.7 Autothermal Thermophilic Aerobic Digestion

A promising technology for meeting the current and proposed U.S. federal requirements for pathogen control and land application of municipal wastewater sludge. Autothermal thermophilic aerobic digestion, or ATAD, has been studied since the 1960s and significantly developed since the mid-1970s. It is currently widely and successfully implemented in Europe.

ATAD systems are normally two-stage aerobic processes that operate under thermophilic temperature conditions (40° to 80°C) without supplemental heat. Typical ATAD systems operate at 55°C and reach 60° to 65°C in the second stage. They rely on the heat released during digestion to attain and sustain the desired operating temperatures.

The ATAD process has many benefits: a high disinfection capability, odor reduction, low space and tankage requirements, and high sludge treatment rate. It is a relatively simple technology that is easy to operate (automatic monitoring or control equipment and full-time staff are not required) and economical, particularly for small facilities. It provides a proven, cost-effective way to achieve aerobic digestion and to produce sludge that can be applied to land in the U.S. without any management restrictions for pathogen control.

Prestage systems also provide thermophilic digestion and are normally incorporated in the treatment process ahead of conventional anaerobic digestion.

ATAD can be used in lieu of anaerobic processes for high strength effluents. It is also an applicable process for organic industrial wastes.

The aerobic thermophilic biological technology is viable and has applicability for destruction of organic bearing wastes. The system can be applied for treatment of organic sludges, high-strength organic wastes and hot streams containing biodegradable organics. For autothermophilic conditions, waste strength must be greater than 30,000 mg COD/ℓ, the reactor must be insulated and covered, and a relatively efficient aeration system (transfer efficiency of approximately 12%) is required

Scientists at Michigan State University are studying the use of thermophilic bacteria for site remediation.

1.5.8 Biological Aerated Filter

Wastewater is filtered downward through a fully submerged bed of small rocks, which help to form the biofilm, and air is forced into the bed. No settler is used, but periodic backwashing is required. This is a compact treatment system.

1.5.9 Biological Tower

The biological tower is similar to the trickling filter, except that plastic media can be stacked to heights of 12 m. The use of lightweight plastic media allows construction of tall towers (thus conserving land) with high specific surface areas (allowing higher volumetric loading than possible in conventional trickling filters).

Many German companies are replacing their lagoons with tower-like reactors that consume less energy and take up less surface area while more than doubling the mass transfer of oxygen in aerobic-treatment reactions. The towers handle chemical oxygen demand (COD) concentrations between 2 and 12 g/ℓ, according to Bayer AG (Leverkusen).

Air is introduced at the bottom of the tower. The configurations of the injectors and the sizes of the air bubbles they provide are customized to ensure an even distribution of bubbles in the reactor, and to prevent the bubbles from coalescing as they move up through the tower.

In the U.S., where energy costs are lower than in Europe, biotowers promise to be economically viable only for the treatment of highly concentrated or toxic waste streams, or in areas where excess land is not available.

1.5.10 Composting

Composting provides a means of achieving high-rate aerobic digestion of organics by mesophilic and thermophilic microorganisms. When waste is applied to the soil in windrow piles, and aerated by spreading or turning the pile, complete oxidation of simple and complex organics is achievable. In addition to microbial degradation, the process encourages adsorption of metals onto the soil–organic media. It is the only biological treatment process which is relatively insensitive to toxicants. Between adsorption of solvent inorganics and precipitation of metals in the alkaline aerobic compost environment, inhibition is prevented.

Energy demand is very low, and is limited primarily to fuel costs to operate earth-moving equipment. Chemical requirements include limestone and nutrient addition. There are no exotic compounds known to be formed. There are also no sludges or brines requiring ultimate disposal. Leachate from composting beds requires aqueous biological treatment, i.e., activated sludge treatment, for decomposition of solvent organics.

The **aerated static pile** process involves mixing dewatered sludge with a bulking agent, such as wood chips, followed by active composting in specially constructed piles. Typically, both recycled bulking agent and new (external) bulking agent are used for mixing. Induced aeration, either positive (blowing) or negative (suction), is provided during the active composting phase. Temperature and oxygen are monitored during active composting as a means of process control. The active composting period lasts at least 21 days, following which alternate pathways to produce finished compost may be employed as described below.

If at the end of the 21 day active composting period, composted material is sufficiently dry, screening may be performed directly to separate bulking agent for recycle. The recycled bulking agent is generally stored prior to reuse in the mixing operation. Screened compost is restacked and cured for at least 30 days and then stockpiled as finished compost prior to distribution.

If at the end of the 21 day active composting period, compost material is not sufficiently dry for screening, a separate drying step is required prior to screening, curing, and stockpiling. Alternatively, unscreened compost may be restacked for the 30 day curing period, after which drying, screening, and stockpiling are performed.

The **conventional windrow process** involves initial mixing of dewatered sludge with a bulking agent such as finished compost, often supplemented with an external amendment, followed by formation of long windrows. Formation of the windrows is generally performed in two steps. Typically, front-end loaders are used to initially stack material in a rough windrow configuration; then a specially designed mobile composter is used to fine mix the material by turning the windrow in place.

An active windrow composting period of 30 days (or more) is provided following initial mixing and formation. During this period, the windrows are periodically turned with a mobile composter (in some cases front-end loaders may be used) to aerate and remix the material. A turning frequency of two to three times per week is typical. Temperature is monitored for process control. Following the active windrow composting period, the composted material is allowed to cure for at least 30 days; then, a portion of the finished compost is recycled and a portion is stockpiled for distribution.

The **aerated windrow process** is similar to the conventional windrow process with one exception: a system for induced aeration is provided in addition to aeration by turning with a mobile composter. Either positive- or negative-induced aeration may be used, which is intended to enhance active composting and drying. Typically, induced aeration is provided using a fan and a fixed arrangement of pipes or channels for delivering air uniformly to the base of the windrow.

A consistent definition of **in-vessel composting** has not been established by the industry or the regulatory agencies. It is sometimes broadly interpreted as composting that takes place in a container of some sort where the material to be composted is aerated and mixed by mechanical means. The OME report classifies vessels used in this type of composting process as either rotating drums or tanks.

The mixing and tumbling of Municipal Solid Waste (MSW) inside a rotating horizontal drum provides particle size reduction and mixing of air and moisture. The drums are similar to a cement kiln in design and are as long as 180 feet with a diameter of up to 12 feet, although much smaller drums are also used. Some rotating drums retain the material inside for about 8 hours, functioning more as a pulping device than a composter since the materials must then be composted by one of the other methods. Some drums retain the waste for several days or weeks and actually function to digest the material, requiring less time in subsequent composting steps. Due to higher capital and operating costs, in-vessel systems are most commonly used with large volumes of MSW and sewage sludge.

Another type of composting vessel is configured with either horizontal or vertical tanks using forced aeration and mechanical agitation for composting sewage sludge and/or MSW.

Most in-vessel systems are followed by a static pile or windrow composting stage since production of stable compost requires more time than is economically feasible in the vessels.

Composting is also being investigated for treatment of industrial wastes.

1.5.11 Contact Process

The wastewater flows into a small contact tank (30 to 90 minutes detention), where colloidal organic contaminants are captured in flocs and soluble contaminants are oxidized. The settled sludge is sent to a reaeration tank (3 to 6 hr detention of sludge) before it is returned to the contact tank. Reaeration provides oxidation of colloidal material and endogenous decay of biomass. It also reduces the tank volume needed for treatment because both reactions are possible when biological solids are highly concentrated.

1.5.12 Fluidized Beds (Expanded Beds)

In fluidized-bed reactors, solid material, which is colonized by microorganisms is suspended by water flowing upward through the tank. The solid material is either inert (e.g., sand, coal, or plastic) or active (i.e., granular activated carbon). Both aerobic and anaerobic types of fluidized beds are in use or under investigation. In aerobic systems, air is diffused from the bottom through the bed.

There are several advantages that fluidized beds have over packed beds. Because gas bubbles can pass through the bed easily, smaller particles can be used. The use of smaller particles results in a larger biofilm surface area, which can handle higher organic loading. Also, the beds expand rather than clog as the biofilm grows. Growth can easily be controlled by removing particles from the top of the bed, washing them, and returning the cleaned particles to the reactor bottom.

Fluidized beds which use granular activated carbon as the solid material are referred to as biological activated-carbon systems. In these systems, adsorption and biodegradation of organics occurs simultaneously. This has advantages over a design in which the mechanisms occur in series. The granular activated carbon protects the system from organic shock loads, extends the retention time of the less readily biodegraded organics and adsorbs refractory compounds, just as it does in the powdered activated-carbon process. The microorganisms not only degrade the organics in the waste liquid, but also have been shown to regenerate the granular activated carbon. A major drawback to the use of biological activated-carbon systems is the large capital investment in granular carbon required.

No information has been located on emissions of organic vapors from fluidized-bed reactors. For anaerobic systems, they are not assumed to be significant. For aerobic reactors, using diffused-air systems, air emissions are likely to be important.

Air Products and Chemicals, Inc., Allentown, Pennsylvania has developed biological fluidized bed systems—oxitron and anitron systems—to treat industrial, municipal and sanitary wastewaters. These systems have been demonstrated for application in hazardous waste streams, especially of metalworking fluids in the automotive industry and petrochemical industries.

Advantages of the oxitron and anitron systems, according to the developer, are: (1) easy installation and operation; (2) high tolerance for hydraulic shock and greatly increased flow can be accommodated without loss of treatment efficiency; (3) high resistance to toxic shock; (4) rapid restart after shutdown; (5) does not air-strip volatile organics and release them to the atmosphere; and (6) anitron systems produce methane for use in boilers or as fuel for the generation of electricity.

Extended demonstration is needed with different types of hazardous waste streams. Only pumpable liquids, slurries, and sludges are acceptable materials in the system; solid wastes with very low moisture contents cannot be treated.

In the Celgene aerobic process wastewater containing dilute amounts of organic contaminants is treated with nutrients, and the pH adjusted to 6.0 to 7.5. The wastestream is then fed into a vertical, fluidized-bed bioreactor containing microbes immobilized on activated carbon. Relying almost exclusively on the target organic for sustenance, the microbes metabolize the material to carbon dioxide, water and a small amount of biomass.

1.5.13 Hybrid Systems

Hybrid reactors, as the name implies, are a combination of suspended growth and fixed-film reactor principles. In these systems, the fixed film is submerged and the reactor contents are continuously stirred. A large amount of biomass is maintained in the system.

Hybrid reactors, depending upon their configuration, can handle high organic loads (i.e., in the range of 50 to 10,000 ppm). Because these reactors are a completely mixed system, they are less affected by shock loadings. Hybrid reactors are designed to compensate for the principal limitations of fixed-film and suspended growth reactors. However, set-up and operation of hybrid reactors will tend to be somewhat technically demanding than either fixed-film or suspended growth systems.

As with the suspended growth and fixed-film systems, some biomass is produced which will require disposal. The pretreatment and disposal practices required will depend on site-specific requirements.

1.5.14 Land Application (Landfarming)

Land application is the oldest method used for treatment and disposal of municipal and sewage wastes. Cities have used this method for more than 400 years. Several major cities have used "sewage farms" for at least 60 years for waste treatment and disposal. Approximately 600 communities in the United States reuse municipal wastewater treatment plant effluent in surface irrigation. Landfarming is also discussed in Section 1.6.4.

Land treatment has also been used in the United States for industrial sludges. Properly designed and operated facilities should be effective for several waste matrices. The release of volatile organic compounds needs to be addressed during system design. Current Federal regulations include a no migration demonstration for land treatment systems. Volatile organics may need to be removed from waste before it is applied to a land treatment system.

Landfarming facilities are designed to encourage the biological degradation of organic wastes incorporated into the upper soil layer. The soil and climatic characteristics of the site, the chemical characteristics of the waste and the operational techniques employed all influence the extent of biodegradation. Under normal conditions, biodegradation is considered to be the primary loss mechanism, followed by volatilization. Although photolysis and other chemical reactions occur at land treatment facilities, they are normally considered insignificant removal mechanisms.

Land application is typically defined as the spreading of sludge on or just below the surface of the land. Land-applied sludge is usually incorporated into the soil after application to minimize odors, runoff, or contact with animals and humans. The sludge can serve as a soil conditioner and as a partial replacement for commercial fertilizers. Sludge is applied on agricultural lands; forest lands; drastically disturbed lands (land reclamation) or land dedicated specifically to sludge disposal (dedicated land disposal); parklands; golf courses; or home gardens and lawns.

Sludge is applied to the land in liquid or dewatered form. Liquid sludges are transported to the application site in tank trucks and sprayed on or injected into the soil. Dewatered sludges (filtered cakes) can be applied to the land with spreading equipment. Sludge is applied to dedicated sites (where it is applied to the soil in periodic repeated applications) at a higher rate than to agricultural lands or lands used for other purposes.

Concurrent with improving soil productivity, land application also functions as a sludge treatment system. Sunlight, soil microorganisms, and desiccation can destroy many pathogens and organic substances in the sludge. Nutrients, which can cause eutrophication and other problems if released into surface waters, are largely converted into useful biomass, such as crops or wood. The capacity of the land to treat sludge constituents is finite, however, and land application systems must be designed and managed to work within the assimilative capacity of the land and the crops grown on it.

There are three methods of land application:

1. Slow Rate
2. Overland Flow

3. Rapid Infiltration

1.5.15 Methanotropic Systems

The methanotrophic bioreactor system is an aboveground remedial technology for water contaminated with halogenated hydrocarbons. Trichloroethylene (TCE) and related compounds pose a new and difficult challenge to biological treatment. Unlike aromatic hydrocarbons, for example, TCE cannot be used as primary substrates for growth by bacteria. Their degradation depends on the process of cometabolism which is attributed to the broad substrate specificity of certain bacterial enzyme systems. Although many aerobic enzyme systems are reported to cooxidize TCE and related compounds, BioTrol claims that the methane monooxygenase (MMO) of methanotrophic bacteria is the most promising.

Mtehanotrophs are bacteria that can use methane as a sole source of carbon and energy.

Although it has been known that certain methanotrophs can express MMO in either a soluble form or a particulate (membrane-bound) form, BioTrol-sponsored research results have led to a patent pending on the discovery that the soluble form is responsible for extremely rapid rates of TCE degradation.

In the ABB process, methanotrophic bacteria are cultivated for a key enzyme they produce (MMO). Such microbes derive their carbon food source from methane, which is added to the system, not from the hydrocarbon pollutants.

Once the secreted MMO enzyme breaks down chlorinated organics by oxidation, a second line of non-methanotrophic bacteria completely consumes these by-products. This system has successfully converted trichloroethylene (TCE), dichloroethylene (DCE) and vinyl chloride (VC) in groundwater to carbon dioxide, water and chloride ions, such as sodium chloride and dilute HCl.

A series of polyethylene discs continuously rotates inside the reactor, to maximize exposure to both the contaminated influent and the feed gases—oxygen and methane. Each disc hosts a microbial colony; discs can number from dozens to hundreds.

1.5.16 Microbial Rock Plant Filter

This emerging and promising technology utilizing natural processes for municipal wastewater treatment is the result of research conducted by the National Aeronautics and Space Administration (NASA) at the Stennis Space Center (SSC) in Southern Mississippi over the past 20 years. This technology utilizes aquatic and semi-aquatic plants, microorganisms, and high surface area support media such as rocks or crushed stone. Communities, consulting engineers, state agencies, and EPA Region 6 have continued the development.

The technology was developed to treat and reclaim wastewater for reuse in space stations. On Earth, it is a low-cost, cost-effective technology for small communities, on-site treatment, individual systems, and industrial wastewater. Haughton, Benton, and Denham Springs, Louisiana, are the first applications of this technology in Region 6. Long shallow rock filters are heated by solar energy maintaining biological activity rate during cold months.

The scientific basis for municipal wastewater treatment in a vascular aquatic plant system combined with a microbial rock filter (MRF) is the cooperative growth of both the plants and microorganisms associated with the plants and rocks. A major part of the treatment process for degradation of organics is attributed to the microorganisms living on and around the plant root systems and the rock filter. Organics are held in place by the rocks and plant roots where microorganisms are given time for assimilation.

This technology grows only selected plants in wastewater. The rock filters the wastewater in conjunction with the plant roots. Hydroponics is defined as "the growing of plants in a nutrient solution and without soil."

This technology combines the application of hydroponics and the MRF technologies. The rocks (inert) support the plants and roots in a nutrient solution (wastewater). Thus the technology is appropriately defined as a microbial rock plant filter (MRPF).

It is defined by some as a subsurface flow constructed wetland. But this technology is not a derivation of the wetland technology. It is a combination of two technologies: the MRF + hydroponics = MRPF. The MRPF uses different size filter media and design philosophy than the surface and subsurface flow systems described in the EPA Design Manual "Constructed Wetlands and Aquatic Plant Systems for Municipal Wastewater Treatment" (EPA/625/1–88/022). The MRPF is a different design concept and requires a different operation and maintenance and plant management program than constructed wetlands.

Aquatic plants have the ability to translocate oxygen from the upper leaf areas into the roots producing an aerobic zone around the roots where aerobic conditions can be maintained. In addition, aquatic plants have the ability to absorb certain organic molecules intact where these molecules are translocated and eventually metabolized by plant enzymes as demonstrated with systemic insecticides.

Biological reactions that take place between environmental pollutants, plants, and microorganisms are numerous and very complex, and to date, are not fully understood. But there is enough information available to demonstrate that aquatic and semi–aquatic plants serve more of a function than simply supplying a large surface area for microorganisms as some have suggested.

1.5.17 Phosphorous Removal

The removal of phosphorus from municipal wastewaters to control receiving water eutrophication has been receiving high priority in many states and may become a significant constraint in the NPDES discharge permit of many municipalities. Technologies exist for removing phosphorus by physical, chemical, and/or biological means. Biological phosphorus removal (BPR) has rapidly emerged as a desirable alternative process because of its relative ease of implementation at existing plants using conventional activated sludge treatment.

Some treatment plants are required to remove phosphorus, although less of these plants must do so today than a few years ago. This is both because of state bans on the use of detergents containing phosphorus and the generally decreased use of phosphorus in household products.

Precipitating phosphorus can as much as double the amount of sludge requiring treatment and disposal. The amount of chemicals that must be added is a function of the amount of phosphorus needing removal. Fortunately, with the quantity of phosphorus in wastewater diminishing, this quantity is decreasing.

Aerobic processes have been modified by including anaerobic and aerobic sequential environments that facilitate biological phosphorous removal.

A Biological Nutrient Removal System (BNR) consists of the following:

1. A conventional suspended growth (activated sludge) biological treatment process;
2. Designed to nitrify;
3. With an anoxic zone added for nitrogen removal; and
4. An anaerobic zone added for phosphorus removal.

Biological phosphorus removal is a recently developed technique of designing suspended growth activated sludge systems to remove soluble phosphorus from wastewater. Variations on this phenomenon are:

1. Phostrip process
2. Modified Bardenpho process
3. A/O process
4. UCT (University of Capetown) process
5. Sequencing Batch Reactor (SBR) process
6. Operationally modified activated sludge processes
7. Mixed chemical/biological processes
8. Simultaneous nitrate and phosphate removal process (German)

1.5.18 Polishing Ponds

Polishing ponds are usually used as an additional solids removal step following biological treatment processes. They are often used in place of secondary clarifiers following aerated lagoons. Where sufficient land is available at low cost, polishing ponds may present an economically attractive alternative to multimedia filtration or microscreening.

1.5.19 Rotating Biological Contactor

In the rotating biological contactor (RBC), a microbial film is built up on a partly submerged support medium which rotates slowly on a horizontal axis in a tank through which the wastewater flows. The microbial film is thus exposed successively to the nutrients in the wastewater and to air as the medium rotates. This motion maintains the biomass in an aerobic condition. The support medium is available in several configurations, such as discs, lattice construction, or a container of plastic balls. The medium is rotated at a speed of about 1 to 7 revolutions per minute using either a mechanical or air–induced drive system. The actual motion of the biological surface is at right angles to the liquid path at most points. This generates turbulence at the solid–liquid interface which permits high mass transfer of nutrients and oxygen into the biological film and enhances sloughing of the excess film into the tank.

The hydraulic retention time is similar to that of trickling filter (e.g., about 20 minutes for a three–stage RBC with a 50 disk array in each stage). However, the RBC requires only 10% of the ground area that is needed for the trickling filter. Because of the relatively low HRT it has good resistance to sudden changes in operating conditions. In addition, the RBC process offers several advantages over other types of biological treatment process such as operational simplicity, low power requirement, and high treatment efficiency.

RBC systems can be run in either batch or continuous–flow mode and either aerobically or anaerobically. Their efficiency is controlled by the hydraulic retention time and the rotation speed of the disks. Like all fixed–film systems, they require a fairly long start–up time while the biofilm grows, and they are sensitive to temperature, shock loading, and extreme dryness.

There is no recycle of sludges or recirculation of treated effluent in an RBC process. Several RBCs are often operated in series, with the effluent from the last RBC being discharged. Biological solids are usually dewatered prior to disposal.

Rotating biological contactors can be used for treatment of leachate containing readily biodegradable organics. Although not as efficient as conventional activated–sludge systems, RBC's are better able to withstand fluctuating organic loadings because of the large amount of biomass they support.

1.5.20 Roughing Filter

The primary function of a roughing filter is to reduce high organic loadings by its use as an intermediate treatment process upstream of an activated sludge, or perhaps secondary trickling filter, process. Although rock or other media may be used, the typical roughing filter uses plastic media. The roughing filter installation commonly requires forced ventilation.

1.5.21 Sequencing Batch Reactor

The sequencing batch reactor technology is similar to the more widely used activated–sludge process. The main difference is that the five–step treatment cycle is carried out in one tank in batch mode. Sequencing batch reactor technology predates that of the continuous–flow activated–sludge process, but was little used for many years. Recent advances in equipment have caused revived interest in sequencing batch reactors, but their use is still in its infancy. Several full–scale facilities are operating in this country, treating both municipal wastes and hazardous waste leachates.

The process basically consists of five unit processes: fill, react, settle, decant, idle. Reactions initiated during fill (when influent enters the tank) are completed during react, with no flow entering or leaving the tank. Solids separation is accomplished during a similar, ideal quiescent settle period. Clarified supernatant is discharged during decant. While waiting for the start of the next fill cycle, the system is in idle. The SBR process differs from conventional systems in that time is used to separate unit process steps as opposed to multiple dedicated process tanks. This provides powerful flexibility with obvious inherent design, process, and operation advantages. Any unit process operation or sequence can be altered after startup by simply changing time allotments to affect an increase, decrease, or restructuring of any part of the process.

Sequencing batch reactors have been shown to handle greater flows and higher loads with better effluent quality than activated–sludge facilities. Also, they provide greater flexibility in operation and can be used intermittently when waste generation is low. Because only one tank is needed, capital and space requirements are less than for the activated–sludge process. The operation of sequencing batch reactors does require a better control strategy than for activated sludge, however. Also, the system does not tolerate rapid changes in temperature. As is the case for activated sludge, the excess sludge, which may contain toxic organics and heavy metals, need to be treated and disposed of.

The Modified Sequencing Batch Reactor (MSBR) process is designed to achieve the effluent quality associated with sequencing batch reactors, while providing continuous flow at a constant level. Because of the continuous flow, the MSBR process does not require separate tanks for receiving and treating wastewater. In addition, because level remains constant, pumping requirements are minimal.

1.5.22 Submerged Packed Beds

In the process, a stationary biofilm grows on a fixed bed, e.g., wastewater, which is held at a constant level in the treatment tank. For aerobic systems, air is sparged up through the bed. A compartmental design is often used to prevent hydraulic short-circuiting of the primary effluent as it flows in the system. The diffused air helps maintain a mixed liquid and also shears excess biofilm off the medium. The system can then be periodically backwashed to remove the excess biomass.

Submerged packed beds require less energy than fluidized beds and avoid the climate problems associated with rotating biological contactors. They also require little space. No information has been located on air emissions from submerged packed beds. Because they are sparged, emissions are likely to occur from aerobic systems. Emissions can be controlled if the reactors are covered.

1.5.23 Surface Impoundments

A surface impoundment is an excavation or diked area typically used for the treatment, storage, or disposal of liquids, e.g., wastewater, or materials containing free liquids, e.g., sludges. The hydraulic barriers in surface impoundments are usually constructed of low-permeability soil or polymeric membranes or both.

Liquids and solids typically separate in a surface impoundment by gravity settling. Liquids can be removed by draining, evaporation, or flow from an outlet structure. Accumulated solids may be removed by dredging during impoundment operation or when it is closed. Alternatively, solids may be left in place, as a landfill, when the surface impoundment is closed.

In the United States, nearly 30,000 are used by industry, including chemical manufacturers, food processors, oil refineries, primary and fabricated metals manufacturers, paper plants, and commercial waste facilities. Most surface impoundments are not used for waste disposal but rather for waste treatment processes, i.e., neutralization, settling, anaerobic or aerobic digestion, pH adjustment, and polishing. The industrial surface impoundments range in size from less than 0.1 acre (29%) to greater than 100 acres (1%), with the majority less than 5 acres. (One acre = 0.405 hectare.)

The EPA national survey categorized surface impoundment applications into five groups with the percentages in each group used for storage, disposal, or treatment. The majority of agricultural surface impoundments were used for waste storage; the majority of oil and gas surface impoundments for disposal; and the majority of municipal, industrial, and mining impoundments were used for treatment.

The type of surface impoundment required depends on waste composition, waste-generation rate, and the purpose of the impoundment. A surface impoundment can be classified as one of three generic impoundment types: (1) treatment; (2) surge or equalization, i.e., storage; and (3) non-discharging (evaporation or disposal). The greatest number in use are of the treatment type.

Waste inputs and treated waste discharges from treatment impoundments may be steady, fluctuating, or intermittent. Except for some surge or equalization impoundments that are intended to collect runoff, the only external water input is direct precipitation on the impoundment surface and interior dike slopes. Non-discharging surface impoundments generally rely strictly on natural evaporation to maintain liquid level.

More than one impoundment may be required where several incompatible liquid wastes are to be stored. Multiple impoundments may also be desirable for single or compatible wastes in some situations.

The term "surface impoundment" is an all inclusive term covering a number of various processes as follows.

Facultative Lagoons: Facultative lagoons, the most common type, treat wastewater by anaerobic fermentation in the lower layer and aerobic stabilization in the upper layer. The key treatment mechanisms comprise oxygen production by photosynthetic algae and surface reaeration. Aerobic bacteria use the oxygen to stabilize the organic material in the upper layer. Facultative lagoons are used to treat raw municipal wastewater (usually from small communities and also to treat primary or secondary effluent (for small or large cities). The facultative lagoon is the easiest to operate and maintain. Large land areas are required to maintain lagoon biochemical oxygen demand (BOD_5) loadings in a suitable range. The lagoon's facultative treatment capability for raw wastewater usually does not exceed secondary treatment.

Aerated Lagoons: In an aerated lagoon, oxygen for breakdown of pollutants is supplied mainly through mechanical or diffused air aeration rather than by photosynthesis and surface reaeration. Many aerated lagoons are modifications of overloaded facultative lagoons that require aerator installation to supply additional oxygen for proper treatment performance.

Aerobic Lagoons: Aerobic lagoons, much shallower than either facultative or aerated lagoons, maintain dissolved oxygen throughout their entire depth. Oxygen, provided by photosynthesis and surface reaeration, is used by bacteria to stabilize the pollutants. Mixing is often provided to expose all algae to sunlight and to prevent anaerobic conditions at the bottom of the lagoon. Use of aerobic lagoons is limited to warm, sunny climates where a high degree of BOD_5 removal is desired but land area is limited. Because of shallow lagoon depths, the bottoms of aerobic lagoons must be paved or covered to prevent weed growth.

Anaerobic Lagoons: Anaerobic lagoons receive such a heavy organic loading that formation of an aerobic zone is prevented. The principal biological reactions comprise acid formation and methane fermentation. Use of anaerobic lagoons is limited principally to treatment of strong industrial and agricultural wastes, or to pretreatment where an industry contributes wastewater to a municipal system.

Waste Stabilization Ponds (Oxidation Ponds): A type of surface impoundment which relies only on natural processes for aeration. Waste stabilization ponds treat dilute aqueous wastewaters (less than 0.1% solids) with low concentrations of organics. Natural biodegradation reactions are allowed to proceed as wastewater passes slowly through large shallow basins, subjected to wind aeration and sunlight energy for photosynthesis. BOD removal efficiencies range from 50 to 90%. Although large land acreages are required, energy and chemical requirements are insignificant.

1.5.24 Trickling Filters

A trickling filter is an aerobic attached-growth biological treatment process. The system consists of an equalization basin, a settling tank, a filter medium, an influent wastewater distribution system, an under drain system, a clarifier, and a recirculation line. The filter medium consists of a bed of an inert material to which the microorganisms attach themselves and through which the wastewater is percolated. Rocks or synthetic material such as plastic rings and saddles are typically used as filter media. Following equalization and settling of settleable solids in the wastewater, it is distributed over the top of the filter medium by a rotating distribution arm or a fixed distributor system. The wastewater forms a thin layer as it flows downward through the filter and over the microorganism layer on the surface of the medium. As the distribution arm rotates, the microorganism layer is alternately exposed to a flow of wastewater and a flow of air. In the fixed distributor system, the wastewater flow is cycled on and off at a specified dosing rate to ensure that an adequate supply of oxygen is available to the microorganisms. Oxygen from air reaches the microorganisms through the void spaces in the medium.

A trickling filter system is typically used as a roughing filter to reduce the organic loading on a downstream activated sludge process. Trickling filters can be used for the treatment of wastewaters that could potentially produce "bulking" sludge (i.e., a sludge with poor settling characteristics and poor compactability in an activated sludge process) because the microbial solids that slough off the trickling filter medium are relatively dense and can be readily removed in a clarifier.

Trickling filters may be used to biodegrade nonhalogenated and certain halogenated organics in leachate. Although not as efficient as suspended-growth biological treatment processes, trickling filters are more resilient to variations in hydraulic and organic loadings. For this reason, trickling filters are best suited to use as "roughing" or pretreatment units that precede more sensitive processes such as activated sludge.

There are both high-rate and low-rate trickling filters. A typical low-rate trickling filter will have rock media with wastewater application by a rotary distributor. Using recirculation to increase the hydraulic loading, the high-rate trickling filter will accept higher organic loadings. There is a continuous sloughing of excess biological growths. The higher organic load precludes the development of nitrifying bacteria in the filter bed.

Trickling filters are generally applicable to the treatment of the same types of hazardous wastes that are treatable by activated sludge. Because of the relatively short residence time of wastewater in contact with microorganisms, however, the percentage removal of organics is not as high as in activated sludge treatment. Greater removals are achieved as the depth of media and the recycle ratio are increased. Trickling filters are reported to have successfully handled the following waste constituents: acetaldehyde, acetic acid, acetone, acrolein, alcohols, benzene, butadiene, chlorinated hydrocarbons, cyanides, epichlorohydrin, formaldehyde, formic acid, ketones, monoethanolamine, propylene dichloride, and resins.

The advantage of the trickling filter process compared to other wastewater treatment systems is that no power is consumed in agitation or aeration for the creation of the gas-liquid contact area. Power is consumed only in transferring liquid to and from the unit and in distributing it over the packed bed, so that the operating cost is low.

The trickling-filter/solids-contact process was developed in 1979 and is gaining in popularity. It produces better quality effluent than trickling filters and is economical and reliable. In the process, a trickling filter is used to degrade soluble BOD. The effluent from the trickling filter is further treated in an aerobic solids-contact tank. The main purpose of the tank is to decrease particulate BOD by means of flocculation. It also is able to remove approximately 75% of the soluble BOD which remains after trickling-filter treatment. Thus, the solids-contact tank serves as a polishing unit with a low retention time. Flocculation is initiated when the effluent contacts biological solids. The system also has two clarification tanks in which the flocculated solids settle. Some of the settled solids are recycled to the solids-contact tank.

1.5.25 Wetlands (Natural)

While the interest in wetlands for wastewater treatment is fairly recent, the term wetlands is also a relatively new expression, encompassing what for years have simply been referred to as marshes, swamps, or bogs. The difference in these wetlands is related to a large extent to the vegetation which dominates the area. Grasses, or forbs are generally dominant in marshes, trees and shrubs characterize swamps, and sedge/peat vegetation occurs in various bogs.

Natural wetlands are effective as wastewater treatment processes for a number of reasons. Natural wetlands support a large and diverse population of bacteria which grow on the submerged roots and stems of aquatic plants and are of particular importance in the removal of BOD_5 from wastewater. In addition, the quiescent water conditions of a wetland are conducive to the sedimentation of wastewater solids. Other aspects of wetlands that facilitate wastewater treatment are the adsorption/filtration potential of the aquatic plants' roots and stems, the ion exchange/adsorption capacity of wetlands' natural sediments, and the mitigating effect that the plants themselves have on climatic forces such as wind, sunlight and temperature.

Most states (except Florida, and a few others considering special wetland standards) make no distinction between the wetland and the adjacent surface waters and apply the same requirements to both. Under these conditions, economics will not favor the utilization of natural wetlands as a major component in a wastewater treatment process as the basic treatment must be provided prior to discharge to the wetland.

Special situations may arise in which natural wetlands may provide further effluent polishing or, if the wetland is isolated from other surface waters, more basic treatment. The use of treated effluent for enhancement, restoration, or creation of wetlands can be a very desirable and environmentally compatible activity.

Ecologists have long understood that soils in wetlands are often foul because they naturally accumulate contaminants. The methods for a wetland to accumulate contaminants include:

1. Filtering of suspended and colloidal material from the water.
2. Uptake of contaminants into the roots and leaves of live plants.
3. Adsorption or exchange of contaminants onto inorganic soil constituents,

organic solids, dead plant material, or algal material.

4. Neutralization and precipitation of contaminants through the generation of HCO_2 and NH_2 by bacterial decay of organic matter.
5. Destruction or precipitation of chemicals in the anaerobic zone catalyzed by the activity of bacteria.

Wastewater treatment by natural and constructed artificial wetland systems is generally accomplished by sprinkling or flood irrigating the wetland area with wastewater or by passing the wastewater through a system of shallow ponds, channels, basins, or other constructed areas where the emergent aquatic vegetation has been planted or naturally occurs and is actively growing. The vegetation produced as a result of the system's operation may or may not be removed and can be utilized for various purposes; e.g., composted for use as a source of fertilizer/soil conditioner, dried or otherwise processed for use as animal feed supplements, digested to produce methane, or eventually harvested as valuable timber.

A wetland is mainly used for polishing treated effluents. Use is highly site-specific and depends upon soil, climate, and wastewater, or contaminated groundwater characteristics. The method is not suited to areas where it is subject to freezing. Use of a wetland for treatment of groundwater or wastewater contaminated with toxic or hazardous materials may not be environmentally acceptable due to the potential risk of spreading dangerous chemicals to a much larger area for a prolonged period of time. However, a wetland may still be considered for use as a polishing treatment after the majority of the toxic compounds have been removed by other treatment methods.

If a natural site is available, a wetland can offer low-cost treatment while requiring a very low level of energy. However, when it is used for treatment of contaminated waters, the system potentially becomes a liability, and is also likely to expose operators to toxic substances.

1.5.26 Wetlands (Constructed)

With so many possible removal processes, a wetland is the typical contaminant treatment system in a natural ecosystem. In addition, it operates in a passive mode requiring no additional reactants and no continuous maintenance.

In the last decade, engineers began to use wetlands for the removal of contaminants from wastewater. In some instances, natural wetlands were used. However, a natural system will accommodate all the above removal processes and probably will not operate to maximize a certain process. If a wetland is constructed, it can be designed to maximize a specific process suitable for the removal of certain contaminants from water. Engineering as well as ecological reasons lead to the choice of constructing a wetland for contaminant removal rather than using an existing natural ecosystem.

Use of wetland wastewater treatment systems based on emergent plant species and their associated microbial communities is more widespread than use of floating aquatic plant systems. Most wetland processes involve the growth of rooted emergent plants such as reeds and bulrushes in an artificial bed and the passage of wastewater either across the surface of the wetland (surface–flow systems), or through the growing medium in which the wetland plants are rooted (subsurface–flow or root zone systems).

The surface–flow wetland approach utilizes the stems of wetland plants as the main site for effluent treatment. In this method, beds of emergent wetland plants, such as reeds or bulrushes, are flooded with pretreated effluent which is retained within the wetland system for a predetermined period prior to discharge. Surface–flow wetland plant stems provide a substratum for the microorganisms which achieve the desired effluent treatment. Wetland processes result in the accumulation of organic material in the bottom of the system where microorganisms also occur in high densities and further enhance effluent treatment, particularly in terms of nitrogen elimination, and anaerobic decomposition of detrital material to carbon dioxide and organic acids.

The principle behind the subsurface–flow wetland treatment system involves passage of wastewater through a specially prepared soil, sand, or gravel medium in which reeds or other emergent plants are grown. Wastewater treatment occurs in the growing medium, principally as a consequence of the growth of wetland plant rhizomes, which are claimed to enhance the hydraulic conductivity of the growth medium and introduce oxygen into adjacent areas of the growing medium.

Wetland treatment effectiveness is a function of retention time and capacity of the vegetation and sediments to retain and/or cycle certain constituents. In using a wetland to polish domestic secondary treated effluent, the following general guidelines are considered reliable. It has been shown that an effluent suspended solids level of 5 to 10 mg/ℓ can be achieved with a retention period of about 1 to 2 days. A longer retention time is required for effective BOD removal. An effluent BOD value of 10 to 15 mg/ℓ can be achieved with 4 to 8 days of retention of a secondary treated effluent. Total nitrogen levels of the order of 4 to 6 mg/ℓ can be achieved with 10 to 12 days of retention. Total phosphorus levels of 2 to 4 mg/ℓ can be achieved with 15 to 20 days of retention. In the case of nitrogen and phosphorus removal vegetation and detritus, harvesting and collection will be necessary prior to decomposition to capture the nitrogen and phosphorus associated with the biomass. This management interval will be a variable depending on the removal requirements, the growing period, and the size of the wetland. In the long run, when steady state conditions are reached, an annual harvesting schedule of a portion of the wetland will be required.

Although using a constructed wetland for wastewater treatment is a relatively new concept, it is of importance to small communities. These communities are attracted to constructed wetlands because they are inexpensive to build and easy to operate.

Most constructed wetlands treat wastewater flows of less than one million gallons per day (mgd), although larger systems can treat flows as high as 20 mgd.

There are two types of constructed wetlands to consider. A free water surface flow (FWS) wetland resembles a natural wetland. Here, the wastewater is exposed as it flows over the surface of the system. This offers benefits in treatment, but also presents concerns about accidental exposure to the wastewater. The wetlands usually must be posted and fenced to prevent accidental exposure to the public.

In the second type, a subsurface flow (SF) wetland, wastewater flows through about one foot of rock media. Because the wastewater is never exposed, there is less concern about exposure, odor problems, and mosquitoes. The rock media, however, represents from 50 to 80% of the cost of a SF wetland.

The major costs and energy requirements for constructed wetlands are associated with preapplication treatment, pumping and transmission to the site, distribution at the site, minor earthwork, and land costs. In addition, a constructed system may require the installation of a barrier layer to limit percolation to groundwater and additional containment structures in case of flooding.

Possible constraints to the use of constructed wetlands for wastewater treatment include the following:

1. Geographical limitations of plant species, as well as the potential that a newly introduced plant species will become a nuisance or an agricultural competitor.
2. Constructed wetlands that discharge to surface water require 4 to 10 times more land area than a conventional wastewater treatment facility. Zero-discharge constructed wetlands require 10 to 100 times the area of conventional wastewater treatment plants.
3. Plant biomass harvesting is constrained by high plant moisture content and wetland configuration.
4. Some types of constructed wetlands may provide breeding grounds for disease producing organisms and insects and may generate odors if not properly managed. Muskrats can also be a problem.

Constructed wetlands, however, offer the engineer greater hydraulic control for general use and are not restricted by many of the environmental concerns and user conflicts associated with natural wetlands. Unlike natural wetlands, which are confined by availability and proximity to the wastewater source, constructed wetlands can be built anywhere, including lands with limited alternative uses. They also offer greater flexibility scope for design and management options and thus may provide superior performance and reliability.

1.5.27 White–Rot Fungus

White–rot fungus *(Phanerochaete chrysosporium)* has the ability to degrade the very complex lignin molecule, and because of this, attention has directed towards utilizing this fungus to destroy hazardous complex organic chemicals, particularly aromatics.

Specifically, white–rot fungus has been shown to degrade lindane, benzo(a)pyrene, DDT, TCDD, and PCBs to innocuous end products. The studies performed, to date, suggest that white–rot fungus may prove to be an extremely useful microorganism in the biological treatment of hazardous organic waste.

The lack of selectivity of the white–rot fungus allows the use of a single organism for treating a mixture of organic compounds, as opposed to the standard use of bacterial consortiums for treating multicomponent contaminants. The primary factors limiting the degradative ability of the fungi in an aqueous phase are the fungi's access to oxygen and the mass transfer resistance, which prevents extensive contact between the organic contaminant and the growing fungi.

Performance of reactors that use an immobilized fungus is superior to reactors in which the fungus was freely suspended. A packed–bed reactor with a porous silica support and a well–mixed reactor with alginate beads as the supporting medium are the best-performing designs.

In a rotating biological contactor, filamentous white–rot fungus attaches to a porous disk that rotates through a contaminated stream. Influent wastewater enters the contactor, contacts the white–rot fungus for a period of time and is then pumped from the contactor.

The contactor can operate in either a continuous or semi–batch mode. Batch operation is not efficient unless nutrients can be added to sustain the growth of fungi, thus making the operation semi–batch.

1.5.28 Flue Gas Treatment

Bacteria are being employed to do what amounts to selective catalytic reduction of oxides of nitrogen (NO_x) in flue gas in a process being developed at Idaho National Engineering Laboratory (INEL); Idaho Falls, Idaho). In tests using a gas stream containing 250 parts per million (ppm) of nitric oxide, the process has converted 99% of the NO to nitrogen and water.

The gas passes through a column that contains *Pseudomonas denitrificans* bacteria immobilized on compost, which also serves as a nutrient source. For a food supplement, a sugar solution is dripped over the bed "every few days."

The bacteria work best at 30° to 45°C, so the system could be put near the end of a stack, where the gases are cooler.

Studies undertaken by the Illinois Department of Energy and Natural Resources have confirmed their preliminary observations that *Botryococcus braunii* can tolerate and grow well in flue gas CO_2 concentrations of 10 to 15%, and produce oil. The highest extracted oil was observed in 10% CO_2 enriched air. Initial pH of the medium at or near 10 pH is favorable to cell growth probably by stimulating the CO_2 solubilization in the medium.

1.6 BIOREMEDIATION

Biological processes are being used to remediate contaminated soils and groundwater, using both ex situ and in situ processes. Sediments, sludges and surface water are also treated by biological processes.

The natural activity of microorganisms is used in the bioremediation process to decontaminate soils and groundwater polluted with organics. Effective microorganisms are often found in small quantities at a contaminated site and, through nutrient enrichment, can be multiplied and encouraged to accelerate the natural degradation process. If the proper organisms are not already present, often they may be introduced.

Bioremediation is the process of using bacteria to biodegrade organic compounds in soils. Under favorable conditions, microorganisms may be capable of completely degrading many organic compounds into carbon dioxide and water or organic acids and methane.

The applicability of bioremediation depends on the biodegradability of site contaminants. Petroleum compounds, such as gasoline and diesel fuel, are known to be readily biodegradable. Other biodegradable contaminants include alcohols, phenols, esters, and ketones. Chlorinated compounds become more difficult to biodegrade as the number of chlorine molecules increases. The biodegradation rate, or half–life, of large, heavily chlorinated compounds such as PCBs is very slow. However, recent work has shown that chlorinated contaminants are more degradable than previously assumed.

Many highly chlorinated aromatics and aliphatics can be destroyed microbiologically most rapidly by sequential anaerobic–aerobic treatment. In general, the biochemical pathway providing the highest rate for the initial steps of microbial destruction of the highly chlorinated organics is anaerobic reductive dechlorination. Once partially dechlorinated, the resulting compounds typically degrade faster under aerobic, oxidizing conditions.

It has recently been reported that of the 130 sites currently being tracked by the Bioremediation Field Initiative, 85 (65%) are planning, implementing, or have recently completed the use of bioremediation technologies on soils. According to collected data, bioremediation of the vadose zone is much more prevalent than of the saturated zone. Groundwater, the next most frequent media receiving biotreatment, is being remediated at 55 sites. Sediments and sludge are third and fourth with 15 and 12 sites, respectively. Only 2 of the sites reporting data are bioremediating surface water at this time. Many sites are conducting or planning bioremediation activities for more than one media.

Ex situ treatment is currently the most popular bioremediation technique, with over 80 sites employing some form of bioreactor, land treatment, pile, or treatment in an aerated lagoon.

Bioremediation and thermal desorption appear to be the favored innovative technologies to treat sites where VOCs occur with SVOCs. Bioremediation has been chosen 22 times to treat VOCs, primarily nonchlorinated VOCs, such as benzene. In all but five cases, SVOCs also are being treated.

Over 50 sites are utilizing in situ treatment including land treatment, pile, bioventing, nutrient addition, confined treatment, and other technologies. The vast majority of sites report operating units under aerobic rather than anaerobic conditions, although several sites are employing both. Thirty-five sites report using indigenous organisms, while 12 reported using exogenous organisms, and a few sites are using or planning to use both.

Any aqueous waste streams associated with processes described in this section can also be treated biologically by the wastewater treatment processes discussed in the previous section.

Treatment trains employing one or more treatment processes may be required for complex waste streams; and bioreclamation can be preceded by, or otherwise used in combination with, other treatments that can reduce toxic concentrations to a tolerable level.

In addition, bioremediation processes can be used for oil spill cleanup on water or land. The EPA successfully investigated bioremediation techniques in a field demonstration project at the Prudhoe Bay oil spill in Alaska.

In situ bioremediation has four distinguishing features:

1. The active agents for the cleanup are microorganisms, usually bacteria, that biodegrade the contaminants; hence the bio part of bioremediation emphasizes the use of microorganisms.
2. The microorganisms are present in the intact aquifer or soils and perform their biodegradation reactions in situ, or in place. Thus the soil and water do not need to be removed for treatment.
3. Naturally occurring or added bacteria are "stimulated" to bring about rapid biodegradation rates. Making the rate of biodegradation as fast as possible makes the bioremediation approach technically and economically attractive.
4. Stimulation means that the number of microorganisms active in biodegradation of the contaminant is increased by many orders of magnitude. This acceleration is brought about by the controlled addition

of the materials that are normally missing from the environment and, when added in the correct amount, allow growth and activity of the desired microorganisms.

1.6.1 Biotreatments—Advantages

(1) Ability to Treat Wastes On Site: Cuts out transportation costs and eliminates risk underlying that operation. For example, even the simple collection of municipal garbage amounts to over half of the total cost of dealing with these solid wastes.

(2) Minimum Disruption of Sites: (a) Since waste can be treated on site, there is no need to dig up and haul away anything unless there are some "untreatable" contaminants in the site. (b) Little disruption is caused by operations such as addition of oxygen and pumping water through the contaminated area in order to dilute the waste for biological treatment.

(3) Faster Than Certain Other Methods: Air stripping or carbon adsorption can take up to 50 years, while bioreclamation could be in the one, two or three year time frame.

(4) Use as Polishing Step with Other Methods: (a) Importance of looking at biotech in combination with other technologies—it is shortsighted to focus on individual tools. (b) Final step that gets rid of trace contaminants and achieves permanent degradation of wastes.

(5) Biosystems Are Not Energy Intensive: Microorganisms can work at ambient temperatures, especially the aerobic species, vs, for example, oxidation through incineration.

(6) Others Include: (a) low capital and operating costs, b) minimal specialized equipment requirements, (c) low technology profile, and (d) availability of trained contractors.

1.6.2 Biotreatments—Disadvantages

(1) Temperature: At temperatures below 50°F, the metabolism of microorganisms slows significantly. This means that in northern countries, biodegradation ex situ processes are seasonal unless the installation is heated.

(2) Dilute Conditions: Microbes can assimilate waste in an aqueous system only. They work better when contaminants are diluted. (a) They work best when there are a few tens to a few thousands of parts per million of pollutants in soil or water. (b) Large molecules are often insoluble.

(3) Specificity of Microorganisms: No one microbe does the job alone; usually a complex mixture of microflora is required. The more complex the component, the more complex the microbial population has to be to handle it.

(4) Black Box Syndrome: (a) The main problem is getting industry people to understand how systems work and how to operate them properly—85% of biological treatment system failures are due to human error rather than system problems. (b) Companies using biosystems must hire specialists to keep the systems working properly.

(5) Fragility of Biosystems: (a) Standard microbial products can have shelf-life problems. (b) Toxic effluents can poison the biomass, creating problems to reactivate the microflora. (c) Biosystems will not survive if not fed properly.

(6) Site Specific Technology: Hydrogeological factors can limit the use of bioremediation methods.

(7) Biodegradation Not Applicable to All Contaminants: Not all contaminants are susceptible to biodegradation; for example, metal contaminants, cyanide complexes, radioactive wastes or inorganic substances. Very large complex molecules do not exist in nature; therefore, bacteria generally lack enzymes that can degrade them.

(8) Time Requirements: Can take up to 2 years to complete.

(9) Space Constraints: Crowded conditions due to buildings or other obstructions can cause difficulties.

1.6.3 Reasons for Failure

Frederickson, et al, and Black, et al, have presented a number of reasons for failure of the process, including:

1. The presence of cotoxicants such as heavy metals that inhibit biodegradation,
2. Physical constraints on electron acceptor—nutrient delivery,
3. Slow reaction rates caused by physical constraints (e.g., low temperature),
4. Biologically unavailable contaminants,
5. Conversion of contaminants to toxic metabolites,
6. Heterogeneous distribution of contaminants,
7. Lack of microorganisms with the necessary biochemistry to degrade target contaminants,
8. Soil with high percentage of clays, can slow the procedure,
9. Sorption of contaminants to organic matter,
10. Aging bonds the contaminant to the soil matrix, reducing the ability of the contaminant to dissolve in the water phase.
11. High salinity soils, and
12. Oil concentrations in excess of 10%.

1.6.4 Soils—Ex Situ

Ex situ biological processes for remediating contaminated soils consist of the solid-phase, and slurry-phase treatment processes. land farming could also be utilized, as long as regulations are adhered to.

Solid-Phase Treatment: Solid-phase soil bioremediation is a process that treats soils in an above grade system using conventional soil management practices to enhance the microbial degradation of contaminants. The system can be designed to contain and treat soil leachate and volatile organic compounds. It has been used to treat pentachlorophenol and creosote wastes, oil field and refinery sludges, pesticide wastewaters, gasoline, PCBs and PAHs.

The system consists of a treatment bed which is lined with a high-density liner. Clean sand is placed on top to provide protection for the liner and proper drainage for contaminated water as it leaches from contaminated soils placed on the treatment bed. Lateral perforated drainage pipe is placed on top of the synthetic liner in the sand bed to collect soil leachate. If volatile contaminants must be contained, the lined soil treatment bed is completely covered by a modified plastic film greenhouse. An overhead spray irrigation system contained within the greenhouse provides for moisture control and a means of distributing nutrients and microbial inocula to the soil treatment bed.

Volatile organic compounds which may be released from the soil during processing are swept through the structure to an air management system. Biodegradable volatile organic compounds can be treated in a vapor phase bioreactor. Non–biodegradable volatile organic compounds can be removed from the effluent gas stream by adsorption on activated carbon or incineration.

Contaminated leachate which drains from the soil is transported by the drain pipes and collected in a gravity–flow lined sump and then pumped to an on–site bioreactor for treatment. Treated leachate can then be used as a source of microbial inocula and reapplied to the soil treatment bed through an overhead irrigation system, after adjusting for nutrients and other environmental parameters.

In another variation, soil is excavated and placed in a treatment area, in mounds resembling extended compost heaps. In the "dry" variation of this process, the soil is periodically turned over to ensure good aeration; in the "wet" technique, a sprinkling system is used to add water and nutrients. An air distribution system, buried in the piles, augments the oxygen supply.

Contained Solid–Phase: Treatment occurs in an enclosure which allows more process control. Treatment in a device which is defined as a "tank" under RCRA is viable means of achieving land ban requirements. Testing is underway on PCB waste in rotating drums which allow control of oxygen levels. No secondary containment is required if the waste does not contain free liquids. It is more costly than land treatment (farming) due to extra expense required to satisfy tank standards and provide more process control, e.g., aeration.

Slurry–Phase Treatment: This biodegradation technology involves the treatment of contaminated soil or sludge in a large mobile bioreactor. This system maintains intimate mixing and contact of microorganisms with the hazardous compounds and creates the appropriate environmental conditions for optimizing microbial biodegradation of target contaminants.

The first step in the treatment process is to create the aqueous slurry. During this step stones and rubble are physically separated from the waste, and the waste is mixed with water, if necessary, to obtain the appropriate slurry density. The water may be contaminated groundwater, surface water, or another source of water. A typical soil slurry contains about 50% solids by weight; a slurried sludge may contain fewer solids. The actual percent solids is determined in the laboratory based on the concentration of contaminants, the rate of biodegradation, and the physical nature of the waste. The slurry is mechanically agitated in a reactor vessel to keep the solids suspended and maintain the appropriate environmental conditions. Inorganic and organic nutrients, oxygen, and acid or alkali for pH control may be added to maintain optimum conditions. Microorganisms may be added initially to seed the bioreactor or added continuously to maintain the correct concentration of biomass. The residence time in the bioreactor varies with the soil or sludge matrix, physical/chemical nature of the contaminant, including concentration, and the biodegradability of the contaminants. Once biodegradation of the contaminants is completed, the treated slurry is dewatered. The residual water may require further treatment prior to disposal.

Depending on the nature and concentration of the contaminants, and the location of the site, any emissions may be released to the atmosphere, or treated to prevent emission. Fugitive emissions of volatile organic compounds, for instance, can be controlled by modifying the slurry–phase bioreactor so that it is completely enclosed.

Aside from the biodegradability of a particular compound, other limiting factors include the presence of inhibiting compounds and operating temperature. Heavy metals and chlorides may inhibit microbial metabolism because of their toxicity. The operating temperature range is approximately 15° to 70°C. Dissolved oxygen is also critical and must be monitored along with pH, nutrients, and waste solubility.

One advantage of treatment in a contained process is that a remediation system can be designed to pretreat waste contaminated with heavy metals as well as biodegradable semivolatile and volatile compounds. Soil washing and extraction of metals using weak acids and chelating agents can be combined with biological treatment by coupling two separate slurry–phase reactors in series.

Advantages:

1. Offers most control of the physical/chemical environment.
2. Most certain process to monitor in terms of effectiveness.
3. Enclosed reactors can capture fugitive volatile emissions.
4. Provides highest biological reaction rates.
5. Offers capability to treat the broadest range of organic compounds and soil types, i.e., treat the most difficult to biodegrade.
6. Treatability testing and engineering scale–up for this technology is relatively simple.

Disadvantages:

1. Considerable energy may be required to keep soil in suspension, thereby adding to cost. Pretreatment may be necessary to remove dense material (gravel, stones, etc.).
2. Tanks or containers need to meet appropriate RCRA standards, including requirements for secondary containment.

FEBD Process: The Institute of Gas Technology (IGT) Fluid Extraction–Biological Degradation (FEBD) Process extracts hydrocarbon contaminants from soil and then biologically degrades the pollutants in aerobic bioreactors. The process consists of three stages; extraction, separation, and biodegradation. Contaminants are first removed from the soil by solubilization in supercritical carbon dioxide in an above ground extraction vessel. The hydrocarbon contaminants are then collected in a separation solvent, and clean CO_2 is recycled to the extraction stage. Separation solvent containing the organic wastes is sent to the biodegradation stage where the wastes are converted to CO_2, water, and biomass.

Landfarming: The controlled application of waste materials to soil for degradation by the resident microflora is called landfarming. Landfarming of petroleum wastes has proven to be a successful alternative to incineration when energy conservation and costs are considered. This alternative to in situ biotreatment may be employed in cases where soil permeability is too low for effective groundwater recirculation. The contaminated soil is spread over the surface of the landfarm and incorporated into the top 8 to 12 inches of clean soil. Nutrients can be added at this time, and the soil can be tilled to increase the oxygen level for enhanced biodegradation. Rototilling equipment vigorously mixes the soil, promoting the aeration and mixing process more effectively than disks or bulldozers. Tilling the waste material into the soil immediately after application will decrease its chance of migration out of the area. This process has been used extensively as a disposal mechanism for oily sludge. EPA "no migration" regulations must be carefully considered.

Advantages:

1. Relatively maintenance-free operation. Soil is applied in lifts which are several inches thick and periodically cultivated (approximately 2 week intervals). Nutrients or manure may be added as supplements. Occasional irrigation may be required to maintain soil moisture.
2. Construction of leachate collection system will minimize chance for off-site migration.

Disadvantages:

1. No control of volatile emissions.
2. Land treatment is defined as a form of land disposal under RCRA 3004(b). If "placement" invokes land ban requirements, a no-migration petition is required.
3. Local government conflicts.
4. Poor public perception.

Populations of bacteria added to soils often decline rapidly and become metabolically inactive. To efficiently degrade contaminants, microorganisms must be metabolically active. Thus, a significant obstacle to the successful use of microorganisms for environmental applications is their long-term survival and the expression of their degradative genes in situ. Rhizosphere microorganisms are known to be more metabolically active than those in bulk soil, because they obtain carbon and energy from root exudates and decaying root matter. Rhizosphere populations are also more abundant, often containing 10^8 or more culturable bacteria per gram of soil, and bacterial populations on the rhizoplane can exceed 10^9/g root.

Many of the critical parameters that influence the competitive ability of rhizosphere bacteria have not been identified, but microorganisms have frequently been introduced into soil (bioaugmentation) as part of routine or novel agronomic practices. However, the use of rhizosphere bacteria and their in situ stimulation by plant roots for degrading organic contaminants has received little attention. Published studies have demonstrated the feasibility of using rhizobacteria *(Pseudomonas putida)* for the rapid removal of chlorinated pesticides from contaminated soil.

Land application is also discussed in section 1.5.14.

Composting: This technique can provide an interesting treatment alternative. It has been utilized for petroleum contaminated soils. An additional benefit of composting is that the end product could be useful as top soil, mulch or fill material. See also 1.5.10.

1.6.5 Soils—In Situ

In situ biodegradation is the term for biological treatment processes that are performed in place and therefore do not require excavation and removal of the contaminated soil. This treatment method includes widely used technologies such as land treatment as well as some emerging technologies that employ subsurface injection of oxygen or nutrients to promote the biodegradation of contaminants.

A major limitation of this system is that it only works when the adapted bacteria are in direct contact with the contaminants. This requires constant turning of the soil and removal of "clean" soil. This process can be time consuming and inexpensive, depending on the depth of contamination.

Proper mixing of bacteria in the soil is essential, as is direct contact between the microbes, the soil and added nutrients. This is accomplished by mixing in thin layers and turning the soil over mechanically. Each layer of treated soil must be removed before the next layer can be treated.

The technology involves the adaptation of naturally-occurring microorganisms to perform specific biodegradation of targeted hazardous wastes. Once adapted, the process involves the accelerated growth of these microorganisms and eventually, inoculation into the soil or other matrix in which the water is contained. Nutrients and catalysts are also added to the matrix to enhance the microbial activity. The matrix is then physically manipulated and subsequent inoculation of microorganisms, nutrients and catalysts are added over time depending on the need. Biodegradation of the contaminants occurs over a relatively short period of time (usually two to four months).

In situ bioremediation in the unsaturated (vadose) zone can be applied as a specialized form of soil vacuum extraction. The air circulation induced by soil vacuum extraction ensures an ample supply of oxygen to the indigenous microbial population. Other vadose zone in situ bioremediation systems use infiltration galleries or injection wells for delivery of oxygen and nutrients.

Since volatilization makes a potentially large contribution to the overall removal achieved by most in situ biotreatment processes, this technology is generally not suitable for remediating sites which are contaminated with volatile fuels or other contaminants, or for remediating sites that are close to sensitive receptors. In situ biotreatment is best suited for volatile fuel sites in remote locations, and sites that are contaminated with less volatile fuels (such as JP–4, JP–5, or diesel fuel).

In situ biodegradation is often used in conjunction with a groundwater pumping and reinjection system to circulate nutrients and oxygen through a contaminated aquifer and associated soils. It can provide substantial reduction in organic contaminant levels in soils without the cost of soil excavation.

Enhanced biodegradation (bioreclamation) is one of the in situ methods that is engineered to create favorable aerobic conditions in unfavorable conditions such as nonhomogeneous soils, delicate geochemical balances, and uncertain organic substrates. A major rate limiting factor in in situ biodegradation is the presence of dissolved oxygen. Hydrogen peroxide is currently the preferred oxygen source; at 40 mg/ℓ of groundwater, it releases enough oxygen to maintain continuous biodegradation. Other sources of oxygen include air, and pure oxygen. Nitrate is also being investigated as an alternate electron receptor.

The presence of iron in the subsurface causes hydrogen peroxide depletion at a faster rate. A prerequisite for the application of hydrogen peroxide as an oxygen source is soil pretreatment, which is necessary to prolong the stability of peroxide in situ. Several phosphate compounds are currently being tested as complexing agents for iron to increase the stability of peroxide. Anaerobic pathways are also available but are generally considered too slow to constitute active cleanup.

In situ bioremediation is applicable to a majority of contaminants found in soils, since most of them are organic compounds derived from agriculture, industry, commerce, and transportation. On the down side, many sites are contaminated with compounds from several organic classes along with metals, salts, and radionuclides. In situ bioremediation cannot, at present, address all aspects of such a site.

Landfarming involves the aeration of oil and other hazardous materials in soil and sludges by tilling or other cultivation methods, with the addition of nutrients. This method has been used by the oil refining industry for the disposal of oily sludges for many years. The methods can be applied in situ, where soil contamination is relatively shallow. Addition of microbial cultures can be used to augment the indigenous microbial population and speed up the rate of biodegradation.

Advantages:

1. Under favorable conditions, this will be the lowest cost bioremediation technology.
2. Ongoing testing concerns degradation of chlorinated compounds (TCE) using methane oxidizing bacteria (methanotrophs).
3. Although the reaction rate is lowest, a large volume of soil may be treated at once.
4. Research in anaerobic processes for reductive dechlorination shows promising results.

Disadvantages

1. Applications limited to favorable site conditions which require soils that are sandy and highly permeable (K greater than 10^{-3} m/s).
2. Extensive treatability studies and site characterization is required. Relevant physical and chemical parameters include pH, redox conditions, temperature, TOC, Fe and Mn concentrations, heavy metals, and nutrients (nitrogen and phosphorous) dissolved oxygen, carbon dioxide, nitrate, and sulfate.
3. Proper design and operation is necessary to avoid groundwater contamination.
4. The precise fate of degraded hydrocarbons, such as gasoline, is not yet known.
5. Most difficult process to conclusively monitor cleanup efficiency since no mixing takes place, i.e., it may be difficult to get characteristic or representative samples if soil concentrations vary widely.
6. Water recirculation may be limited by biofouling or biological growth which reduces permeability.
7. Difficulties may arise in the dissemination of oxygen and nutrients in low permeability or highly heterogeneous regimes.
8. Some states may not allow reinjection of treated groundwater, therefore, amendments must be delivered to the injection point in clean water.
9. May be relatively ineffective for LNAPL and DNAPL.

1.6.6 Groundwater–Ex Situ

Ex situ groundwater processes are termed Pump–and–Treat, which is discussed in Chapter 6. Almost all remediation of groundwater at heavily contaminated sites is based on groundwater extraction by wells or drains, usually accompanied by treatment of the extracted water prior to disposal. This often causes initial decrease in contaminant concentrations in the extracted water, followed by a leveling of concentration, and sometimes a gradual decline that is generally expected to continue over decades. In such cases, the goal of reaching stringent health-base cleanup standards is very remote, and the ultimate cost of cleanup very high.

After the groundwater has reached the surface it can be remediated by any number of physical, chemical, and biological methods, and reinjected, if allowed. Biological methods include: activated sludge, aerated surface impoundments, land treatment, anaerobic digestion, trickling filters, and rotating biological discs.

1.6.7 Groundwater–In Situ

An emerging technology for the in-situ remediation of groundwater is the use of microorganisms to degrade contaminants which are present in aquifer materials. Although in situ bioremediation has been used for a number of years in the restoration of groundwater contaminated with petroleum hydrocarbons, its application to other classes of contaminants is relatively recent.

Most biological in situ treatment systems are carried out by stimulating indigenous microorganisms to degrade those organic contaminants dissolved in groundwater and attached to aquifer solids. The process, which is an adaptation of earlier attempts to remediate gasoline-contaminated aquifers, involves the circulation of oxygen and nutrients through a contaminated aquifer using extraction and injection wells. The placement of the wells depends on the size and configuration of the affected area, and the hydraulic conductivity of the groundwater formation.

Research is under way to test the use of nitrate instead of oxygen during in situ treatment systems to promote the anaerobic degradation of organic contaminants. Investigations into additional methods to enhance in situ bioremediation include the addition of a readily degradable substrate to aid in the degradation of more recalcitrant molecules, and the addition of a nontoxic substitute for a specific contaminant in order to induce degradative enzyme activity that will affect both the substitute and the specific contaminant.

In addition to the stimulation of indigenous microbial populations to degrade organic compounds in a contaminated aquifer, another technique, which has not been fully demonstrated, is the addition of microorganisms with specific metabolic capabilities. These microbial populations have been altered to degrade specific compounds by enrichment culturing or genetic manipulation. Enrichment culturing involves exposure of microorganisms to increasing concentrations of a contaminant. Genetic manipulation is accomplished by exposure of organisms to a mutagen, followed by enrichment culturing, or by the use of DNA recombinant technology to change the genetic structure of the microorganism. It is important to note that the inoculation of specialized microbial populations into the subsurface may not result in degradation for a number of reasons including the concentration of the contaminant, geochemistry of the formation, or other organisms that are toxic or inhibitory to the inoculated organisms.

There are a number of advantages to the use of in situ bioremediation. Unlike other aquifer remediation techniques, it can often be used to treat contaminants that are sorbed to aquifer material or trapped in pore spaces. The time required using in situ bioremediation can often be faster than extraction and treatment processes. For example, a gasoline spill was remediated in 18 months using in situ bioremediation, while pump–and–treat techniques were estimated to require 100 years. In situ bioremediation often costs less than other remediation options. There are also disadvantages to in situ bioremediation. Many organic compounds are resistant to degradation as are heavy metals. In addition, organic compounds that otherwise might be subject to degradation may be toxic or inhibit the growth of microorganisms at concentrations often found at contaminated sites. Injection wells may also become clogged from profuse microbial growth resulting from the addition of nutrients and oxygen.

In situ bioremediation is difficult to implement in low permeability acquifers that do not permit the transport of adequate supplies of nutrients and electron acceptors to active microbial populations. Aquifers with hydraulic conductivities of 10^{-4} cm/sec (100 ft/yr) or more usually considered good candidates for in situ bioremediation.

In most contaminated hydrogeologic systems, the remediation process may be so complex, in terms of contaminant behavior and site characteristics, that no single system or unit is capable of meeting all requirements. Consequently, several unit operations may be combined in series or in parallel to effectively restore groundwater quality to the required level. Barriers and hydrodynamic controls may serve as temporary plume control measures, however, hydrodynamic processes are integral parts of any withdrawal and treatment or in situ treatment process.

Although it is difficult to quantify the importance of contaminant distribution to project feasibility, project success will clearly require movement of nutrient–enriched water through those areas of the site which contain the highest concentrations of contamination. Sites which contain a few point sources of contamination, whether a lagoon or a leaking tank, can generally be treated fairly reliably with an in situ treatment method. However, at sites which contain multiple and undefined sources of contamination treatment methods become much more difficult to design and operate in a predictable fashion. The probability of successful remediation is definitely influenced by ones understanding of the sources and transport mechanisms for the contaminants.

A material is considered easily degradable if the genetic and enzymatic equipment required for the degradation of a compound is widely distributed in nature and if bacteria can obtain sufficient energy from the compound to use the material as a sole carbon source. Although newer innovative techniques may lead to ways of treating the more recalcitrant materials, these processes are likely to be more complex than those currently being used on gasoline contaminants.

(1) Simple hydrocarbons and light petroleum distillates such as gasoline, kerosene, diesel, jet fuel and light mineral oils are generally degradable. Their rate of degradation decreases with increasing molecular weight and decreasing solubility. Increased branching and cyclic structures also slow the degradation process.

(2) Aromatic hydrocarbons with up to two rings (including benzene, toluene, xylene, ethylbenzene and naphthalene) are readily degradable. The rate of degradation of larger polyaromatic hydrocarbons decreases as size increases and solubility decreases (a 3 ring PAH contains up to 14 carbons).

(3) Alcohols, amines, esters, mercaptans, carboxylic acids, and nitriles are also usually degradable, but these compounds also tend to be toxic to unacclimated bacteria at high concentrations. Nitro substitution and ether linkages usually make degradation more difficult.

(4) Chlorinated hydrocarbons (both straight chain and aromatic) become increasingly difficult to degrade as the degree of chlorine substitution increases. As a result polychlorinated biphenyls (PCBs) and other polychlorinated hydrocarbons (chloroform, carbon tetrachloride, tetrachloroethylene, trichloroethylene and dichloroethylenes) are not readily biodegraded aerobically, and are toxic at ppm levels.

(5) Pesticides are another very complex set of organic contaminants. In general, those which are found at hazardous waste sites (DDT, Lindane, Aldrin, Chlordane, etc.) are not readily biodegraded.

The degradation of viscous organics materials like number 6 fuel oil, creosote and refinery wastes are often controlled by their physical condition in the soil/water matrix. If they are present as small droplets of oil occluded within the pores of a soil, there will be very little exposed surface area for degradation and the process will be inhibited.

In situ bioreclamation is a valuable technique for removing a large portion of soil and groundwater receptors. However, if remediation objectives require complete destruction of small concentrations of organics within isolated pockets of the site, the probability of success will be very dependent on the homogeneity of the formation.

Characteristics of the ideal candidate site for successful implementation of in situ bioremediation include: (1) a homogeneous and permeable aquifer; (2) a contaminant originating from a single source; (3) a low groundwater gradient; (4) no free product; (5) no soil contamination; and (6) an easily degraded, extracted, or immobilized contaminant. Obviously, few sites meet these characteristics. However, development of information concerning site specific geological and microbiological characteristics of the aquifer, combined with knowledge concerning potential chemical, physical, and biochemical fate of the wastes present, can be used to develop a bioremediation strategy for a less–than–ideal site.

Lack of fundamental scientific knowledge and incompleteness in the corresponding databases limit the reliable application of in situ bioremediation. Although many fundamental areas deserve research attention, four areas stand out as truly essential to the advancement of in situ bioremediation: (1) the means to quantify the biodegradation kinetics, (2) dissolution/desorption kinetics of poorly soluble substrates, (3) biologically induced clogging, and (4) transport of colloids.

In situ biorestoration of aquifers contaminated by halogenated aliphatic compounds requires a unique approach, since in most cases the halogenated aliphatic compounds can not be utilized by native microorganisms as primary substrates for growth. However, they can be degraded as secondary substrates by microorganisms which utilize another primary substrate for growth. The in situ degradation of these compounds is therefore promoted by the stimulation of a particular class of native microorganisms through the introduction of the appropriate primary substrate for growth (electron donor) and electron acceptor into the treatment zone.

One method relies on the transformation of the chlorinated aliphatic compounds by methane–utilizing bacteria (methanotrophs). These bacteria grow on methane as a sole carbon source under aerobic conditions. The chlorinated aliphatic compounds are thought to be transformed by the methane monooxygenase enzyme, an enzyme with a broad range of specificity, that is produced by the methanotrophic bacteria.

Lawrence Livermore National Laboratory (LLNL), Livermore, CA, has just completed a feasibility study on an in situ microbial filter. The idea is to let the forces of nature bring the soluble pollutant to the microorganism rather than incur the cost and energy of bringing a dense fluid like water to the surface for treatment.

In this microbial filter method, relatively thin vertical filters are installed in the subsurface to intercept contaminant plumes that are being transported by the flowing groundwater. The filters are created by injecting into the subsurface methanotrophic microbes grown in surface bioreactors and allowing them to become attached to the soil. The microbes produce an enzyme (MMO) that fortuitously catalyzes the degradation of TCE into carbon dioxide, water and chloride ions. Because no external energy or carbon source is provided, the microbes remain metabolically active only for a limited time so that the filter needs to be periodically replenished with fresh microbes. Two methods of establishing and maintaining the microbial filter using either vertical or horizontal wells were studied.

The method involves injecting nonindigenous bacteria into a sand-filled trench that bisects the leading edge of a contaminated groundwater plume. As the groundwater flows through the trench, the contaminants are metabolized by the bacteria. Unlike other in situ bioremediation methods, nutrients are not injected along with the bacteria to stimulate subsurface growth. Instead, the new concept relies on the use of resting (nondividing) microbial cells to break down the volatile organic contaminants and achieve more efficient degradation. The approach will be most suitable to aquifers with rapidly moving, very dilute plumes of contaminants, such as chlorinated ethenes and chloroform.

Many water-table aquifers contain oxygen, which can support aerobic microorganisms that can degrade a wide variety of organic contaminants. The extent of biodegradation of these compounds in groundwater is limited by the concentration of oxygen. Roughly two parts of oxygen are required to completely metabolize one part of organic compound. obviously, the prospects for aerobic metabolism of these compounds will depend on their concentration as well as on the concentration of other degradable organic materials in the aquifer. Concentrated plumes of organic contaminants cannot be degraded aerobically until dispersion or other processes dilute the plume with oxygenated water.

Many of the commonly encountered organic pollutants in aquifers are synthetic organic solvents that are very persistent in oxygenated waters. This important class of organic contaminants commonly enters groundwater as spills from underground storage tanks. Groundwater contamination in the Santa Clara Valley of California (Silicon Valley) is a good example. Recent research has shown that this class of organic contaminants can be cometabolized by bacteria that grow on gaseous aliphatic hydrocarbons like methane or propane. The potential use of cometabolism for in situ restoration is under evaluation.

When the concentration of organic contaminants is high, oxygen in the groundwater will be totally depleted and aerobic metabolism will stop. However, further biotransformations often will be mediated by a variety of anaerobic bacteria. Anaerobes that produce methane, called methanogens, are only active in highly reduced environments. Molecular oxygen is very toxic to them. Methane can be produced by the fermentation of a few simple organic compounds such as acetate, formate, methanol, or methylamines. Molecular hydrogen can also be used in the reduction of inorganic carbonate to methane. Although the microorganisms that actually produce the methane can use a very limited set of organic compounds, they can act in consort with other microorganisms which break more complex organic compounds down to substances that the methanogenic organisms can use. These partnerships or consortia can totally degrade a surprising variety of natural and synthetic organic compounds.

Contaminants in solution in groundwater as well as vapors in the unsaturated zone can be completely degraded or can be transformed to new compounds. Undoubtedly, thousands of contamination events are remediated naturally before the contamination reaches a point of detection. However, methods are needed to determine when natural biorestoration is occurring, the stage the restoration process is in, whether enhancement of the process is possible or desirable, and what will happen if natural processes are allowed to run their course. A number of researchers are presently working in this area.

1.6.8 Enhancement of Biochemical Mechanisms

There are a number of treatments that could enhance microbial activity in hazardous waste contaminated soil.

Colloidal Gas Aphrons: The introduction of microscopic bubbles of gas (gas aphrons) into the soil can enhance aerobic biodegradation of dissolved and dispersed organic contaminants by delivering gases at greater than their solubility limits. In laboratory experiments, colloidal gas aphrons have been shown to increase the concentration of gases present within the soil matrix. The use of CGAs at uncontrolled hazardous waste sites depends on the microdispersion as a source of oxygen for in situ bioreclamation. The contaminated medium retains the CGAs for much longer periods of time than it does air directly injected into the contaminated matrix because directly injected air moves rapidly toward the unsaturated zone and allows little oxygen retention.

Soil Moisture: Moisture control may take the form of supplemental water to the site (irrigation), removal of excess water (drainage, wellpoints), a combination of techniques for greater moisture control, or other methods (e.g., soil additives). Furthermore, the addition of vegetation to a site will increase evapotranspiration of water and, therefore, assist in retarding the downward migration of water (i.e., leaching).

Oxygen Control: Aerobiosis can be maintained by the addition of air, oxygen, or other oxidants or oxygen sources (such as hydrogen peroxide, ozone, and nitrates). Gas injection or infiltration of water containing these alternative oxygen sources is being used for the reclamation of soil contaminated with hazardous wastes. Both ozone and hydrogen peroxide have been demonstrated to enhance dissolved oxygen levels in soil/groundwater systems and, consequently, to stimulate microbial activity. Ozone and hydrogen peroxide can also chemically degrade (oxidize) the contaminants completely or partially.

The application of soil venting or air sparging technology is also appropriate. Pneumatic fracturing can also be utilized. Tillage is another soil venting technology. Bioventing is discussed in Section 1.6.10. Air sparging is used in conjunction with vacuum extraction.

Oxygen enhancement with microbubbles technology is designed to carry oxygen and other nutrients to subsurface microorganisms to stimulate in situ bioremediation of organic contaminants in groundwater. Oxygen is mixed with an inexpensive, biodegradable surfactant to produce highly stable microbubbles in the 40 per micron size range. Bubbles this size can remain dispersed in a coarse sand matrix in the saturated zone without significant coalescence. Partially supported by EPA and the Air Force, researchers at Virginia Polytechnic Institute and State University (Virginia Tech) in Blacksburg, Virginia, have taken the lead in developing this technology.

Magnetic Fields: Research at the New Jersey Institute of Technology has revealed the effect of magnetic fields on the speed of biological activity during bioremediation. Positively charged magnetic energy increases life, growth and development within the bioremediation process, while negatively charged energy inhibits growth, according to the research.

Funnel–and–Gate System: The Waterloo Centre for Groundwater Research has developed Funnel–and–Gate systems that isolate contaminant plumes in groundwater and funnel the plumes through in situ bioreactors. The Funnel–and–Gate consists of low hydraulic conductivity cutoff walls with gaps that contain in situ reactors (such as reactive porous media), which remove contaminants by abiotic or biological processes. The cutoff walls (the funnel) modify flow patterns so that groundwater flows primarily through high conductivity gaps (the gates).

Soil pH: Depending on the nature of the hazardous waste components contaminating the soil, it may be advantageous to optimize the soil pH for a particular segment of the microbial community because both structure and activity are affected by the soil pH. Some fungi have a competitive advantage at slightly acidic pH, whereas actinomycetes flourish at slightly alkaline pH. Soil pH has also been shown to be an important factor in determining the effect various pesticides have on soil microorganisms. Near neutral pH values are probably most conducive to microbial functioning in general.

Soil Nutrients: As in the case of all living organisms, microorganisms must have specific inorganic nutrients (e.g., nitrogen, phosphorus, potassium, calcium, magnesium, sulfur, iron, and trace metals) and a carbon and energy source to survive. The organic contaminants present in the soil may provide the carbon and energy source and serve as primary substrates. If the compound of interest is only degraded cometabolically, however, a primary substrate must be made available for the microbial population. The primary source of carbon may already be present in the soil or it may be added e.g., glucose, acetate, citrate). Carbon sources also could be added if the concentration of contaminants present in the soil are not sufficient to support an active microbial population; however, the addition of these compounds could inhibit the biodegradation of the compound(s) of interest as a result of preferential degradation.

Soil Temperature: Soil temperature is one of the most important factors controlling microbiological activity and the rate of decomposition of organic matter. It also influences the rate of volatilization of compounds from soil. Soil temperature can be modified by regulating the oncoming and outgoing radiation or by changing the thermal properties of the soil.

Addition of Nonspecific Organic Amendments: Stimulating general soil microbial activity and population size through the addition of organic matter increases the opportunity to select organisms that can degrade hazardous waste components. High microbial activity allows cometabolic processes to act on recalcitrant hazardous waste components. The addition of manures, plant materials, or wastewater treatment digestor sludge at levels characteristic of composting may prove valuable to biological treatment of soil contaminated with hazardous wastes.

Cometabolism: Thomas and Ward define cometabolism as the biodegradation of an organic substance by a microbe that cannot use the compound for growth and hence must rely on other compounds for carbon and energy. Three mechanisms of cometabolism are: (1) analogue enrichment; (2) nonanalogue enrichment with methane; and (3) other nonanalogue hydrocarbon substrates.

Augmentation With Acclimated or Mutant Microorganisms: Biological treatment methods described thus far have relied on the stimulation of microbial activity in the soil or on the natural selection of populations of microorganisms capable of degrading toxic waste constituents. Although these approaches show considerable promise for treating many kinds of organic hazardous waste constituents, the metabolic range of the natural soil microbiota may not be capable of degrading certain compounds or classes of compounds. Also, microbial metabolic specialists may not develop large enough populations under limited substrate conditions to degrade xenobiotic compounds rapidly enough to meet treatment criteria. In situations such as these, it may be advisable to add exogenously grown microorganisms to the soil. These microorganisms can be selected by enrichment culturing or genetic manipulation, and they can be acclimated to the degradation of different organic contaminants by repeatedly exposing them to the compound of interest. Microbial inoculants with a broad range of metabolic capabilities are available commercially, and experience with their use in both soil and aquatic systems contaminated with waste chemicals is expanding.

Application of Cell–Free Enzymes: Enzymes produced by microorganisms, which can transform hazardous compounds to nonhazardous or more labile products, could be harvested from cells grown in mass culture and applied to contaminated soils. Industry commonly uses crude or purified enzyme extracts, either in solution or immobilized on glass beads, resins, or fibers, to catalyze a variety of reactions, including the breakdown or transformation of carbohydrates and proteins.

Encapsulated Microorganisms: Encapsulate degradative microorganisms, together with necessary nutrients, in a polymer matrix, then dehydrate the encapsulated microorganisms. Encapsulated microorganisms applied to a site can be released from the capsules by various regulating mechanisms, such as water dissolution of the polymer matrices. The release of encapsulated microorganisms is manipulated by using different polymer matrix materials, encapsulation configurations, and manufacturing processes.

Microbial Suppression: There are situations where microbial activity actually contributes to the contamination problem. In the worst scenario, microbes may create additional contaminants that may have even more serious environmental and health consequences than the original contaminant.

Thus, the proper bioremediation approach may actually be focused on microbial activity suppression rather than enhancement. An example would be the creation of vinyl chloride, in the biodegradation of chlorinated hydrocarbons.

1.6.9 Vegetative Uptake

The ability of higher plants (i.e., seed–producing) to remove and accumulate compounds from the soil has resulted in studies for their potential use as an in situ treatment technique for both organic and inorganic compounds. The potential method of treatment by plants may occur through bioaccumulation, transformation (i.e., metabolizing the compound to nontoxic metabolites), or by adsorbing to plant roots for microbial degradation. Plant uptake of both organics and inorganics in the soil environment is influenced by numerous physical and chemical factors, including pH, clay content, cation exchange capacity, soil texture and compaction, organic matter content, plant species, and toxicity of the compound.

Uptake of compounds by plants occurs through chemical partitioning onto the external root surfaces leading to accumulation into the root with subsequent access to the vascular system of the plant. In general, plant uptake of a chemical in the soil can be accomplished through the following main pathways:

1. Root uptake into conduction channels.
2. Uptake of vapor in the surrounding air by the vegetative parts of the plant.
3. Uptake by external contamination of shoots by soil and dust, followed by retention in the cuticle or penetration through it.
4. Uptake and transport in the oil cells of oil–containing plants (carrots, parsnips).

Most uptake by the plant will occur through the first two pathways, although the second two pathways may apply under specific conditions, e.g., uptake and transport of highly lipophilic compounds into the oil cells of oil–containing plants.

Several differences occur between the plant uptake of organic versus inorganic compounds. Uptake of elements can take place if the element exists as either a cation or anion. Several variables may influence the concentration of metals found in plants, including species, cultivar, maturity, and plant part. Leafy vegetables (lettuce, cress) accumulate significantly greater amounts of cadmium than do plants such as corn or wheat.

A major consideration to be addressed when assessing the uptake of inorganics is toxicity. Plant species differ significantly in their tolerance of metals, which could affect their use as an in situ treatment technique.

Plant uptake of organic compounds has also been investigated. Nonionic (organic) adsorption in the soil is largely to the organic matter that coats most particles in the soil. Several studies have shown that plant roots adsorb high levels of lipophilic pollutants from the soil which compete with existing soil organic matter.

Advantages:

1. Soils can be treated without excavating large quantities of material.
2. Worker exposure is minimized.
3. Cost of this technology would be relatively low.

Disadvantages

1. Toxicity of pollutants may have adverse effects on the plant or on animals eating the plant.
2. Plants will, in most cases, only remove small amounts of the contaminant.
3. Plants would need to be disposed of, e.g., incinerated, after uptake of the contaminants.

1.6.10 White Rot Fungus

The use of white rot fungus to degrade complex organic chemicals in wastewater has been discussed earlier. It is also useful for bioremediation.

The white rot fungus is not naturally found in the soil and does not compete well when alone with the native microflora of nonsterile soils. To aid the fungus, wood chips are commonly used as both a support and growth media for the fungi. Wood by–products or other lignin–containing materials may also be used.

Mixing of fungi–impregnated material with contaminated soils is the current state of technology for both in situ and ex situ bioremediation of contaminated soils. Development of an in situ treatment that does not require mixing is an attractive, cost–reducing alternative and may be required in some instances (e.g., situations involving explosives at levels of contamination that are close to their explosive limits, making the use of heavy equipment unsafe). Such a treatment should provide a delivery system for the oxygen, microorganisms, and nutrients that makes minimal use of machinery and provides an opportunity for good contact of the organic contaminants with the degrading organisms.

Certain non–white rot fungi have also shown considerable promise.

The fungal treatment process involves inoculation of the contaminated soil with selected fungal strain(s) followed by addition of nutrients (if needed), irrigation, and aeration of the soil through tilling/cultivating to provide optimal fungal growth conditions. Inoculation is accomplished by physically mixing the soil and the inoculum. Mixing can be performed in solids mixing equipment, or in situ by placing the inoculum on the contaminated soil and tilling until the two are thoroughly mixed. In the case of ex situ mixing, the soil–inoculum mix must then be spread over the ground. Land farming procedures such as irrigation, aeration and nutrient addition are then implemented periodically to sustain the fungal activity within the soil matrix. As a result of the fungal activity, the hazardous compounds are transformed and become irreversibly bound to soil organic matter, in which state they are not biologically active and thus do not present toxicity problems. The fungal treatment can take several weeks to several months to achieve the desired level of contaminant reductions.

This fungal treatment has been tested for treatment of soils contaminated with organic wood preserving compounds such as pentachlorophenol (PCP) and select polynuclear aromatic hydrocarbons (PAHs) found in creosote. Warm temperatures (greater than 80°F) and sufficient moisture (greater than 30%) in the target matrix are desirable for the optimal growth of the fungus and, thus, for the degradation of the contaminants.

1.6.11 Bioventing

Vacuum extraction can enhance biodegradation of volatile and semi–volatile chemicals in the soil by providing oxygen to the soil for use by microorganisms. Larger amounts of oxygen can be supplied per volume of air than per volume of water. This use of vacuum extraction to enhance biodegradation is also known as bioventing.

Bioventing systems are composed of hardware identical to that of conventional soil vacuum extraction (SVE) systems, with vertical wells and/or lateral trenches, piping networks, and a blower or vacuum pump for gas extraction. They differ significantly from conventional systems, however, in their configuration and philosophy of design and operation. The primary purpose of a bioventing system is to employ moving air to transfer oxygen to the subsurface where indigenous organisms can utilize it as an electron acceptor to carry out aerobic metabolism of soil contaminants. As such, bioventing system extraction wells are not placed in the center of the contamination as in conventional SVE systems, but on the periphery of the site, where low flow rates maximize the residence time of vent gas in the soil to enhance in situ biodegradation and minimize contaminant volatilization.

Because it is a biological treatment approach, however, bioventing does require the management of environmental conditions to ensure maintenance of bioactivity at the site. Management of soil moisture and soil nutrient levels to avoid inhibition of microbial respiration within the vadose zone can be accomplished fairly easily, and has been used to optimize contaminant biodegradation at field sites when other variables, i.e., toxicity, do not limit microbial activity.

The two major design considerations for bioventing systems are (1) whether the contaminants of concern are biodegradable under prevailing site conditions, i.e., whether inhibition or toxicity is evident at the site, and (2) whether the required terminal electron acceptor, i.e., oxygen, can be effectively transported within the soil to encourage aerobic contaminant biodegradation. The first question can be answered using soil–gas composition and in situ respiration measurements, while the second question is answered from in situ air permeability measurements.

1.6.12 Biosparging

Biosparging is a variant of air sparging where oxygen stimulated biodegradation is the aim, rather than volatilization. As with air sparging, soil venting is used to recover gas discharged through the water table.

1.7 METALS REMOVAL

Biological treatment is a separation process rather than a destruction technology for metal–containing wastes.

Biological treatment processes do not alter or destroy inorganics. In fact, concentrations of soluble inorganics should be kept low so that enzymatic activity is not inhibited. Trace concentrations of inorganics may be partially removed from the liquid waste stream during the biological treatment, because of adsorption into the microbial cell coating. Typically, microorganisms have a net negative charge and are therefore able to perform cation exchange with metal ions in solution. Anionic species, such as chlorides and sulfates, are not affected by biological treatment.

High concentrations of heavy metals are toxic to most microorganisms and often cause serious upsets in biological systems. Thus, influent heavy metal concentrations which can be tolerated and removed is the major criterion on which these technologies are evaluated. In addition, factors such as type of influent, its strength, and the extent of system acclimation are also used to evaluate the viability of biological treatment as a technology for the removal of heavy metals from wastes. There are direct interaction (redox), or indirect interaction processes.

Several mechanisms can affect the removal of heavy metals during biological treatment including sulfide precipitation, adsorption, and bioflocculation. The first mechanism, hydrogen sulfide precipitation, is initiated by the pH dependent generation of hydrogen sulfide by bacteria. Soluble metal ions react with the hydrogen sulfide and are precipitated as insoluble metallic sulfides. The second mechanism, adsorption of cationic metallic ions, may result from the anionic nature of certain cellular material, clay particles, and industrial wastes constituents. Also, the organic part of organo–metallic complexes may be adsorbed through the cell walls of the biological organisms, thus trapping the metals. The third mechanism, bioflocculation, is related to the synthesis of insoluble extracellular polymer strands. These extracellular polymers can act as non-specific sorbers for metal ions.

Typically, the removal of heavy metals in a biological system and the type of mechanism which dominates are dependent on the species of heavy metal present. The distribution of a particular heavy metal among various chemical forms, however, largely depends upon the physical and chemical properties of the environment established by the treatment process itself. Upon introduction into the biological treatment system, species of heavy metal make adjustments toward a new equilibrium state defined by chemical environment parameters such as pH, oxidation reduction potential (ORP), the presence of complexing agents, and concentrations of precipitant ligands. At this point, adsorption to solid phases or biomass, and intracellular storage can occur.

It has been found that the microbial removal of heavy metals consists of initial rapid uptake followed by slow, but consistent long-term uptake. The race of uptake is greatly affected by solution pH. Sludge age, as well as the extent of acclimation, can also affect the extent of metal removal in an activated sludge system.

Microorganisms (bacteria, fungi, and microalgae) can accumulate relatively large amounts of toxic heavy metals and radionuclides from the environment. These organisms often exhibit specificity for particular metals. The metal content of microbial biomass can be a substantial fraction of total dry weight with concentration factors (metal in dry biomass to metal in solution) exceeding one million in some cases. Both living and inert (dead) microbial biomass can be used to reduce heavy metal concentrations in contaminated waters to very low levels—parts per billion and even lower. In many respects (e.g., specificity, residual metal concentrations, accumulation factors, and economics) microbial bioremoval processes can be superior to conventional processes, such as ion exchange and caustic (lime or hydroxide) precipitation for heavy metals removal from waste and contaminated waters.

However, the potential advantages of bioremoval processes must still be developed into practical operating systems. There is great variability from one biomass source to another in bioremoval capabilities. Bioremoval is affected by pH, other ions, temperature, and many other factors. The biological (living vs dead) and physical (immobilized vs dispersed) characteristics of the biomass also greatly affect metal binding. Even subtle differences in the microbial biomass, such as the conditions under which it was cultivated, can have major effects on heavy metal binding.

Many microbes produce both intra- and extra-cellular metal complexing agents which could be considered in practical metal removal processes. Bioremoval processes are greatly affected by the microbial species and even strain used, pH, redox potential, temperature, and other conditions under which the microbes are grown. Development of practical applications of bioremoval requires applied research using the particular waste solutions to be treated, or close simulations thereof. From a practical perspective, the selection of the microbial biomass and the process for contacting the microbial biomass with the metal containing solutions are the key issues. Much of the recent commercial R&D has emphasized commercially available, inert, microbial biomass sources as these can be acquired in sufficient quantities at affordable costs.

Algae are particularly well suited for metal bioremoval. A recent commercial application of bioremoval utilizes inert (dead) immobilized microalgae biomass as ion exchange materials for the removal of heavy metals from industrial wastewaters. Also, living microalgal cultures have been used to remove metals from mine effluents. Microbial cells and biomass can bioaccumulate metals and radionuclides by a large variety of mechanisms, both dependent and independent of cell metabolism. Microbial cell walls can act as ion exchange and metal complexation agents. Heavy metals can precipitate and even crystallize on cell surfaces. Metabolically produced hydrogen sulfide or other metabolic products can bioprecipitate heavy metals.

Typically, algae is contacted with the influent metals–containing wastewater in an aerated lagoon. The lagoon is usually a lined, flat–bottom pond enclosed by earthen dikes. Oxygen transfer between the air and water is accomplished through algae photosynthesis, although platform–mounted mechanical aerators can be used to enhance transfer. Influent wastewater enters near the center of the lagoon and effluent discharges at the windward side.

Advantages of this type of system relative to other biological processes include lower capital and operating costs. In addition, operational flexibility is increased since the effluent flow can be regulated. Disadvantages include extensive physical space requirements, poor industrial waste treatment capacity, and seasonal performance variations.

Tank–like effluent treatment equipment can also be utilized, and columns have been designed for a range of flows between 1 and 100 gallons per minute.

Passive systems have been used extensively for coal mine drainage. Three systems include: (1) aerobic wetlands, (2) wetlands constructed with an organic substrate, and (3) noxic limestone drains.

For the past 10 years, there has been considerable research undertaken in various countries, in the development of specific organisms designed priority for the bioaccumulation of heavy metals. While the available literature emphasizes activated sludge treatment, anaerobic and algal systems have been increasingly explored in recent years.

Recent research and development has included:

1. Evaluating microbial systems for their ability to quantitatively leach heavy metals from sludge.
2. Evaluating microbial system for quantitatively precipitating heavy metals from solution.
3. Removal and recovery of metals in waste streams with metallothioneins which are metal binding proteins.
4. Removal of sulfur from coal.
5. Researchers at the Hebrew University in Rehovot, Israel, have developed a method for removing metals from wastewater using water ferns. Azolla, a water fern found in Asia, East Africa and Central America, can be used to remove metals such as copper, zinc, chromium, cadmium, nickel, silver, titanium and uranium from industrial waste. It can be grown in settling ponds and, when harvested and dried, used as filtering material in paint and metals–processing plants. Regulatory considerations are extremely important for any plant uptake of hazardous material.

6. An anaerobic/citric acid process for metal removal has been developed at Brookhaven National Laboratory.
7. Biotreatment Limited (U.K.) is currently undertaking a laboratory research program to assess the potential of various forms of microbial biomass as matrices for the accumulation of metal cations that are of significance in nuclear reprocessing effluents. This study focuses on the biosorption of three of the cations most significant in nuclear effluent streams (strontium, ruthenium, and cobalt).
8. Phytoremediation by Cunningham at duPont: Plants take up the metals in their roots and translocate them to their shoots, which are harvested, burned in a kiln, and the metals recovered and recycled. The challenge is finding or engineering plants that can absorb, translocate, and tolerate heavy metals, while producing enough foliage to make the process efficient.
9. The U.S. Bureau of Mines has developed porous beads containing immobilized biological materials such as sphagnum peat moss for extracting metal contaminants from wastewaters.

1.7.1 Processes Include

(1) Heap Leaching for Cyanide Removal: A bacterial treatment system can provide alternative rinsing technology for decommissioning precious metals heap leach facilities. This alternative increases the rate of cyanide degradation in heaps by activating natural populations of cyanide–oxidizing bacteria indigenous to the site and/or introducing additional populations of natural bacteria with known cyanide–degrading capabilities.

The bacteria–enhanced process increases the rate of cyanide rinsing from the heaps and enables complete water recycling. This has three major advantages: it eliminates the need for toxic or corrosive chemicals to destroy the cyanide in process solutions; it diminishes the amount of fresh water needed for cyanide rinsing; and it eliminates the water balance problem caused by the large volumes of contaminated wastewater generated during conventional rinsing that must be evaporated. Ideally, the bacteria–enhanced rinsing will completely and permanently destroy the cyanide in the process solutions as well as in the heaps.

(2) Polymeric Beads: Porous polymeric beads containing immobilized biological materials have been developed to extract toxic metals from water. The beads, designated as BIO–FIX beads, are prepared by blending biomass such as sphagnum peat moss or algae into a polymer solution and spraying the mixture into water. The beads have distinct advantages over traditional methods of utilizing biological materials in that they have excellent handling characteristics and can be used in conventional processing equipment or low–maintenance systems. Cadmium, lead, and mercury are a few of the many metals readily removed by BIO–FIX beads from acid mine drainage (AMD) waters, metallurgical and chemical industry wastewaters, and contaminated ground waters. Because of their affinity for metal ions at very low concentrations, National Drinking Water Standards and other discharge criteria are frequently met. Adsorbed metals are removed from the beads using dilute mineral acids. In many cases, the extracted metals are further concentrated to allow recycle of the metal values.

(3) Algae Process: The AlgaSORB™ sorption process (Bio-Recovery Systems, Inc.) is designed to remove heavy metal ions from aqueous solutions. The process is based on the natural affinity of the cell walls of algae for heavy metal ions.

The sorption medium is comprised of algal cells immobilized in a silica gel polymer. This immobilization serves two purposes: (1) it protects the algal cells from decomposition by other microorganisms, and (2) it produces a hard material that can be packed into chromatographic columns that, when pressurized, still exhibit good flow characteristics.

The system functions as a biological ion-exchange resin to bind both metallic cations (positively charged ions, such as mercury, Hg^{+2}) and metallic oxoanions (large, complex, oxygen-containing ions with a negative charge, such as selenium oxide, SeO_4^{-2}). Anions such as chlorides or sulfates are only weakly bound or not bound at all.

Like ion-exchange resins, the algae-silica system can be recycled. However, in contrast to current ion-exchange technology, the components of hard water (calcium, Ca^{+2}, and magnesium, Mg^{+2}) or monovalent cations (sodium, Na^+, and potassium, K^+) do not significantly interfere with the binding of toxic heavy metal ions to the algae-silica matrix.

After the media are saturated, the metals are stripped from the algae by using acids, bases, or other suitable reagents. This produces a small volume of solution containing highly concentrated metals that must undergo treatment.

(4) Current Projects: As reported by Mattison, the following are examples of large scale projects.

1. 6,000 m^3/day of zinc- and sulfate-contaminated groundwater are being treated by sulfate-reducing bacteria to precipitate zinc sulfide for recycle to a zinc smelter (Belgium).
2. Streams of 1,700 m^3/day and nearly 30,000 m^3/day of acid mine drainage are being treated by iron-oxidizing bacteria to allow easy removal of iron and other heavy metals (Japan).
3. Leachate from treating 80 tons/day of copper smelter flue dust is being treated by iron-oxidizing bacteria to facilitate metal removal; savings due to bio-oxidation estimated at $360,000/year (Japan).
4. Over 400 "constructed wetlands" are treating acid coal mine drainage utilizing sulfate-reducing and iron-oxidizing bacteria to consume acidity and render iron and other metals easily precipitated. Many pay for themselves in the first year of operation (USA).
5. 21,000 m^3/day of dilute cyanide mine solutions are treated to degrade free and metal-complexing cyanide and to entrap heavy metals prior to discharge to a trout stream. This biological approach is economically and environmentally superior to chemical treatment alternatives (USA).

1.8 BIOFILTRATION/BIOSCRUBBING

Biofiltration is used extensively in Europe, particularly Germany and the Netherlands to treat gases with low concentrations of VOCs (under 1,000 ppm), and for odor control. One type of biofilter consists of a packed column containing biologically active mass. The support material can be of the following four types: (1) nonbiodegradable inactive material, such as glass or sand, which has no significant adsorption potential for the organics; (2) biodegradable inactive material, such as peat, with low adsorption potential for the organics, but has organic matter; (3) nonbiodegradable active material, such as activated carbon, which has high adsorption potential for organics; and (4) biodegradable active material, such as polymeric adsorbent, which has adsorption potential for the organics and has biodegradable organic groups.

The biologically active matter (biomass) can exist either as a uniform biofilm on the support medium, or as a biomass particle trapped in the void spaces between the support material. In the case of a biofilm, the biomass is attached to the support material with simultaneous diffusion and degradation of the organics. In the case of a biomass particle, the organics degrade as they diffuse through the active biomass.

The process conducts raw gas from the stripper through a radial blower and spray humidifier before entering the biofilter. Humidified gas enters the lower section, which contains the aeration system. The upper section contains biologically active filter material that can be derived from composts based on municipal solid waste, wood waste or peat. As they rise through the filter bed, target pollutants are removed by diffusion into a wet film covering the filter particles, and then by aerobic degradation. In the case of non-chlorinated VOCs the by-products are carbon dioxide and water.

Control efficiencies of more than 90% for the target pollutants can be achieved if the filter is sized to provide the required degradation capacity for a given pollutant load. However, higher chlorinated organics show significantly reduced biodegradation rates.

Another type is the soil biofilter. Traditionally, the term "biofilter" has been used to define processes that use compost, peat, bark, soil, or mixtures of these substances as the filter medium. These media serve as a support system for a microbial population. Filter media is underlain with a gas distribution system, commonly perforated pipe. Gases flow through the bed where the pollutants are adsorbed to the filter media. After contact with the microorganisms the pollutants are broken down thus regenerating the adsorption capacity of the bed. Water is sprayed over the bed's surface or by humidifying the influent gases. The terms "soil filters," "soil biofilters," or "soil beds" delineate processes where the filter media is soil. Soil biofilters are generally less permeable to gas flow than biofilters that use compost, peat, or bark media thus a larger soil biofilter area is required for the same back pressure.

Biofilters and soil filters have been applied to control odors from wastewater treatment plants and industrial processes since 1953. Recently, these processes have been used for volatile organic compound emissions removal from chemical and process industries.

Other processes mentioned in the literature that employ biological treatment of waste gases include bioscrubbers and trickling filters. Bioscrubbers are generally used when the biological degradation products, such as acids from H_2S and ammonia removal, would harm the biofilter bed, or when contaminants are insoluble in water. Treating a wider variety of contaminants than biofilters do, bioscrubbers come in two forms: activated sludge scrubbers, and trickling filters, which can move beyond simple organics to treat chlorinated waste streams.

An engineered biofilter using synthetic media, such as activated carbon, has been developed by Alcoa that shows improvements in removal efficiency, biodegradation, and space requirements over existing filters.

Advantages of biofiltration include:

1. Vertical stratification of the microorganisms, with different predominant organisms existing at various levels of the biofilter bed height. Through the process of natural selection, microorganisms of a certain type will dominate at a specific height which maximizes their growth due to the existence of the optimum conditions, such as, concentration of organics, pH, temperature, humidity, etc.
2. No breakthrough of the organic(s) due to continuous degradation as compared to breakthrough in an activated carbon system when its capacity is reached. Initially, the concentration of the organic(s) in the support material will increase until a steady state is established, when the rate of transport of the organic(s) from the gas phase to the support material is balanced by the rate of biodegradation of the organic(s).
3. Higher rate of biodegradation than in activated sludge systems due to the existence of an immobilized biofilm, which can contain a significantly higher concentration of the microorganisms than found in conventional activated sludge. Since the rate of biodegradation is dependent on the concentration of the microorganisms, a significantly higher concentration in the biofilm will result in an increased rate of biodegradation.
4. Potential of using a variety of organisms, either under aerobic or anaerobic conditions. Mixed cultures that have been acclimated to specific organics can be used as easily as pure cultures, which are capable of degrading certain organics only. Aerobic and anaerobic biofilters can be used sequentially to degrade a mixture of organics containing components that are recalcitrant under aerobic conditions.
5. Less potential for contamination of support material by nonbiodegradable organics or high molecular weight contaminants, which is likely in the case of completely mixed continuous systems, such as activated sludge plants or fluidized bed reactors, handling aqueous waste streams. For the biofilter, the organic contaminants that are introduced through the gas phase would not have a high molecular weight or be recalcitrant compounds that can accumulate in the support material.

1.9 BIOCONVERSION

A number of biological processes are being developed to convert industrial wastes, as well as other raw materials, into fuel and chemicals. Examples include:

1. Digestion of municipal solid and industrial wastes to methane by anaerobic digestion. This process occurs naturally in municipal waste landfills.
2. Production of ethanol from paper mill sludge, cellulosics, waste paper, yard waste, and other organic wastes. The waste material is first broken down with acids or enzymes into the component sugars, then treats the sugars with a bioengineered bacteria. Other chemicals such as furfural could be produced from the hemicellulose hydrolysate.

3. Conversion of the CO, CO_2, and H_2 in industrial waste gas streams into acetic acid.
4. Conversion of sodium oxalate (a hazardous by-product of the alumina process) into sodium carbonate and bicarbonate.
5. The production of linear alkanes, olefins, alcohols and esters by a process involving anaerobic digestion and electrolytic oxidation. In the process, waste material is fed to an anaerobic digester in which methane formation is inhibited. This enhances formation of linear aliphatic carboxylic acids from acetic to hexanoic acid. The organic acids can be removed from the fermenter by liquid-liquid extraction and then converted to the final product by electrolytic oxidation. The product is dependent on the organic acid produced in the digestion step and the conditions of the electrolytic oxidation.
6. Liquid fuel production from biomass can be accomplished by any of several different processes including hydrolysis and fermentation of the carbohydrates to alcohol fuels, thermal gasification and synthesis of alcohol or hydrocarbon fuels, direct extraction of biologically produced hydrocarbons such as seed oils or algae lipids, or direct thermochemical conversion of the biomass to liquids and catalytic upgrading to hydrocarbon fuels.
7. In the fossil fuels industry, suggested areas of R&D include enhanced oil recovery, in situ bitumen extraction, site remediation, basic biogeochemical and transport phenomena studies, bioconversion of natural gas, wastewater and sludge treatment, sulfur processing, and carbon dioxide removal.

REFERENCES

1. Antonopoulos, A.A., *Biotechnological Advances in Processing Municipal Wastes for Fuels and Chemicals,* Noyes Data, 1987.
2. Benedict, A.H., et al, *Composting Municipal Sludge,* Noyes Data, 1988.
3. Berkowitz, J.B., et al, *Unit Operations for Treatment of Hazardous Industrial Wastes,* Noyes Data, 1978.
4. Bioremediation Report, 10/92.
5. Block, R., et al, *Bioremediation—Why Doesn't it Work Sometimes?,* Chem. Eng. Prog. 8/93.
6. Bowker, R.P. G., et al, *Phosphorus Removal from Wastewater,* Noyes Data, 1990.
7. Brubaker, G., *Screening Criteria for In Situ Bioreclamation of Contaminated Aquifers.*
8. *Bugs Digest Chlorinated Organics,* Chem. Eng. 2/93.
9. Burton, D.J., et al, *Treatment of Hazardous Petrochemical and Petroleum Wastes,* Noyes Data, 1989.
10. Chambers, C.D., et al, *In Situ Treatment of Hazardous Waste-Contaminated Soils—Second Edition,* Noyes Data, 1991.
11. Corbitt, R.A., *Standard Handbook of Environmental Engineering,* McGraw-Hill, 1990.
12. Davis, M.L., et al, *Introduction to Environmental Engineering—Second Edition,* McGraw-Hill, 1991.
13. Dean, N., Kremer, F., "Advancing Research For Bioremediation," *Environmental Protection,* 9, 1992.
14. EPA, *Biological Treatment of Wood Preserving Site Groundwater by BioTrol, Inc.—Applications*

Analysis Report, RREL, EPA/540/A5-91/001, 9/91.
15. EPA, *Bioremediation Action Committee—Summary of February 12, 1992 Meeting,* ORD, 1992.
16. EPA, *Bioremediation in the Field,* various issues, 1990-1992.
17. EPA, *Bioremediation of Hazardous Wastes,* EPA/600/R-92/126, 8/92.
18. EPA, *Bioremediation of Hazardous Wastes,* ORD, EPA/600/9-90/041, 12/90.
19. EPA, *Corrective Action: Technologies and Applications,* EPA/625/4-89/020, 9/89.
20. EPA, *Design Manual—Constructed Wetlands and Aquatic Plant Systems for Municipal Wastewater Treatment,* ORD, CERI, EPA/625/1-88/022, 9/88.
21. EPA, *Eighteenth Annual Risk Reduction Engineering Laboratory Research Symposium,* EPA/600/R-92/028, 4/92.
22. EPA, *Evaluation of Oxidation Ditches for Nutrient Removal,* EPA 832-R-92-003, 9/92.
23. EPA, *Fungal Treatment Bulletin,* EPA/540/MR-93/514, 6/93.
24. EPA, *Handbook: Groundwater and Contamination,* EPA/625/6-90/0162, 9/90.
25. EPA, *Handbook: Stabilization Technologies for RCRA Corrective Actions,* EPA/625/6-91/026, 8/91.
26. EPA, *Innovative Hazardous Waste Treatment Technologies, Domestic and International, 2nd Forum,* EPA/540/2-90/009, 9/90.
27. EPA, *Innovative Treatment Technologies: Overview and Guide to Information Sources,* OSWER, EPA/540/9-91/002, 10/91.
28. EPA, *In Situ Remediation of Groundwater,* EPA/540/S-92/017, 1/93.
29. EPA, *In Situ Treatment of Contaminated Groundwater,* OSW, EPA, 9/16/92.
30. EPA, *Municipal Wastewater Treatment Technology,* Noyes Data, 1993.
31. EPA, *RCRA Corrective Action Stabilization Techniques,* EPA/625/R-92/014, 10/92.
32. EPA, *The Superfund Innovative Technology Evaluation Program: Technology Profiles—Fourth Edition,* RREL, EPA/540/5-91/008, 11/91.
33. EPA, *Summary Report—High-Priority Research on Bioremediation,* Bioremediation Research Needs Workshop, Washington, DC, April, 1991.
34. EPA, *Synopses of Federal Demonstrations of Innovative Site Remediation Technologies, 2nd ed.,* EPA/542/B-92/003, 8/92.
35. Frederickson, J., et al, *In Situ and On-Site Bioreclamation,* Env. Sci. Tech., 9/93.
36. Foster, M.H., et al, "Bioaugmentation Aids Wastewater Systems," *Environmental Protection,* 10/92.
37. Fouhy, K., *Cleaning Waste Gas, Naturally,* Chem. Eng., 12/92.
38. Fouhy, K., *Mighty Microbes,* Chem. Eng., 11/92
39. Govind, R., et al, *Development of a Novel Biofilter for Aerobic Biodegradation of VOCs.*
40. Gupto, B., et al, *Data Summary of Municipal Solid Waste Management Alternatives,* NREL, DOE, DE 92016433.
41. Hartley, R.P., *Surface Impoundments,* Noyes Data, 1992.
42. Hastic, B., *The Use of Aquatic Plants in Wastewater Treatment, A Literature Review,* Univ. of Texas, 12/92.
43. Holden, T., et al, *How to Select Hazardous Waste Treatment Technologies for Soils and Sludges,* Noyes Data, 1989.
44. Hydroqual, Inc., *Investigation and Field Testing of Anaerobic Biological Treatment of Pharmaceutical Wastewaters,* U.S. Department of Energy, DOE/CH/10239-1 (DE 90010576), 3/90.
45. Jackman, A.P., et al, *Hazardous Waste Treatment Technologies,* Noyes Data, 1991.
46. Kamnikar, B., "Bioremediation of Contaminated Soil," *Pollution Engineering,* 11/92.
47. Kingsley, M., et al, *Environmental Restoration Using Plant-Microbe Bioaugmentation,* Battelle Pacific Northwest Labs, PNL-SA-22251, 4/93.
48. Kleinmann, R., et al, *Treat Mine Water Using Passive Methods,* Poll. Eng., 8/93.
49. Lawrence Livermore National Laboratory, *In-Situ Microbial Filter,* Haz. Waste Cons., S-6/93.
50. Luey, J., et al, *Biodegradation of Hazardous Waste Using White Rot Fungus: Project Planning and Concept Development Document,* USDOE, PNL-7534, 11/90.
51. Mattison, P., *Bioremediation of Metals—Putting It to Work,* COGNIS, Santa Rosa, CA, 1992.

52. Noyes, R., *Handbook of Pollution Control Processes,* Noyes Data, 1991.
53. Nunno, T., et al, *International Technologies for Hazardous Waste Site Cleanup,* Noyes Data, 1990.
54. Palmer, S.A.K., et al, *Metal/Cyanide Containing Wastes,* Noyes Data, 1988.
55. Quinlan, B., et al, *Pure Oxygen-Enhanced Biodegradation for Contaminated Groundwater,* Remediation, 8/92.
56. Reuter, R.H., *Biotechnology Workgroup for Department of Defense Soil and Groundwater Decontamination Applications,* Naval Civil Engineering Laboratory, NCEL CR-91.007, 6/91.
57. Riser-Roberts, E., *In Situ/On-Site Biodegradation of Refined Oils and Fuels (a Technology Review)—Volumes 1-3,* Naval Civil Engineering Laboratory, NCEL CR 92.008, 6/92.
58. Rittman, B., *Aerobic Biological Treatment,* ES&T, Vol. 21, No. 2, 1987.
59. Rittman, B., et al, *In Situ Bioremediation, 2nd Ed.,* Noyes Publications, 1994.
60. Rosengrant, L., Lopez, L., *Treatment Technology Background Document,* USEPA, PB91-160556HDT, 1/91.
61. Rue, D., et al, *Fluid Extraction-Biological Degradation of Organic Wastes.*
62. Scholze, R.J. Jr., et al, *Biotechnology for Degradation of Toxic Chemicals in Hazardous Wastes,* Noyes Data, 1988.
63. Secor Inc., *Strategic Industry Analysis: Biotechnology in the Waste Treatment Industry,* Montreal, Quebec, 9/89.
64. Semprini, L., et al, *A Field Evaluation of In-Situ Biodegradation Methodologies for the Restoration of Aquifers Contaminated with Chlorinated Aliphatic Compounds: Results of a Preliminary Demonstration,* EPA, Stanford Univ., Technical Report No. 302, 11/87.
65. Semprini, L., et al, *Methodologies for Evaluating In-Situ Bioremediation of Chlorinated Solvents,* EPA/600/R-92/042, 3/92.
66. Sims, J.L., et al, *Bioremediation of Contaminated Surface Soils,* EPA-600/9-89/073, 8/89.
67. Surprenant, N., et al, *Halogenated-Organic Containing Wastes,* Noyes Data, 1988.
68. Torpy, M.F., *Anaerobic Treatment of Industrial Wastewaters,* Noyes Data, 1988.
69. Usinowicz, P., et al, *Thermophilic Process Cuts Biomass Wastes,* Env. Prot., 3/93.
70. Vance, D., *Remediation By In-Situ Aeration,* Nat. Env. Jnl., 7-8/93.
71. Worthy, W., "Anaerobic Digestion of Industry Waste Sought," *Chemical & Engineering News,* 10/28/91.
72. Zitomer, D., et al, *Sequential Environments for Enhanced Biotransformation of Aqueous Contaminants,* ES&T, Vol. 27, No. 2, 1993.

2

Chemical Technology

Chemical treatment is a class of processes in which specific chemicals are added to wastes or to contaminated media in order to achieve detoxification. Depending on the nature of the contaminants, the chemical processes required will include pH adjustment, lysis, oxidation, reduction or a combination of these. Thus, chemical treatment is used to effect a chemical transformation of the waste to an innocuous or less toxic form. In addition, chemical treatment is often used to prepare for or facilitate the treatment of wastes by other technologies.

In general, chemical treatment processes are applicable to a broad range of organic and inorganic wastes. For example, they can be used for the oxidation of organics, for pH adjustments to precipitate heavy metals, and for lysis of chlorinated organics to cleave chlorine atoms from organic molecules in preparation for subsequent oxidative processes. It should be remembered that chemical processes are very specific as to the waste that they treat. Thus, it is frequently necessary to link several unit operations together to effect the desired removal objectives. Other waste components must also be carefully considered because they can affect the chemical process by consuming more reagents, generating unwanted precipitates, inhibiting the reaction or creating safety issues when their presence is not recognized.

The function of pH adjustment is to neutralize acids and bases and to promote the formation of precipitates (especially of heavy metal precipitates) which can subsequently be removed by conventional settling techniques. These purposes are not mutually exclusive, precipitates can be formed as the result of neutralizing a waste. Conversely, neutralization of the waste stream can result when adjusting the pH to effect chemical precipitation. Typically, pH adjustment is effective in treating inorganic or corrosive wastes.

In-situ chemical treatment uses the same principles employed for above-ground chemical processes. Materials are added to neutralize, oxidize or remove contaminants in groundwater or soils in order to avoid digging or pumping of the contaminated waste above ground for treatment.

In-situ treatment can be used when it is uneconomical to haul or when infeasible or uneconomical to dig or pump the contaminated waste matrix for treatment in a reactor. This approach should be used whenever excavation or removal causes an increased threat to human health. It can reduce the cost of a remediation program. Because chemicals are applied to the contaminated waste matrix, specifically soil and groundwater, a potential exists for reaction with the soil. Permeability problems can occur as the result of precipitate formation. This can result in inadequate mixing of the contaminant with the treatment chemical. Gas generation may also occur.

Catalysis should be considered as a tool for use in chemical reactions. Catalysis is essentially applicable to all types of organic compounds including hydrocarbons containing halogens, nitrogen, oxygen, sulfur, and phosphorus. It is also applicable to oxidation and reduction reactions involving heavy metals, the oxidation of cyanides, carbonyls, and sulfites. Catalysis should be considered for waste treatment when a chemical transformation process has been shown to be applicable or when the chemistry of the components in the waste stream indicates that a chemical transformation is possible. If the transformation can be successfully catalyzed, lower operating costs and lower energy costs can be anticipated.

Catalysts are extremely versatile; they can provide selective detoxification as in the dechlorination of chlorinated pesticides or can provide complete destruction as in the air oxidation of cyanides. Also, their appropriate use in a chemical manufacturing process can serve as a powerful pollution prevention tool.

2.1 ACID AND ALKALINE LEACHING

Acid leaching is a treatment technology used to treat wastes in solid or slurry form containing metal constituents that are soluble in a strong acid solution or can be converted by reaction with a strong acid to a soluble form. This process has been used to recover metals such as mercury, copper, nickel, silver, and cadmium from inorganic wastes generated in the primary metals and inorganic chemicals industries.

The acid leaching process is most effective with wastes having high (over 1,000 ppm) levels of metal constituents. Wastes containing lower levels of such contaminants are more difficult to process because the low metal concentrations require longer contact times.

Acid leaching can also be used to extract heavy metals or radionuclides from mixed wastes, principally soils, separating the material into its hazardous and radioactive components. This process has greater utility when combined with unit operations such as ion exchange, solvent extraction, chemical precipitation, and filtration.

Another type of leaching process, alkaline leaching, is used to treat wastes containing metal constituents that are soluble in a strong caustic or alkaline solution. This process is mainly useful in recovering aluminum from bauxite ore.

Leaching is a process in which a solid material is contacted with a liquid solvent for selectively dissolving some components of the solid into the liquid phase. Leaching can sometimes be used to extract various metals from sludges. The goals of this process are as follows: (1) to dissolve the metals in a liquid phase to produce a solution that can be reused directly in a process or from which the metal can be recovered by other techniques, such as electrowinning; and (2) to produce a secondary sludge (leach residue) that is nonhazardous or from which additional metals can be reclaimed by other processes. A leaching process selected for metals recovery should be sufficiently flexible to remove a mixture of elements from sludge with a variety of characteristics (e.g., extent of aging, solids content).

Several leaching agents can potentially be used, including sulfuric acid, ferric sulfate, ammonia or ammonium carbonate, hydrochloric acid, sulfur dioxide, ferric chloride, nitric acid, or a caustic solution. Selection of a suitable solvent and unit processes depend on the chemical state and physical environment of the metals. Each reagent has its advantages and disadvantages. Sulfuric acid, the most commonly used and developed reagent, has the advantages of low cost, minor corrosion problems, and the ability to dissolve many metal compounds. Acid leaching has been used to recover metals such as copper, nickel, silver and cadmium from inorganic wastes generated in the primary metals and inorganic chemicals industries. The acid leaching process is most effective with wastes having high concentrations (greater than 1,000 ppm) of metal constituents. Ferric sulfate can be obtained from spent pickle liquors and used to provide a sulfate solution for metal removal.

Ammonia and ammonium carbonate offer better selectivity for dissolving metals, but these reagents are expensive and must be recovered for the process to be economical. In ammonia leaching, ammonium carbonate is used to convert nickel, copper, zinc, and cadmium to water–soluble amines. Iron, chromium, and calcium remain in the sludge as insoluble hydroxides. The remaining chromium hydroxide in the sludge can then be oxidized by air to the soluble dichromate species in the presence of added caustic. The leach solution of sodium dichromate can be crystallized for recovery of chromium. Sulfur dioxide must be added to the leached sludge to reduce chromium to the trivalent state, and the residual sludge must be adjusted to a pH of 7 or 8 to minimize chromium solubility. The sludge resulting after ammonia leaching can be leached with sulfuric acid to produce the hydrated trivalent form of chromium, which can then be oxidized to dichromate and recovered.

Leaching of hydroxide sludges with caustic has also been evaluated. Because of the low solubility of trivalent chromium in caustic and the amphoteric character of zinc, proper control of pH and sludge conditions could allow for dissolution of zinc and separation from the insoluble chromium. In particular, calcination can be used to convert the hydroxide to more inert oxides to avoid solubilization of chromium; however, the calcination step would add significantly to process costs. After zinc is extracted, oxidation of the remaining sludge would produce dichromates that are readily soluble in an additional caustic leach. Remaining metals could be recovered with an acid leach process.

2.1.1 Acid Leaching

The basic principle of operation for acid leaching is that solubilities of various metals in acid solutions aid in their removal from a waste. The process concentrates the constituent(s) leached by the acid solutions. These constituents can then be filtered to remove residual solids and neutralized to precipitate solids containing high concentrations of the constituents of interest, which can be further treated in metals recovery processes. Alternatively, the acid solutions can be electrolyzed to recover pure metals. An acid leaching system usually consists of a solid/liquid contacting unit followed by a solid/liquid separator. The most frequently used acids include sulfuric (H_2SO_4), hydrochloric (HCl), and nitric (HNO_3). Although any acidic pH can theoretically be used, acid leaching processes are normally run at a pH from 1 to 4.

Acid leaching processes can be categorized into two major types: (a) treatment by percolation of the acid through the solids, or (b) treatment by dispersion of the solids within the acid solution. Both treatments are followed by subsequent separation of the solids from the liquid. In both types of systems, sufficient acid must be supplied to keep the pH at the level needed to effectively leach the metals from the waste.

Percolation is typically conducted in batch tanks. Batch percolators are large tanks ranging in size up to 50,000 gal. First, the solids are placed in the tank, and then acid is added. The acid percolates through the solids and drains out through screens or a porous medium in the tank bottom. Following treatment, the solids are removed and further treated using stabilization, and/or they are land disposed.

Acid leaching by **dispersion** of fine solids into the acid is performed in batch tanks. The untreated waste and the acid are mixed in the reaction tank to ensure effective contact between the solids and the acid. Following mixing, the suspension may be pumped to stirred holding tanks, where the leaching is allowed to proceed to completion. The treated solids are then usually separated from the acid by filtration and further treated using stabilization, and/or they are land disposed.

Solid Waste Particle Size: Both the solubility reaction rate of the acid with the hazardous metal constituents in the waste and the rate of transport of acid to and from the site of the hazardous constituents are affected by the size of the solid waste particles. The smaller the particles, the more rapidly they will leach because of the increased surface area of the waste that is exposed to the acid. If the particle size of the untested waste is greater than that of the tested waste, the system may not achieve the same performance. Grinding the untested waste may be required to reduce the particle size and achieve the same treatment performance, or it may be necessary to consider other, more applicable treatment technologies for treatment of the untested waste.

Alkalinity of the Waste: The neutralizing capacity (or alkalinity) of the waste solids affects the amount of acid that must be added to the waste to achieve and/or maintain the desired reactor pH. In addition to dissolving the waste contaminants, the acid will also dissolve some of the alkaline bulk solids; therefore, highly alkaline wastes require more acid or a stronger acid to maintain pH during treatment. If the alkalinity in an untested waste is greater than that in a tested waste, the system may not achieve the same performance. Use of additional acid or a stronger acid may be required to compensate for the increased alkalinity and achieve the same treatment performance, or other, more applicable treatment technologies may need to be considered for treatment of the untested waste.

Solubility of the Metal Constituents in the Acid: The metal constituents must dissolve in the acid to form soluble salts for the process to be effective. Thus, the acid selected should be one that forms soluble salts for all of the constituents to be removed. If the solubility of a metal constituent(s) of concern in an untested waste is less than that of another constituent(s) of concern in a previously tested waste in the same acid, or less than the solubility of the same metal(s) tested with a different acid, the system may not achieve the same performance. Use of another acid may be required to increase the solubilities of the metal constituent(s) of concern and achieve the same treatment performance, or other, more applicable treatment technologies may need to be considered for treatment of the untested waste.

Concentration of Leachable Metals: The amount of leachable metals is a measure of the maximum fraction of the waste that can be expected to leach in the acid leaching system. A relatively low concentration of leachable metals implies that most of the waste will remain in the solid or slurry waste residues, i.e., it is nonleachable. If the concentration of leachable metals in the untested waste is significantly less than that in the tested waste, the system may not achieve the same performance. Use of a higher concentration of acid or a stronger acid may be required to leach less leachable components and achieve the same treatment performance, or other, more applicable treatment technologies may need to be considered for treatment of the untested waste.

Residence Time: The extent of reaction and dissolution of the contaminants in the acid is directly related to the contact time.

Type and Concentration of Acid Used: If the hazardous constituents to be removed in the acid leaching system are already present in the waste in soluble form, or are solubilized by pH reduction, then any acid that will reduce the pH to the desired value may be used. However, if chemical reaction is necessary to form soluble species, then the appropriate acid, as well as the appropriate concentration of the acid, must be used to ensure effective leaching of the metal constituents.

pH: For dispersed solids systems, the feed of acid to the leaching reactor is based on pH monitoring and control because the reaction rate is highly pH dependent. Therefore, a pH should be determined, based on the residence time and amount of hazardous metal constituents in the waste, that provides for complete dissolution of metal constituents. For percolation systems, pH monitoring of the acid percolating through the tank i.e., leaving the system) should ensure that enough acid is being added.

Degree of Mixing: Mixing provides greater contact between the acid and the solid waste particles, ensuring more rapid leaching of metal contaminants from the waste. The quantifiable degree of mixing is a complex assessment that includes, among other factors, the amount of energy supplied, the length of time the material is mixed, and the related turbulence effects of the specific size and shape of the tank. The exact degree of mixing is beyond simple measurement.

2.2 CHELATION

A chelating molecule contains atoms which can form ligends with metal ions. If the number of such atoms in the molecule is sufficient and if the final molecular shape is such that the metal atom is essentially surrounded then the metal will not be able to form ionic salts which can precipitate out. Thus, chelation is used to keep metals in solution and to aid in dissolution for subsequent transport and removal, e.g., soil washing.

Chelating chemicals can be chosen for their affinity to particular metals, e.g., EDTA and calcium. The presence of fats and oils can interfere with the process.

Chelation could also be described as a fixation process in that a hazardous material is bonded by the chelation agent in more than one position (making it unable to react chemically and thus less toxic). Chelation is applicable to aqueous solutions contaminated with heavy metal ions. There are two types of chelation agents: (1) sequestrants which bind the metal ion but remain soluble within the water column, and 2) precipitants which simply cause precipitation of the chelate–metal ion complex (and in this manner remove it from the water column). Sequestrants have the disadvantage that, even though the metal ion has been detoxified to a certain degree, the complex remains in the water column, and can present a hazard in itself. Ethylenediamine tetra–acetic acid (EDTA), for example, is a commonly used sequestrant and even has medicinal application as a treatment for metal poisoning. However, its aquatic toxicity threshold is in the range of 100 to 1,000 ppm. Treatments requiring addition of EDTA in concentrations greater than 1,000 ppm should, therefore, be carried out with concern for human health considerations.

The Japanese have been investigating the use of chelating resins, and dialkyl thiocarbonate, for removal of arsenic from wastewater.

The Toronto Harbor Commissioners (THC) have developed a soil treatment train designed to treat inorganic and organic contaminants in soils.

The THC's treatment train consists of three soil remediation technologies: a soil washing technology, a technology that removes inorganic contamination by chelation, and a technology that utilizes chemical and biological treatment to reduce organic contaminants.

The process utilizes an attrition soil wash plant to remove relatively uncontaminated coarse soil fractions using mineral processing equipment while concentrating the contaminants in a fine slurry which is routed to the appropriate process for further treatment. The wash process includes a trommel washer to remove clean gravel, hydrocyclones to separate the contaminated fines, an attrition scrubber to free fines from sand particles, and a density separator to remove coal and peat from the sand fraction.

If only inorganic contaminants are present, the slurry can be treated in the inorganic chelator unit. This process uses an acid leach to free the inorganic contaminant from the fine slurry and then removes the metal utilizing solid chelating agent pellets in a patented countercurrent contactor. The metals are recovered by electrowinning from the chelating agent regenerating liquid.

Organic removal is accomplished by utilizing a chemical pretreatment of the slurry from the wash plant or the metal removal process and biological treatment in upflow slurry reactors utilizing the bacteria which have developed naturally in the soils being treated. The treatment soil is dewatered utilizing hydrocyclones and transported back to the site from which it was excavated.

The metals removal process equipment and chelating agent were fouled by free oil and grease contamination, forcing the curtailment of sampling prematurely. This establishes a limitation for this technology since biological treatment or physical separation of oil and grease will be required to void such fouling.

The IT Corporation is developing a mixed waste treatment process designed to address one of the more difficult waste treatment problems—soils contaminated with both hazardous and radioactive constituents.

Inorganic contaminants are removed by three physical and chemical separation techniques: (1) gravity separation of high density particles, (2) chemical precipitation of soluble metals, and (3) chelant extraction of chemically bound metals.

Some of the radionuclides that are not in soluble form remain with the soil through the gravity separation process. These radionuclides are removed from the soil via extraction with a chelant. The chelant solution then passes through an ion exchange resin to remove the radionuclides. The chelant solution is recycled back to the soil extraction step.

Chelant extraction has been successfully applied to surface contamination in the nuclear industry for more than 20 years. Application of this chemistry using soil washing equipment is expected to achieve similar results on subsurface contamination.

Argonne National Laboratory is evaluating chelating agents (EDTA) and (NTA for removing lead from soil. Maximum removal of lead occurred with EDTA, which is pH–insensitive.

2.3 DEHALOGENATION

Chemical dechlorination processes use specially synthesized chemical reagents to destroy hazardous chlorinated molecules or to detoxify them to form other compounds that are considered less harmful and environmentally safer. In recent years, several dechlorination processes using different reagents have been developed to detoxify PCBs and many chlorinated organic compounds. The residue structures are generally nontoxic or lower in toxicity than the original compound. These processes were first developed for the treatment of PCB–containing oils, but several have potential application to in situ treatment of contaminated soils.

The reaction mechanism is nucleophilic substitution. Nucleophilic substitution removes chlorine from aromatic compounds by two mechanisms: the intermediate complex mechanism, and the benzene mechanism.

2.3.1 Glycolate Dehalogenation

One of the first major PCB–contaminated oil dechlorination processes was developed by the Goodyear Tire and Rubber Company. This process was intended to remove PCBs from heat transfer fluids using sodium naphthalene as the reagent. The sodium naphthalene reagent is prepared by complexing naphthalene and metallic sodium with the solvent tetrahydrofuran (THF). The reagent mix is reacted with the PCB–contaminated fluid at ambient temperature and at a reagent to chlorine ratio of 50 to 100:1 (under a nitrogen blanket). Under these conditions, the PCB molecule is stripped of chlorine to form sodium chloride and polyphenyls which, after quenching, are vacuum distilled in order to recover the THF and naphthalene. The use of the priority pollutant, naphthalene, proved to be a source of concern. Subsequent dechlorination processes were designed to utilize alternate reagents, such as glycols.

In 1978, the EPA sponsored research which led to the development of the first of the series of APEG (Alkylene Polyethylene Glycolate) reagents, which were shown to effectively dechlorinate PCBs and oils. Essentially, these reagents were alkali metal/polyethylene glycols which react rapidly to dehalogenate halo–organic compounds of all types, under both ambient and high temperature conditions. In the APEG reagents, the alkali metal ion is held in solution by the large polyethylene glycol anion. PCBs and other halogenated molecules are uniquely soluble in APEG reagents. These qualities combine to get a single–phase system in which the anions readily displace the halogen atoms. The reaction of halogenated aromatics with PEGs result in the substitution of the PEG for the chlorine atom to form a PEG ether. The PEG ether, in turn, may then decompose to a phenol. The Franklin Research Institute was also instrumental in research and development.

These processes use polyethylene glycols (PEG) or their derivatives that have been reacted with alkali (usually potassium) metals or their hydroxides to dechlorinate. When alkali metals are used, the reagents are susceptible to decomposition by water. The use of alkali metal hydroxides solve this problem.

KPEG (potassium hydroxide/polyethylene glycol) is the most commonly used type of APEG reagent. Sodium hydroxide/tetra–ethylene glycol (ATEG), is another variation of the reagent. However, sodium hydroxide has also been used in the past and most likely will find increasing use in the future because of patent applications that have been filed for modification to this technology.

The glycolate dehalogenation process is potentially effective in detoxifying specific types of aromatic organic contaminants, particularly dioxins and polychlorinated biphenyls (PCBs). The process involves heating and physically mixing contaminated soils, sludges, or liquids with an alkali metal hydroxide–based polyethylene glycol reagent in a mobile batch reactor.

Before treatment, soils are sieved to remove any large rocks and/or debris. The contaminated media are commingled with a reagent to form a homogeneous slurry. The reagent primarily consists of potassium or sodium hydroxide (KOH or NaOH) and polyethylene glycol (PEG); other reagents such as dimethylsulfoxide (DMSO) or sulfolane (SFLN) may be added to improve the efficiency of the process. The slurry is simultaneously heated (25° to 150°C) and mixed, consequently decomposing halogenated contaminants into less toxic, water–soluble compounds (glycol–ethers and chloride salts).

Treatment time in the reactor ranges from 0.5 to 5 hours, depending on the contaminant type, initial concentration of the contaminant, water content, humic and clay content (for soils), and the level of treatment desired. Water is vaporized in the reactor and collected in a condensate receiver. A carbon adsorption filter traps any volatile compounds that are not condensed.

Additional treatment of soils is required to desorb reaction by–products and reagent from the dechlorinated soil. This treatment includes physically mixing the dehalogenated soil with water in successive washing cycles. The treated soil is then dewatered and redeposited on–site, while the reagent and wash waters are recycled and ultimately treated and/or delisted.

The major technology consideration is determining how a large volume of residual wastewater generated from the soil washing/dewatering process will be managed. The residual effluent may require treatment prior to disposal; however, if the volume of wastewater is extremely high (i.e., volumes generated from greater than 30,000 yd^3 of washed soil), it may be more cost–effective to petition EPA to delist the residual effluent, whereby it may be disposed without further treatment. Post–treatment options commonly employed when treating residual wastewaters may include chemical oxidation, biodegradation, carbon adsorption, precipitation, or incineration.

Glycolate dehalogenation operations require no special handling (although special handling of contaminated media, e.g., dioxin contaminated waste may be required) and energy requirements are not extreme; therefore, operation and maintenance costs are relatively low. A full–scale dehalogenation unit with a capacity of 80 yd^3 per batch requires an average of 670 kW, with 930 kW peak. A sufficient power source is required and may present additional costs if a source is not readily accessible. Preconstruction engineering controls, to guard against accidental spills, include leveling and lining (synthetic) the areas under and adjacent to the treatment facility and diking the area surrounding the facility.

The presence of reactive metals, e.g., aluminum, under alkaline conditions may be deleterious to the dechlorination process. If these contaminants are present in the waste matrix, a suitable pretreatment may be necessary to remove them prior to the dechlorination treatment.

Dechlorination may also be effective in treating halogenated aliphatic (W04) and nonpolar halogenated organic compounds (W01). However, the degradation products from the APEG treatment of halogenated aliphatic compounds include acetylene. Acetylene is an extremely volatile and potentially explosive compound which may also form acetylides. Several of these metal acetylides, particularly those of copper and silver, are very unstable and can explode upon drying.

As for halogenated cyclic alipatics, ethers, esters, and ketones (W05), data for compounds with similar physical and chemical characteristics suggest that dechlorination is potentially effective on these compounds.

The APEG solutions may also be potentially effective on other halogenated polar aromatic compounds (W03). Wastes contaminated with nonhalogenated organic compounds and inorganic compounds cannot be treated by dechlorination, because of the absence of chlorine in these treatability groups. It is possible that contaminants in these treatability groups may volatilize into the air stream or transfer into the reagent during treatment of halogenated compounds. Dechlorination may be used as part of a treatment train to treat waste mixtures which contain halogenated and nonhalogenated organic compounds and inorganic compounds.

Factors limiting the effectiveness of glycolate dehalogenated include highly concentrated contaminants, high water content, low pH, high humic content (soil), and the presence of other alkaline–reactive materials (e.g., aluminum, other metals).

Design and Operating Parameters Affecting Performance: The design and operating parameters which influence the effectiveness of the dechlorination process are soil moisture and type, degree of mixing, reaction time, PEG:soil:contaminant ratio, pH control, temperature, and solvent and PEG recoveries. These parameters are summarized as follows:

(1) *Soil Moisture and Type*—Water bound in the soil deactivates the alkali polyethylene glycolate reagent. The soil should be dried before reagent is added. The type of soil will also affect the amount of time necessary to treat contaminated soils with the APEG reagents. Contaminants can be bound tightly to certain soils, hence reaction times and temperatures may have to be varied to effectively remove the contaminants.

(2) *Degree of Mixing*—The degree of mixing between the soil and the polyethylene glycolate (APEG solution) is a critical factor. The degree of mixing needs to be thorough to ensure that the contaminant in the soil makes intimate contact with the dechlorination reagent. Creating a slurry may improve removal efficiencies.

(3) *Reaction Time*—Reaction time between the contaminant and the dechlorination reagent must be sufficient to ensure that all possible reactions occur. This must be determined experimentally depending on the type and amount of solvent and physical and chemical characteristics of the soil.

(4) *APEG:Soil:Contaminant Ratio*—This ratio is dependent on the type of APEG, the initial contaminant concentration, the solvent used (usually dimethylsulfoxide), and the soil type. As a general rule, two and a half times more APEG reagent than the maximum number of chlorine atoms on contaminant molecules is required to effectively dechlorinate (des Roslers, U.S. EPA, February 1989).

(5) *pH Control*—The pH is extremely important to the operation of the process. It has

been reported that the pH should be raised above 12 before adding APEG reagent. Continuous monitoring to ensure that the pH is maintained in the proper range during the treatment process is recommended. Contaminated soils containing organic esters, such as n-butyl esters of 2,4-D and 2,4,5-T, must be pretreated with potassium hydroxide to a pH above 12 prior to adding APEG reagent to avoid wasting the special dechlorinating reagent.

(6) *Temperature*—The optimum temperature range for dechlorination varies depending on the compounds present. Typically, the reaction is run at about 100°C, but dechlorination of some compounds may require temperatures up to 150°C.

(7) *Solvent and APEG Recoveries*—The degree of recovery of solvent following the dechlorination treatment is important to the overall economics of the process. Recovery of APEG reagents by distillation is typically only about 50%, while washing is significantly more successful, yielding 94 to 99% recovery of the reagent.

Advantages:

1. It has greater public acceptance than incineration.
2. Dehalogenation has been used successfully to treat contaminant concentrations of PCBs reported as high as 45,000 ppm to less than 2 ppm.
3. Uses standard reactor equipment to mix/heat soil and reagents.
4. Energy requirements are moderate, and operation and maintenance costs are relatively low.
5. Treatment units can be mobile.
6. Short treatment time.

Limitations:

1. Most effective with aromatic halides when APEG and KPEG reagents are used, although ATEG reportedly works with aliphatic halides. The presence of other pollutants, such as metals and other inorganics, can interfere with the process.
2. Wastewater will be generated from the residual washing process. Treatment may include chemical oxidation, biodegradation, carbon adsorption, or precipitation.
3. Engineering controls, such as a lined and bermed treatment area and carbon filters on gas effluent stacks, may be necessary to guard against releases to the environment.

Firms offering processes include:

1. Acurex Corporation (through Chemical Waste Management).
2. Golson Research Corporation.
3. PPM, Inc.
4. Sea Marconi Technologies Group (CDP Process).
5. Sun Ohio (PCBX Process).

2.3.2 Alkaline Processes

Battelle–Northwest has developed an aqueous phase process in which organic matter is digested at 250° to 400°C (480° to 750°F) with mild alkali under pressure (500 to 3,000 psig), and in the absence of oxygen. The result is the conversion of organic solids into an oil, which under some circumstances can have a heating value of up to 90% that of diesel oil. The product oil is easy to separate from the wastewater, and is generated in up to 70% of theoretical energy yield. Residence time is 0.5 to 5 hours.

Lindane and chloroform were almost totally destroyed, and a 2,4,5–T analog was partially degraded, demonstrating the ability of alkali under these conditions to remove both aromatic and aliphatic chlorine atoms from organic molecules.

Rulkens, Assink, & Van Gemert (TNO) has developed a process for extraction of organic bromine compounds from a contaminated site in the Netherlands using NaOH.

Extraction with NaOH solution consists of the following process steps:

1. Soil pretreatment to separate large objects, e.g., stones, and reduce the size of large clods of soil (crushing and wet sieving).
2. Intimate mixing of soil with extracting agent (approximately 0.2% NaOH solution); the soil–to–water ratio is about 3 to 1 on a weight basis.
3. Extraction and washing of the soil with clean extracting agent in countercurrent flow in two modified screw classifiers in line.
4. Dewatering of soil before redeposition. The remaining alkalinity of the soil will be largely neutralized by absorption of CO_2 from the ambient air.
5. The overflow of the first modified screw classifier is led through a settling tank for fine mineral particles dragged out from the screw classifier by the extracting liquid. The particles that settle, with diameters between approximately 35 and 60 μm (approximately 1% of the total soil), are collected from time to time and washed separately by mixing them with NaOH–solution.
6. Sludge forming by adding lime as coagulant and polyelectrolyte as flocculant. The sludge formed can be separated in a tiltable plate separator.
7. Dewatering of the sludge in a solid bowl centrifuge with scroll discharge.
8. Effluent polishing by deep bed filtration, activated–carbon adsorption, and finally anion exchange to remove any bromides formed by hydrolysis. The cleaned extracting agent can be recycled to the extraction process in the screw classifiers.

Experiments showed that it is possible to remove the bromine compounds from the soil down to a level of 1 mg Br/kg or less. The cleaned extracting agent contains less than approximately 0.6 mg Br/kg, the main part of which is bromide.

2.3.3 Catalytic Dechlorination

Researchers are studying the use of metal catalysts in the degradation of halogenated organic compounds in aqueous solutions. Based on batch and column tests, the catalyst performs two functions: (1) it produces highly reducing conditions, and (2) it participates in the degradation process. The catalysts have been effective in degrading a range of halogenated methanes, ethanes and ethenes.

GARD, a division of Chamberlain Manufacturing Corporation, Niles, Illinois has developed a treatment system using a platinum–based catalyst to break down halogenated (containing chlorine, fluorine or bromine) hydrocarbons into acids and simple hydrocarbons. GARD has established two different processes of the catalytic dehalogenation technology: one for liquids containing low concentrations of halogenated compounds (e.g., silvex) and another for liquids present as pure halogenated compounds or their solutions in high concentrations (e.g., ethylene dibromide, EDB).

The process system for low concentrations of halogenated compounds features a waste storage tank, a pump, a preheater, a catalytic reactor, a cooling system, a vapor/liquid separator and a scrubber. The process begins by pumping dilute halogenated liquid wastes into the reactor where they are heated to 300° to 400°F at 50 to 150 psi. Hydrogen is then added, which reacts with wastes, some of which are converted to acids.

Each halogen atom of the halogenated compound is replaced by a hydrogen atom, resulting in a fully hydrogenated aromatic hydrocarbon. Typically, the solvent is not affected by the catalytic dehalogenation process. Thus, the final product will consist of a mixture of the original solvent and the dehalogenated compound. The excess hydrogen is recycled; in addition, the product recycle is built into the system to safeguard against incomplete reaction due to upset conditions or cases where a single pass would not achieve the desired degree of halogenation.

The reaction product leaving the reactor is cooled, then, following a vapor/liquid separation stage, the dehalogenated hydrocarbon product and halogen acid collected in a scrubber are transferred to storage.

In the second process, for liquids present as pure halogenated compounds or their solutions in high concentrations, ethylene dibromide (EDB) is usually used directly instead of dissolved in solvent; therefore, there is no solvent to recover. Debromination of EDB is possible either by catalytic hydrodebromination described above or by catalytic oxidation. Ethylene is oxidized to carbon dioxide and catalyst temperature is maintained in the range of 800° to 1000°F required for a complete removal of EDB. Bromine can be easily condensed and recovered, while ethylene and bromide are difficult to capture.

Halogenated liquid solvents, oils, and halogenated wastes of any concentration are acceptable feed materials for the catalytic dehalogenation process. The temperature range of the process is from 300° to 1000°F, while the pressure ranges are 50 to 150 psi. Typically, the residence time for dehalogenation varies from 10 to 20 minutes.

Advantages: The Catalytic Dehalogenation process treats halogenated compounds by replacing the halogen atoms with hydrogen atoms. Typically, these compounds are found as liquid solvents for reuse. The economic value of such materials is retained, while at the same time, the halogenated compounds are rendered harmless. The method takes very little pretreatment, and is highly mobile. It uses conventional equipment and has relatively mild operating conditions.

Limitations: The process is capable of treating only liquids. Some liquid residues require further treatment.

The Hungarian Academy of Science has developed a new type of supported palladiam catalyst that activates hydrodehalogenation. In other words, the catalyst removes the chlorine atoms and replaces them with hydrogen atoms. For example, tetra– and penta–chlorobenzene with supported paladium catalyst at atmospheric pressure and at temperatures of 120° to 140°C will react to form benzene and chlorobenzene.

The Waterloo Centre for Groundwater Research in Ontario, Canada, consistently is getting good results in the development of a permeable reaction wall that degrades halogenated organic compounds in situ. The wall consists of a porous medium, containing an iron–based catalyst that degrades the contaminants as they pass through the wall. This passive method of remediation thus prevents further downstream migration of contamination and degrades contaminants many times faster than the natural rate of degradation. Because the degradation occurs in situ, the contaminants are not transferred from the water to a different medium.

Other processes include the use of potassium hydroxide reagent with tetraethylene glycol as a catalyst, investigated by Radian Corporation (Research Triangle Park, North Carolina); and the CHLOROFF process developed by Kinetics Technology International (KTI–Zoetermeer, Netherlands).

2.3.4 Light Activated Reduction

Another approach to dechlorination of halogenated organics is employed in the light activated reduction of chemicals (LARC). The LARC processes, which uses UV light and hydrogen gas to degrade extracted chlorinated hydrocarbons, was developed by **Atlantic Research Corporation.** However, despite its initial promise as demonstrated by its ability to destroy halogenated organics such as Aroclor 1254, kepone, and tetrabromophthalic anhydride, the process has not been actively pursued, primarily because of economic considerations.

2.3.5 KGME Process

Chemical Waste Management's (CWM) DeChlor/KGME process involves the dechlorination of liquid–phase halogenated compounds, particularly polychlorinated biphenyls (PCB). KGME, a CWM proprietary reagent, is the active species in a nucleophilic substitution reaction, in which the chlorine atoms on the halogenated compounds are replaced with fragments of the reagent. The products of the reaction are a substituted aromatic compound, which is no longer a PCB aroclor, and an inorganic chloride salt.

KGME is the potassium derivative of 2–methoxyethanol (glyme) and is generated in situ by adding stoichiometric quantities of potassium hydroxide (KOH) and glyme. The KOH and glyme are added to the reactor vessel, along with the contaminated waste. The KGME is formed by slowly raising the temperature of the reaction mixture to about 110°C (230°F), although higher temperature can be beneficial.

The DeChlor/KGME technology is preferable to the older sodium (Na) dispersion treatment method because it is less expensive and because the KGME reagent is much more tolerant of water in the reaction mixture; the water can cause a fire or explosion in the presence of Na metal. One advantage of the DeChlor/KGME process over KPEG or APEG methods is that only about one–quarter the weight of KGME is required for dehalogenation as would be required if KPEG were used. Also, considerably less waste is produced, and no polymeric treatment residue, which is difficult to handle, is formed.

The reaction product mixture is a fairly viscous solution containing reaction products and the unreacted excess reagent. After this mixture has cooled to about 93°C (200°F), water is added to help quench the reaction, improve the handling of the mixture, extract the inorganic salts from the organic phase for disposal purposes, and help clean out the reaction vessel for the next batch of material to be treated. The two resulting phases, aqueous and organic, are separated, analyzed, and transferred to separate storage tanks, where they are held until disposal.

2.3.6 Base Catalyzed Decomposition

This process was developed by EPA at the **Risk Reduction Engineering Laboratory.** Charles Rogers and coworkers at the Risk Reduction Engineering Laboratory have been working on detoxifying chlorinated organics for over ten years. They focused their efforts on contaminated soils, and developed several chemical treatment processes that are described as base-catalyzed dechlorination (BCD) processes. Unlike the earlier versions that use polyethylene glycol (PEG), the latest version of this technology uses no PEG, and represents new chemistry for dechlorination. This new mechanism is a breakthrough in treatment technology, and provides a clean and inexpensive reaction.

Unlike incineration, BCD processes offer lower cost of decontamination, reduced air pollution risk, and greater public acceptance.

The Base-Catalyzed Decomposition Process (BCDP) is a dehalogenation/dechlorination process that strips off chlorine in the PCB molecule and forms sodium chloride and biphenyls. The BCDP uses a rotary reactor in which most of the decomposition takes place. The contaminated soil is screened, processed with a crusher and pug mill, and stockpiled.

Next, in the main treatment step, this stockpile is mixed with sodium bicarbonate ($NaHCO_3$). The sodium bicarbonate is used in an amount equal to about 10% of the weight of the stockpile. The mixture is then heated for about one hour at 630°F in the rotary reactor. PCBs are decomposed and partially volatilized in this step.

The clean soil removed from the reactor can be returned to the site. Off-gases from this reactor, which contain dust and trace amounts of PCBs, are filtered, scrubbed, and vented to the atmosphere. PCBs in the vapor condensate, residual dust, spent carbon, and filter cake are decomposed in a stirred-tank slurry reactor. The resulting sludge can be disposed of in the same manner as municipal sewage sludge.

2.4 HYDROLYSIS

Hydrolysis is the process of breaking a bond in a molecule (which is ordinarily not water-soluble) so that it will go into ionic solution with water. Hydrolysis can be achieved by the addition of chemicals (e.g., acid hydrolysis), by irradiation (e.g., enzymatic bond cleavage). The cloven molecule can then be further treated by other means to reduce toxicity.

Hydrolysis is suitable for pretreating difficult-to-treat wastes and for organics with substituents, such as phenols or chlorinated organics with reactive chlorine atoms. Hydrolysis is specific for only a limited number of contaminants.

The resulting residuals from a hydrolysis process are an aqueous effluent and insoluble organics.

Hydrolysis was favorably applied to a site in which the wastewater contained very soluble, refractory organics. In addition, tars were being produced in high quantities on this site. Both of these problems were solved using a hydrolyzer. As a result, the wastewater treatment goals were achieved, and the production of tar was reduced.

The primary design parameter considered for hydrolysis is the half-live, which is the time required to react 50% of the original compound. The half-life of a reaction is generally dependent on the reaction pH and temperature and the reactant molecule. Hydrolysis reactions can be catalyzed at low pH, high pH, or both, depending on the reactant. In general, an increase in temperature will increase the hydrolysis rate. Improving the conditions for the hydrolysis reaction results in a shorter half-life, and therefore, the size of the reaction vessel required is reduced.

Hydrolysis is a treatment technology which should be strongly considered for wastewaters which contain carbamate, phosphate, phosphorothioate, phosphorodithioate, and phosphonothioate PAIs. For virtually all PAIs in these structural groups for which treatability testing was performed, a half-life less than 30 minutes was achieved at high temperature (60°C) and high pH (pH 12). Literature data shows that many of the PAIs in fact react even faster than EPA's study demonstrated. Study conditions were such that the "zero" reaction time was in fact at least 15 minutes (i.e., 15 minutes had elapsed between the time the initial sample was taken and analyzed). In some cases, the PAI had been completely destroyed within that 15 minute period (i.e., the PAI was not detected in the sample). In such cases, the half-life was estimated to be at less than 30 minutes, and a 30 minute half-life was used in calculating reactor sizes and retention times, hence cost, for treatment. Literature data, however, confirms that for PAIs such as malathion the half-life is less than one minute.

For many compounds high pH and ambient temperature were enough to result in a half-life less than an hour, especially for the carbamates. Acid hydrolysis was only effective for a small number of compounds tested. However, for organophosphorus and carbamate pesticide hydrolysis, alkaline hydrolysis is usually faster than acid hydrolysis. The urea PAIs tested were not hydrolyzed effectively, so long reaction times would be necessary to treat most urea PAIs.

Acid hydrolysis of dithiocarbamate PAIs can achieve short half-lives; however, this reaction results in evolution of carbon disulfide gas; therefore, hydrolysis is not considered to be feasible for dithiocarbamate PAIs. Hydrolysis has also been used to treat triazine PAIs, but only at high temperature with catalyst because this reaction proceeds very slowly in the normal range of conditions used in wastewater treatment.

Hydrolysis has been identified as the most effective technology for achieving high levels of destruction of pesticide active ingredients in the carbamates and organophosphate structural groups. This technology has been demonstrated at a number of manufacturing facilities, and in both EPA and industry-supplied treatability studies.

Hydrolysis may be applied to a wide range of waste types, primarily for destruction of nitriles, amides, esters, and some chlorinated hydrocarbons. Familiarity with the specific reaction chemistry is necessary before this technique should be used to treat wastes, due to the fact that the hydrolysis reaction products are often as toxic, or even more toxic, than the original waste component. Because of this, in situ hydrolysis is not a recommended course of action. For example, hydrolysis of the pesticide parathion yields as one of the reactions p-nitrophenol which is also quite toxic.

Hydrolysis is a common commercial process which can be conducted with relatively simple equipment (e.g., batch-wise in open tanks) or in more elaborate equipment (e.g., countercurrent towers). Capital costs will vary considerably depending upon the equipment and operating temperature and pressure. Raw material costs are usually small.

Acid Hydrolysis: The agents for acid hydrolysis are most commonly hydrochloric and sulfuric acids, but others such as formic and oxalic acids are also potential reagents. Muriatic acid (30% hydrochloric acid solution) is readily available at hardware stores. Sulfuric acid (20%) is available in the form of drain cleaner products and can be used.

It must be remembered, however, that these materials are extremely corrosive and must be handled with care.

Acid hydrolysis usually proceeds more rapidly in solution. Thus, stirring in an open container with excess acid is feasible though potentially dangerous. Another method is to mix the waste with sand or other adsorbent in a pit or trench and then add the acid. Disposal of wastes by this method cannot be performed in an area where groundwater contamination is possible. Treated waste streams will generally require pH adjustment before they can be discharged.

One final word of caution: Do not treat cyanides or phosphides with strong acids as hydrogen cyanide gas and phosphine gas, both highly toxic, will be given off as reaction products.

Alkaline Hydrolysis: Alkaline hydrolysis mostly utilizes sodium hydroxide, but alkali carbonates (limestone), alkaline hypochlorite, calcium compounds lime, slaked lime), and magnesium and aluminum compounds are also used.

As with acid hydrolysis, alkaline hydrolysis can be performed in batch processes in open tanks with stirring, although this may be dangerous. Treatment in a trench filled with adsorbent material is recommended for treatment of small quantities of wastes, especially pesticide wastes. This should be done only in areas where groundwater contamination will not occur.

If a large volume of waste is to be discharged after treatment, it will probably be necessary to adjust the pH.

2.5 ION EXCHANGE

Ion exchange is a treatment technology applicable to (a) metals in wastewaters where the metals are present as soluble ionic species (e.g., Cr^{+3} and CrO_4^{-2}); (b) nonmetallic anions such as halides, sulfates, nitrates, and cyanides; and (c) water–soluble, ionic organic compounds including (1) acids such as carboxylics, sulfonics, and some phenols, at a pH sufficiently alkaline to yield ionic species, (2) amines, when the solution acidity is sufficiently acid to form the corresponding acid salt, and (3) quaternary amines and alkysulfates.

Ion exchange, when used in hazardous waste treatment, is a reversible process in which hazardous cations and/or anions are removed from an aqueous solution and are replaced by nonhazardous cations and/or anions such as sodium, hydrogen, chloride, or hydroxyl ions. Ion exchange resins are cationic if they exchange positive ions (cations) and anionic if they exchange negative ions (anions). When the waste stream to be treated is brought into contact with a bed of resin beads (usually in a packed column), an exchange of hazardous ions for nonhazardous ions occurs on the surface of the resin beads. Initially, a nonhazardous ion is loosely bound to the surface of the resin. When a hazardous ion is near the resin, it is preferentially adsorbed to the surface of the resin (based on the differences in ionic potential), releasing the nonhazardous ion.

Cation exchange resins contain mobile positive ions, such as hydrogen (H^+) or sodium (Na^+), which are attached to immobile functional acid groups, such as sulfonic (SO_3^-) and carboxylic (COO^-) groups. Anion exchange resins have immobile basic ions, such as amine (NH_2^-), to which the mobile anions, such as hydroxyl (OH^-) or chloride Cl^-), are attached.

Ion exchange material is contacted with the solution containing the ion to be removed until the active sites in the exchange material are partially or completely used up ("exhausted") by that ion.

After exhaustion, the resin is then contacted with a relatively low volume of a very concentrated solution of the exchange ion to convert ("regenerate") it back to its original form.

For instance, in the case of a sodium-based resin, a strong solution of sodium chloride is typically the regenerant solution. The regenerant solution forces the previously removed ions back into solution. This relatively low volume solution, now highly concentrated with the contaminant ions, must then be treated prior to disposal for recovery or removal of the hazardous cation or anion contaminants. There will continue to be a high concentration of the regenerant ion (sodium in the above example) in the used regenerant solution because excess regenerant ion is necessary to force the contaminant ions back into solution. The direction and extent of the completion of the exchange reaction depend upon the equilibrium that is established between the ions in the solution (M^+X^-) and those in the exchange material (R^-N^+).

2.5.1 Ion Exchange Process

Most ion exchange operations are conducted in packed columns. The aqueous solution to be treated is continuously fed to either the top or the bottom of the column. A typical fixed-bed ion exchange column consists of a vertical cylindrical pressure vessel with corrosion-resistant linings. If appropriate, a filter is installed at the inlet of the column to remove suspended particles because they may plug the exchange resin. Spargers are provided at the top and bottom of the column to distribute waste flow. Frequently, a separate distributor is used for the regenerant solution to ensure an even flow. The resin bed, usually consisting of several feet of ion exchange resin beads, is supported by a screen near the bottom distributor or by a support bed of inert granular material. Externally, the unit has a valve manifold to permit downflow operation, upflow backwashing (to remove any suspended material), injection of the regenerant solution, and rinsing of any excess regenerant.

A typical process for a basic two-step cation/anion ion exchange system includes a series treatment with separate cation and anion exchange systems. Some systems contain both anion and cation exchange resins in the same vessel. The pressure vessels used for ion exchange generally range in size from 2 to 6 feet in diameter for prepackaged modular systems, which typically handle 25 to 300 gpm flow rates, to a maximum custom size of 12 feet in diameter, which can handle flow rates up to 1,150 gpm. The height of these vessels varies between 6 and 10 feet to provide adequate resin storage, distribution nozzle layout, and freeboard capacity for bed expansion during backwashing. The nominal surface loading area of the ion exchange vessels ranges from 8 to 10 gpm/ft^2.

Ion exchange is used to remove a broad range of ionic species from water including:

1. All metallic elements when present as soluble species, either anionic or cationic.
2. Inorganic anions such as halides, sulfates, nitrates, cyanides, etc.
3. Organic acids such as carboxylics, sulfonics, and some phenols, at a pH sufficiently alkaline to give the ions.

4. Organic amines when the solution acidity is sufficiently acid to form the corresponding acid salt.

Sorptive resins can remove a wide range of polar and non–polar organics.

A practical upper concentration limit for ion exchange is about 2,500 to 4,000 mg/ℓ. A higher concentration results in rapid exhaustion of the resin and inordinately high regeneration costs. Suspended solids in the feed stream should be less than 50 mg/ℓ to prevent plugging the resins, and waste streams must be free of oxidants.

This technology is used to treat metal wastes including cations (e.g., Ni, Cd, Hg) and anions (e.g., CrO_4, SeO_4, $HAsO_4$). Limitations are selectivity/competition, pH and suspended solids. Highly concentrated waste streams (greater than about 25,000 mg/ℓ contaminants) can usually be separated more cost effectively by other means. High solid concentrations (greater than about 50 mg/ℓ) should be avoided to prevent resin binding.

Ion exchange is used to extract specific metals from solution. To date, some 30 synthetic "metal loving" compositions of resins that attract specific metals have been developed. The method of attraction employed by the synthetic compositions is similar to that of living cells. Natural cells have a built–in survival mechanism that is highly selective for the capture and transport of certain metals necessary for cellular nutrition, specifically iron, cobalt, zinc, copper, sodium, nickel, potassium, magnesium, and manganese. The synthetic compounds are patterned after the high efficiency and natural metal extraction capability of living cells.

Specific ion exchange and sorptive resins systems must be designed on a case–by–case basis. It is useful to note that although there are three major operating models (fixed bed cocurrent, fixed bed countercurrent, and continuous countercurrent), fixed bed countercurrent systems are most widely used. The continuous countercurrent system is suitable for high flows. Complete removal of cations and anions ("demineralization") can be accomplished by using the hydrogen form of a cation exchange resin and the hydroxide form of an anion exchange resin. For removal of organics as well as inorganics, a combination adsorptive/demineralization system, can be used. In this system, lead beds would carry sorptive resins which would act as organic scavengers, and the end beds would contain anion and cation exchange resins. By carrying different types of adsorptive resins (e.g., polar and non–polar), a broad spectrum of organics could be removed.

Capacities of resins vary greatly with the manufacturer of the resin. The amount of resin needed must be determined by chemical tests using the wastewater to be treated. A resin manufacturer should also be contacted to ensure the correct choice of resins. In order to facilitate the proper selection, the following items of information should be available: (1) name of hazardous material to be removed, (2) concentration (approximate) of hazardous substance, (3) amount of wastewater to be treated, and (4) chemical analysis of ions.

In a new development, Rohm and Haas has developed a family of six carbonaceous adsorbents produced by the pyrolysis of fully activated IERs. After pyrolysis, the beads acquire a carbon shell that combines the high adsorbent activity of activated carbon with the physical strength and controlled pore size of the polymeric bead. A major application for this carbonaceous resin is groundwater remediation, where both high reactivity and selectivity are required to capture selected contaminants to the ppm and ppb concentration range.

Applicability: The list of pollutants for which the ion exchange system has proved effective includes aluminum, arsenic, cadmium, chromium (hexavalent and trivalent), copper, cyanide, gold, iron, lead, manganese, nickel, palladium, platinum, selenium, silver, tin, zinc, and more. Thus, it can be applied to a wide variety of industrial concerns. Because of the heavy concentrations of metals in their wastewater, the metal finishing industries utilize ion exchange in several ways. As an end-of-pipe treatment, ion exchange is certainly feasible, but its greatest value is in recovery applications. It is commonly used as an integrated treatment to recover rinse water and process chemicals. Some electroplating facilities use ion exchange to concentrate and purify plating baths. Also, many industrial concerns, including a number of nonferrous metals manufacturing plants, use ion exchange to reduce salt concentrations in incoming water sources.

The ion exchange process may be used to remove cyanide in a ferrocyanide complex from wastewater. The process generates a concentrated stream of the complex, which may be treated using cyanide precipitation. Ion exchange is applicable to cyanide removal when the cyanide is complexed with iron. Experimental data have shown that a specific resin is very selective to the removal of iron cyanide complexes.

Ion exchange has been used in hydrometallurgical processing for recovery of valuable metals. Uranium processing and extraction is an active field for both liquid and solid ion exchange. It has also been used for removal and isolation of radioactive wastes. Ion exchange can also be used to remove certain organic compounds, however, in general there are other more competitive processes available for this use.

With the exception of occasional clogging or fouling of the resins, ion exchange has proved to be a highly dependable technology. Only the normal maintenance of pumps, valves, piping and other hardware used in the regeneration process is required. Few, of any, solids accumulate within the ion exchangers, and those which do appear are removed by the regeneration process. Proper prior treatment and planning can eliminate solid build-up problems altogether. The brine resulting from regeneration of the ion exchange resin must usually be treated to remove metals before discharge. This can generate solid waste.

Advantages and Limitations: Ion exchange is a versatile technology applicable to a great many situations. This flexibility, along with its compact nature and performance, makes ion exchange a very effective method of wastewater treatment. However, the resins in these systems can prove to be a limiting factor. The thermal limits of the anion resins, generally in the vicinity of 60°C, could prevent its use in certain situations. Similarly, nitric acid, chromic acid, and hydrogen peroxide can all damage the resins, as will iron, manganese, and copper when present with sufficient concentrations of dissolved oxygen. Removal of a particular trace contaminant may be uneconomical because of the presence of other ionic species that are preferentially removed. The regeneration of the resins presents its own problems. The cost of the regenerative chemicals can be high. In addition, the waste streams originating from the regeneration process are extremely high in pollutant concentrations, although low in volume. These must be further processed for proper disposal.

Xanthates: The use of xanthates for the removal of metals from waste streams appears to be a new, promising technology for treating metal-bearing wastewaters. Xanthates contain functional groups capable of forming insoluble complexes with metals, and the sludge so formed can be separated by conventional means.

Xanthates can be generated by mixing starch or cellulose with carbon disulfide in a caustic medium. Three types of xanthates have been proven in bench pilot scale studies to be effective in removing cadmium, chromium, copper, iron, lead, mercury, nickel, silver and zinc from industrial wastewaters. These are:

1. Soluble starch xanthate with a cationic polymer,
2. Insoluble starch xanthate, and
3. Fibrous cellulose xanthate.

Insoluble starch xanthate is essentially an ion exchange medium used to remove dissolved heavy metals from wastewater. The water may then either be reused (recovery application) or discharged (end–of–pipe application). In a commercial electroplating operation, starch xanthate is coated on a filter medium. Rinse water containing dragged out heavy metals is circulated through the filters and then reused for rinsing. The starch–heavy metal complex is disposed of an replaced periodically. Laboratory tests indicate that recovery of metals from the complex is feasible, with regeneration of the starch xanthate. Besides electroplating, starch xanthate is potentially applicable to any other industrial plants where dilute metal wastewater streams are generated. Its present use is limited to one electroplating plant.

Unlike hydroxide precipitation, this process is reported to be effective in removing metals over a wide pH range of 3 to 11, with an optimum range between 7 and 9. Cellulose xanthate is as effective as starch xanthate in removing trace metals, and is superior to starch xanthate in terms of sludge settling characteristics, filterability, and handling.

Xanthate may also be used as a complexing agent to prevent the formation of soluble anions from insoluble amphoteric metal hydroxides.

Xanthate treatment offers advantages over hydroxide precipitation, including the following:

1. A higher degree of metal removal;
2. Less sensitivity to pH fluctuation (metal xanthates do not exhibit amphoteric solubilities;
3. Less sensitivity to the presence of complexing agents;
4. Improved sludge dewatering properties; and
5. The capability of the selective removal of metals.

DeVoe–Holbein Technology: DeVoe–Holbein technology uses coordinating compounds covalently bonded to the surface of an inert carrier material to capture metal ions. In waste treatment applications, the reactants are used in equipment similar to that employed for ion exchange resins.

The technology was originally developed by DeVoe–Holbein as an adaptation of biological mechanisms in which living cells selectively extract a variety of metal nutrients from their environment. Cells can acquire target metals by means of specialized molecular sites on their surfaces that recognize and bind only that species. Examples of such selective reactants are the nonprotein iron–binding molecules, collectively known as "siderophores."

One of the reported advantages of the DeVoe–Holbein system is that it is capable of yielding a more highly concentrated regenerant than ion exchange. Several options for downstream utilization of the concentrated metal regenerant are therefore possible. The process appears most applicable to the selective removal of valuable metals, e.g., silver, from waste streams.

2.5.2 Waste Characteristics Affecting Performance (WCAPs)

In determining whether ion exchange will achieve the same level of performance on an untested waste as on a previously tested waste and whether performance levels can be transferred, EPA examines the following waste characteristics: (a) the concentration and valence of the contaminant, (b) the concentration of competing ionic species, (c) the concentration of interfering inorganics and organics, (d) the concentrations of dissolved and suspended solids and oil and grease, and (e) the corrosiveness relative to the resin material.

Concentration and Valence of the Contaminant: As the concentration and valence of adsorbable ions in the wastewater increase, the size of the resin bed required will increase as well, or, alternatively, the bed will become exhausted more rapidly. This is because a given amount of ion exchange resin has a limited number of sites to adsorb charged ions. If, for example, the valence is doubled or the concentration of the adsorbed ions is doubled, the sites will be exhausted twice as quickly. Hence, very high concentrations of the waste may be inappropriate for ion exchange because of rapid site exhaustion, which could conceivably require regenerant volumes to be essentially equal to waste flow volumes. If the concentration and/or the valence of the contaminant in an untested waste is significantly higher than that of the tested waste, the system may not achieve the same performance. A larger exchange bed or more frequent regeneration may be required to exchange higher concentrations and/or higher valences of the contaminant and achieve the same treatment performance, or other, more applicable treatment technologies may need to be considered for treatment of the untested waste.

Concentration of Competing Ionic Species: The presence of other contaminants or ions in the wastewater can affect the performance of the ion exchange unit in removing the hazardous contaminant of concern. Other ions in the wastewater with the same charge as the contaminant of concern will compete for exchange sites on the resin. Also, ions with a higher valence will be preferentially adsorbed. While a low concentration of the contaminant of concern may be readily removed from a solution with a low concentration of other similarly charged ionic species, the contaminant may not be removed as efficiently from solutions where high concentrations of similarly charged ions exist, especially if those ions have a higher valence than that of the contaminants. If the ions of concern are removed from a solution with high concentrations of other similarly charged ions, the resin will become exhausted more rapidly because most resins cannot selectively adsorb one contaminant in a solution containing other similarly charged ionic species. If the concentration of competing ionic species in an untested waste is significantly higher than that in the tested waste, the system may not achieve the same performance. A larger exchange bed or more frequent regeneration may be required to exchange higher concentrations of competing ionic species and achieve the same treatment performance, or other, more applicable treatment technologies may need to be considered for treatment of the untested waste.

Concentration of Interfering Inorganics and Organics: Interfering inorganics, such as iron precipitates, can accumulate in the pores of anion exchangers; these inorganics will physically break down or block the resin particles. Some organic compounds, particularly aromatics, can be irreversibly adsorbed by the exchange resins. Also, some ions tend to oxidize after they are removed from solution. For instance, Mn^{+2} (manganese) may oxidize to the insoluble Mn^{+4} state, thereby permanently fouling the exchange sites and requiring the premature replacement of the resin. If the concentration of interfering inorganics and organics in an untested waste is significantly higher than that in the tested waste, the system may not achieve the same performance and other, more applicable treatment technologies may need to be considered for treatment of the untested waste.

Concentrations of Dissolved and Suspended Solids and Oil and Grease: High concentrations of dissolved and suspended solids and oil and grease can affect the performance of ion exchange sites. Conventional ion exchange systems are usually downflow; i.e., the wastewater flows down through the resin bed. Regeneration is accomplished in either the downflow or upflow mode. If excessive concentrations of dissolved and suspended solids and/or oil and grease are present in the wastewater, the bed may clog and require backwashing prior to exhausting its exchange capacity. Backwashing may prove ineffective in the removal of some solids or oils. If the concentration of dissolved and suspended solids and/or oil and grease in an untested waste is significantly higher than that in the tested waste, the system may not achieve the same performance and other, more applicable treatment technologies may need to be considered for treatment of the untested waste.

Corrosiveness: Some wastewaters are extremely corrosive to ion exchange resin materials, reducing efficiency or increasing downtime for maintenance and repair. For instance, strong solutions of chromates may oxidize many resins, requiring premature replacement. If the corrosiveness of the untested waste is significantly higher than that of the tested waste, the system may not achieve the same performance and other, more applicable treatment technologies may need to be considered for treatment of the untested waste.

Design and Operating Parameters: In assessing the effectiveness of the design and operation of an ion exchange system, EPA examines the following parameters: (a) the amount and type of resin, (b) the amount and type of regenerant solution, (c) the hydraulic loading, and (d) the exchange temperature.

For many hazardous organic constituents, analytical methods are not available or the constituent cannot be analyzed in the waste matrix. Therefore, it would normally be impossible to measure the effectiveness of the ion exchange system. In these cases EPA tries to identify measurable parameters or constituents that would act as surrogates to verify treatment.

For organic constituents, each compound contains a measurable amount of total organic carbon (TOC). Removal of TOC in the ion exchange system will indicate removal of organic constituents. Hence, TOC analysis is likely to be an adequate surrogate analysis where the specific organic constituent cannot be measured. However, TOC analysis may not be able to adequately detect treatment of specific organics in matrices that are heavily organic–laden (i.e., the TOC analysis may not be sensitive enough to detect changes at the milligrams per liter (mg/ℓ) level in matrices where total organic concentrations are hundreds or thousands of milligrams per liter). In these cases other surrogate parameters should be sought. For example, if a specific analyzable constituent is expected to be treated as well as the unanalyzable constituent, the analyzable constituent concentration should be monitored as a surrogate.

Amount and Type of Resin: The main design parameter that affects the performance of ion exchange systems is the amount and type of resin used. Numerous cationic and anionic resins are commercially available. The selection of a resin is based on a variety of factors. Different resins have different exchange capacities, and some have greater affinity than others for specific ions. Certain resins are designed to tolerate corrosive, oxidizing, or high temperature solutions, so their exchange capacity does not degrade as rapidly with use. Most resins will effectively remove contaminant ions from solution until they become exhausted. However, if resin bed exhaustion occurs too frequently, or if regeneration requires excessive volumes of the regenerant, the type and/or amount of resin might need to be changed. In some instances, pretreatment technologies may be required prior to ion exchange. For most metals removal, cation resins are usually required. However, some metal complexes, such as copper cyanide [$Cu(CN)_4^{-2}$], chromates (CrO_4^{-2}), and arsenates (AsO_4^{-3}), are anionic and require the use of anion exchange resins.

Amount and Type of Regenerant Solution: For hydrogen–based cation exchangers, acid regenerant solutions are used (e.g., sulfuric, nitric, or hydrochloric acids). For sodium–based cation resins, sodium chloride is generally used. For anion exchange resins, alkali (commonly sodium hydroxide) is used to regenerate hydroxide–based resins. Sodium chloride is used for chloride–based anion resins.

Hydraulic Loading Rate: The amount of time that the wastewater contaminants are in contact in the ion exchange resin (i.e., residence time) impacts the extent to which ion exchange occurs. Higher residence times generally improve exchange performance, but require larger ion exchange beds to maintain the same overall throughput. For a given size ion exchange bed, the residence time can be determined by the hydraulic loading rate. Typical hydraulic loading rates for ion exchange systems range from 600 to 15,000 gal/day–ft^2.

Exchange Temperature: High temperatures reduce resin life, requiring premature replacement. The high temperature limit for anionic resins may be approximately 60°C.

2.6 NEUTRALIZATION

Neutralization is a technique of accepted technical and economic feasibility by which certain hazardous wastes can be treated. It is in widespread use in various industries, in many applications. The basic principle behind the process is simple—combining an acid with an alkali to adjust the pH of the product to an acceptable level. In the case of effluent wastes, either excess acidity or excess alkalinity is corrected, bringing the pH to a value between 6.0 and 9.0. There are numerous methods by which the admixture can be accomplished and a multitude of waste streams with diverse chemical and physical characteristics which are treatable by this technique.

However, it is sometimes only necessary to adjust the pH to approximately 5 to 6 (i.e., partial neutralization) to achieve certain treatment objectives. In other applications it may be necessary to neutralize an acid to pH 9 or higher to precipitate metallic ions or to completely clarify a waste for acceptable discharge. These techniques are called under– and over–neutralization, respectively.

Neutralization is used to treat waste acids and waste alkalies (bases) in order to eliminate or reduce their reactivity and corrosiveness. Neutralization can be a very inexpensive treatment, especially if waste alkali can be used to treat waste acid and vice/versa. Residuals include a neutral effluent containing dissolved salts and any precipitated salts.

The process should be performed in a well-mixed system to ensure completeness. Care should be taken to ensure compatibility of the waste and treatment chemicals to prevent the formation of more toxic or more hazardous compounds than were originally present.

Limitations include the concentration (because of the heat of the exothermic reaction), the physical form (such as sludges or solids), and the need for corrosion-resistant equipment. The resulting effluent may contain dissolved inorganic salts at concentrations which may be unacceptable for discharge. Based on the chemical composition of the waste stream, a precipitate may be formed which may require removal and disposal.

Neutralization of acidic or basic waste streams is used in the following situations.

1. To enhance precipitation of dissolved heavy metals;
2. To prevent metal corrosion and damage to other construction materials;
3. As a preliminary treatment allowing effective operation of the biological treatment process;
4. To provide neutral pH water for recycle uses;
5. To reduce detrimental effects on a facilities receiving water; and
6. Spills on land or water.

Neutralization may be accomplished in either a collection tank, rapid mix tank, or equalization tank by commingling acidic and alkaline wastes, or by the addition of chemicals. Alkaline wastewaters are typically neutralized by adding sulfuric or hydrochloric acid, or compressed carbon dioxide. Acidic wastewaters may be neutralized with limestone or lime slurries, soda ash, or caustic soda. The selection of neutralizing agents depends upon cost, availability, ease of use, reaction by-products, reaction rates, and quantities of sludge formed. The most commonly used chemicals are lime (to raise the pH) and sulfuric acid (to lower the pH).

Reagents used for neutralization include:

1. Other acid/alkali wastes
2. Limestone
3. Lime slurry (lime, waste carbide lime, cement kiln dust)
4. Caustic (sodium hydroxide, sodium carbonate)
5. Mineral acids (hydrochloric and sulfuric acids)
6. Carbonic acid (carbon dioxide, boiler flue gas, submerged combustion)
7. Magnesium hydroxide

The selection of the appropriate reagent for wastewater neutralization processes is site-specific and dependent on the following considerations: wastewater characteristics, reagent costs and availability, speed of reaction, buffering qualities, product solubility, costs associated with reagent handling, and residual quantities and characteristics. Typically, the first step in reagent selections is to characterize the wastewater. General parameters of interest include flow (rate, quantity), pH, pollutant loading, physical form of waste, and waste/reagent compatibility. These characteristics narrow the range of reagents and treatment configurations available for consideration.

Following the selection of candidate reagents, the quantity of reagent required to neutralize the waste to the desired end point must be determined. Reagent quantity is usually calculated by developing a titration curve for each candidate reagent using representative wastewater samples. These data determine the quantity of reagent required to bring the sample volume of wastewater to the desired pH.

In the final selection, the optimal reagents and reagent/waste feed ratio will be those which incur the least overall cost, including not only the cost of the reagent itself, but also the cost of purchasing and maintaining the reagent and neutralization systems, and the costs associated with residual handling. The combination of all such factors may make a slightly more expensive reagent less expensive overall.

One common system for neutralizing acidic or basic waste streams utilizes a multiple compartmental basin usually constructed of concrete. This basin is lined with acid brick or coated with a material resistant to the expected environment.

In order to reduce the required volume of the neutralization basin, mixers are installed in each compartment to provide more intimate contact between the waste and neutralizing reagents, thus speeding up reaction time. Stainless steel plates mounted on the floor of the pit and directly below the mixers will reduce corrosion damage to the structure. Basin inlets are baffled to provide for flow distribution, while effluent baffles can help to prevent foam from being carried over into the receiving stream. In some cases, neutralization may be accomplished in a discharge sewer.

Magnesium hydroxide is assuming increased importance for neutralization of acidic waste streams. The advantages of magnesium hydroxide include (a) ease of bringing effluent discharges to a pH of 9, (b) less corrosive than other alkalies, allowing use of less expensive equipment, (c) does not produce calcium sulfate as a by–product, (d) better degree of control as compared to caustic soda, and (e) forms a more easily filterable sludge, as compared to caustic soda.

Neutralization tanks are fabricated from a wide range of construction materials such as masonry, metal, plastic, or elastomers. Corrosion resistance can be enhanced with coatings or liners which prevent the premature decomposition of tank walls. For example, concrete reactors susceptible to corrosion can be installed with a two–layer coating of a 6.3 mm base surface (glass–reinforced epoxy polyamide) covered by a 1.0 mm coating of polyurethane elastomer to extend service lifetimes.

Vessel geometrics can be either cubical or cylindrical in nature with agitation provided overhead in line with the vertical axis. While cubical tanks need no baffling, cylindrical vessels are typically constructed with suitable ribs to prevent swirling and maintain adequate contact between the reactants. A general rule of thumb in the design of neutralization reactors is that the depth of the liquid should be roughly equivalent to the tank diameter or width.

Reactors can be arranged in either single– or multi–stage configurations and operate in either batch or continuous mode. Multi–stage continuous configurations are typically required to neutralize concentrated wastes with variable feed rates. In these units most of the reagent is added in the first vessel with only final pH adjustments (polishing) made in the remaining reaction vessels. This is particularly true when using sluggish reagents which require extensive retention time. Single–stage continuous or batch neutralization is suitable for most applications with highly buffered solutions or dilute wastewaters not subject to rapid changes in flow rate or pH.

A holdup period is required to provide time for the neutralization reaction to go to completion. This factor is especially critical where a dry feed (lime or limestone) or slurry is used as the control agent. In these systems, the solids must dissolve before they react, increasing the required holdup time and tank capacity. For example, liquid reagents used in continuous flow operations generally require 3 to 5 minutes of retention time in the first tank. Three minutes corresponds to the absolute minimum size that will not cause considerable splashing or trapping of air. In comparison, solid–based reagent systems such as lime or limestone typically require 30 and 45 minute retention time, respectively.

Agitation serves the purpose of equalizing the hydrogen or hydroxide concentration profile within the reaction vessel as the influent is dispersed in the tank. Vessels with large stagnant areas provide little mixing between reactants and causes large disturbances when concentrated materials are released into the system. For accurate pH control, Hoyle has suggested that agitator capacity should be measured as a ratio of the system dead time (the interval between the addition of a reagent and the first observable pH change) to the retention time (volume of the vessel divided by the flow through the vessel). A ratio of dead time to retention time of 0.05 approaches an optimum value.

The pH control systems for batch neutralization processes can be quite simple with only on–off control provided via solenoid or air activated valves. Control system designs for continuous flow neutralization systems are more complicated because the wastewater feeds often fluctuate in both flow and concentration. Systems currently available include: proportional, cascade, feed–forward, or feedback pH control. Each system has distinct advantages and disadvantages. The pH control equipment usually consists of a pH probe, monitor, and recorder. In addition, there is typically a control panel with an indicator, starters and controls for metering pumps, all relays, high/low pH alarms, switches, and mixer motor starters.

Neutralization is a relatively simple unit treatment process which can be performed using readily available equipment. Only storage and reaction tanks with accessory agitators and delivery systems are required. Because of the corrosivity of the wastes and treatment reagents, appropriate materials of construction are needed to provide a reasonable service–life for equipment. The process is reliable provided pH monitoring units are used. The feed of the neutralization agent may be regulated automatically by the pH monitoring unit thereby ensuring effective neutralization and minimizing worker contact with corrosive neutralizing agents.

Neutralization of hazardous wastes has the potential of producing air emissions. Acidification of streams containing certain salts, such as sulfide, will produce toxic gases. Feed tanks should be totally enclosed to prevent escape of acid fumes. Adequate mixing should be provided to disperse the heat of reaction if wastes being treated are concentrated. The process should be controlled from a remote location is possible.

Spills: Neutralization is applicable to spills of acidic or alkaline hazardous materials (or substances which when spilled into water form an acid or a base) and can be applied in situ under the proper circumstances. In emergency field situations, the factors limiting the use of neutralization often are the volume of the spill, the violence of the neutralization reactions, and the production of potentially toxic gases. The final pH of the neutralized hazardous substance would generally be environmentally acceptable if in the range of 6.0 to 9.0.

The use of strong bases as neutralizing agents should be limited to spills of strong acids on land where effective containment is possible and control of the reaction is somewhat achievable. Treatment of water spills should be limited to the use of weak bases. For land spills, aqueous sodium bicarbonate, sodium carbonate, or sodium hydroxide solution is often added to spilled material contained in a lined trench. This technique has been successfully used to neutralize several chlorine spills. Acid spills can also be treated with lime, sodium carbonate, soda ash, and magnesium hydroxide. A technique which has been successfully employed to prevent frothing and sputtering when neutralizing an acid spill with limestone, is to mix flyash with the acid before adding the limestone.

Spills of acids in water or substances which react with water to form acids should be treated with weak bases. Examples of applicable neutralizing agents are: (1) sodium bicarbonate, (2) sodium carbonate, and (3) calcium carbonate.

Lime or calcium hydroxide are also used, but addition of these materials in excess will cause large changes in the pH. Although reaction rates will generally be slower for calcium carbonate than for the sodium-based agents, calcium carbonate, on the other hand, has several distinct advantages for use in neutralizing water spills.

After recovery of as much material as possible for recycle or reuse, neutralization of a spill of a strong base is done most economically with a strong acid. For land spills where the remainder, of the spilled strong base, is contained (i.e., lined trench, concrete), neutralization with strong acids (i.e., hydrochloric, sulfuric) can be accomplished. In a non-porous soil, mechanical mixing to depth of penetration can be done and then the neutralized soil may be excavated for disposal. Only qualified personnel should handle the neutralization process and small scale field tests should first be performed to determine if any negative (i.e., heat generation) reactions take place.

Aqueous spills of base compounds that react with water to form bases should be treated with weaker acids to prevent harmful pH changes caused by an overdose of the acid neutralizing agent. Examples of these are acetic acid, sodium dihydrogen phosphate, and even gaseous CO_2. Sodium dihydrogen phosphate is applicable in all situations because it is a buffer salt which tends to keep the pH within a certain range. Overdosing, at most, would cause the pH to decrease to approximately 4.0. The user should be aware that the neutralized products of acetic acid and sodium dihydrogen phosphate, namely, acetates and phosphates, may be unacceptable for discharge. CO_2 also acts as a buffer and has been used to neutralize ammonia spills in water, but it is difficult to dissolve and disperse properly.

Mixing of Acid and Alkali Wastes: The process of acid/alkali mixing (mutual neutralization) may be the most economical method of neutralization available, particularly in cases where compatible acid/alkaline wastes are present in the same plant. Prior to implementation, data are typically collected on the volume and concentration of each waste stream and their respective flow patterns (batch or continuous). In addition, waste stream mixing characteristics are usually investigated in order to predetermine possible waste incompatibilities that would prevent or limit the use of the technology. For example, the precipitation of metal hydroxides or other insoluble species (e.g., calcium sulfate) may result in increased sludge generation or plugging of the transport or dewatering equipment. If the sludge generation is considerable, the increased dewatering, disposal, and maintenance costs could possible outweigh the benefit of any savings realized on reagent costs. Also, if incompatible wastes produce a reaction that is too sluggish or difficult to control, or generate reaction products that are toxic (i.e., hydrogen cyanide) or highly exothemmic, then implementation may not be feasible.

As with most neutralization processes, acid/alkali mixing can be operated in either a batch or continuous mode. Operational type depends primarily on the variation inflow rate or concentration of the divergent influent streams. Batch operations are typically utilized in treating concentrated bath dumps or intermittent flow applications. Reactor configurations can be either single– or multi–stage. However, unit processing of corrosive wastes (pH less than 2 or greater than 12.5) generally require multi–stage continuous operation.

Limestone: Limestone treatment is a well–developed and established technology for the neutralization of acidic waste streams. Limestone is a particularly effective reagent for the neutralization of dilute acid waste streams containing low concentrations of acid salts and suspended solids. With modifications, it may also be a good neutralizing agent for many corrosive acidic waste streams, either as a primary treatment for weak acids or as a pretreatment for other processes, i.e., partial neutralization. However, in most applications, limestone has been replaced by more cost–effective reagents such as lime slurry and caustic soda which eliminate solids handling problems. In addition, caustic soda results in reduced sludge generation.

Limestone is available in either high calcium ($CaCO_3$) or dolomitic ($CaCO_3MgCO_3$) form. Both types of limestone are available as either a powder or crushed stone. Crushed stone diameters are typically 0.074 mm (200 mesh) or less since both the reactivity and completeness of the reaction increase proportionately to the available surface area. High calcium is most commonly used because of its greater reaction rate and its more widespread availability. Dolomitic limestone reactivity will increase if finely ground and sludge production will be minimal due to the formation of soluble magnesium sulfate. However, its reactivity is generally too slow even with grinding, and hence not suitable for most applications.

Lime Slurry: Lime slurry treatment of corrosive waste streams is analogous to that of limestone neutralization. It is one of the oldest and perhaps most prevalent of all industrial waste treatment processes. It is used extensively as an alkaline reagent in the neutralization of pickling wash waters, plating rinses, acid mine drainage, and process waters from chemical and explosive plants. Lime slurry has replaced limestone in many applications as a low–cost alkali due to its greater available surface area, pumpable form, continuous application, and greater effectiveness in removing Ca salts from the process. However, similar to the use of limestone, a major disadvantage of the process is the formation of a voluminous sludge product.

Limes are formed by the thermal degradation of limestone (calcination), and are available in either high calcium (CaO) or dolomitic (CaO–MgO) form. The pure oxidized calcium product is referred to as quicklime. Quicklime varies in physical form and size, but can generally be obtained in lump (63 to 255 mm), pebble (6.3 to 63 mm), ground (1.45 to 2.38 mm) or pulverized (0.84 to 1.49 mm) form. As with limestone, experimental evidence has shown an increase in dissolution as the size of a lime particle diminishes. For example, a 100% quicklime of 100 mesh (0.149 mm) will dissolve twice as fast as one of 48 mesh (0.35 mm).

Although lime can be fed dry, for optimal efficiency it is slaked (hydrated) and slurried before use. Slaking is usually carried out at temperatures of 82° to 99°C with reaction times varying from 10 to 30 minutes. Following slaking, a wet plastic paste is formed (lime putty) and then slurried with water to a concentration of 10 to 35%.

Caustic Soda: Pure anhydrous sodium hydroxide (NaOH) or caustic soda is a white crystalline solid manufactured primarily through the electrolysis of brine. Caustic soda is a highly alkaline, water–soluble compound especially useful in reactions with weakly acidic materials where weaker bases such as sodium carbonate are less effective. It is also useful in the precipitation of heavy metals and in neutralizing strong acids through the formation of sodium salts.

Although available in either solid or liquid form, NaOH is almost exclusively used in water solutions of 50% or less. The solution is marketed in either lined 55 gallon drums or in bulk, i.e., tank car or truck. As a solution, caustic soda is easier to store, handle, and pump relative to either lime or limestone. In comparison to lime slurries, caustic soda will not clog valves, form insoluble reaction products, or cause density control problems. However, when sodium hydroxide is stored in locations where the ambient temperature is likely to fall below 12°C, heated tanks should be provided to prevent reagent freezing.

After lime, sodium hydroxide is the most widely used alkaline reagent for acid neutralization systems. Its chief advantage over lime is that, as a liquid, it rapidly disassociates into available hydroxyl (OH−) ions. Holdup time is minimal, resulting in reduced feed system and tankage requirements. Caustic soda's main disadvantage is reagent cost. As a monohydride, in neutralizing diprotonated acids such as sulfuric, two parts base are required per part of acid neutralized. In contrast, dihydroxide bases such as hydrated lime, only require one part base per part of acid neutralized.

Mineral Acids: Mineral acid treatment of corrosive waste streams is the most widespread of neutralization processes for alkaline wastes. The two primary mineral acid reagents are sulfuric and hydrochloric which are characterized by their highly reactive nature, complete miscibility with water, and rapid disassociation rates. In concentrated form, application may result in the generation of an acid mist or toxic fumes. Therefore, the choice of an acidic reagent is typically based on ease of handling, as well as cost per unit basicity, and in some cases such as food processing, end product characteristics.

Sulfuric acid (H_2SO_4) is the most widely used of all mineral acid reagents. Ease of manufacture, diprotonated reaction chemistry and concentrated nature, combine to make it the least expensive mineral acid on a neutralization equivalent basis. It is supplied in concentration liquid form (93 to 98%), is highly reactive, strongly hygroscopic and presents a burn hazard to personnel. Dilute solutions are highly corrosive to iron and steel, whereas concentrated solutes (greater than 93%) are not corrosive. Protection from freezing during storage and transport is required since sulfuric acid exhibits a maximal freezing point of 8°C (47°F) at a concentration of 85%. In addition, pH control overshoot and fuming characteristics frequently require diluting the acid to 30% concentration prior to application. In diluting operations, the acid should be slowly added to the water, with provisions for agitation, adequate ventilation and protective clothing.

When sulfuric acid is uneconomical or otherwise inapplicable as a neutralization reagent, hydrochloric acid (HCl) is often used as a substitute. Hydrochloric (Muriatic) acid is supplied in aqueous solutions of 35 to 37% acid. Although it is somewhat more reactive than sulfuric, the most concentrated commercial grade contains at least 63% water, increasing transportation costs and limiting most major uses to a radius of 300 to 500 km from the producing source. In addition, its higher unit cost and monoprotonated reaction chemistry result in an overall reagent cost approximately double that of sulfuric acid on a neutralization equivalent basis. Hydrochloric acid is highly disassociated (a 10,000 mg/ℓ HCl solution will result in pH of 0.9), and extremely corrosive, attacking most metals through surface dissolution. Commonly used plastics and elastomers are recommended as materials of construction when designing neutralization systems using hydrochloric acid as reagent.

The main advantage of using HCl is that it generates soluble end products. However this attribute may not be beneficial in some cases, since it may cause the waste to exceed dissolved solids and metal effluent standards. As with sulfuric, hydrochloric acid will react vigorously with water, sometimes evolving an acid mist which can destroy the mucous membrane and cause choking, coughing, headache and dizziness. In addition, hydrochloric acid will decompose in the presence of heat into toxic hydrogen chloride gas. Therefore, caution must be exercised when storing or handling concentrated hydrochloric acid to minimize splashes, spills or mist generation.

Carbonic Acid Treatment: Carbonic acid neutralization of alkaline waste streams is a relatively old, but as of yet, undeveloped treatment technology. As early as 1931, Curtis and Copson patented a process using a reaction product carbonic acid to neutralize a cotton waste (Kier liquor) treated with caustic soda. The inherent problem with carbonic acid treatment is that carbonic acid, a weak acid disassociates slowly in solution, retards reaction rates, and limits pH reduction applications to the pH 7 to 8 range. In addition, carbonic acid reaction products are slightly alkaline in nature and tend to act as buffers in the neutralization of concentrated alkaline wastes.

Typically, carbonic acid is generated directly in the neutralization chamber by injecting carbon dioxide into the wastewater solution, Upon hydration, the carbon dioxide will form carbonic acid and neutralize excess alkalinity. Carbon dioxide is available as either a compressed gas or the by-product of a combustion process.

The primary advantages of compressed carbon dioxide are minimal capital requirements, uncomplicated piping, and the inability to over-acidify the wastewater. Its primary disadvantages are a low dissolved oxygen content at the point of injection, and a high reagent cost on a neutralization equivalent basis.

2.7 OXIDATION

The processes described in this section are based on chemical oxidation as differentiated from thermal (including wet air and supercritical water oxidation), electrolytic, radiative, and biological oxidation, which are discussed in other chapters.

Liquids are the primary waste form treatable by chemical oxidation. The most powerful oxidants are relatively non-selective; therefore, any easily oxidizable material in the waste stream will be treated. If for instance an easily oxidizable organic solvent were used, little of the chemical effect of the oxidizing agent could be used on the hazardous constituent. This, therefore, essentially limits the use of the most commonly used oxidants to aqueous wastes.

Gases have been treated by scrubbing with oxidizing solutions for the destruction of odorous substances, such as certain amines and sulfur compounds. Potassium permanganate, for instance, has been used in certain chemical processes, in the manufacture of kraft paper and in the rendering industry. Oxidizing solutions are also used for small-scale disposal of certain reactive gases in laboratories.

Oxidation has limited application to slurries, tars, and sludges. Because other components of the sludge, as well as the material to be oxidized, may be attacked indiscriminately by oxidizing agents, careful control of the treatment via multi-staging of the reaction, careful control of pH, etc. are required.

Chemical oxidation has been found very effective in the treatment of industrial and domestic wastewater. In particular, oxidation offers one of the few methods for removing odor, color, and various potentially toxic organic substances like phenols, pesticides and industrial solvents. It also disinfects drinking water by killing or inactivating pathogenic microorganisms that may be present.

Oxidation reactions can be carried out using simple, readily available equipment; only storage vessels, metering equipment, and contact vessels with agitators are required. However, implementation is complicated because every oxidation/reduction reaction system must be designed for the specific application. Laboratory- and/or pilot-scale testing are essential to determine the appropriate chemical feed rates and reactor retention times in accordance with reaction kinetics. Oxidation and reduction has not been widely used in treating hazardous wastestreams.

A major consideration in electing to utilize oxidation technology is that the treatment chemicals are invariably hazardous, and great care must be taken in their handling. In particular, the handling of many oxidizing agents is potentially hazardous and suppliers' instructions should be carefully followed.

In some cases, undesirable by-products may be formed as a result of oxidation. For example, addition of chlorine can result in formation of bio-resistant end products which can be odorous and more toxic than the original compound. The possibility of this undesirable side reaction needs to be considered when using chlorine for oxidation of wastewaters.

Chemical oxidation can be an effective way of pretreating wastes prior to biological treatment; compounds which are refractory to biological treatment can be partially oxidized making them more amenable to biological oxidation.

One of the major limitations with chemical oxidation is that the oxidation reactions frequently are not complete (reactions do not proceed to CO_2 and H_2O). Incomplete oxidation may be due to oxidant concentration, pH, oxidation potential of the oxidant, or formation of a stable intermediate. The danger of incomplete oxidation is that more toxic oxidation products could be formed. Chemical oxidation is not well suited to high-strength, complex waste streams. The most powerful oxidants are relatively non-selective and any oxidizable organics in the waste stream will be treated. For highly concentrated waste streams this will result in the need to add large concentrations of oxidizing agents in order to treat target compounds. Some oxidant such as potassium permanganate can be decomposed in the presence of high concentrations of alcohols and organic solvents.

Chemical oxidation should be considered as a first treatment step when the waste contains cyanide. Chemical oxidation should be evaluated as a first treatment for wastes that have constituents or concentrations of constituents that are not amenable to other treatment methods. Chemical oxidation should be considered as a final polishing step for residual traces of contaminants remaining after certain other treatments.

Chemical oxidation is a treatment technology used to treat wastes containing organics. It is also used to treat sulfide wastes by converting the sulfide to sulfate. The destruction of cyanides in wastes is usually accomplished by chemical oxidation. Chemical oxidation can also be used to change the oxidation state of metallic compounds to valences that are less soluble, such as converting arsenic in wastes to the relatively insoluble pentavalent state.

Chemical oxidation is applicable to dissolved cyanides in aqueous solutions, such as wastewaters from metal plating and finishing operations, or to inorganic sludges from these operations that contain cyanide compounds. For cyanides, chemical oxidation is most applicable to solutions containing less than 500 mg/ℓ of cyanides when the cyanides are in a form that can be easily disassociated in water to yield free cyanide ions. If cyanides are present in water as a tightly bound complex ion, e.g., ferrocyanide, only limited treatment may occur. If the waste contains greater than 500 mg/ℓ of cyanide, but no more than about 100,000 mg/ℓ, electrolytic oxidation may be more appropriate.

Chemical oxidation may also be used for destruction of the organic component of organometallic compounds in wastes, thus freeing the metal component for treatment by chemical precipitation or stabilization. Organic compounds such as EDTA, NTA, citric acid, glutaric acid, lactic acid, and tartrates are often used as chelating agents to prevent metal ions from precipitating out in electroless plating solutions. When these spent plating solutions require treatment for metals removal by chemical precipitation, the organic chelating agents must first be destroyed. Chemical oxidants, potassium permanganate in particular, are effective in releasing metals from complexes with these organic compounds.

The basic principle of operation for chemical oxidation is that inorganic cyanides, some dissolved organic compounds, and sulfides can be chemically oxidized to yield carbon dioxide, water, salts, simple organic acids, and, in the case of sulfides, sulfates. Metallic ions such as arsenites can be oxidized to higher, less soluble valences such as arsenates. The principal chemical oxidants used are hypochlorite, chlorine gas, chlorine dioxide, hydrogen peroxide, ozone, and potassium permanganate.

2.7.1 Waste Characteristics Affecting Performance (WCAPs)

In determining whether chemical oxidation will achieve the same level of performance on an untested waste that it achieved on a previously tested waste and whether performance levels can be transferred, EPA examines the following waste characteristics: (a) the concentration of other oxidizable contaminants, and (b) the concentration of metal salts.

Concentration of Other Oxidizable Compounds: The presence of other oxidizable compounds in addition to the constituents of concern will increase the demand for oxidizing agents and, hence, potentially reduce the effectiveness of the treatment process. Inorganic reducing compounds such as sulfide may also create a demand for additional oxidizing agent. If TOC and/or inorganic reducing compound concentrations in the untested waste are significantly higher than those in the tested waste, the system may not achieve the same performance. Additional oxidizing agent may be required to effectively oxidize the waste and achieve the same treatment performance, or other, more applicable treatment technologies may need to be considered for treatment of the untested waste.

Concentration of Metal Salts: Metal salts, especially lead and silver salts, will react with the oxidizing agent(s) to form metal peroxides, chlorides, hypochlorites, and/or chlorates. These reactions can cause an excessive consumption of oxidizing agents and potentially interfere with the effectiveness of treatment.

An additional problem with metals in cyanide solutions is that metal–cyanide complexes are sometimes formed. These complexes are negatively charged metal–cyanide ions that are extremely soluble. Cyanide in the complexed form may not be oxidizable, depending on the strength of the metal–cyanide bond in the complex and the type of oxidizing agent used. Iron complexes [for example, the ferrocyanide ion, $Fe(CN)_6^{-4}$) are the most stable of the complexed cyanides.

If the concentrations of metal salts and/or metal–cyanide complexes in the untested waste are significantly higher than those in the tested waste, the system may not achieve the same performance. Additional oxidizing agent and/or a different oxidizing agent may be required to effectively oxidize the waste and achieve the same treatment performance, or other, more applicable treatment technologies may need to be considered for treatment of the untested waste.

2.7.2 Design and Operating Parameters

In assessing the effectiveness of the design and operation of a chemical oxidation system, EPA examines the following parameters: (a) the residence time, (b) the amount and type of oxidizing agent, (c) the degree of mixing, (d) the pH, (e) the oxidation temperature, and (f) the amount and type of catalyst.

For many hazardous organic constituents, analytical methods are not available or the constituent cannot be analyzed in the waste matrix. Therefore, it would normally be impossible to measure the effectiveness of the chemical oxidation treatment system. In these cases EPA tries to identify measurable parameters or constituents that would act as surrogates to verify treatment.

For organic constituents, each compound contains a measurable amount of total organic carbon (TOC). Removal of TOC in the chemical oxidation treatment system will indicate removal of organic constituents. Hence, TOC analysis is likely to be an adequate surrogate analysis where the specific organic constituent cannot be measured.

However, TOC analysis may not be able to adequately detect treatment of specific organics in matrices that are heavily organic–laden (i.e., the TOC analysis may not be sensitive enough to detect changes at the milligrams per liter level in matrices where total organic concentrations are hundreds of thousands of milligrams per liter). In these cases other surrogate parameters should be sought. For example, if a specific analyzable constituent is expected to be treated as well as the unanalyzable constituent, the analyzable constituent concentration should be monitored as a surrogate.

Residence Time: The residence time impacts the extent of volatilization of waste contaminants. For a batch system, the residence time is controlled by adjusting the treatment time in the reaction tank. For a continuous system, the waste feed rate is controlled to make sure that the system is operated at the appropriate design residence time.

Amount and Type of Oxidizing Agent: Several factors influence the choice of oxidizing agents and the amount to be added. The amount of oxidizing agent required to treat a given amount of oxidizable constituent(s) will vary with the agent chosen. Enough oxidant must be added to ensure complete oxidation; the specific amount depends on the type of oxidizable compounds in the waste and the chemistry of the oxidation reactions. Theoretically, the amount of oxidizing agent to be added can be computed from oxidation reaction stoichiometry; in practice, an excess of oxidant should be used. Testing for excess oxidizing agent will determine whether the reaction has reached completion. In continuous processes, oxidizing agent is added by automated feed methods. The amount of oxidizing agent needed is usually measured and controlled automatically by an oxidation–reduction potential (ORP) sensor.

Degree of Mixing: Process tanks must be equipped with mixers to ensure maximum contact between the oxidizing agent and the waste solution. Proper mixing also limits the production of any solid precipitates from side reactions that may resist oxidation. Mixing also provides an even distribution of tank contents and a homogeneous pH throughout the waste, improving oxidation of wastewater constituents. The quantifiable degree of mixing is a complex assessment that includes, among other factors, the amount of energy supplied, the length of time the material is mixed, and the related turbulence effects of the specific size and shape of the tank. This is beyond the scope of simple measurement.

pH: Operation at the optimal pH maximizes the chemical oxidation reactions and may, depending on the oxidizing agent being used, limit the formation of undesirable reaction by-products or the escape of cyanide from solution as HCN, CNCl, or C_2N_2 gas. The pH is controlled by the addition of caustic, lime, or acid to the solution. In most cases, a slightly or moderately alkaline pH is used, depending on the type of oxidizing agent being used and the compound being treated. In alkaline chlorination treatment of organics, a slightly acidic pH may be selected as an optimum. In permanganate oxidation, a pH of 2 to 4 is often selected.

Oxidation Temperature: Temperature affects the rate of reaction and the solubility of the oxidizing agent in the waste. As the temperature is increased, the solubility of the oxidizing agent, in most instances, is increased and the required residence time, in most cases, is reduced.

Oxygen Versus Air: The fundamental difference between air and pure oxygen is that for equivalent total pressure, much higher oxygen partial pressures are obtained with oxygen. This an obvious consequence of the diluent effect of the nitrogen contained in air. Thus, when the reaction order with respect to oxygen is greater than zero, substantially more air at higher pressure must be used to obtain the same process result as that which can be obtained with oxygen. This translates into the following advantageous process characteristics for oxygen based systems:

1. Higher reaction rate at equivalent total pressure.
2. Equivalent reaction rate at lower pressure.
3. Lower total gas throughput.
4. Better selectivity (system specific as discussed above).

These process characteristics can be translated into economic benefits as follows:

1. Increased productivity when oxygen is used to replace air in an existing reactor.
2. Lower capital investment for new reactor systems.
3. Lower environmental compliance costs.
4. Lower compression costs.
5. Greater chemical efficiency due to selectivity improvement.

2.7.3 Catalytic Oxidation

Adding a catalyst that promotes oxygen transfer and thus enhances oxidation has the effect of lowering the necessary reactor temperature and/or improving the level of destruction of oxidizable compounds. For waste constituents that are more difficult to oxidize, catalyst addition may be necessary to effectively destroy the constituent(s) of concern. Catalysts typically used for this purpose include copper bromide and copper nitrate.

One of the earliest investigations of catalytic oxidation was conducted by Battelle Laboratories in 1971 to study the adsorption of free and complexed cyanide onto activated carbon in the presence of copper. Subsequent efforts were undertaken by the Calgon Corp. to also develop a cyanide detoxification method utilizing catalytic oxidation on granular activated carbon. Cupric ions are added to the wastewater along with oxygen prior to passing the cyanide–bearing waste through a granular activated carbon column. According to Calgon, "cupric ions are added to the water to accelerate and increase the efficiency of the catalytic oxidation of cyanide by granular activated carbon." In addition to improving the catalytic oxidation of the cyanide, "the presence of cupric ions results in the formation of copper cyanides, which have a greater adsorption capacity than copper or cyanide alone."

Hydrogen bromide wastes from pharmaceutical, agrichemical and other plants can be oxidized into bromine, using a solid catalyst from Catalytica Inc. The process uses oxygen instead of chlorine for the oxidation, thus generating water as the only by–product. With chlorine, bromides and sodium chloride brine waste must be disposed of.

Chemical Waste Management, Inc. has developed a process to treat wastewaters, such as leachates, ground water, and process waters, containing mixtures of salts, metals, and organic compounds. The proprietary technology is a combination of evaporation and catalytic oxidation processes. Wastewater is concentrated in an evaporator by boiling off most of the water and the volatile contaminants, both organic and inorganic. Air or oxygen is added to the vapor, and the mixture is forced through a catalyst bed, where the organic and inorganic compounds are oxidized. This stream, composed of mainly steam, passes through a scrubber, if necessary, to remove any acid gases formed during oxidation. The stream is then condensed or vented to the atmosphere. The resulting brine solution is either disposed of or treated further, depending on the nature of the waste.

This technology can be used to treat complex wastewaters that contain volatile and nonvolatile organic compounds, salts, metals, and volatile inorganic compounds. Suitable wastes include leachates, contaminated groundwaters, and process waters. The system can be designed for any capacity, depending on the application and the volume of the wastewater. Typical commercial systems range from 10 to 10,000 gal/min.

2.7.4 Chlorine Oxidation

Chlorination has been used in drinking water treatment since the turn of the century, however, the formation of trihalomethanes will probably deter further expansion of chlorine for this use. Oxidized organic compounds are produced by the chlorination process. Chlorine is used in a number of ways for detoxification including alkaline chlorination, chlorine dioxide, chloroiodides, hypochlorites, and chlorinolysis.

Alkaline Chlorination: When chlorine is added to wastewaters, under alkaline conditions, reactions occur which lead to oxidation (chlorination) of the contaminant. This oxidation process, which is widely used in the treatment of cyanide wastes, is generally referred to as the "alkaline chlorination" process. Cyanides can be oxidized with chlorine to the less toxic cyanates. Additional chlorine will then oxidize the cyanates to nontoxic nitrogen gas, carbon dioxide, and bicarbonates.

Alkaline chlorination is used to treat free cyanides and complex cyanides although combinations with Fe or Ni will take a longer time. Limitations include the exothermic heat of the reaction, nonselective competitions with other species and additional chlorine demands. Fairly close pH control (7.5 to 9.0) is required to avoid toxic volatiles release.

The oxidation of cyanides in wastewater by chlorine has been the most widely acceptable method of cyanide treatment for the past thirty years. In an alkaline solution of pH 8.5 or higher, chlorine reacts with cyanide to form cyanate.

The oxidation of cyanide wastes by chlorine is a widely used process in plants using cyanide in cleaning and metal processing baths. Alkaline chlorination is also used for cyanide treatment in a number of inorganic chemical facilities producing hydrocyanic acid and various metal cyanides.

This reaction requires 6.8 pounds of chlorine to convert 1 pound of cyanide to cyanate. Usually more than 6.8 pounds of chlorine is used due to the presence of other oxidizable components. If the pH is controlled to between 8.5 and 9.0, at ambient temperature and pressure it takes 10 to 30 minutes for 100% conversion of free cyanide to cyanate. If the pH is increased, the reaction time diminishes. However, if pH drops to as low as 8.0, not only does the conversion decrease tremendously, but also highly toxic cyanogen chloride, CNCl, begins to form. It is therefore very important to maintain the pH level above 8.5 when chlorinating cyanides.

The cyanate which is formed by the reaction is further converted to bicarbonate and nitrogen at a pH of 8.5 to 9.5. Again, the pH must be maintained above 8.5 to inhibit the formation of toxic by-products, in this case, chloroamines. Chlorination reactions are commonly carried out at ambient temperature and pressure. Besides pH, the effectiveness of destruction of cyanide depends on factors such as initial cyanide concentration, presence of metal ions like nickel, iron and cobalt, and mass transfer effects on the chlorine in solution.

Alkaline chlorination can be accomplished by either batch or continuous processes. For batch treatment, the wastewater is transferred to a reaction tank, where the pH is adjusted and the oxidizing agent is added. In some cases, the tank may be heated to increase the reaction rate. It is important that the tank be well mixed for effective treatment to occur. After treatment, the wastewater is either directly discharged or transferred to another process for further treatment.

In the continuous process, automatic instrumentation is used to control pH, reagent addition, and temperature. An oxidation-reduction potential (ORP) sensor is usually used to measure the extent of reaction.

Chlorination is typically used to treat dilute cyanide solutions and rinse waters having cyanide concentrations below 1,000 mg/ℓ. For spent process solutions (which have cyanide concentrations well above 1,000 mg/ℓ), however, the chlorination process becomes quite inefficient. Other forms of treatment, such as wet air oxidation are required to obtain a high percentage of cyanide removal. It is also possible to combine such treatment with chlorination in a two stage process which can treat a wide range of wastes to a relatively high degree of destruction.

The presence of metal ions such as nickel, iron and cobalt in cyanide rinse water interfere with the destruction of cyanides to cyanates. Nickel forms a cyanide complex which cannot be completely transformed to cyanate in a relatively short time by chlorine.

It is necessary to remove the nickel ion first by metal ion precipitation and then proceed with a chlorination step to remove all the cyanides.

All chlorination technologies require that the chlorine dissolves in the aqueous phase prior to the oxidation step. The rate of cyanide destruction therefore depends on the dissolution rate of chlorine in aqueous solutions. To minimize mass transfer limitations, a well-designed bubbler or dispersion system is usually used to increase the gas-liquid interfacial transfer area.

Chlorine, which is produced in the chlorinator, is mixed with alkali to raise the pH of the solution. The pH of the solution is maintained at an optimal preset value by the use of a pH meter and an oxidation-reduction-potential (ORP) controller. The liquid effluent from the first reactor (which contains mostly cyanates) then enters the second reactor where it is further converted to nitrogen and carbon dioxide (from bicarbonate decomposition). The reaction time needed in each stage of the process is approximately 25 to 30 min. The degree of conversion of cyanides can be evaluated from samples of the effluent stream by analyzing for the cyanide content using gas chromatography as well as from the dissolved chlorine concentration by the amperometrical method (i.e., by measuring the current in the solution and converting it to concentration using a calibration curve). For many cyanide containing wastestreams, such a process will typically result in an effluent which can be safely discharged to municipal sewers under most existing regulations.

Treatment of cyanides by chlorination offers the cheapest possible treatment of cyanide containing aqueous wastes from the electroplating and metal finishing industries, provided that the wastestreams contain only a small amount of organics. For high concentrations of organics in the wastewater, chlorination often becomes undesirable due to the formation of organohalides, especially trihalomethanes. Under such conditions, other oxidizing agents such as ozone will be needed in order to prevent the formation of organohalides.

There are a few companies which manufacture and sell custom-designed chlorination units for treatment of cyanide wastewater for metal finishers. The available units are capable of removing cyanides as well as some heavy metals. The volume of wastewater that the units can handle ranges from 100 to 1,000 gal/min. Because of the large volume of cyanide containing water which usually requires treatment at a typical metal finishing operation, most cyanide wastes are treated on-site. The chlorination unit is located at the source of the wastewater, thereby eliminating the cost of transporting the waste off-site for treatment.

Advantages:

1. Proven technology with documented cyanide destruction efficiencies.
2. Operates at standard operation temperatures and pressures and is well suited to automatic control.
3. Modular design allows for plant expansion and can be used in different configurations.
4. When treating dissolved HCN, calcium, potassium, or sodium cyanide no sludges are generated.
5. Low cost.

Disadvantages:

1. Need for careful pH and ORP control.

2. Possible chemical interference in the treatment of mixed wastes (i.e., large oxidation chemical excesses required for complete reactions).
3. Process is not selective and therefore restricted to specific product wastestreams.
4. Potential hazard of shipping, storing, and handling of chlorine gas, hypochlorite salts, sodium hydroxide, and concentrated sulfuric acid.
5. Unable to treat ferro and ferricyanides and has difficulty treating nickel cyanide.
6. Potential for creating toxic residue which will require post–treatment (i.e., fixation/solidification/encapsulation).

Chlorine Dioxide Oxidation: Oxidation by chlorine dioxide has substantial advantages over oxidation by chlorine in the treatment of drinking water. The oxidation power of chlorine dioxide is not impaired over a wide range of pH, and is therefore not very sensitive to fluctuations in pH. The effectiveness of chlorine oxidation is very much dependent on the pH value, and at typical operating values of pH from 6 to 10, chlorine partially loses its oxidizing power. Chlorine dioxide does not react with ammonia or nitrogenous compounds, whereas these react with chlorine to form chloramines. Finally, chlorine dioxide does not form trihalomethanes during oxidation of organic chemicals. All of these factors make oxidation by chlorine dioxide one of the techniques of choice for removing low levels of organic chemicals from water.

Most of the oxidizing capacity of chlorine dioxide comes from the reduction of ClO_2 to a chlorite ion. As with chlorine, wastewater treatment by chlorine dioxide is usually done at the site where the waste is produced. The chlorine dioxide is usually produced by reacting sodium chlorite with acids, such as, hydrochloric or sulfuric acids. Extreme caution must be exercised when mixing $NaClO_2$ with acids. The reaction between sulfuric acid and solid $NaClO_2$ is explosive. Therefore, the reaction must be carried out exclusively using solutions of $NaClO_2$ in water.

Typically, a packed bed reactor is used to enhance mass transfer. The flowhseet is arranged so that the acid for both reactions is added simultaneously, with the excess HCl from the first reaction carrying over to react with HOCl in the second reaction. The $NaClO_2$ is added immediately prior to the point where the combined solution enters the packed bed. The performance of the system can be determined by measuring the concentrations of the chlorine dioxide in the entering and exit streams. These concentrations can be found by amperometric titration or spectrophotometry. Monitoring them on–line is essential for establishing an effective control system.

The by–products produced by the oxidation of most organic compounds with chlorine dioxide tend to be less troublesome in downstream uses of the wastewater than when chlorine is used. Phenol oxidation by chlorine dioxide forms quinones and chloroquinones, which in excess chlorine dioxide, decompose to oxalic and maleic acids. Both products are odorless. This makes chlorine dioxide preferable to chlorine which reacts with phenols to form mono–, di– and trichloro–derivatives. These compounds are highly odorous and only slowly decompose with excess chlorine. Chlorine and chlorine dioxide react similarly with unsaturated aliphatics to form dichloro–compounds, chloroketones, chlorohydrins, and epoxides. Chlorine dioxide oxidizes primary and secondary aliphatic alcohols to acids whereas chlorine does not attack alcohols at all.

Exxon chemicals, Inc., and Rio Linda Chemical Co. have developed a process that uses chlorine dioxide, generated on–site by a patented process, to oxidize organically contaminated aqueous waste streams, and simple and complex cyanide in water or solid media. Chlorine dioxide is an ideal oxidizing agent because it chemically alters contaminants to salts and nontoxic organic acids.

Chlorine dioxide gas is generated by reacting sodium chlorite solution with chlorine gas, or by reacting sodium chlorite solution with sodium hypochlorite and hydrochloric acid. Both processes produce at least 95% pure chlorine dioxide.

In aqueous treatment systems the chlorine dioxide gas is fed into the waste stream via a venturi, which is the driving force for the generation system. The amount of chlorine dioxide required depends on the contaminant concentrations in the waste stream and the concentration of oxidizable compounds, such as sulfides.

In soil treatment applications, the chlorine dioxide may be applied in–situ via conventional injection wells or surface flushing. The concentration of chlorine dioxide would depend on the level of contaminants in the soil.

Chlorine dioxide treatment systems have been applied to drinking water disinfection, food processing sanitation, and as a biocide in industrial process water. Since chlorine dioxide reacts via direct oxidation rather than substitution (as does chlorine), the process does not form undesirable trihalomethanes.

This technology is applicable to aqueous wastes, soils, or any leachable solid media contaminated with organic compounds. It can also be applied to groundwater contaminated with pesticides or cyanide; sludges containing cyanide, PCPs or other organics; and industrial wastewater similar to refinery wastewater.

Chloroiodides: A method for the degradation of substances containing both aromatic rings and ether bonds was reported in 1979 (Botre, 1979; des Rosiers, 1983; Esposito, 1980). This is of current interest because 2,3,7,8–TCDD contains two aromatic rings connected by two ether bonds. The method utilized chloroiodides attached to quaternary ammonium salt surfactant molecules to rupture the ether bonds, and thus split the 2,3,7,8–TCDD molecule into smaller fragments. End products are chlorophenols and related compounds. The mechanism of the ether bond rupturing is thought to be the loss of an iodine atom from the surfactant, and subsequent formation of reactive hydrogen iodide at a location in an aqueous solution near the 2,3,7,8–TCDD molecule. Hydrogen iodide by itself is known to rupture ether bonds, but usually only in a strongly acidic environment. However, the formation of the hydrogen iodide in close proximity to the 2,3,7,8–TCDD molecule seems to be the key factor.

One method of 2,3,7,8–TCDD degradation described involves the extraction of 2,3,7,8–TCDD from soil using aqueous solutions of surfactants containing chloroiodide groups. The aqueous residues from the soil washings are extracted with benzene, methanol, or methylene chloride. These extracted liquids containing the 2,3,7,8–TCDD may require evaporation under reduced pressure to concentrate the solution and thus enhance the reaction by bringing the 2,3,7,8–TCDD molecule and chloroiodide–bearing surfactant molecules into more frequent contact.

The chloroiodide derivatives producing the most promising results for the cleavage of ether bonds are alkyldimethylbenzylammonium (benzalkonium) chloroiodide, and 1–hexadecylpyridinium (cetylpyridinium) chloroiodide (CPC). The low solubilities of these chloroiodides in water can be increased with the addition of micellar solutions of the same surfactants with chloroiodide groups. Micellar solutions consist of large polymeric particles (clusters) of the surfactants. Common solubilizing agents are benzalkonium chloride, used to enhance the solubility of benzalkonium chloroiodide, and cetylpyridinium chloride, used to enhance the solubility of cetylpyridinium chloroiodide.

Surfactant micellar solutions of 2,3,7,8–TCDD without chloroiodides are stable when stored in the dark, but decompose when exposed to sunlight or UV irradiation. This form of treatment seems to be appropriate and effective for the decontamination of buildings, furniture, etc., where surfactant contact accompanied by exposure to UV is possible. However, the use of chloroiodides has been shown to be effective in the decomposition of 2,3,7,8–TCDD without irradiation. This latter method utilizing chloroiodides is therefore more suitable for degrading bulk solutions.

Hypochlorite Oxidation: Hypochlorite oxidation consists of adding sodium or calcium hypochlorite (bleaching agents) to oxidize organic wastes. Such technology will be recognized as the common method of disinfecting home swimming pools. This method may produce toxic chlorinated organic by–products and it must be done under controlled (not in situ) conditions, i.e., batch reactors. It is a nonspecific reaction.

Hypochlorites can be used for clean up of certain chemical spills. Sodium hypochlorite is available in an aqueous solution (household bleach). Sodium hypochlorite has been shown to be effective in the treatment of cyanides and nitriles, aliphatic amines, and phenolic compounds. In the treatment of phenols, an elevated pH (>10) is usually necessary in order to ensure a complete reaction and limit the formation of incomplete reaction products, e.g., chlorinated phenols. Free chlorine, either as a gas (land spill applications) or as chlorine ions (water spills), is a product of the oxidation reaction. Therefore, extreme care must be utilized. Chlorine gas generated by the application of sodium hypochlorite to a land spill can be allowed to dissipate in the atmosphere. Chloride produced in an aqueous solution, however, must be removed by further treatment such as activated carbon adsorption.

Note that an alkaline pH is required in aqueous solution containing cyanides before treatment with chlorine. Acid pH will facilitate the formation of hydrogen cyanide and/or cyanogen chloride, both of which are highly toxic.

Chlorinolysis: In the process termed chloronolysis (or chlorolysis in Germany), chloride is introduced to the waste at high temperatures and pressures. At temperatures above 500°C, under excess chlorine conditions, the carbon–carbon bonds of hydrocarbons can be broken and the molecular fragments can react with chlorine to form low molecular weight chlorinated hydrocarbons. It is essentially a pyrolytic process carried out in the presence of chlorine.

Chlorinolysis is primarily used as a manufacturing process to produce carbon tetrachloride. Its potential as a waste treatment process is limited by dependency on the carbon tetrachloride market.

2.7.5 Hydrogen Peroxide Oxidation

This process is based on the addition of hydrogen peroxide to oxidize organic compounds. Hydrogen peroxide is not the stable oxide of hydrogen and since it readily give up its extra oxygen, it is an excellent oxidizing agent. The process is a nonspecific reaction. It may be exothermic/explosive or require addition of heat and/or catalysts. Oxidation by hydrogen peroxide is being developed by Steinfeld and Partner in Germany for in situ treatment. It may also be used for soil surface treatment.

The peroxide oxidation process is run under similar conditions, and with similar equipment, to those used in the alkaline chlorination process. Hydrogen peroxide is added as a liquid solution. Oxidation with H_2O_2 is generally performed in the presence of a metal catalyst. Typical catalysts include iron sulfate, iron wool, nickel salts, and aluminum salts. The waste is heated and then treated with H_2O_2 while being agitated. The H_2O_2 oxidation tends to proceed quickly under basic conditions.

Ozone and hydrogen peroxide, can be used together, and has been found by USAE Waterways Experiment Station to be as effective, and less costly than ultraviolet based systems.

Cyanide Oxidation: Hydrogen peroxide oxidation removes both cyanide and metals in cyanide containing wastewaters. In this process, cyanide bearing waters are heated to 49° to 54°C (120° to 130°F) and the pH is adjusted to 10.5 to 11.8. Formalin (37% formaldehyde) is added while the tank is vigorously agitated. After 2 to 5 min, a proprietary peroxygen compound (41% hydrogen peroxide with a catalyst and additives) is added. After an hour of mixing, the reaction is complete. The cyanide is converted to cyanate, and the metals are precipitated as oxides or hydroxides. The metals are then removed from solution by either settling or filtration.

The main equipment required for this process is two holding tanks equipped with heaters and air spargers or mechanical stirrers. These tanks may be used in a batch or continuous fashion, with one tank being used for treatment while the other is being filled. A settling tank or a filter is needed to concentrate the precipitate.

The hydrogen peroxide oxidation process is applicable to cyanide–bearing wastewaters, especially those containing metal–cyanide complexes. In terms of waste reduction performance, this process can reduce total cyanide to less than 0.1 mg/ℓ and the zinc or cadmium to less than 1.0 mg/ℓ.

Chemical costs are similar to those for alkaline chlorination using chlorine, and lower than those for treatment with hypochlorite. All free cyanide reacts and is completely oxidized to the less toxic cyanate state. In addition, the metals precipitate and settle quickly, and they may be recoverable in many instances. However, the process requires energy expenditures to heat the wastewater prior to treatment.

Ultraviolet Enhancement: Ultraviolet enhancement of hydrogen peroxide oxidation processes are discussed in Chapter 7.

Air Pollution Control: Hydrogen peroxide can be effective in solving air pollution problems involving NO_x, SO_2, H_2S, mercaptans, aldehydes, and odor control.

Hydrogen peroxide's strong oxidizing properties offer two primary benefits in gas scrubbing systems:

1. Depending on the application, hydrogen peroxide may enhance the rate of absorption and thus reduce the scrubber size; and
2. As an environmentally–compatible chemical, hydrogen peroxide eliminates liquor disposal problems since the scrubber effluent can be safely disposed of or recycled.

2.7.6 Ozonation

Ozone is a powerful oxidizing agent, having oxidizing potential greater than either chlorine or chlorine dioxide. Oxidation by ozone instead of chlorine or chlorine dioxide has been found to eliminate entirely the formation of undesirable end-products such as organohalides. It is effective for disinfection, odor and color removal, and destruction of cyanides and toxic organic compounds in water. In addition, ozonation coupled with ultraviolet radiation destroys some trihalomethanes and their precursors. The unconsumed dissolved ozone after treatment quickly decomposes to oxygen, such that no secondary pollutant is formed which might directly or indirectly (through side reactions) cause environmental health problems.

Ozone can be used to pretreat wastes to break down refractory organics or as a polishing step after biological or other treatment processes to oxidize untreated organics. Ozone is usually produced by high-voltage ionization of atmospheric oxygen.

Ozone is currently used for treatment of hazardous wastes to destroy cyanide and phenolic compounds. The rapid oxidation of cyanides with ozone offers advantages over the slower alkaline chlorination method. Limitations include the physical form of the waste (i.e., sludges and solids are not readily treated) and nonselective competition with other species.

Ozonation can be conducted in a batch or continuous process. The ozone for treatment is produced on site because of the hazards of transporting and storing ozone as well as its short shelf life. The ozone gas is supplied to the reaction vessels by injection into the wastewater. The batch process uses a single reaction tank. As with alkaline chlorination, the amount of ozone added and the reaction time used are determined by the type and concentration of the oxidizable contaminants, and vigorous mixing should be provided for complete oxidation.

In continuous operation, two separate tanks may be used for reaction. The first tank receives an excess dosage of ozone. Any excess ozone remaining at the outlet of the second tank is recycled to the first tank, thus ensuring that an excess of ozone is maintained and also that no ozone is released to the atmosphere. As with alkaline chlorination, an ORP control system is usually necessary to ensure that sufficient ozone is being added.

Ozone is usually generated at the point of use in a flowing air or oxygen stream by an electric discharge process. Mixtures of 1 to 3% of ozone/air and 3 to 5% ozone/oxygen can be produced. These are then mixed with water in a contactor to ensure efficient transfer into the liquid. Ozone product is energy intensive, with only 10% of the power supplied to the ozone generator producing ozone. The remainder of the power produces light, sound and heat which are undesirable by-products and, hence, represent process inefficiencies. Therefore, the electricity cost for the ozonation process makes up a considerable percentage of the operating cost.

Using oxygen instead of air as the gas feed to the electric discharge unit roughly doubles the ozone production for the same amount of input of electrical energy. However, since only 4% (on the average) of the oxygen feed is converted to ozone, in order to minimize the loss of oxygen, the oxygen coming out from the generator must be recycled to the generator after augmentation with make-up oxygen.

Because ozone is only slightly soluble in water, it is important to design a good contacting system to maximize the mass transfer between the gaseous ozone and the liquid. Some of the factors which affect the mass transfer of ozone are pressure, temperature, bubble size, and method and time of contact.

There are no definite limits on the concentrations of the wastes for the ozonation process. Any limitations depend upon operating conditions as well as desired removal efficiencies. However, the removal efficiencies of contaminants are high if the aqueous solutions contain small amounts (up to 1,000 mg/ℓ) of contaminants such as organics, cyanides, sulfites, etc. Sludges, slurries and tars are unlikely candidates for ozonation, unless they could be greatly diluted and dispersed in water. Such highly viscous materials pose special mixing problems. Furthermore, they would likely be opaque to ultraviolet radiation limiting the possibility of assisting the oxidation process in this manner.

Ozonation can be used for:

1. Oxidation of organic compounds in wastewater and groundwater.
2. Removal of cyanides.
3. Oxidation of natural organics as an aid to floc formation.
4. Destruction of THM precursors prior to chlorination.
5. Destruction of THMs and other toxic organic chemicals.
6. Oxidation of natural and synthetic organic materials to enhance their biodegradability for removal on biological filters.
7. Oxidation of inorganic ions in wastewaters, followed by hydrolysis and precipitation.

Advantages: There are several factors which suggest that ozonation may be a viable technology for treating certain dilute aqueous waste streams:

1. Capital and operating costs are not excessive when compared to incineration provided oxidizable contaminant concentration levels are less than 1%.
2. The system is readily adaptable to the on site treatment of hazardous waste because the ozone can and must be generated on site.
3. Ozonation can be used as a final treatment for certain wastes since effluent discharge standards can be met.
4. It can be used as a preliminary treatment for certain wastes (e.g., preceding biological treatment).
5. Easier and safer to use than chlorine.
6. Generally more effective than chlorine.

Limitations: There are limitations which often will preclude use of ozonation as a treatment technology. These include:

1. Ozone is a nonselective oxidant; the waste stream should contain primarily the contaminants of interest.
2. Certain compounds because of their structure are not amenable to ozonation, e.g., chlorinated aliphatics.
3. Ozone systems are generally restricted to 1% or lower levels of toxic compounds. The system is not amenable to bulky wastes.
4. Toxic intermediates may persist in the waste stream effluent.
5. Ozone decomposes rapidly with increasing temperature, therefore, excess heat must be removed rapidly.
6. More expensive than chlorination.

Removal of Cyanides by Ozonation: Ozone effectively oxidizes cyanides to cyanates. The reaction is normally carried out at ambient temperature and pressure. Ozone oxidation beyond the cyanate level is quite slow and requires another form of oxidation treatment like alkali chlorination in order to convert cyanates to nitrogen and bicarbonates.

Ozone oxidation of cyanide is strongly influenced by pH. The reaction has been found to be first order with respect to the dissolved ozone concentration at all values of pH, but the order for cyanide decreases with increasing pH. The reaction orders with respect to cyanide for pH values of 11, 9.5 and 7 are, respectively, 0.55, 0.83 and 1.06. The optimal pH level for the reaction is seven, with 1.2 mols of ozone being consumed for each mol of cyanide being oxidized. Ozone can usually achieve removal rates of nearly 99% except at the highest concentrations of cyanides. In such cases it is reasonable to expect that a two stage process will be required.

Some advantages of ozone oxidation for handling process effluents are its suitability to automatic control and on–site generation and the fact that reaction products are not chlorinated organics and no dissolved solids are added in the treatment step. Ozone in the presence of activated carbon, ultraviolet, and other promoters shows promise of carbon, reducing reaction time and improving ozone utilization, but the process at present is limited by high capital expense, possible chemical interference in the treatment of mixed wastes, and an energy requirement of 25 kWh/kg of ozone generated. Cyanide is not economically oxidized beyond the cyanate form.

Enhancement by ultraviolet and ultrasonic techniques are discussed elsewhere in this book.

2.7.7 Permanganate Oxidation

Permanganate oxidation is a chemical reaction by which wastewater pollutants can be oxidized. When the reaction is carried to completion, the by–products of the oxidation are not environmentally harmful. A large number of pollutants can be practically oxidized by permanganate, including cyanides, hydrogen sulfide, and phenol. In addition, the chemical oxygen demand (COD) and many odors in wastewaters and sludges can be significantly reduced by permanganate oxidation carried to its end point. Potassium permanganate can be added to wastewater in either dry or slurry form. The oxidation occurs optimally in the 8 to 9 pH range

One of the by–products of this oxidation is manganese dioxide (MnO_2), which occurs as a relatively stable hydrous colloid usually having a negative charge. These properties, in addition to its large surface area, enable manganese dioxide to act as a sorbent for metal cation, thus enhancing their removal from the wastewater.

Commercial use of permanganate oxidation has been primarily for the control of phenol and waste odors. Several municipal waste treatment facilities report that initial hydrogen sulfide concentrations (causing serious odor problems) as high as 100 mg/ℓ have been reduced to zero through the application of potassium permanganate. A variety of industries (including metal finishers and agricultural chemical manufacturers) have used permanganate oxidation to totally destroy phenol in their wastewaters.

Permanganate oxidation has several advantages as a wastewater treatment technique. Handling and storage are facilitated by its nontoxic and noncorrosive nature. Performance has been proven in a number of municipal and industrial applications. The tendency of the manganese dioxide by–product to act as a coagulant aid is a distinct advantage over other types of chemical treatment.

The cost of permanganate oxidation treatment can be limiting where very large dosages are required to oxidize wastewater pollutants. In addition, care must be taken in storage to prevent exposure to intense heat, acids, or reducing agents; exposure could create a fire hazard or cause explosions. Of greatest concern is the environmental hazard which the use of manganese chemicals in treatment could cause. Care must be taken to remove the manganese from treated water before discharge.

The oxidation of wastewater pollutants by potassium permanganate is a proven treatment process in several types of industries. It has been shown effective in treating a wide variety of pollutants in both municipal and industrial wastes.

A process has been developed in Germany for in situ oxidation of arsenic in groundwater using potassium permanganate. It involves accelerated oxidation of trivalent arsenic into pentavalent arsenic and precipitation of complex arsenic–iron–manganese compounds.

The process consists of injecting a solution of $KMnO_4$ and water into the ground through injection wells and piezometers. The solution naturally mixes with contaminated groundwater and the natural oxidation process of the arsenic is accelerated. The source of the arsenic was the residue from a zinc ore smelter located near Cologne.

2.7.8 Ruthenium Tetroxide

Ruthenium tetroxide is a powerful oxidizing agent. It is more effective than either hypochlorite or permanganate in attacking aromatic substances. The reagent can be used in solution with water or with organic solvents which demonstrate no nucleophilic character such as chloroform, methylene chloride, acetic acid, fluorotrichloromethane, and nitromethane.

Degradation using ruthenium tetroxide is by aromatic ring cleavage. In tests where chlorophenols were treated with ruthenium tetroxide, all of the aromatic ring carbons were accounted for as carbon dioxide, and the aromatic chlorosubstituents gave rise to chloride ions. A similar analysis of the degradation products of TCDD was not carried out due to analytical difficulties related to the low solubility of the compound. It was inferred, however, that because of the close chemical and structural similarities between TCDD and chlorophenols that they would be degraded in a similar manner.

One factor affecting the rate of pollutant degradation using ruthenium tetroxide (RuO_4) is temperature. In one set of experiments, the rate of pollutant degradation increased 2.4-fold per 10°C rise within the test temperature range.

This technology will require considerable work before it can be applied in the field. The high cost of ruthenium tetroxide and the toxicity of process residuals may limit application of this technology.

2.7.9 Sulfur–Based Processes

Sulfur Dioxide: In 1982 Inco Metals Company announced the development of a technology for the destruction of cyanide in gold mill waste streams. The process involves the selective oxidation to cyanate of both free and complexed cyanide species using a mixture of SO_2 and air at controlled pH in the presence of copper as a catalyst. Metals are precipitated from solution as hydroxides. The process also removes iron cyanide, not by oxidation, but by precipitation as an insoluble copper or zinc ferrocyanide. The SO_2/air oxidation process destroys the metal cyanide complexes typically present in metal finishing and gold mining effluents.

SO_2/air oxidation of cyanide depends on efficient mixing of air with the waste to ensure an adequate supply of oxygen. Because of this factor, the equipment requirements for this process are similar to those for ozonation. SO_2 is sometimes supplied with the air by using flue gas containing SO_2 as the air source. Otherwise, sulfur in the +4 oxidation state can be fed as gaseous sulfur dioxide, liquid sulfurous acid, sodium sulfite solution, or sodium bisulfite solution. Sodium bisulfite solution, made by dissolving sodium metabisulfite in water, is the most frequently used source of SO_2. This process is usually run continuously, with the addition of oxidizing agent and acid/alkali being controlled through continuous monitoring of ORP and pH, respectively.

Polysulfide Treatment: In the cyanide–polysulfide reaction it has been postulated that 1 mol cyanide reacts with 1 mol of polysulfide to produce 1 mol of less toxic thiocyanate. However, it should be noted that in sufficient quantities, thiocyanates can cause toxic inhibition to biological treatment systems.

Despite the large quantity of experimental data available, industrial applications of polysulfide treatment of cyanide bearing wastewaters has been mostly limited to fluid catalytic cracking and coal gasification effluents. Another industrial application reported in the literature is a large commercial waste treatment facility in California. The cyanide treatment process at this facility is batch in nature and consists of two 18,000 gal storage/treatment tanks into which cyanide wastes (greater than 100 ppm CN_T) are pumped. The treatment reagent used is calcium polysulfide. This reagent is stored in an adjacent fiberglass tank. The amount of reagent required to complete the cyanide to thiocyanate reaction is predetermined by on site laboratory analysis of incoming waste for reactive CN. The process typically handles approximately 40,000 gal of waste per month at 50% of its capacity.

Ferrous Sulfate: The formation of less toxic cyanide complexes such as ferro and ferric–cyanides also has been used as a method for detoxifying of cyanide wastewaters. This process involves the use of iron salts to form complex compounds with the free cyanide in the wastes. Eventually these cyanide complexes are precipitated and removed as a sludge.

The major advantage of this treatment method is that it is relatively inexpensive in locations where waste ferrous sulfate is available. However, considerable quantities of sludge may be formed and the treated solutions are strongly colored. There also is evidence that ferrocyanides may be decomposed to free cyanide by sunlight. The regeneration of the cyanide under these conditions would contaminate the receiving stream.

This method has received very little acceptance by industry in this country, but appears to be used in Europe. The complexing process apparently does not completely destroy cyanide under practical operating conditions. Cyanide levels in treated solutions may be as great as 5 to 10 ppm. Thus, the sludges formed would appear to be toxic and will require substantial post–treatment prior to final disposal.

2.8 PRECIPITATION

Chemical precipitation is a physicochemical process whereby a substance in solution is transformed into a solid phase and driven out of solution. Precipitation entails altering the chemical equilibrium affecting the solubility of the hazardous component by either adjusting the pH of the solution (often involving a redox reaction and sometimes floc formation), or by adding a substance that will react with the dissolved substance to form a less soluble product.

Precipitation reactions can also follow as a result of a temperature change. Current application of precipitation with respect to hazardous waste treatment include the removal of heavy metals from wastewater by lime treatment, treatment of dye manufacturing wastes, and removal of organic colloids from pulp and paper mill wastewater effluents. Precipitation may be applied to almost any liquid waste stream containing hazardous solids which can be settled out of solution.

There are many factors which affect the efficiency of precipitation (pH, nature and concentration of hazardous substances in water, precipitant dosage, temperature, water turbulence, etc.). In practice, the optimum precipitant and dosage for a particular application are determined by a "trial and error" approach using jar tests.

Precipitation can be followed by coagulation and flocculation, in order to enhance sedimentation. Coagulation and flocculation are discussed elsewhere. Xanthate precipitation is discussed under "Ion Exchange."

This technology is used to treat aqueous wastes containing metals. Limitations include the fact that not all metals have a common optimum pH at which they precipitate. Chelating and complexing agents can interfere with the process. Organics are not removed except through adsorptive carryover. The resulting sludge may be hazardous by definition but often may be delisted by specific petition.

The principal precipitation reagents include:

1. Lime—least expensive, generates highest sludge volume.
2. Caustic and carbonates—more expensive than lime, generates smaller amount of sludge, applicable for metals where their minimum solubility within a pH range is not sufficient to meet clean-up criteria.
3. Sulfides—effective treatment for solutions with lower metal concentrations.
4. Sodium borohydride—expensive reagent, produces small sludge volumes which can be reclaimed.

Chemical precipitation normally depends on several variables:

1. Maintenance of an alkaline pH throughout the precipitation reaction and subsequent settling.
2. Addition of a sufficient excess of treatment ions to drive the precipitation reaction to completion.
3. Addition of an adequate supply of sacrificial ions (such as iron or aluminum) to ensure precipitation and removal of specific target ions.
4. Effective removal of precipitated solids.

Installation of a metals precipitation system inevitably results in the problem of sludge disposal. The cost of hauling the sludge to a licensed hazardous waste landfill will depend on the volume of sludge, the distance hauled, and the sludge composition. Sometimes it is possible to dispose of calcium-based reagent sludges through agricultural or acid pond liming.

Another option is to treat the waste by immobilizing the waste constituents for as long as they remain hazardous. This method of treatment, based on fixation or encapsulation processes, is a possibility for some metals containing wastes. Certain of these residuals could be found hazardous; their heavy metal content may lead to positive tests for EP toxicity. In such cases, encapsulation may be needed to eliminate this characteristic.

Chemical precipitation of metals has long been the primary method of treating metal–laden industrial wastewaters. Due to the success of metals precipitation in such applications, the technology is being considered and selected for use in remediating groundwater containing heavy metals. In groundwater treatment applications, the metal precipitation process is often used as a pretreatment for other treatment technologies (such as chemical oxidation or air stripping) where the presence of metals would interfere with the other treatment processes.

2.8.1 Applicability

Chemical precipitation is a treatment technology applicable to wastewaters containing a wide range of dissolved and other metals, as well as other inorganic substances such as fluorides. This technology removes these metals and inorganics from solution in the form of insoluble solid precipitates. The solids formed are then separated from the wastewater by settling, clarification, and/or polishing filtration.

For some wastewaters, such as chromium plating baths or plating baths containing cyanides, the metals exist in solution in a very soluble form. This solubility can be caused by the metal's oxidation state, for example, high cyanide–containing wastewaters. In both cases, pretreatment, such as hexavalent chromium reduction or oxidation of the metal–cyanide complexes, may be required before the chemical precipitation process can be applied effectively. In the case of arsenic, the arsenic–containing solution is normally treated first with oxidizing agents such as alkali hypochlorite solution to convert the lower–valence arsenic compound to arsenate. The arsenic ion is then typically precipitated out as ferric arsenate. Some compounds must be reduced prior to precipitation. For instance, selinites and selenates are oxidants and are readily reduced to elemental selenium, which is insoluble in aqueous solutions. Sulfur dioxide, sulfides, sulfites, and ferrous ion are all effective for this reduction reaction.

Chemical precipitation may also be applicable to mixed waste for separating radionuclides from other hazardous constituents in wastewaters. Specific conditions of pH, temperature, and precipitating reagent addition are required to selectively remove part or all of the radioactive component as a precipitate.

2.8.2 Principles of Operation

The basic principle of operation of chemical precipitation is that metals and inorganics in wastewater are removed by the addition of a precipitating agent that converts the soluble metals and inorganics to insoluble precipitates. These precipitates are settled, clarified, and/or filtered out of solution, leaving a lower concentration of metals and inorganics in the wastewater. The principal precipitation agents used to convert soluble metal and inorganic compounds to less soluble forms include lime [$Ca(OH)_2$], caustic ($NaOH$), sodium sulfide (Na_2S), and, to a lesser extent, soda ash (Na_2CO_3), phosphate ($PO_4^{=}$), and ferrous sulfide (FeS).

The solubility of a particular compound depends on the extent to which the electrostatic forces holding the ions of the compound together can be overcome. The solubility changes significantly with temperature, with most metal compounds becoming more soluble as the temperature increases. Additionally, the solubility is affected by other constituents present in the wastewater, including other ions and complexing agents. Regarding specific ionic forms, nitrates, chlorides, and sulfates are, in general, more soluble than hydroxides, sulfides, carbonates, and phosphates.

Once the soluble metal and inorganic compounds have been converted to precipitates, the effectiveness of chemical precipitation is determined by how successfully the precipitates are physically removed. Removal usually relies on a settling process; that is, a particle of a specific size, shape, and composition will settle at a specific velocity, as described by Stokes' Law. For a batch system, Stokes' Law is a good predictor of settling time because the pertinent particle parameters essentially remain constant. In practice, however, settling time for a batch system is normally determined by empirical testing. For a continuous system, the theory of settling is complicated by such factors as turbulence, short-circuiting of the wastewater, and velocity gradients, thus increasing the importance of empirical tests to accurately determine appropriate settling times.

2.8.3 Chemical Precipitation Process

The equipment and instrumentation required for chemical precipitation vary depending on whether the system is batch or continuous. Both systems are discussed below.

For a batch system, chemical precipitation requires a feed system for the treatment chemicals and a reaction tank where the waste can be treated and allowed to settle. When lime is used, it is usually added to the reaction tank in a slurry form. The supernatant liquid is generally analyzed before discharge to ensure that settling of precipitates is adequate.

For a continuous system, additional tanks are necessary, as well as the instrumentation to ensure that the system is operating properly. In this system, wastewater is fed into an equalization tank, where it is mixed to provide more uniformity, thus minimizing the variability in the type and concentration of constituents sent to the reaction tank.

Following equalization, the wastewater is pumped to a reaction tank where precipitating agents are added. This is done automatically by using instrumentation that senses the pH of the system for hydroxide precipitating agents, or the oxidation–reduction potential (ORP) for nonhydroxide precipitating agents, and then pneumatically adjusts the position of the treatment chemical feed valve until the design pH or ORP value is achieved. (The pH and ORP values are affected by the concentration of hydroxide and nonhydroxide precipitating agents, respectively, and are thus used as indicators of their concentrations in the reaction tank.)

In the reaction tank, the wastewater and precipitating agents are mixed to ensure commingling of the metal and inorganic constituents to be removed and the precipitating agents. In addition, effective dispersion of the precipitating agents throughout the tank is necessary to properly monitor and thereby control the amount added.

Following reaction of the wastewater with the stabilizing agents, coagulating or flocculating compounds are added to chemically assist the settling process. Coagulants and flocculants increase the particle size and density of the precipitated solids, both of which increase the rate of settling. The coagulant or flocculating agent that best improves settling characteristics varies depending on the particular precipitates to be settled.

Settling can be conducted in a large tank by relying solely on gravity or can be mechanically assisted through the use of a circular clarifier or an inclined plate settler. Following the addition of coagulating or flocculating agents, the wastewater is fed to a large settling tank, circular clarifier, or inclined plate settler, where the precipitated solids are removed. These solids are generally further treated in a sludge filtration system to dewater them prior to disposal.

The supernatant liquid effluent can be further treated in a polishing filtration system to remove precipitated residuals both in cases where the settling system is underdesigned and in cases where the particles are difficult to settle.

2.8.4 Waste Characteristics Affecting Performance (WCAPS)

In determining whether chemical precipitation will achieve the same level of performance on an untested waste that it achieved on a previously tested waste and whether performance levels can be transferred, EPA examines the following waste characteristics: (a) the concentration and type of metals, (b) the concentration of total dissolved solids (TDS), (c) the concentration of complexing agents, and (d) the concentration of oil and grease.

Concentration and Type of Metals: For most metals, there is a specific pH at which the metal precipitate is least soluble. As a result, when a waste contains a mixture of many metals, it is not possible to operate a treatment system at a single pH or ORP value that is optimal for the removal of all metals. The extent to which this affects treatment depends on the particular metals to be removed and their respective concentrations. One alternative is to operate multiple precipitations, with intermediate settling, when the optimum pH occurs at markedly different levels for the metals present. If the concentration and type of metals in an untested waste differ from and are significantly higher than those in the tested waste, the system may not achieve the same performance. Additional precipitating agents, alternate pH/ORP values, and/or multiple precipitations may be required to achieve the same treatment performance, or other, more applicable treatment technologies may need to be considered for treatment of the untested waste.

Concentration of Total Dissolved Solids: High concentrations of total dissolved solids (TDS) can interfere with precipitation reactions, as well as inhibit settling. Poor precipitate formation and flocculation are results of high TDS concentrations, and higher concentrations of solids are found in the treated wastewater residuals. If the TDS concentration in an untested waste is significantly higher than that in the tested waste, the system may not achieve the same performance. Higher concentrations of precipitating agents may be required to achieve the same treatment performance, or other, more applicable treatment technologies may need to be considered for treatment of the untested waste.

Concentration of Complexing Agents: A metal complex consists of a metal ion surrounded by a group of other inorganic or organic ions or molecules (often called ligands). In the complexed form, metals have a greater solubility. Also, complexed metals inhibit the reaction of the metal with the precipitating agents and therefore may not be removed as effectively from solution by chemical precipitation. However, EPA does not have analytical methods to determine the concentration of complexed metals in wastewaters. The best indicator for complexed metals is to analyze for complexing agents, such as cyanide, chlorides, EDTA, ammonia, amines, and methanol, for which analytical methods are available. Therefore, the concentration of complexing agents is used as a surrogate waste characteristic for the concentration of metal complexes. If the concentration of complexing agents in an untested waste is significantly higher than that in the tested waste, the system may not achieve the same performance. Higher concentrations of precipitating agents may be required to achieve the same treatment performance, or other, more applicable treatment technologies may need to be considered for treatment of the untested waste.

Concentration of Oil and Grease: The concentration of oil and grease in a waste inhibits the settling of the precipitate by creating emulsions that require a long settling time. Suspended oil droplets in water tend to suspend particles such as chemical precipitates that would otherwise settle out of solution. Even with the use of coagulants or flocculants, the settling of the precipitate is less effective. If the concentration of oil and grease in an untested waste is significantly higher than that in the tested waste, the system may not achieve the same performance. Pretreatment of the waste may be required to reduce the oil and grease concentration and achieve the same treatment performance, or other, more applicable treatment technologies may need to be considered for treatment of the untested waste.

2.8.5 Design and Operating Parameters

In assessing the effectiveness of the design and operation of a chemical precipitation system, EPA examines the following parameters: (a) the pH/ORP value; (b) the precipitation temperature; (c) the residence time; (d) the amount and type of precipitating agents, coagulants, and flocculants; (e) the degree of mixing; and (f) the settling time.

pH/ORP Value: The pH/ORP value in continuous chemical precipitation systems is used as an indicator of the concentration of precipitating agents in the reaction tank and thus is used to regulate their addition to the tank. The pH/ORP value also affects the solubility of metal precipitates formed and therefore directly impacts the effectiveness of their removal.

Precipitation Temperature: The precipitation temperature affects the solubility of the metal precipitates. Generally, the lower the temperature, the lower the solubility of the metal precipitates and vice versa.

Residence Time: The residence time impacts the extent of the chemical reactions to form metal precipitates and, as a result, the amount of precipitates that can be settled out of solution. For batch systems, the residence time is controlled directly by adjusting the treatment time in the reaction tank. For continuous systems, the wastewater feed rate is controlled to make sure that the system is operating at the appropriate design residence time.

Amount and Type of Precipitating Agents, Coagulants, and Flocculants: The amount and type of precipitating agent used to effectively treat the wastewater depends on the amount and type of metal and inorganic constituents in the wastewater to be treated. Other design and operating parameters, such as the pH/ORP value, the precipitation temperature, the residence time, the amount and type of coagulants and flocculants, and the settling time, are determined by the selection of precipitating agents. The addition of coagulants and flocculants improves the settling rate of the precipitated metals and inorganics and allows for smaller settling systems (i.e., lower settling time) to achieve the same degree of settling as a much larger system. Typically, anionic polyelectrolyte flocculating agents are most effective with metal precipitates, although cationic or nonionic polyelectrolytes also are effective. Typical doses range from 0.1 to 10 mg/ℓ of the total influent wastewater stream. Conventional coagulants, such as alum (aluminum sulfate), are also effective, but must be dosed at much higher concentrations to achieve the same result. Therefore, these coagulants add more to the settled sludge volume requiring disposal than do the polyelectrolyte flocculants.

Degree of Mixing: Mixing provides greater uniformity of the wastewater feed and disperses precipitating agents, coagulants, and flocculants throughout the wastewater to ensure the most rapid precipitation reactions and settling of precipitate solids possible. The quantifiable degree of mixing is a complex assessment that includes, among other things, the amount of energy supplied, the length of time the material is mixed, and the related turbulence effects of the specific size and shape of the tank. This is beyond the scope of simple measurement.

Settling Time: Adequate settling time must be provided to make sure that removal of the precipitated solids from the wastewater has been completed.

2.8.6 Hydroxide Precipitation

Precipitation of many heavy metals is accomplished by adjusting the pH of the wastewater to alkaline, which causes the soluble metal ions to form insoluble metal hydroxides. This pH adjustment is usually achieved by the addition of caustic (sodium hydroxide), limestone (calcium carbonate) or lime (calcium hydroxide). Recently, the usage of magnesium oxide (MgO) and magnesium hydroxide has increased [$Mg(OH)_2$] because they generate lower sludge volumes and are more effective in treating low-metal-concentration (≤50 ppm) effluents. For metal hydroxide species, the solubilities are known to increase with both rising and falling pH values outside the 7 to 10 pH range. Because of the need for the discharge to meet the proper pH range (6 to 9) and because of differences in industrial waste compositions, actual industrial effluent concentrations from precipitation systems will tend to be greater than their minimal solubilities. Sand filtration of the effluent following sedimentation can further reduce the metal concentrations in the discharge. Coprecipitation with other metals can enhance the removal of a given metal species. Because of the relatively low cost of lime (or caustic), hydroxide precipitation is the metal precipitation system most commonly used by industry.

Hydroxide precipitation is a widely used and well developed technology for reducing metals effluent concentrations to acceptable levels. The process operates at ambient temperature and pressure and is well suited to automatic control. Its ability to treat a wide variety of industrial waste streams has been well demonstrated in bench, pilot, and full-scale systems. Environmental impacts can result from emissions during the precipitation process and the production of large volumes of potentially hazardous sludge. Exit gases can be scrubbed by using a control system, however, sludge reduction methods (seeding, dilution, vacuum filtration, etc.), have only partially offset the problems associated with sludge generation. Therefore, new methods of sludge disposal and reduction and recycle/reuse options (such as agricultural liming) should be considered.

The high costs attributed to sludge disposal demonstrates the main drawback to hydroxide precipitation. As land disposal costs increase, treatment processes such as hydroxide precipitation, which generate large quantities of hazardous sludge will lose their cost advantage over the more expensive recovery technologies.

2.8.7 Sulfide Precipitation

Chelated wastewater (e.g., those containing chelating agents such as EDTA) can severely hinder metals removal in hydroxide precipitation systems. Sulfide precipitation, however, which drops out metal ions as very insoluble sulfides, has been shown to yield high metal removals even in highly chelated wastewater. This process is used as an alternative to hydroxide precipitation or as a polishing step following hydroxide precipitation. Two different processes are used for sulfide precipitation: (1) the soluble sulfide process, and (2) the insoluble sulfide process. The soluble sulfide process uses sodium sulfide as the treatment reagent, whereas the insoluble sulfide process uses ferrous sulfide as the reagent. Both processes generate metal sulfide sludges that tend to be difficult to dispose of properly because of potential sulfide reactivity. The insoluble sulfide process generates a larger volume of sludge than does hydroxide precipitation because of the liberation of ferrous ions during treatment and the subsequent conversion of these ions to ferrous hydroxide. For arsenic, cadmium, lead, mercury, nickel, silver, and zinc, removal levels are significantly better with sulfide precipitation than with hydroxide precipitation. Again, as in hydroxide precipitation, sand filtration will help reduce metal discharges.

A commercially developed insoluble sulfide process known as the Sulfex System (Permutit Company) shows promise as a treatment for metal–containing wastes. This system uses a ferrous sulfide slurry in greater than stoichiometric amounts to precipitate metals.

For effective metals removal, the pH of the wastewater must be maintained within the neutral to slightly alkaline range. Contact with acidic wastewaters will result in poorer removal and can cause the emission of hydrogen sulfide (H_2S) gas.

Because of operational problems and odor generation, the soluble sulfide process had shown little promise. With the recent development of ion–specific probes, however, control of sodium sulfate addition to match demand is possible, which has sparked new interest in the process. Polyelectrolyte developments have eliminated previous separation difficulties by flocculating the fine metal sulfide particles generated.

The major advantage of the sulfide precipitation process is that the extremely low solubility of most metal sulfides promotes very high metal removal efficiencies; the sulfide process also has the ability to remove chromates and dichromates without preliminary reduction of the chromium to its trivalent state. In addition, sulfide can precipitate metals complexed with most complexing agents. The process demands care, however, in maintaining the pH of the solution at approximately 10 in order to restrict the generation of toxic hydrogen sulfide gas. For this reason, ventilation of the treatment tanks may be a necessary precaution in most installations. The use of insoluble sulfides reduces the problem of hydrogen sulfide evolution. As with hydroxide precipitation, excess sulfide ion must be present to drive the precipitation reaction to completion. Since the sulfide ion itself is toxic, sulfide addition must be carefully controlled to maximize heavy metals precipitation with a minimum of excess sulfide to avoid the necessity of additional wastewater treatment. At very high excess sulfide levels and high pH, soluble mercury–sulfide compounds may also be formed.

Where excess sulfide is present, aeration of the effluent stream can aid in oxidizing residual sulfide to the less harmful sodium sulfate (Na_2SO_4). The cost of sulfide precipitants is high in comparison to hydroxide precipitants, and disposal of metallic sulfide sludges may pose problems. An essential element in effective sulfide precipitation is the removal of precipitated solids from the wastewater and proper disposal in the appropriate site. Sulfide precipitation will also generate a higher volume of sludge than hydroxide precipitation, resulting in higher disposal and dewatering costs. This is especially true when ferrous sulfide is used as the precipitant.

Sulfide precipitation may be used as a polishing treatment after hydroxide precipitation–sedimentation. This treatment configuration may provide the better treatment effectiveness of sulfide precipitation while minimizing the variability caused by changes in raw waste and reducing the amount of sulfide precipitant required.

2.8.8 Carbonate Precipitation

Carbonate precipitation with soda ash (sodium carbonate) or calcium carbonate has proven to be an effective process for removal of cadmium, lead, nickel, and zinc. Precipitation as insoluble carbonates tends to occur at more neutral pH conditions than with hydroxide precipitation. Carbonate solubilities tend to be less than those of the corresponding hydroxide. The carbonate–based reaction mechanism, however, proceeds at a slower pace than the hydroxide–based system. The solubility of soda ash also limits its use because a chemical feed of only 20% by weight can be maintained at room temperature without recrystallization. An advantage of soda ash is low sludge generation; however, these sludges can be difficult to filter. Calcium carbonate sludges show much better filtration properties.

Carbon dioxide has also been used to treat metal–containing waste streams. Carbon dioxide gas is injected into the wastewater, and upon hydration it will form carbonic acid. The carbonic acid will then react with the available hydroxides to form the less–soluble carbonates.

High reagent costs associated with CO_2 systems make this treatment technique less attractive than conventional systems. Carbonate has proven to be more effective at removing lead than hydroxide, and it is the method of choice if nickel reclamation is deemed appropriate.

The main advantages of carbonate technology are buffering capability, superior handling characteristics (i.e., little dust, good flow, and no arching in the feeder), and widespread availability. Main disadvantages are slow reaction time (typically a minimum of 45 minutes retention) and low solubility (20% by weight). Since carbonates are not particularly corrosive and soda ash generates less sludge than comparable calcium–based technologies, environmental impacts are few.

2.8.9 Sodium Borohydride Precipitation

Sodium borohydride ($NaBH_4$), a strong reducing agent, can precipitate heavy metals in their elemental form from an alkaline solution (pH 8 to 11). This technology was pioneered in the late 1960's. Precipitation of single metal waste streams can produce metal sludges suitable for recycling or reclamation. Sodium borohydride offers the advantage of low sludge volumes compared with those produced by the conventional hydroxide precipitation method. This chemical also has the added benefit of removing metals to lower concentration levels than possible with conventional treatment. Sodium borohydride is commonly available as a stabilized water solution of 12% $NaBH_4$ in caustic soda. Sodium borohydride is capable of reducing chromates to the trivalent state before deposition as chromium hydroxide. The high cost of the reagent currently limits its use in industrial metal–bearing wastewaters.

2.8.10 Phosphate Precipitation

Pilot studies have been conducted to evaluate phosphate precipitation as a treatment alternative for recovering trivalent metals such as chromium, iron, and aluminum from mixed metal solutions. Under low pH conditions, phosphate, in the form of phosphoric acid (H_3PO_4) or sodium phosphate, can effectively strip trivalent cations from solution in preference to divalent cations. Phosphate products filter easily and can be compacted to a high solids content. Work conducted on metal sludge leachates has shown good results.

2.8.11 Differential Precipitation

Differential precipitation is a wastewater treatment technique in which multistep titration is used to form and precipitate out specific metal salts at selected titration points. This process may be followed by a recovery process (e.g., electrowinning) for the metals remaining in solution. The process can be designed to precipitate specific metals targeted for recovery. Thermodynamic modeling is required for complex wastewaters to identify key points in the titration process.

2.8.12 Zinc Cementation

Powdered zinc can be used to precipitate elements from wastewaters that are more electronegative than zinc such as chrome or copper. This precipitation technique is termed cementation. The cementation process has been shown to be effective in precipitating lead and cadmium from wastewaters. Zinc has been used to treat mercury in sludges. These sludges were then retorted as a final treatment step.

2.8.13 Coprecipitation

Coprecipitation is the process of precipitating a given metal species in association with other metal species. For some metal ionic forms, such as arsenate (AsO_4^{-3}), coprecipitation is the treatment method of choice. Coprecipitation involves both adsorption of the soluble ion onto a bulk solid and coagulation of fine solids by the bulk precipitate.

A process has been developed in Japan for the removal of heavy metals from acidic wastewater. The process, known as ferrite coprecipitation, has the potential for producing a marketable residual by converting the metal ions in solution into insoluble ferromagnetic oxides or ferrites which can be removed magnetically or by filtration. The treatment is applied by adding a ferrous salt to the metal–bearing wastewater, then neutralizing and oxidizing the complex heavy metal–ferrite coprecipitate. Particle sizes are reported to be relatively large and sludges formed can be safely disposed of by landfilling.

Another Japanese process involves coprecipitation of arsenic and gallium from wastewater of a semiconductor plant, achieved through the addition of $Fe(OH)_3$, $FeCl_3$, and NH_4OH, at a pH of 7. Filtration is used to separate out the precipitate, which is then resuspended in water and NaOH, at a pH of 9 to 13, to redisolve the gallium. Arsenic remains in solid form which is filtered out and the sludge is prepared for final disposal.

The presence of substantial quantities of iron in metal–bearing wastewaters before treatment has been shown to improve the removal of toxic metals. In some cases this iron is an integral part of the industrial wastewater; in other cases iron is deliberately added as a preliminary treatment or first step of treatment. The iron functions to improve toxic and other metals (such as molybdenum) removal by three mechanisms: (1) the iron coprecipitates with toxic metals forming a stable precipitate which desolubilizes the toxic metal; (2) the iron improves the settleability of the precipitate; and (3) the large amount of iron reduces the fraction of toxic metal in the precipitate. Coprecipitation with iron has been practiced for many ears incidentally when iron was a substantial constituent of raw wastewater, and intentionally when iron salts were added as a coagulant aid. Aluminum or mixed iron–aluminum salt also have been used.

Iron coprecipitation is a process that is used in the Uranium Mill Tailing Remedial Action (UMTRA) program to remove radium, uranium, and other contaminants from the surface runoff wastes generated during remedial action. It is also used at the Oak Ridge Y–12 Plant to remove uranium from nitrate–containing wastes. Iron is added to the stream and then precipitated with the contaminants when the pH of the solution is raised by the addition of lime or sodium hydroxide. Once the precipitation has occurred, the contaminant–containing solids must be separated from the water. This can be done using microfiltration as at the UMTRA site at Lakeview, Oregon, or by settling as used at the Oak Ridge Y–12 Plant. Coprecipitation is a process that removes metal ions, however, it will not remove nitrate ions, which are a serious contamination problem in some of the groundwaters at Hanford.

2.8.14 Lignochemicals and Humic Acids

The Bureau of Mines investigated the feasibility of removing heavy metals from mineral–process waste streams by precipitation with lignochemicals and humic acids. Lignochemicals are by–products from the paper industry, while humic acids are obtained by caustic treatment of peat, subbituminous coal, and lignite. These high–molecular–weight organic materials have many functional groups which can coordinate and form inner complex salts with heavy metals that are crystalline precipitates. Filtering of a humic–acid– or lignochemical–treated solution or waste stream containing these precipitates removes the heavy metal sequestrates.

2.8.15 Titanic Acid Process

The Japanese have developed a process for removal of arsenic from wastewaters that contain several metal ions. Wastewaters are treated with a titanium compound to form titanic acid, which forms a coprecipitate with arsenic. After coprecipitation and filtration, sludge containing arsenic can be disposed of in a landfill.

2.8.16 Xanthate Precipitation

This is a process in which metals are removed through precipitation. Metal contaminants in the water exchange with Na^+ ions contained by the xanthated material to form an insoluble complex. The heavy metals–laden material can then be removed from solution by sedimentation and filtration.

Currently, hydroxide precipitation is used extensively in the treatment of heavy metal–contaminated groundwaters and wastewater. Xanthate treatment offers many advantages over hydroxide precipitation, including the following:

1. A higher degree of metal removal;
2. Less sensitivity to pH fluctuation (metal xanthates do not exhibit amphoteric solubilities);
3. Less sensitivity to the presence of complexing agents;
4. Improved sludge dewatering properties; and
5. The capability of the selective removal of metals.

2.8.17 Cyanide Precipitation

Cyanide precipitation, although a method for treating cyanide in wastewaters, does not destroy cyanide. The cyanide is retained in the sludge that is formed. Reports indicate that during exposure to sunlight, the cyanide complexes can break down and form free cyanide. For this reason, the sludge from this treatment method must be disposed of carefully.

Cyanide precipitation can be used when cyanide destruction is not feasible because of the presence of cyanide complexes which are difficult to destroy. Effluent concentrations of cyanide well below 0.15 mg/ℓ are possible. Cyanide precipitation is an inexpensive method of treating cyanide. Problems may occur when metal ions interfere with the formation of the complexes.

Cyanide may be precipitated and settled out of wastewaters by the addition of zinc sulfate or ferrous sulfate. In the presence of iron, cyanide will form extremely stable cyanide complexes. The addition of zinc sulfate or ferrous sulfate forms zinc ferrocyanide or ferro ferricyanide complexes.

Cyanide precipitation occurs in two steps: reaction with ferrous sulfate or zinc sulfate at an alkaline pH to form iron or zinc cyanide complexes followed by reaction at a low pH with additional ferrous sulfate to form insoluble iron cyanide precipitates. Cyanide precipitation is applicable to all cyanide containing wastewater and, unlike many oxidation technologies, is not limited by the presence of complexed cyanides. Cyanide precipitation has been selected as the technology basis for cyanide control because of the presence of iron, nickel, and zinc in wastewaters. These metals are known to form stable complexes with cyanide.

2.8.18 Crystallization

Crystallization is a recovery technique in which metal contaminants in spent solutions are precipitated through temperature reduction and then are removed by settling or centrifugation. The applicability of crystallization as a treatment alternative for metal-bearing hazardous wastes is limited to liquid waste with appropriate solubility characteristics. As such, crystallization is most applicable to spent acid wastes from pickling, plating, etching, or other types of metal finishing operations, such as caustic soda etching of aluminum.

Crystallization systems have been applied on a commercial scale, primarily by generators of large volumes of spent solutions (e.g., iron and steel plants). There are several different commercially applied processes for recovery of sulfuric acid from spent pickle liquor. All processes, however, rely upon the basic principles of crystallization of iron salts (mainly ferrous sulfate) from the spent pickle liquor and the addition of enough fresh sulfuric acid to return the pickling solution to its original acid strength. These commercial acid recovery systems allow the free sulfuric acid remaining in the spent pickling solution to be reused. The processes differ in the methods used to crystallize the ferrous sulfate.

2.9 PYROMETALLURICAL PROCESSES

Most of the pyrometallurgical processes identified for metal waste treatment are classified as "calcination" or "smelting" operations. Calcination processes are generally those which form metal oxides, while smelting produces pure metal.

Drying and calcination are usually carried out in various types of kilns such as rotary kilns, shaft furnaces, and rotary hearths. Smelting operations are conducted in blast or reverberatory furnaces as described in reports and texts dealing with metal processing. Many nonferrous metals can be extracted by reduction smelting: copper, tin, nickel, cobalt, silver, antimony, bismuth, and others. Blast furnaces are sometimes used for the smelting of copper or tin, but reverberatory furnaces are more common for most metals.

One of the newer pyrometallurgical processes to be developed is one which employs the ultra-high temperatures of a plasma arc furnace. Waste dusts from furnace operations may be fed to a plasma burner operating at temperatures as high as 5000°C. The high heat will pyrolyze (break apart) the molecules of the waste mixture. Recovery may then be effected through selectively precipitating metals at their appropriate condensation point. This is a proposed method for handling solid, metallic wastes, particularly those in which a variety of metals are contained, such as dusts from specialty steelmaking furnaces.

Overall, the key element in evaluating the economic attractiveness of pyrometallurgical systems is the value which may be derived from recovery of metals. However, systems which can not produce reusable materials may be attractive in terms of providing good volumetric reduction of wastes, but may not be viable economically.

2.10 REDUCTION

Chemical reduction is a chemical reaction in which electrons are transferred to the chemical being reduced from the chemical initiating the transfer (the reducing agent). The function of reduction processes is to convert inorganics to a less toxic and/or more easily treated form. It also serves as a pretreatment step for inorganics in which chemical precipitation is used to remove the metal hydroxide from solution. Organics can also be reduced.

The basic principle of chemical reduction is to reduce the valence of the oxidizer in solution. Reducing agents used to effect the reduction include the sulfur compounds sodium sulfite (Na_2SO_3), sodium bisulfite ($NaHSO_3$), sodium metabisulfite ($Na_2S_2O_5$), sulfur dioxide (SO_2), and sodium hydrosulfide (NaHS). The ferrous form of iron (Fe^{+2}) is a popular reducing agent in many cases. Elemental magnesium (Mg), zinc (Zn), and copper (Cu) are also effective reducing agents. Frequently, hydrazine (N_2H_2) is used as a reducing agent also. The reactions are usually accomplished at pH values between 2 and 3. Precipitation techniques have been discussed earlier.

Redox Reactions: Reduction–oxidation (redox) reactions involve the chemical transformation of reactants in which the oxidation state of one reactant is raised while the other is lowered. The process destroys or reduces the toxicity of many toxic organics and heavy metals. Use of reducing agents for treatment is less common than oxidizing agents because of the high reactivity of the reducing agents.

Redox reactions are applied to a number of different contaminants; either oxidizing agents or reducing agents are applied to the waste in separate reaction vessels. Redox treatment has most commonly been applied to aqueous wastes containing heavy metals. For example, water used to flush source material from soils may be treated via redox reactions.

Efforts have recently focused on applying redox treatment to slurries, sludges, and soils. Applying a water–reagent mixture to sludges and soils will aid in mixing. In addition, combining this treatment with a soil flushing system may improve performance.

The process is nonspecific. Solids must be in solution. Reactions can be explosive. Waste composition must be well known to prevent the inadvertent production of a more toxic or more hazardous end product.

2.10.1 Chemical Reduction Process

The chemical reduction treatment process can be operated in a batch or continuous mode. A batch system consists of a reaction tank, a mixer to homogenize the contents of the tank, a supply of reducing agent, and a source of acid and base for pH control. A continuous chemical reduction treatment system usually includes a holding tank upstream of the reaction tank for flow and concentration equalization. It also typically includes instrumentation to automatically control the amount of reducing agent added and the pH of the reaction tank. The amount of reducing agent is controlled by the use of a sensor called an oxidation–reduction potential (ORP) cell. The ORP sensor electronically measures, in millivolts, the level to which the oxidation/reduction (redox) reaction has proceeded at any given time. It must be noted, however, that the ORP reading is very pH dependent. Consequently, if the pH is not maintained at a steady value, the ORP will vary somewhat, regardless of the level of chemical reduction.

When chemical reduction is used for treating hexavalent chromium, the trivalent chromium that is formed is either reused or further treated by stabilization and land disposed. Likewise, for selenium reduction, the precipitated elemental selenium may be recovered or stabilized and disposed of.

2.10.2 Waste Characteristics Affecting Performance (WCAPs)

In determining whether chemical reduction will achieve the same level of performance on an untested waste that it achieved on a previously tested waste, and whether performance levels can be transferred, EPA examines the following waste characteristics: (a) the concentration of other reducible compounds and b) the concentration of oil and grease.

Concentration of Other Reducible Compounds: The presence of other reducible compounds (also called oxidizers) in addition to the BDAT list constituents of concern will increase the demand of reducing agents, thereby potentially reducing the effectiveness of the treatment process. As a surrogate for the amount of organic oxidizers present in the waste, EPA might analyze for total organic carbon (TOC) in the waste. Inorganic oxidizing compounds such as ionized metals (e.g., silver, selenium, copper, mercury) may also create a demand for additional reducing agent. EPA might attempt to identify and analyze for these metal constituents. If TOC or inorganic oxidizer concentration in the untested waste is significantly higher than that in previously tested wastes, the system may not achieve the same performance as that achieved previously. Additional reducing agent may be required to effectively reduce the untested waste and achieve the same treatment performance, or other, more applicable treatment technologies may need to be considered for treatment of the untested waste.

Concentration of Oil and Grease: EPA believes that oil and grease compounds could cause monitoring problems because of fouling of instrumentation (e.g., electrodes for pH and ORP sensors). If the concentration of oil and grease in an untested waste is significantly higher than that in a tested waste, the untested system may not achieve the same performance. Therefore, other, more applicable treatment or pretreatment technologies may need to be considered for treatment of the untested waste.

2.10.3 Design and Operating Parameters

In assessing the effectiveness of the design and operation of a chemical reduction system, EPA examines the following parameters: (a) the residence time, (b) the amount and type of reducing agent, (c) the degree of mixing, (d) the pH, and (e) the reduction temperature.

For many hazardous organic constituents, analytical methods are not available or the constituent cannot be analyzed in the waste matrix. Therefore, it would normally be impossible to measure the effectiveness of the chemical reduction treatment system. In these areas EPA tries to identify measurable parameters or constituents that would act as surrogates to verify treatment.

For organic constituents, each compound contains a measurable amount of total organic carbon (TOC). Removal of TOC in the chemical reduction treatment system indicates removal of organic constituents. Hence, TOC analysis is likely to be an adequate surrogate analysis where the specific organic constituent cannot be measured.

However, TOC analysis may not be able to adequately detect treatment of specific organics in matrices that are heavily organic–laden (i.e., the TOC analysis may not be sensitive enough to detect changes at the milligrams per liter level in matrices where total organic concentrations are hundreds or thousands of milligrams per liter). In these cases other surrogate parameters should be sought. For example, if a specific analyzable constituent is expected to be treated as well as the unanalyzable constituent, the analyzable constituent concentration should be monitored as a surrogate.

Residence Time: The residence time affects the extent of reaction of waste contaminants with reducing agents. For a batch system, the residence time is controlled by adjustment of the treatment time in the reaction tank. For a continuous system, the waste feed rate is controlled to make sure that the system is operated at the appropriate design residence time.

Amount and Type of Reducing Agent: Several factors influence the choice of reducing agents and the amount to be added. The amount of reducing agent required to treat a given amount of reducible constituent will vary with the agent chosen. Enough reducing agent must be added to ensure complete reduction; the specific amount will depend on the type of reducible compounds in the waste and the chemistry of the reduction reactions. Theoretically, the amount of reducing agent to be added can be computed from reduction reaction stoichiometry. In practice, however, an excess of reducing agent should be used. Testing for excess reducing agent, if possible, will determine whether the reaction has reached completion.

In continuous processes, the addition of reducing agent is usually accomplished by automated feed methods. The amount of reducing agent needed is usually metered and controlled automatically by an oxidation–reduction potential (ORP) sensor.

Degree of Mixing: Process tanks must be equipped with mixers to ensure maximum contact between the reducing agent and the waste oxidizing solution. Proper mixing also homogenizes any solid precipitates that may be present, or that may form from side reactions, so that they can also be reduced if necessary. Mixing provides an even distribution of tank contents and a homogeneous pH throughout the waste, improving reduction of wastewater constituents. The quantifiable degree of mixing is a complex assessment that includes, among other factors, the amount of energy supplied, the length of time the material is mixed, and the related turbulence effects of the specific size and shape of the tank. This is beyond the scope of simple measurement.

pH: For batch and continuous systems, the pH affects the reduction reaction. The reaction speed is usually significantly reduced at higher pH values (typically above 4.0). It is worth noting that some reduction reactions may proceed better under alkaline conditions, in which case pH must be properly controlled to the appropriate alkaline range. For a batch system, the pH can be monitored intermittently during treatment. For a continuous system, the pH must be continuously monitored because it affects the ORP reading.

Reduction Temperature: Temperature affects the rate of reaction and the solubility of the reactants in the waste. As the temperature is increased, the solubility of the reactants, in most instances, is increased and the required residence time, in most cases, is reduced.

2.10.4 Chromium Reduction

The major application of chemical reduction involves the treatment of chromium wastes.

Sulfur dioxide, sodium bisulfite, sodium metabisulfite, and ferrous sulfate form strong reducing agents in aqueous solution and are often used in industrial waste treatment facilities for the reduction of hexavalent chromium to the trivalent form. The reduction allows removal of chromium from solution in conjunction with other metallic salts by alkaline precipitation. Hexavalent chromium is not precipitated as the hydroxide.

Gaseous sulfur dioxide is a widely used reducing agent and provides a good example of the chemical reduction process. Reduction using other reagents is chemically similar.

The reaction is favored by low pH. A pH of from 2 to 3 is normal for situations requiring complete reduction. At pH levels above 5, the reduction rate is slow. Oxidizing agents such as dissolved oxygen and ferric iron interfere with the reduction process by consuming the reducing agent.

A typical treatment consists of 45 minutes retention in a reaction tank. The reaction tank has an electronic recorder–controller device to control process conditions with respect to pH and oxidation reduction potential (ORP). Gaseous sulfur dioxide is metered to the reaction tank to maintain the ORP within the range of 250 to 300 millivolts. Sulfuric acid is added to maintain a pH level of from 1.8 to 2.0. The reaction tank is equipped with a propeller agitator designed to provide approximately one turnover per minute.

Chromium reduction is most usually required to treat electroplating and metal surfacing rinse waters, but may also be required in nonferrous metals manufacturing plants. A study of an operational waste treatment facility chemically reducing hexavalent chromium has shown that a 99.7% reduction efficiency is easily achieved. Final concentrations of 0.05 mg/ℓ are readily attained, and concentrations of 0.01 mg/ℓ are considered to be attainable by properly maintained and operated equipment.

The major advantage of chemical reduction to reduce hexavalent chromium is that it is a fully proven technology based on many years of experience. Operation at ambient conditions results in minimal energy consumption, and the process, especially when using sulfur dioxide, is well suited to automatic control. Furthermore, the equipment is readily obtainable from many suppliers, and operation is straightforward.

One limitation of chemical reduction of hexavalent chromium is that for high concentrations of chromium, the cost of treatment chemicals may be prohibitive. When this situation occurs, other treatment techniques are likely to be more economical. Chemical interference by oxidizing agents is possible in the treatment of mixed wastes, and the treatment itself may introduce pollutants if not properly controlled. Storage and handling of sulfur dioxide is somewhat hazardous.

Maintenance consists of periodic removal of sludge; the frequency of removal depends on the input concentrations of detrimental constituents.

Pretreatment to eliminate substances which will interfere with the process may often be necessary. This process produces trivalent chromium which can be controlled by further treatment. However, small amounts of sludge may be collected as the result of minor shifts in the solubility of the contaminants. This sludge can be processed by the main sludge treatment equipment.

The reduction of chromium waste by sulfur dioxide or sodium bisulfite is a classic process and is used by numerous plants which have hexavalent chromium compounds in wastewaters from operations such as electroplating, conversion coating and noncontact cooling.

Because of its abundancy and low cost, ferrous sulfate ($FeSO_4$) is frequently used for chromium reduction. For rapid reduction, $FeSO_4$ requires the pH of the chromate waste stream to be between 2 and 3.

Subsequent neutralization results in large volumes of iron hydroxide sludges, which makes disposal costly. An alkaline ferrous sulfate reduction process has been tried. Because this process requires maintaining the pH between 7 and 10, it can be accomplished in the neutralization/precipitation tankage. One deficiency of this alkaline process is the difficulty encountered in accurately controlling ferrous sulfate additions to the wastewater.

2.10.5 Other Inorganic Reduction Processes

Other uses for chemical reduction include oxidizing wastes containing reducible organics and inorganic oxidizers such as calcium hypochlorite, hydrogen peroxide (and other peroxides), and nitric acid. Because this technology frequently requires that the pH be in the acidic range, it would not normally be applicable to wastes that contain significant amounts of cyanide or sulfide. In such wastes, lowering of the pH can result in the release of toxic gases such as hydrogen cyanide or hydrogen sulfide. Hexavalent selenium can be reduced to elemental selenium.

Hexavalent chromium is usually present in wastes from the plating industry, metal surface preparation processes, the chromium pigments industry, and leather tanning processes. Selenates are frequently found in some mining and ore processing wastes. Organic and inorganic oxidizers are found in propellant explosives and in the chemical manufacturing industries.

One other common reduction process is the application of sodium borohydride to reduce metals in waste streams. Sodium borohydride is a mild but effective reducing agent, and is currently used in some chlor–alkali plants to reduce the soluble mercury ion to metallic mercury which is removed from solution by carbon adsorption.

Sodium borohydride is also reported to be effective in removing silver, mercury, gold, lead, and cadmium. However, this technology is only being applied in limited cases, the cost of the chemical being the major drawback.

Chemical reduction can be used for reduction of lead compounds in wastewater. Alkali metal hydrides are being used to recover silver from photographic waste effluents. Reduction processes are also applicable to chelated–metal–bearing wastes.

2.10.6 Reduction of Organics

Chemical reduction through the use of catalyzed metal powders and sodium borohydride has been shown to degrade toxic organic compounds. Reduction with catalyzed iron, zinc, or aluminum effect treatment through mechanisms such as hydrogenolysis, hydroxylation, saturation of aromatic structures, condensation, ring opening, and rearrangements to transform toxic organics to innocuous forms. Reductive dehalogenation of a variety of chlorinated organics, unsaturated aromatics, and aliphatics has been demonstrated in laboratory studies in which catalyzed metal powders were used.

In the pesticides industry, chemical reduction has been used to treat wastewaters containing an alkyl halide PAI (Pesticide Active Ingredient). The PAI is reduced with the addition of sodium bisulfite and ultraviolet light, i.e., sunlight.

Thermal gas–phase reduction of organic hazardous wastes in aqueous matrices is an alternative to incineration. The reaction is conducted in a hydrogen–rich reducing atmosphere with a complete absence of oxygen, resulting in virtually complete dechlorination of organic molecules and production of lighter recoverable hydrocarbons. After scrubbing the HCl from the gas stream, these hydrocarbons may be used as fuel in the boiler that preheats the waste. No chlorinated dioxins or furans are formed since the chlorine has been removed. ELI Eco Technologies is currently building a mobile system based on its patented thermochemical process as a demonstration project for the Canadian

Department of National Defense.

The Eco Logic system is a gas–phase thermochemical process which employs a reduction reaction of hydrogen with organic and chlorinated organic compounds at elevated temperatures. The reduction reaction breaks the large–chain molecules into less problematic hydrocarbons. Approximately 95% of the reformed gaseous product is recirculated back to the reactor, with the remaining 5% used to cofire a propane fired preheat boiler. The boiler stack emissions are not significant.

2.10.7 Amalgamation

The term amalgamation refers to two types of processes that can be used as treatment for mercury wastes. Neither amalgamation process significantly reduces the leachability of mercury; both processes only convert it into a more easily processed form. In both processes, a solid alloy of mercury and a base metal, such as zinc, is formed. This alloy can be subsequently processed by retorting to recover mercury.

The two processes differ from each other in the types of wastes managed. The first process is applicable only to solutions containing dissolved mercury salts. The principal current use for the process is in treatment of wastewaters containing organomercury salts. The second process is usable only for wastes rich in elemental mercury. Since the use of this second process is merely a convenience to avoid handling mercury in liquid form, its use is negligible. Both processes are applicable only to wastes containing mercury where selective recovery of the mercury is deemed viable.

The amalgamation processes depend on the ability of mercury to form low–melting–point solid alloys with metals such as copper and zinc, which have the thermodynamic capability of simultaneously reducing mercuric and mercurous salts to elemental mercury. Basically, an excess of the less noble metal (zinc or copper) is contacted with a waste containing mercury or mercury salts. The chemical reaction reducing the mercury in the mercury salts occurs and the elemental mercury liberated forms an alloy with the excess metal added.

2.10.8 High Temperature Metals Recovery (HTMR)

High temperature metals recovery (HTMR) is a technology applicable to wastes containing metal oxides and metal salts (including cadmium, chromium, lead, nickel, and zinc compounds) at concentrations ranging from 10 to 70% with low levels (i.e., below 5%) of organics and water in the wastes. There are a number of different types of high temperature metals recovery systems, which generally differ from one another in the source of energy used and the method of recovery. These HTMR systems include the rotary kiln process, the plasma arc reactor, the rotary hearth electric furnace system, the molten slag reactor, and the flame reactor.

HTMR is generally not used for mercury–containing wastes even though mercury will volatilize readily at the process temperatures present in the high temperature units. The retorting process is normally used for mercury recovery because mercury is very volatile and lower operating temperatures can be used. Thus, the retorting process is more economical than HTMR for mercury–bearing wastes.

The HTMR process has been demonstrated on wastes such as baghouse dusts and dewatered scrubber sludge from the production of steels and ferroalloys. Zinc, cadmium, and lead are the metals most frequently recovered. The process has not been extensively evaluated for use with metal sulfides. The sulfides are chemically identical to natural minerals ordinarily present in ores used as feedstocks by primary smelters. Some sulfide-bearing wastes from the chrome pigments industry have been sent to such primary smelters. However, with sulfides, a possibility exists for formation of either carbon disulfide from reaction with carbon or sulfur dioxide from reaction with oxygen in the HTMR processes.

Metal halide salts are also not directly used in HTMR processes. They, however, may be converted to oxides or hydroxides, which are acceptable feedstocks for HTMR processes.

The basic principle of operation for HTMR is that metal oxides and salts are separated from a waste through a high temperature thermal reduction process that uses carbon, limestone, and silica (sand) as raw materials. The carbon acts as a reducing agent and reacts with metal oxides to generate carbon dioxide and free metal. The silica and limestone serve as fluxing agents. This process yields a metal product for reuse and reduces the concentration of metals in the residuals and, hence, the amount of waste that needs to be land disposed.

The HTMR process consists of a mixing unit, a high temperature processing unit (kiln, furnace, etc.), a product collection system, and a residual treatment system.

The mixing unit homogenizes metal–bearing wastes, thus minimizing feed variations to the high temperature processing unit. Before the wastes are fed into the high temperature processing unit, fluxing agents and carbon can be added to the mixing unit and mixed with the wastes. The fluxes used (sand and limestone) are often added to react with certain metal components, preventing their volatilization and resulting in an enhanced purity of the desired volatile metals removed.

The blended waste materials are fed to a furnace, where they are heated to temperatures ranging from 1100° to 1400°C (2012° to 2552°F), resulting in the reduction and volatilization of the desired metals. The combination of temperature, residence time, and turbulence provided by rotation of the unit or addition of an air or oxygen stream helps ensure the maximum reduction and volatilization of metal constituents.

The product collection system can consist of either a condenser or a combination condenser and baghouse. The choice of a particular system depends on whether the metal is to be collected in the metallic form or as an oxide. Recovery and collection are accomplished for the metallic form by condensation alone, and for the oxide by reoxidation, condensation, and subsequent collection of the metal oxide particulates in a baghouse. There is no difference in these two types of metal recovery and collection systems relative to the kinds of waste that can be treated; the use of one system or the other simply reflects the facility's preference relative to product purity. In the former case, the direct condensation of metals allows for the separation and collection of individual metals in a relatively uncontaminated form; in the latter case, the metals are collected as a combination of several metal oxides.

The treated waste residual slag, containing higher concentrations of the less–volatile metals than the untreated waste, is sometimes cooled in a quench tank and (a) reused directly as a product (e.g., a waste residual containing mostly iron can be reused in steelmaking); (b) reused after further processing (e.g., a waste residual containing oxides of iron, chromium, and nickel can be reduced to metallic form and then recovered for use in the manufacture of stainless steel); (c) stabilized (material has no recoverable value) to immobilize any remaining metal constituents and then land disposed; or (d) directly land disposed as a slag.

The high temperature metals recovery (HTMR) process may be applicable to mixed wastes if the volatility of the radioactive metal component is significantly different from that of the nonradioactive portion. This technology is similar to distillation of organics, but instead "distills" more volatile inorganic components from less volatile ones.

A process has been developed by Environscience, Inc.

2.10.9 Nitrogen Oxides Reduction

Dry NO_x removal techniques involve the reduction of NO_x into molecular nitrogen and other nonpolluting gases, water and/or carbon dioxide, depending on which process is used. These products are exhausted from the NO_x removal process as a gas and do not normally require additional treatment. In some situations, the quantity of unreacted additives or the incomplete conversion of carbon monoxide into carbon dioxide may warrant additional treatment of the flue gas.

In order to accomplish NO_x reduction, a reducing agent is mixed with the flue gas and induced to react with the NO_x present in the flue gas. Consequently, a key element in each technology is the degree of dispersion and mixing of the reductant in the flue gas.

The **selective catalytic reduction process** was developed nearly 30 years ago. In this process, ammonia (NH_3) gas is reacted with NO_x in the presence of a catalyst to form molecular nitrogen (N_2) and water. The process is referred to as "selective," because the ammonia preferentially reduces the NO_x to N_2 rather than participating in other reactions with other flue–gas constituents. The process takes place inside a reactor that contains the catalyst. For maximum reduction of NO_x, the ammonia must be evenly distributed throughout the gas flow. To accomplish this, an injection grid is used, which injects precisely controlled amounts of ammonia through a bank of small but numerous nozzles. The ammonia is then normally mixed in the flow with baffles, turns, etc. Gas–flow distributors are used to provide a uniform flow through the catalyst.

Originally, this process required a narrow temperature band, but improvements have now made it possible to operate between approximately 500° and 800°F. Below 500°F, the catalyst activity is too low, and above 800°F the catalyst can be damaged by sintering. It has also been reported that at higher temperatures the process may actually increase the amount of NO_x present. Other important process parameters include initial NO_x concentration, removal efficiency required, amount of catalyst in the reactor, residence time in the presence of the catalyst, and the molar ratio of ammonia to nitrogen oxides (NH_3/NO_x).

Selective noncatalytic reduction was developed approximately 20 years ago. As it was originally developed, this technology was similar to the selective catalytic reduction process. Ammonia was injected through an injection grid into the off–gas to react with NO_x to reduce it to nitrogen and water. As with selective catalytic reduction, this process relies upon temperature control and effective reductant dispersion in the flue gas to maximize NO_x removal and to minimize releases of unreacted reductant. However, unlike selective catalytic reduction, this process does not utilize a catalyst and must be operated at a higher temperature, typically 1600° to 2200°F, to provide sufficient energy for the reaction.

As the technology has matured, variations of the process have evolved. Lower operating temperatures can be utilized by injecting hydrogen gas. Other reductants besides ammonia are now available. For example, urea can be used, as it will break down when heated to form the same type of reactive species as ammonia. Urea can be mixed with water and handled as a liquid, which typically requires less equipment. Liquid injection is also reported to have smaller energy requirements. The disadvantage of using urea is it may increase carbon monoxide emissions in the flue gas by 5 to 25 ppm.

One of the newer NO_x reduction technologies being investigated is a **nonselective noncatalytic reduction process** also referred to as reburning, methane injection, or natural gas injection. This process is similar to selective noncatalytic reduction except a less selective reductant is used. As such, this process is referred to as nonselective, as the reductant will not preferentially reduce NO_x but can easily react with other combustion gases, especially oxygen.

In nonselective noncatalytic reduction, natural gas or some other combustible gas is injected into the flue gas. This creates a reducing environment where the NO_x is reduced to molecular nitrogen and water or carbon monoxide. If other nitrogen compounds, such as ammonia or cyanide are present, the nitrogen in these compounds can also be converted to molecular nitrogen. After sufficient mixing and time to react, air is injected into the flue gas to burn the remaining combustible gases.

In **nonselective catalytic reduction,** a nonselective reductant, such as methane, hydrogen, or some other gaseous fuel, is added to the flue gas and then passed over a catalyst at approximately 570° to 850°F. The nonselective nature of the process means that the reductants will react with oxygen, so extra reductant is needed. The reaction of the reductant with oxygen also generates excess heat, resulting in a temperature rise of approximately 270°F for every percent of oxygen in the flue gas. Therefore, if the flue gas has 2 to 3% oxygen, multiple reactors may be necessary to effectively remove the heat.

In one variation of this technology that is commercially available, nonselective catalytic reduction is combined with other technologies to make a hybrid process.

2.11 SCRUBBING/ABSORPTION

Absorption can be either a physical or chemical process. Physical absorption as applied to scrubbing of a gas with a solvent is discussed in Chapter 6. Chemical absorption where the absorbed (or dissolving) gas reacts with a material in a scrubbing operation is discussed in this section.

2.11.1 Sulfur Dioxide

Wet Scrubbing: The dominant type of Flue Gas Desulfurization (FGD) in the U.S. today is wet scrubbing based on either limestone or lime. Lime scrubbing is similar to limestone processes except that the feed slurry is prepared by slaking lime rather than by grinding limestone. High (over 90%) SO_2 removals are easier to obtain with lime than with limestone and more complete sorbent utilization is realized, but the higher cost of the raw sorbent offsets some of the advantages. The sorbent slurry is recirculated between an absorber (typically a spray tower) and a hold tank where fresh sorbent is added, and a bleed stream of reaction products is pumped to the waste disposal system or settling pond. The outlet gas from the scrubber passes through a slurry entrainment separator ("demister") and is reheated for buoyancy before discharge to the stack.

The reaction products from the first generation of installed scrubbers form a $CaSO_3/CaSO_4$ sludge that is difficult to dewater, although fixation techniques involving fly ash and/or proprietary additives have been developed to give the waste stability. Second generation scrubbers have greatly reduced this problem by operating in either an inhibited or forced oxidation mode. The inhibited oxidation mode of operation uses colloidal sulfur to retard calcium sulfite oxidation resulting in larger product crystals, which improves dewatering of the product. In the forced oxidation mode, the calcium sulfite is oxidized to produce calcium sulfate (gypsum), which can be dewatered and landfilled or, in some cases, sold as a by-product.

Sodium-based systems use a scrubbing solution that avoids many of the operational problems that have hampered lime/limestone scrubbers, particularly erosion and scaling. Sulfur dioxide removals well over 90% are easily achieved. Typically, sodium carbonate is dissolved in a mixing tank and pumped to the absorber vessel. A bleed stream of scrubber effluent removes reaction products such as sodium sulfite and sodium bisulfite. This stream is sent to an evaporation pond for disposal. Due to the soluble nature of the products, groundwater contamination may be a concern.

An important variation on sodium scrubbing is the double-alkali process. It combines the high removals and good operability of the sodium process with the waste disposal characteristics of a lime/limestone scrubber. In this case, the bleed stream is pumped to a reactor tank where it is blended with slaked lime or finely ground limestone. Calcium sulfite and sulfate crystals are formed and removed for disposal while the sodium sulfite scrubbing liquor is regenerated.

A magnesium oxide scrubbing process is used by the Philadelphia Electric Company. Reagents used in other processes include sodium sulfite, ammonium sulfite, fly ash/alkali, sodium carbonate, elemental sulfur/Glauber's salt, hydroxide, and calcium silicate.

There are four types of wet acid-gas removal devices—packed-bed, tray, wet fluidized-bed, and spray scrubbers. Acid-gas absorption is more efficient when the gases are saturated. Thus, a quencher is used to saturate the gases upstream of the acid removal device. This quencher also serves to cool the gases and eliminate temperature problems. Because of the moisture involved, mists are frequently produced; consequently, mist eliminators are used at the exit or as a second stage.

To contact the liquid caustic reagent with the off-gas, the gas flow can be countercurrent to the liquid (i.e., in the opposite direction), crossflow to the liquid (i.e., perpendicular to the liquid flow), or cocurrent (i.e., in the same direction). Countercurrent flow, the most common configuration, has larger concentration gradients that increase the degree of acid absorption into the liquids and, therefore, increase the degree of acid conversion to a salt. Crosscurrent flow is also effective and is frequently used. Cocurrent flow is rarely used.

It is important to maintain the proper pH of the caustic liquor. If the liquor pH is below 7, then acid gases may be passing through the unit. If the pH is approximately 10 or higher, carbon dioxide may be reacting with and unnecessarily consuming the caustic reagent. To minimize the use of the caustic reagent and maximize the conversion of the acids, it is good practice to keep the pH between 7 and 9.

Wet acid–gas removal devices will accumulate dissolved and suspended solids. To keep these solids from being entrained by the gas flow, liquids must be periodically removed from the unit and replaced with fresh water. The liquid blowdown may be high in salts and, in the case of radioactive waste treatment, may also contain some nuclides, including tritiated water that condenses in the unit.

Most wet acid–gas removal devices can be fouled by particulate, so the devices are normally used after the particulate has been removed.

Dry Scrubbing: Dry acid–gas removal technology has been developed in an effort to lower the cost of operations as well as secondary waste disposal. Less costly alkali reagents, such as limestone, can be used without the need of alkali preparations, such as lime slaking, which provides savings in both initial and operating costs. Secondly, because a dry secondary waste is generated, there is no need for wastewater treatment facilities or drying equipment. The result is additional savings in initial costs and operating costs.

In the **dry sorbent injection (DSI) process,** a dry powder sorbent, (i.e., an alkali reagent) is injected in the off–gas at a point where the sorbent reacts with the acid gases that are present. A common method of injection is via pneumatic transport through a nozzle, but mechanical feeders, such as augers, are also frequently used. For this process to perform as required, the flue gas velocity must be sufficient to carry the dry powder in the gas stream. It is also important that the powder be uniformly dispersed in the flue gas. In addition, because the alkali reagent is a solid carried by the flue gas, the process must be combined with an effective particulate removal device that is downstream from the sorbent injection point.

The possible locations to inject the alkali range from injection into the thermal unit to injection into the flue gas duct upstream of gas cooling to injection in the particulate removal device. Because of this flexibility, there is a great variety in process parameters that are used in DSI. In general, the process is more effective when the gas stream is cooled to approximately 250° to 300°F and the gas stream is nearly saturated. It is important, however, to keep the temperature above the dew point and not let the gases become saturated, as wet surfaces from condensation will cause problems, such as cake buildup and corrosion.

Dry fluidized–bed scrubbing can be separated from the combustion unit by using a separate vessel for the fluidized–bed scrubber. By using a separate vessel, dry fluidized scrubbing can be used for all types of thermal systems. In this manner, optimum conditions for combustion can be used in the thermal device, and optimum conditions for dry scrubbing can be used in the fluidized scrubber.

In a fluidized–bed system, gas is passed through a bed of small, solid particles. The velocity of the gas through the solids bed is sufficient so that the drag force on the particles is greater than the force of gravity on the particles. As a result, the bed will expand, and the particles will move through the bed, carried by the gas flow. Under these conditions, the bed will act like a fluid, taking the shape of the container and leveling out on the surface rather than forming a pile of solids.

There are different types of fluidization depending on the gas velocity through the bed. For slower velocities (i.e., slightly above the point of minimum fluidization), the particles will remain close together, and the bed will have a boiling motion as small bubbles of gas pass through the bed. This dense–phase fluidization, referred to as a bubbling bed, allows the particles to move around within the bed.

Semidry Scrubbing: There is one technology that falls in the category of semidry scrubbers, namely **spray dryer absorption.** This technology is referred to as semidry because a wet caustic feed material is used but the residue produced is dry. The dry residue has the advantage of being easier to dispose in comparison to a wet residue, which is a major reason for the development and use of this technology.

Spray dryer absorption utilizes a cylindrical reactor vessel in which the flue gas and a liquid alkali are intimately mixed to form a salt. There are many variations of this technology using cocurrent or countercurrent designs in up– or down–flow arrangements. Some reactors have one gas entrance and others have multiple gas entrances. Semidry acid–gas removal is similar to dry sorbent injection in that it produces a dry secondary waste. As such, semidry acid removal also requires a particulate removal device.

In the spray dryer absorption process, water and alkali reagent are mixed to form a solution. The solution is then injected into the reactor as a finely atomized spray. Acid gases present in the flue gas are absorbed into the small liquid droplets and reacted with the alkali reagent to form a salt. As the droplets pass through the reactor, heat from the flue gas evaporates the water, cooling the flue gas and forming solid particles of salt, unreacted alkali, and possibly fly ash. The dry solids are then collected downstream of the reactor in a particulate removal device.

The liquid scrub solution is normally atomized in one of two ways: a two–fluid nozzle or a rotary atomizer. The two–fluid nozzle is advantageous in that it has a large scrub liquor passageway that is not easily plugged by solids. These nozzles use a second fluid, either steam or air, at high pressure to atomize the scrub liquor. A second advantage of these nozzles is that they can be easily serviced without cooling the reactor by simply unbolting a flange and removing the spray nozzle assembly.

Combustion Cleaning: Fluidized–bed combustion, for example, reduces emissions by controlling combustion parameters and by injecting a sorbent or a pollutant absorbent (such as crushed limestone), into the combustion chamber along with the coal. Pulverized coal mixed with crushed limestone is suspended on jets of air (or fluidized) in the combustion chamber. As the coal burns, sulfur is released, and the limestone captures the sulfur before it can escape from the boiler. The sulfur chemically combines with the limestone to form a new solid waste product, a mixture of calcium sulfite and calcium sulfate. Some of the solid waste is removed with the bed ash through the bottom of the boiler. Small ash particles, or fly ash, that escape the boiler is captured with dust collectors (cyclones and baghouses). More than 90% of the sulfur released from coal can be captured in this manner.

The fluidized mixing of fuel and sorbent enhances both the coal–burning and sulfur–capturing processes and allows for reduced combustion temperatures of 1400° to 1600°F, or almost half the temperature of a conventional boiler. This temperature range is below the threshold where most of the NO_x forms. Thus, fluidized–bed combustors substantially reduce both SO_2 and NO_x emissions.

Fluidized–bed combustors can be either atmospheric or pressurized. The atmospheric type operates at normal atmospheric pressure while the pressurized type operates at pressure 6 to 16 times higher than normal atmospheric pressure. The pressurized fluidized–bed boiler offers potentially higher efficiency, reduced operating costs, and less waste products than does the state–of–art atmospheric fluidized–bed boiler. A new type of atmospheric fluidized–bed boiler offers circulating (entrained) fuel flow instead of the bubbling bed or fixed bed used in earlier approaches. Circulating fluidized beds allow for finer coal feed, better fuel mixing, higher efficiencies, and increased SO_2 capture.

2.11.2 Nitrogen Oxides

Although NO_x is an acid gas, relatively poor NO_x removal is obtained using traditional acid–gas removal techniques. The reason for this poor removal is the low solubility of NO. For most thermal systems, NO makes up approximately 90% of the NO_x. Because the NO has a low solubility in water, it is not readily absorbed from the gas phase into the liquid phase and, therefore, will not readily react with the caustic scrubbing solutions.

To alleviate this problem, the NO must be converted to NO_2 which is more soluble in water, or a scrubbing agent that can collect NO must be used. These are the approaches that are used in wet NO_x abatement. However, the processes of this type that have been commercialized are proprietary and may utilize proprietary oxidizers and scrubbing solutions. Proprietary additives may also be used. These proprietary agents may have specialized requirements for equipment that are also considered proprietary. The confidential nature of these processes makes it very difficult to ascertain whether anything else is needed to make the process work, what conditions are required for successful operation, and whether this technology is more appropriate for particular applications.

When using oxidation and absorption, an oxidizer, such as ozone or hypochlorite, is first mixed with the flue gas. This oxidizer will convert the NO to NO_2, which can then be absorbed in an acid gas scrubber. Depending on the process, it may be possible to recover nitric acid from the scrubber and use this elsewhere, or the NO_2 can be scrubbed with a caustic solution (such as sodium hydroxide) and converted to a nitrate salt.

Even though the NO_2 is soluble, it is not as soluble as other acid gases such as HCl. For this reason, different parameters are used when scrubbing NO_2 compared to other gases. The gas velocity may be slower (i.e., larger scrubber diameter), the packing section may be much longer (i.e., deeper beds or multiple scrubbers), and higher liquid–to–gas ratios may be used.

As with the oxidation/absorption process, scrubbing with **chelating agents** was developed in order to overcome the extremely low solubility of NO, which is the major form of NO_x from most thermal systems. In this process, a caustic scrubber is used as in other liquid scrubbers; however, a chelating agent is also added to the scrubbing liquor. The chelating agent will form a complex with NO so that the NO will pass from the gas stream to the liquid stream. Once in the liquid solution, the NO will react with the caustic agent in the same manner as other acid gases. The chelating agent is then free to complex with another NO molecule so the process can be repeated.

2.11.3 Others

Hydrogen Sulfide Scrubbing: Hydrogen sulfide in off–gases can be scrubbed with a number of chemicals in various chemical absorption processes:

1. Sodium carbonate
2. Limestone
3. Dolomite
4. Cobalt titanate and zinc oxide (Illinois Gas Institute)
5. Sodium carbonate, vanadium salt, and a catalyst (disodium salt of

anthroquinone disulfonic acid) (Stretford)
6. Zinc ferrite (Kellogg–Rust–Westinghouse)
7. Mixed lithium, sodium, potassium and calcium carbonates (Battelle PNL)
8. Sodium carbonate and naphthaquinone sulfonic acid (Takahax)
9. Iron salt chelated with an organic compound, plus buffering salts (LO–CAT)
10. Iron chelate process (Shell/Dow)
11. Oxides of zinc, iron, vanadium and copper (Hampton Univ.)

Mercury Scrubbers: By reacting mercury with chemicals such as sodium hypochlorite or with a chelating agent and cupric chloride, water–soluble species of mercury are formed. As such, mercury can be removed from the flue gas using conventional wet scrubbing technologies such as packed beds or spray towers. In this manner, mercury removal efficiencies of 90 to 95% have been obtained. While this technology is relatively new and little information is known, it has excellent potential and may also be useful in removing other pollutants.

Active Metals Scrubbing: Active metals scrubbing is a process which uses sodium, zinc, or aluminum metal to rapidly react with halogenated compounds. It has only been commercialized for destruction of PCB–containing wastes.

REFERENCES

1. Ainsworth, S., *Resurgence in Demand Reving Market for Sodium Chlorite,* C&EN, 3/22/93.
2. Arienti, M., et al, *Dioxin–Containing Wastes,* Noyes Data, 1988.
3. Bergos, B., *Control of Organic Substances in Water and Wastewater,* Noyes Data, 1987.
4. Berkowitz, J., et al, *Unit Operations for Treatment of Hazardous Industrial Wastes,* Noyes Data, 1978.
5. Breton, M., et al, *Treatment Technologies for Solvent Containing Wastes,* Noyes Data, 1988.
6. Burton, D., et al, *Treatment of Hazardous Petrochemical and Petroleum Wastes,* Noyes Data, 1989.
7. Chambers, C., et al, *In Situ Treatment of Hazardous Waste Contaminated Soils,* Noyes Data, 1991.
8. Cushnie, G., *Electroplating Wastewater Pollution Control Technology,* Noyes Data, 1985.
9. Dalton, J., *An Assessment of Off–Gas Treatment Technologies for Application to Thermal Treatment of Department of Energy Wastes,* DOE/MWIP–1, 9/92.
10. EPA, *Corrective Action: Technologies and Applications,* EPA/625/4–89/020, 9/89.
11. EPA *Development Document for Effluent Limitations Guidelines and Standards for the Inorganic Chemicals Manufacturing Point Source Category,* EPA 440/1–82/007, 6/82.
12. EPA, *Development Document for Effluent Limitations Guidelines and Standards for the Nonferrous Metals Manufacturing Point Source Category,* EPA 440/1–89–019.1, 5/89.
13. EPA, *Development Document for Effluent Limitations Guidelines and Standards for the Organic Chemicals, Plastics, and Synthetic Fibers Point Sound Category,* EPA–440/1–87/009, 10/87.
14. EPA, *Forum on Innovative Hazardous Waste Treatment Technologies: Domestic and International,* EPA/540/2–89/056, 9/89.
15. EPA, *Forum on Innovative Hazardous Waste Treatment Technologies: Domestic and International (Second),* EPA/540/2–90/009, 9/90.
16. EPA, *Forum on Innovative Hazardous Waste Treatment Technologies: Domestic and International (Third),* EPA/540/2–91/015, 9/91.
17. EPA, *Innovative Technology/Glycolate Dehalogenation,* 9200 5–254FS, 11/89.
18. EPA, *Innovative Treatment Technologies, Overview and Guide to Information Sources,* EPA/540/9–91/002, 10/91.
19. EPA, *In Situ Treatment of Contaminated Ground Water,* OSW, 9/16/92.
20. EPA, *Proposed Technical Development Document for the Pesticide Chemicals Manufacturing Category Effluent Limitations Guidelines, Pretreatment Standards, and New Source Performance Standards,* EPA 821 R–92–005, 4/92.

21. EPA, *Risk Reduction Engineering Laboratory Research Symposium, (Eighteenth),* EPA/600/R–92/028, 4/92.
22. EPA, *Summary of Treatment Technology Effectiveness for Contaminated Soil,* EPA 9355.4–06, 6/90.
23. EPA, *Synopses of Federal Demonstrations of Innovative Site Remediation Techniques, Second Edition,* EPA/542/B–92/003, 8/92.
24. EPA, *The Superfund Innovative Technology Evaluation Program, Technology Profiles,* EPA/540/5–90/006, 11/90.
25. EPA, *Synopses of Federal Demonstrations of Innovative Site Remediation Technologies, 2nd Edition,* EPA/542/B–92/003, 9/92.
26. EPA, *Treatment Technology Background Document,* OSW, 1/91.
27. Freeman, H., *Innovative Thermal Hazardous Organic Waste Treatment Processes,* Noyes Data, 1985.
28. Holden, T., *How to Select Hazardous Waste Treatment Technologies for Soils and Sludges,* Noyes Data, 1989.
29. *Ion Exchange: A New Sphere of Action,* ChEng, 9/92.
30. Jackman, A., *Hazardous Waste Treatment Technologies,* Noyes Data, 1991.
31. Jamshidi, M., *Environmentally Conscious Manufacturing,* Rio Grande Technology Foundation, 1992.
32. Krishnan, E., et al, *Recovery of Metals from Sludges and Wastewaters,* Noyes Data, 1993.
33. Meltzer, M., et al, *Metal–Bearing Waste Streams,* Noyes Data, 1990.
34. Noyes, R., *Handbook of Pollution Control Processes,* Noyes Data, 1991.
35. Nunno, T., et al, *International Technologies for Hazardous Waste Site Cleanup,* Noyes Data, 1990.
36. Palmer, S., et al, *Metal/Cyanide Containing Wastes,* Noyes Data, 1988.
37. Roberts, R., et al, *Hazardous Waste Minimization Initiation Decision Report,* Naval Civil Engineering Laboratory, 6/88.
38. Robinson, S., *Hydrogen Peroxide: First Aid for Air Pollution,* Nat. Env. Jnl. 5–6/93.
39. Suprenant, T.N., et al, *Halogenated–Organic Containing Wastes,* Noyes Data, 1988.
40. Unterberg, W., *How to Respond to Hazardous Chemical Spills,* Noyes Data, 1988.
41. Wagner, K., et al, *Remedial Action Technology for Waste Disposal Sites,* Noyes Data, 1986.
42. Wilk, L., et al, *Corrosive–Containing WAstes,* Noyes Data, 1988.

3

Containment and Barrier Technology

This chapter includes those containment and barrier concepts utilized to prevent the migration of wastes from hazardous waste landfills, surface impoundments, and waste piles; as well as from municipal waste landfills. There are regulatory requirements for design of these facilities.

Also discussed are barriers used to contain wastes or contamination in the cleanup of hazardous waste sites. Additional information will be found regarding natural underground barriers, and barriers for contaminated dredged material.

3.1 HAZARDOUS WASTE FACILITIES

On November 8, 1984, Congress enacted the Hazardous and Solid Waste Amendments (HSWA to the Resource Conservation and Recovery Act (RCRA), placing stringent new requirements on the land disposal of hazardous waste. Among other requirements, Congress amended section 3004 of RCRA and added section 3015 to impose specific design standards for land disposal units. It has been amended since that date.

A final rule was issued on 1/29/92 (57FR3462) that modifies the existing double-liner and leachate collection and removal system requirements for new and replacement surface impoundments and landfills and for lateral expansions of these units, including those units at interim status facilities.

New surface impoundment and landfill units for which construction commences after January 29, 1992, and replacement units reused after and lateral expansions of existing units for which construction commences after July 29, 1992 must have a double liner consisting of a top liner designed to prevent the migration of hazardous constituents into the liner during the active life and post-closure period (e.g., a geomembrane), and a composite bottom liner consisting of a geomembrane underlain by at least three feet of compacted soil material having a hydraulic conductivity of no more than 1×10^{-7} cm/sec.

EPA is also extending the revised landfill double-liner and leachate collection and removal system requirements to new waste pile units for which construction commences after January 29, 1992, and replacement units reused after and lateral expansions of waste pile units for which construction commences after July 29, 1992.

Under current RCRA and CERCLA regulations, there are four distinct waste containment systems: surface impoundments, waste piles, landfills, and on–site isolation. Each containment system is built of components with distinct engineering functions, e.g., moisture barrier, reinforcement, drainage, etc. In turn, each component is composed of elements, i.e., individual materials or products, that have particular field inspection requirements.

Surface Impoundments: These are basins used to store or dispose of primarily liquid wastes. If the system is planned to be removed after the operating life and the site cleaned of all contamination, then it is considered a storage unit. If the waste is stabilized, the free liquid is removed, and the system is closed and monitored, then it is a disposal unit. Surface impoundments can include liners, leachate collection systems, leachate detection systems, and gas collection systems. If the facility is designed as a disposal unit, then a closure system is necessary.

Landfills: These are final disposal units for solid and hazardous wastes. Landfills have the same components and elements as the surface impoundment disposal units.

Waste Piles: These are structures in which waste can be treated and/or stored temporarily. A waste pile system must have a similar bottom liner system as the surface impoundment but will not have a final cover since it is only a temporary structure. However, waste piles must be covered by a structure which keeps precipitation, wind and surface water run–on away from the waste. Typically, these protective structures are simple metal buildings, although other protective cover may be used. For example, a geomembrane can be used to cover the waste.

On–Site Waste Isolation: Waste isolation systems include caps built over waste to minimize infiltration of rainwater, and both horizontal barriers under the waste and vertical barriers at the lateral extent of the waste to prevent uncontrolled release of leachate. These are systems which are constructed on remediation sites to isolate and allow for the treatment of an uncontained waste.

Waste facility caps reduce water infiltration, control gas and odor emissions, improve the aesthetics, and provide a stable surface over the waste. A composite barrier capping system is required for the closure of hazardous waste storage facilities. A hardened cap is typically required in an arid climate where a vegetative cover will not survive, in urban areas where vegetation may be undesirable, or at industrial facilities where it would be advantageous to continue using the site. The hardened cap integrates the vegetative layer, protective layer (biotic) and drainage layer into one layer. The hardened cap can be constructed using "hard" elements, such as graded stone, asphalt, or concrete. The use of a hardened surface layer does not eliminate the need for the geomembrane/clay moisture barrier components in the cap.

Horizontal barriers are installed below an existing waste mass to contain the waste and prevent the movement of contaminate into the surrounding soil and water. Horizontal barriers are very difficult to inspect due to the overlying waste.

Horizontal barrier techniques involve the injection of grout under the waste using one of the following methods:

1. Vertical borings drilled through the waste and pressure injection or rotary jetting of the grout beneath the waste.
2. Horizontal borings drilled below the waste from trenches and pressure injection of the grout beneath the waste.

3. Block displacement method which surrounds the waste with a vertical grout wall and then injects low pressure grout beneath the waste. The entire waste block is raised in this process.

All of these methods require that the borings be spaced close enough together that the grout bulbs or lenses overlap and form a continuous barrier. Verification of the overlap is critical but very difficult. Potential inspection methods are limited to test excavations, exploratory borings, and observed impact on ground water if the barrier is placed beneath the ground water table.

Vertical barriers are wall–like systems used to isolate any contaminates that have leached from the waste and that are moving laterally. To be effective, these barriers should intercept a continuous impervious horizontal layer below the waste. This bottom layer can be a naturally occurring layer such as an aquiclude, or a horizontal barrier. Several types of vertical barriers are commonly used, including:

1. **Slurry wall:** A trench surrounding the waste, filled with a soil bentonite and/or concrete–bentonite slurry.
2. **Grout curtain:** Grout is injected in a series of vertical columns that surround the waste, creating a continuous curtain.
3. **Geomembrane curtain:** Interlocking geomembrane panels are placed in a vertical trench surrounding the waste. In some cases the geomembrane is used in conjunction with the slurry wall to form a composite liner system.

3.1.1 Bottom Containment Designs

The basic bottom liner design, for hazardous waste landfills, is two or more liners with a leachate collection system above and between the liners. The redundancy aspect of the design is that if the top liner does not perform as designed, then the second leachate collection system will alert appropriate personnel while corrective actions are implemented. The bottom liner in this design is assumed to contain the waste until the corrective action is in place. The design was reviewed and modeled in saturated and unsaturated hydraulic flow conditions. The result of these studies is the current recommended design of a double liner which has a bottom composite liner and a top geomembrane. The composite bottom liner is one that consists of a geomembrane in intimate contact with a compacted, low permeability natural soil. The composite liner design has been determined to be more hydraulically efficient than the geomembrane or natural soil liner working independently.

The liner system currently being used by most hazardous waste management facilities incorporate in descending order a filter layer, followed by a primary leachate collection and removal system (LCRS), a primary geomembrane, a leak detection, collection and removal system (LCDRS), and a composite liner above the native soil foundation (EPA, 1987). The composite liner is defined as a geomembrane and a compacted, low hydraulic conductivity ($k \leq 1 \times 10^{-7}$ cm/sec) natural soil.

In bottom liner systems for construction and field seaming purposes, the geomembrane is to be at least 0.75 mm (30 mils) thick or 1.12 mm (45 mils) thick if left exposed to the elements for more than 30 days. These thicknesses may not be suitable for all geomembrane materials. The required geomembrane thickness will depend on the site–

specific design, installation/construction concerns, seamability, and long-term durability.

3.1.2 Top Cover System Designs

Proper closure is essential to complete a landfill. Research has established minimum requirements needed to meet the stringent, necessary, closure criteria for both hazardous and nonhazardous waste landfills in the United States. In designing the landfill cover, the objective is to limit the infiltration of water to the waste so as to limit creation of leachate that might possibly escape to groundwater sources.

The cover system must be devised at the time the site is selected and the plan and design of the landfill containment structure is chosen. The location, the availability of low-hydraulic conductivity soil, the stockpiling of good topsoil, the availability and use of geosynthetics to improve performance of the cover system, the height restrictions to provide stable slopes, and the use of the site after the postclosure care period are typical considerations. The goals of the cover system are to minimize further maintenance and to protect human health and the environment.

The closure of a hazardous waste landfill will normally have as its main criteria the minimization of moisture into the facility. Allowing moisture into a hazardous waste facility will subject the waste to leaching of potentially toxic pollutants into the leachate. Minimizing leachates in a closed waste management unit requires that liquids be kept out and that the leachate that does exist be detected, collected, and removed. Where the waste is above the groundwater zone, a properly designed and maintained cover can prevent (for practical purposes) water from entering the landfill and, thus, minimize the formation of leachate.

The current recommended design is a multilayered system consisting of, from bottom to top:

1. **A Low-Hydraulic Conductivity Geomembrane/Soil Layer:** A 60 cm 24 in) layer of compacted natural or amended soil with a hydraulic conductivity of 1×10^{-7} cm/sec in intimate contact with a minimum 0.5 mm (20 mil) geomembrane liner.
2. **A Drainage Layer:** A minimum 30 cm (12 in) soil layer having a minimum hydraulic conductivity of 1×10^{-2} cm/sec, or a layer of geosynthetic material having the same hydraulic characteristics.
3. **A Top, Vegetation/Soil Layer:** A top layer with vegetation (or an armored top surface) and a minimum of 60 cm (24 in) of soil graded at a slope between 3 and 5%.

Because the design of the final cover must consider the site, the weather, the character of the waste, and other site-specific conditions, these minimum recommendations may be altered. Design innovation is encouraged to meet site specific criteria. For example, in extremely arid regions, a gravel top surface might compensate for reduced vegetation, or the middle drainage layer might be expendable. Where burrowing animals might damage the geomembrane/low-permeability soil layer, a biotic barrier layer of large-sized cobbles may be needed above it. Where the type of waste may create gases, soil or geosynthetic vent structures would need to be included.

3.2 MUNICIPAL WASTE LANDFILLS

Municipal solid waste landfills are defined as publicly or privately owned landfills that receive household waste. In addition to household wastes, an MSWLF may also receive other types of RCRA Subtitle D wastes, such as commercial solid waste, nonhazardous sludge, small quantity generator waste, and industrial solid waste.

On October 9, 1991, EPA promulgated standards for new and existing municipal solid waste landfills (MSWLFs) under RCRA Subtitle D (56 *FR* 50978). The new rule establishes minimum national criteria for the location, design, operation, cleanup, and closure of MSWLFs under 40 *CFR* Part 258. States that obtain authorization for their own programs are allowed to exercise flexibility in implementing the new criteria. Owners/operators located in states without approved programs will have to strictly comply with the federal requirements.

The new rule also establishes six locational restrictions that deal with airport safety, floodplains, unstable areas, wetlands, seismic impact zones, and fault areas.

In addition a number of operating criteria were established, one of which states that six inches of daily cover must be provided.

Under Part 258, new MSWLF units and lateral expansions will be required to meet certain design criteria. The standards are not applicable to existing units. Two design criteria are presented, depending on whether the facility is located in an approved or unapproved state. A composite liner system will be required at landfills located in states without EPA approved permit programs. The liner system is designed to be protective in all locations and consists of a flexible membrane liner, a 2 foot compacted soil layer, and a leachate collection system.

In approved states, a site–specific performance standard is applicable. The performance standard requires that the design ensure that maximum contaminant levels (MCLs) will note be exceeded at the relevant point of compliance, which will be determined by the state.

When evaluating whether designs meet the performance standard, approved states must consider several site–specific factors, including climate, hydrogeology, and groundwater use.

3.2.1 Bottom Containment Designs

Liner systems for municipal solid wastes may have different designs based on site specific considerations including geology, hydrology and climatic conditions. Two basic approaches are used in the United States. The first is a generic design. This design has a composite liner system that is designed and constructed to maintain less than 30 cm (12 in) depth of leachate over the liner. The second approach based on performance consists of liners and leachate collection systems to ensure that the concentration values of selected chemicals will not be exceeded at some point on the owner/operator's property.

Generic Design: A composite liner is defined as consisting of two components; the upper component is a geomembrane with a minimum of 0.75 mm (30 mil) thickness, the lower component consists of at least a 60 cm (24 in) layer of compacted soil with a hydraulic conductivity less than or equal to a 1×10^{-7} cm/sec. The required geomembrane thickness will depend upon the site–specific design, installation/construction concerns, seamability and long–term durability. The geomembrane must be installed in direct and uniform contact with the compacted soil component so as to minimize the migration of leachate through potential defects in the geomembrane. A leachate collection and removal system (LCRS) should be located immediately above the composite liner to control the level of leachate on the liner.

Performance Based Design: The second design allows the owner operator of the proposed municipal solid waste landfill (MSWLF) to demonstrate that the design is protective of human health and the environment with respect to groundwater quality downgradient from the landfill. The nature of the demonstration is essentially an assessment of the landfill leachate characteristics, the potential for leakage from the landfill of that leachate to groundwater and an assessment of the anticipated fate and transport of those constituents to the proposed point of compliance at the facility. Inherent to this type of approach, is the need to obtain sufficient site specific data to adequately characterize the existing groundwater quality, the pre–existing groundwater regime (flow direction, horizontal and vertical gradients, hydraulic conductivity, specific yield and aquifer thickness). The assessment should consider the effects construction of the MSWLF may have on the groundwater system. The major consideration here, for shallow groundwater systems, is the local capturing of precipitation that normally would have infiltrated as a source of groundwater recharge. An assessment of leakage from the proposed liner and leachate collection design should be based on empirical data from other existing operational facilities of similar design that have the capability of leak detection monitoring. In lieu of the existence or availability of such information, analytical approaches based on conservative assumptions may need to be conducted to estimate anticipated leakage rates. Given known source concentrations, groundwater and soil parameters, and the hydraulic gradients, a simple and hopefully conservative assessment of downgradient concentration at specific times and distances from the source can be conducted. Either one dimensional or two dimensional advection/dispersion contaminant transport methods may be used. The analysis should be performed by qualified professionals and may entail hypothetical computer simulations of groundwater flow and transport.

3.2.2 Cover Systems for Nonhazardous Wastes

The cover system in nonhazardous waste landfills will be a function of the bottom liner system and the liquids management strategy for the specific site. If the bottom liner system contains a geomembrane, then the cover system should contain a geomembrane to prevent the "bathtub" effect. Likewise, if the bottom liner system is a natural soil liner, then the cover system barrier should be hydraulically equivalent to, or less permeable than the bottom liner system. A geomembrane used in the cover will prevent the infiltration of moisture to the waste below and may contribute to the collection of waste decomposition gases, therefore necessitating a gas collection layer.

There are at least two options to consider under a liquids management strategy, mummification and recirculation. In the **mummification** approach the cover system is designed, constructed, and maintained to prevent moisture infiltration to the waste below. The waste will eventually approach and remain in a state of "mummification" until the cover system is breached and moisture enters the landfill. A continual maintenance program is necessary to maintain the cover system in a state of good repair so that the waste does not decompose to generate leachate and gas.

The **recirculation** concept results in the rapid physical, chemical, and biological stabilization of the waste. To accomplish this, a moisture balance is maintained within the landfill that will accelerate these stabilization processes. This approach requires geomembranes in both the bottom and top control systems to prevent leachate from getting out and excess moisture from getting in. In addition, the system needs a leachate collection and removal system on the bottom and a leachate injection system on the top, maintenance of this system for a number of years (depending on the size of the facility), and a gas collection system to remove the waste decomposition gases. In a modern landfill facility, all of these elements, except the leachate injection system, would probably be available. The benefit of this approach is that, after stabilization, the facility should not require further maintenance. A more important advantage is that the decomposed and stabilized waste may be removed and used like compost, the plastics and metals could be recycled, and the site used again. If properly planned and operated in this manner, fewer landfill cells could serve much of a community's waste management needs for many years.

In nonhazardous municipal solid waste landfills natural soils have been used for daily and final covers. However, the use of man–made materials such as foams, recycled paper mixed with polymers, geosynthetics, etc., are gaining popularity for use as daily cover soils. When using natural soils as either the daily or final cover material, it is sometimes necessary to consider different material characteristics to satisfy site specific criteria. A matrix of soil characteristics and health, aesthetics, and site usage characteristics can be developed to provide information on which soil or combination of soils will be the most beneficial.

Health considerations demand the evaluation of each soil type to minimize vector breeding areas and attractiveness to animals. The soil should minimize moisture infiltration (best accomplished by fine grain soils) while allowing gas movement (coarse grain soils are best). This desired combination of seemingly opposite soil properties suggests a layered system. The soil should also minimize fire potential.

Aesthetic considerations include minimizing blowing of paper and other waste, controlling odors, and providing a sightly appearance. All landfill operators strive to be good neighbors and these considerations are very important for community relations.

The landfill site may be used for a variety of activities after closure. For this reason, cover soils should minimize settlement and subsidence, maximize compaction, assist vehicle support and movement, allow for equipment workability under all weather conditions, and allow healthy vegetation to grow. The future use of the site should be considered at the initial landfill design stages so that appropriate end use design features can be incorporated into the cover during the active life of the facility.

3.3 CONTAINMENT AND BARRIER SYSTEMS

3.3.1 Hydraulic Barriers

There are two types of hydraulic barriers, horizontal and vertical. The performance of a barrier system can exceed the sum of the elements that comprise it. A well constructed composite liner system, for example, will have less infiltration than what is expected from independent evaluations of the clay liner and geomembrane.

Geomembranes: Geomembranes are used as low permeability barriers in both the bottom liners and caps of waste containment systems. In a hazardous waste landfill, geomembranes can be used alone as the upper or primary liner, and in conjunction with a low permeability soil layer to form the lower or secondary composite liner.

Increased emphasis is being placed on multilayered cover systems for waste management facilities to minimize the need for further maintenance and to minimize the infiltration of moisture into the unit. These cover systems contain geosynthetic materials that perform specific design, drainage and barrier functions. A major concern is long–term (greater than 30 years) performance of materials used, especially barrier materials. Grasses and woody plants planted or growing naturally on the cover may produce root systems capable of penetrating the barrier materials.

The USEPA believes that waste disposal structures must include the use of geomembranes. They are considered by the Agency to be the only practical means of preventing the migration of chemicals into and through the liners during the active life and post–closure periods. However, they must be carefully selected for waste compatibility, strength, and constructability in the specific design situation. Geomembranes are not intended to be stress–bearing members, and the design should avoid stressing the material as much as possible.

Since hazardous waste regulations and proposed regulations for municipal solid waste require that the final cover be as impermeable as the liner, geomembranes are also recommended in the final cover for surface impoundments, when they are closed as landfills. In this case, a geomembrane is used as the barrier to prevent entry of percolating water to the underlying contaminated material. USEPA's minimum technology guidance for landfill covers describes cover designs using a geomembrane as the principle barrier. The geomembrane need not necessarily be of the same composition in both the liner and cover.

Geomembranes are manufactured using low–permeability synthetic polymers, either non–reinforced or reinforced with fabric material. Geomembranes are made with various base polymers and additives, thus varying considerably in their physical and chemical properties and their interactions with different wastes. No geomembrane is applicable to all wastes; a particular type must be selected for each application.

The base materials of geomembranes are high–molecular weight compounds (polymers). Some common polymers presently in use as base products for geomembranes follow:

1. Thermoplastics, e.g., polyvinyl chloride (PVC).
2. Crystalline thermoplastics, e.g., high density polyethylene (HDPE).
3. Thermoplastic elastomers, e.g., chlorosulfonated polyethylene (CSPE).
4. Elastomers, e.g., butyl rubber.

USEPA has provided detailed technical information about geomembrane materials used to contain specific types of hazardous wastes. Geomembranes and their polymers continue to improve as their manufacturers and users gain experience in waste containment. Indestructability and constructability are two general goals difficult to achieve in the same material.

The compatibility of a liner material with a specific waste is an important consideration in planning a surface impoundment. Liquid wastes in surface impoundments may vary from complex mixtures of variably concentrated constituents to highly concentrated contaminants contained in a relatively simple matrix. Additionally, waste constituents may change over time. During the liner design process, a representative sample of the liquid waste should be analyzed for waste properties that could potentially cause damage to liner material. Several methods for obtaining samples of hazardous wastes have been discussed by USEPA. Potentially detrimental waste properties include the following:

1. Acidity;
2. Alkalinity (greater than pH 10);
3. Temperature extremes; and
4. Hydrocarbons.

If the bottom liner system contains a geomembrane, the cover system should also contain one, in order to "prevent" a bathtub effect.

At one time polyvinyl chloride (PVC) had the majority of the geosynthetics market, however, at this time, high density-polyethylene is in the lead. A number of other specialty synthetic materials are being offered in woven and non-woven forms.

Geomembranes can be utilized in various combinations with soil:

(1) The geomembrane/composite double-liner system consists of two liners—a top primary geomembrane liner and a bottom secondary geomembrane/soil composite liner, separated by a leak-collection system. USEPA's minimum technology guidance recommends that the primary geomembrane be at least 30 mils (0.76 mm) thick where covered by a protective soil and/or geotextile layer. For an uncovered geomembrane, a thickness of at least 45 mils (1.14 mm) is recommended. The guidance suggests that thicknesses of 60 to 100 mils (1.52 to 2.54 mm) may be required to resist various stresses. In any case, the design engineer should recognize that some geomembrane materials may require greater thicknesses to prevent failure or to accommodate unique seaming requirements. It may be possible to use combinations of geotextiles and geomembranes in lieu of increasing the geomembrane thickness in some instances.

(2) The geomembrane/compacted soil double-liner design consists of a primary geomembrane liner placed above a secondary low-permeability-soil liner, separated by a leak-collection system. It is very similar to the geomembrane/composite double-liner system described above, except that it does not contain the geomembrane component in the secondary liner. The primary geomembrane liner has the same recommended minimum thickness and chemical compatibility as the primary geomembrane liner of the geomembrane/composite liner system. The low-permeability-soil secondary liner thickness depends on site- and design-specific conditions, but should not be less than 36 inches (90 cm) thick, according to the guidance.

(3) A further variation of the double liner is a system comprised of two composite geomembrane/compacted soil layers. Many owners and operators of landfills and surface impoundments have indicated that they planned to use a composite liner for the top liner as well as the bottom liner.

In a double-composite-liner system, the liners are separated by a leak-collection layer, no different than those of other double-liner designs, except for the insertion of a separation layer at the top of the leak–collection layer. The separation layer would prevent the migration of the primary soil liner component material into the underlying leak–collection layer. The layer also serves to stabilize the surface of the drainage layer, facilitating compaction of the overlying soil component.

While a double–composite liner should offer further assurance against the escape of contaminated liquid, it is not without its own risks. It appears possible for the primary geomembrane liner to develop a leak that would not drain freely through the soil component to the leak–collection layer. Liquid might thus accumulate between the geomembrane and soil components and go undetected for a long period. On the other hand, this is not altogether contrary to the purpose of the top composite liner's soil component, which would be to impede leakage to the drainage layer.

Geomembranes used in hazardous waste surface impoundments are susceptible to failure during facility operation. There are basic types of geomembrane failure: (1) excessive geomembrane displacement (i.e., vertical or lateral movement), and (2) an unacceptable leak. Geomembrane liner design should specify material components and construction practices to prevent failures by:

1. Protecting the geomembrane from puncture, scratching, abrasion, or other damage (from above and below);
2. Demanding that great care be taken to prevent damage to the geomembrane sheet during all installation processes;
3. Providing gas venting in the drainage layer;
4. Avoiding bridging, rippling, stretching, or other stressed conditions;
5. Allowing slack for shrinkage.
6. Avoiding nonessential penetrations;
7. Eliminating tensile stresses as much as possible, by design;
8. Providing detailed and rigorous seaming instructions;
9. Providing slip–preventing anchorage at the tops of slopes;
10. Requiring well–trained and experienced installation personnel; and
11. Providing a detailed quality assurance plan.

A protective layer of soil or a soil/geotextile layer may be used on the surface of the liner system to protect the underlying geomembrane from construction damage during installation, loads imparted by the waste, weathering, erosion and abrasion, to increase friction, and to dampen potential chemical attack. Above the liquid level of the impoundment, coarser–grained (sometimes rubble) material placed over a geotextile protective layer is often used. The coarse material will generally be more stable on steeper slopes and will dampen wave action and run–up. An even coarser layer, e.g., riprap, may be applied over the geomembrane–protective layer to prevent erosion in larger impoundments. Below the liquid level, sand is often used to protect the geomembrane against puncture, and to dampen the effects of strong chemical or high–temperature waste inputs.

Geomembrane Interlocking Panels: Geomembrane interlocking panels are installed in vertical trenches to construct a low permeability barrier. The panels consist of membrane panels that connect along their lengths much like conventional steel sheet piles. The panels are pre–fabricated and assembled at the site by locking the panels together and placing them into the trench.

Geomembrane interlocking panels have two physical elements; high density polyethylene (HDPE) panels with interlock fittings along their vertical edge, and a soft plastic sealing medium within the interlock that provides a hydraulic seal to the interlock fittings. The HDPE interlock closely resembles that used in conventional steel sheet files: a female channel on one panel that engages a male edge of an adjacent panel.

Geomembrane interlocking panels can be installed in open trenches, slurry trenches, or vibrated into sandy soils using a steel mandrel. Care must be taken to insure that the hydraulic interlock and seal material properly seat along the entire length of the interlock channel. This verification can be difficult in slurry wall and vibrated installations.

Grouts: The grout used in a waste containment system is usually a mixture of cement and bentonite. Grouts are, however, available with silica, acrylate, urethane, and Portland cement binders. Grout can be injected in horizontal or vertical borings using a variety of pressures.

Grouting, including jet grouting, employs high pressure injection of a low–permeability substance into fractured or unconsolidated geologic material. This technology can be used to seal fractures in otherwise impermeable layers or construct vertical barriers in soil through the injection of grout into holes drilled at closely spaced intervals, i.e., grout curtain. However, concerns surround the use of grouting for the construction of vertical barriers in soils because it is difficult to achieve and verify complete permeation of the soil by the grout. Therefore, the desired low permeabilities may not be achieved as expected.

Particulate grouts consist of water plus Portland cement, bentonite, or a mixture of the two which solidifies within the soil matrix. Chemical grouts consist of two or more liquids which gel when mixed together. Often, particulate grouts are used as "pre–grouts" with a second injection of a chemical grout to seal the finer voids.

Using a horizontal or directional drilling method, **Bottom Sealing** involves grout injection techniques to place horizontal or curved barriers beneath a hazardous waste site to prevent downward migration of contaminants. Once in place, the barrier acts as a floor and seals the bottom of the waste site. This technology has possible applications in all soils, including silts, clays, and weak rocks. It can be used with most contaminants including inorganics, organics, metals, mixed, high–level, low–level, and TRU waste. It is used in soils that are contaminated with liquid waste that have the potential of migrating downward.

Block displacement is a plume management method where a slurry is injected in such a manner that it forms a subsurface barrier around and below a specific mass or "block" of earth. Continued pressure injection of the slurry produces an uplift force on the bottom of the "block" which results in a vertical displacement proportional to the slurry volume pumped, thus the name block displacement.

There are many available grouts; however, the selection of grout material depends on site specific factors such as soil permeability, soil grain size, rate of groundwater flow, chemical constituents of soil and groundwater, required grout strength, and cost. Because a grout curtain can be as much as three times as costly as a slurry wall, it is rarely used when groundwater has to be controlled in soil or loose overburden. The major use of grout curtains is to seal voids in porous or fractured rock where other methods of groundwater control are impractical.

Another variation of a grout curtain is the vibrating beam technique for emplacing thin (approximately 4 in) curtains or walls. Although it is sometimes called a slurry wall technique, it is more closely related to a grout curtain since the slurry is injected through a pipe similar to grouting.

The **Fluidized-Bed Zeolite System** utilizes zeolite and particulate/solution polymer based grouts for in situ stabilization and isolation of radioactive and hazardous chemical waste materials that have been disposed in or near underground waste disposal and containment structures. The fluidized-bed will provide chemical fixation by mechanically homogenizing and incorporating waste tank residuals (tank bottoms and sludges) with granular zeolite (or equivalent) materials. Then particulate and solution polymer-based materials are incorporated into the interstitial void volume of the granular zeolite and surrounding geologic media to provide chemical isolation and physical stabilization. This system could be used for remediation of subsurface waste storage/disposal structures such as underground storage tanks, cribs, caissons, piping, and buried sites. This technology will produce a physically stable structure, wherein contaminated materials are anticipated to be isolated from the environment over hundreds to thousands of years.

Compacted Clay Liners (CCL): A CCL may be used as the primary moisture barrier in both waste liner and cover systems. Achieving a low permeability in a clay liner requires a suitable clayey soil and proper preparation and compaction of the soil. Test data clearly demonstrate that the permeability of a clay liner can be increased 100 to 1,000 times if a single parameter in preparation or compaction is neglected. Soil selected for use as a clay liner is specified using soil plasticity (Atterburg Limits) and grain size distribution. Both parameters can be easily monitored during actual field placement of the liner.

Construction of a clay liner requires proper soil preparation and correct compaction equipment and technique. Soils used for clay liners should be processed to ensure that the soil water content is as specified, the soil clods are no larger than 1 to 2 inches, and the maximum particle size is less than required by the project specifications. For composite liners, the maximum rock size in the last lift is frequently less than 1.0 inch (2.54 cm) to minimize potential damage from rocks to the overlying geomembrane. Liner soil preparation is usually done as the soil liner material is placed in a stockpile. Significant moisture adjustment should not be attempted in the 24 hours preceding placement of the soil.

A clay liner is very susceptible to damage due to either desiccation or freezing. Clay liners left exposed must be protected from desiccation using a surface sealant (acrylic sprays or light membranes) or an additional layer of "sacrificial" soil. Soil cover also serves to prevent freezing of the clay liner. Even a single cycle of freezing can significantly increase the permeability of a clay liner.

Clay is the most important component of soil liners because the clay fraction of the soil ensures low hydraulic conductivity. In the United States, however, there is some ambiguity in defining the term "clay" because two soil classification systems are widely used. One system, published by the American Society of Testing and Materials ASTM), is used predominantly by civil engineers. The other, the U.S. Department of Agriculture's (USDA's) soil classification system, is used primarily by soil scientists, agronomists, and soil physicists.

EPA requires that soil liners be built so that the hydraulic conductivity is equal to or less than 1 x 10^{-7} cm/sec. To meet this requirement, certain characteristics of soil materials should be met. First, the soil should have at least 20% fines (fine silt and clay sized particles). Some soils with less than 20% fines will have hydraulic conductivities below 10^{-7} cm/sec, but at such low fines content, the required hydraulic conductivity value is much harder to meet.

Second, plasticity index (PI) should be greater than 10%. Soils with very high PI, greater than 30 to 40%, are sticky and, therefore, difficult to work with in the field. When high PI soils are dry, they form hard clumps that are difficult to break down during compaction.

The USEPA uses the term "low-permeability soil" so as not to imply that there is a narrow restriction on the type of soil that may be used. The term "clay liner" is used casually to refer to any soil liner, but the liner may, in fact, be comprised for the most part of materials other than clay. Also, a sandy soil may be made to meet very low permeability requirements with the addition of only a small percentage of bentonite.

The purposes of the low-permeability soil component of the secondary liner are (1) to minimize the migration of hazardous liquids through the geomembrane component if a breach in the geomembrane should occur, and (2) to attenuate constituents that might leak through the membrane.

Soil liners are not impervious, but they do control seepage and have been used, because of that and their low cost, in the past as the sole liner in surface impoundments and landfills. Soil may be treated, remolded, and/or compacted to achieve prescribed flow-impeding and contaminant-attenuating specifications. However, the USEPA does not believe that a soil liner can ordinarily be constructed to meet, by itself, the requirement of no contaminant breakthrough during the active life and post-closure monitoring period for a surface impoundment.

Despite USEPA's wariness of the impermeability of soil liners, they are still recommended in USEPA's minimum technology guidance as backup to the geomembrane component of the secondary liner in the required double-liner system. Further, low-permeability soil may be used as the secondary liner without the geomembrane if it can be shown that it will provide equivalent performance and not allow contaminant breakthrough during the active life and post-closure monitoring period. The Agency believes that such a showing will be very difficult.

USEPA's minimum technology guidance recommends that the soil component of the secondary liner have a thickness of at least 36 inches (90 cm) and be chemically resistant to the impounded waste. The guidance also states that the soil should be compacted in lifts of 6 inches (15 cm) or less, after compaction, and be free of rocks, roots, and rubbish.

Bentonite and Bentonite Amended Soil: Bentonite provides an effective moisture barrier. Available as either sodium bentonite mined in several western states or a less active calcium bentonite from Georgia, commercial bentonite is purchased in powder or pellet form. The permeability of bentonite ranges from 10^{-8} cm/sec for calcium bentonite to as low as 10^{-10} cm/sec for sodium bentonite. Bentonite can be used by itself to form a moisture barrier or blended with on-site soils to form an acceptable soil liner.

When there is not an adequate supply of low permeability soil on site for the construction of a soil barrier layer, a bentonite soil amendment can be used to lower the permeability of the on–site soils. Typically, a 3 to 6% bentonite amendment by dry weight is sufficient to achieve a permeability of 1 x 10^{-7} cm/sec using on–site soils. Sands, however, may require as much as 10 to 15% bentonite amendment.

Calcium bentonite, though more permeable than sodium bentonite, has also been used for soil blends. Approximately twice as much calcium bentonite typically is needed, however, to achieve a hydraulic conductivity comparable to that of sodium bentonite. One problem with using sodium bentonite, however, is its vulnerability to attack by chemicals and waste leachates.

On–site sandy soils also can be blended with other clay soils available in the area, but natural clay soil is likely to form chunks that are difficult to break down into small pieces. Bentonites, obtained in dry, powdered forms, are much easier to blend with on–site sandy soils than are wet, sticky clods of clay. Materials other than bentonite can be used, such as atapulgite, a clay mineral that is insensitive to attack by waste. Soils also can be amended with lime, cement, or other additives.

Geosynthetic Clay Liner (GCL): A GCL is essentially a pre–fabricated low permeability soil layer. The bentonite is usually sandwiched between two geotextiles or adhered to a geomembrane. GCL's are manufactured in 12 to 14 ft wide sheets and are shipped to the site in rolls. Typically, the weight of bentonite (or bentonite and adhesive) per square foot is specified and must be field verified.

The GCL barrier is installed by simply unrolling the sheets over a prepared subgrade. The subgrade should be free of large rocks, ruts, and objects that could penetrate through the GCL. Additionally, the subgrade must be stable, which can be checked by proof–rolling. Seaming of the GCL barrier is typically limited to a minimum 6 inch overlap of adjacent sheets. The water seal at the seam forms when the bentonite hydrates and "oozes" out from between the geotextiles. Additional dry bentonite powder may be spread between the overlaps to improve the seal.

Geosynthetic clay liners dominate the current market for alternatives to several feet thickness of compacted clay soil. They are represented by five known products, each having distinctive characteristics within the GCL family. Three GCLs are being manufactured in the United States and two in Europe. Because of the differences among these GCLs in materials, design, construction and other characteristics, selecting among them for specific applications requires careful consideration.

GCLs that contain pure sodium bentonite have been shown to retain their original low hydraulic conductivity even after being subjected to numerous freeze–thaw cycles. This property is much more favorable for application to cover systems in cold climates than use of compacted soil materials that must be maintained below the maximum depth of frost action.

Various other alternative materials have been proposed. Fly ash–bentonite–soil mixtures show promise in terms of providing low hydraulic conductivity and high strength. Super–absorbent geotextiles, such as Fibersorb, have been proposed. Sprayed–on geomembranes, applied to a bentonitic blanket material, have been manufactured.

The potential advantages of alternative barriers (compared to low–permeability, compacted soil) are (1) rapid and simple installation of the alternative barrier; (2) the potential for a more predictable end–product with manufactured, alternative barriers; (3)

a more predictable cost for alternative barriers that in some instances may be much lower than for compacted soil; (4) the possibility of utilizing lightweight equipment to install alternative barriers, which minimizes the risk of damaging underlying materials, e.g., FMLs; and (5) the possibility of developing a detailed data base on an alternative barrier material so that the data base does not have to be recreated for every project.

The main disadvantages of alternative barrier materials are: (1) vulnerability to puncture; (2) the possibility of chemical attack of the bentonite; (3) uncertainties about performance of seams; and (4) low shear strength of bentonite.

Slurry Walls: Bentonite slurries are used in vertical barrier walls to displace the natural soils and construct a low permeability barrier. Bentonite slurries are mixtures of 4 to 7% bentonite and water. The bentonite slurry can be used "as is" to prevent the collapse of cut–off wall excavations or blended with on–site soils to form a soil–bentonite SB) slurry used in permanent barrier walls. The soils should have 20 to 40% fines and are blended with the bentonite slurry until a paste is formed. This paste should have the consistency of fresh mortar or concrete and flow easily.

Concrete–bentonite (CB) slurries are used on sites where adequate soils to form soil–bentonite slurries are not available or when increased strengths are needed. The concrete–bentonite slurries are mixtures of approximately 18% concrete, 6% bentonite.

Slurry walls are often used where a waste mass is too large for practical treatment, where residuals from the treatment are landfilled, and where soluble and mobile constituents pose an imminent threat to a source of drinking water. Slurry walls can generally be implemented quickly, and the construction requirements and practices associated with their installation are well understood. These structures are often used in conjunction with covers and treatment technologies such as in situ treatment and groundwater collection and treatment systems.

In the construction of most slurry walls it is important that the barrier is extended and properly sealed into a confining layer (aquitard) so that seepage under the wall does not occur. For a light, non–aqueous phase liquid a hanging slurry may be used. Similarly, irregularities in the wall itself, e.g. soil slumps, may also cause increased hydraulic conductivity.

Slurry walls also are susceptible to chemical attack if the proper backfill mixture is not used. Compatibility of slurry wall materials and contaminants should be assessed in the project design phase.

Slurry walls also may be affected greatly by wet/dry cycles which may occur. The cycles could cause excessive desiccation which can significantly increase the porosity of the wall.

Once the slurry walls are completed, it is often difficult to assess their actual performance. Therefore, long–term groundwater monitoring programs are needed at these sites to ensure that migration of waste constituents does not occur.

The construction of slurry walls involves the excavation of a vertical trench using a bentonite–water slurry to hydraulically shore up the trench during construction and seal the pores in the trench walls via formation of a "filter cake." Slurry walls are generally 20 to 80 feet deep with widths 2 to 3 feet. These dimensions may vary from site to site. There are specially designed "long stick" backhoes that dig to 90 foot depths. Generally there will be a substantial cost increase for walls deeper than 90 feet. Clam shell excavators can reach depths of more than 150 feet. Slurry walls constructed at water dam projects have extended to 400 feet using specialized milling cutters. Depending on the site conditions and contaminants, the trench can be either excavated to a level below the water table to capture chemical "floaters" (this is termed a "hanging wall") or extended ("keyed") into a lower confining layer (aquitard). Similarly, on the horizontal plane the slurry wall can be constructed around the entire perimeter of the waste material/site or portions thereof (e.g., upgradient, downgradient).

As the slurry trench is being excavated it is simultaneously being backfilled with an engineered material that forms the final wall. The three major types of slurry trench backfill mixtures are (1) soil bentonite, (2) cement bentonite, and (3) concrete.

A relatively new development in the construction of slurry walls is the use of mixed-in-place walls (also referred to as soil-mixed walls). The process was originally developed in Japan. A drill rig with multi-shaft augers and mixing paddles is used to drill into the soil. During the drilling operation a fluid slurry or grout is injected and mixed with the soil to form a column. In constructing a mixed-in-place wall the columns are overlapped to form a continuous barrier. This method of vertical barrier construction is recommended for sites where contaminated soils will be encountered, soils are soft, traditional trenches might fail due to hydraulic forces, or space availability for construction equipment is limited. Both this method and a modified method termed "dry jet mixing" are usually more expensive than traditional slurry walls.

Another application of traditional slurry wall construction techniques is the construction of permeable trenches called bio-polymer slurry drainage trenches. This will typically involve the use of a landfill cover in conjunction with the wall. Rather than restricting groundwater flow, these trenches are constructed as interceptor drains or extraction trenches for collecting or removing leachate, groundwater, and groundwater-borne contaminants. These trenches also can be used as recharge systems. The construction sequence is the same as the traditional method described above. However, a biodegradable material (i.e., biopolymer) with a high gel strength is used in the place of bentonite in the slurry, and the trench is backfilled with permeable materials such as sand or gravel.

Diaphragm Walls: Diaphragm walls are barriers composed of reinforced concrete panels (diaphragms), which are emplaced by slurry trenching techniques. They may be cast-in-place or pre-cast, and are capable of supporting great loads. This degree of strength is seldom if ever called for at a hazardous waste site and their use is rare. Because diaphragm walls are constructed in slurry-filled trenches, it is possible to include them in cement-bentonite or soil-bentonite walls for short sections, such as road or rail crossings, that require their greater strength. Provided the joints between the cast panels are made correctly, diaphragm walls can be expected to have permeabilities comparable to cement-bentonite walls. The same compatibility concerns that apply to cement-bentonite, apply to diaphragm walls.

Sheet Piling Cutoff Walls: Sheet piling cutoff walls can be made of wood, reinforced concrete or steel; however, steel sheet piles represent the most effective material for constructing a groundwater barrier. Construction of a steel sheet pile cutoff wall involves driving lengths of steel sheets through unconsolidated deposits with a pile driver. The individual steel sheet piles are connected along the edge of each pile through various types of interlocking joints. These joints provide permeable pathways for groundwater movement which may or may not become watertight naturally depending on the soil characteristics. It may be necessary to fill these joints with an impermeable material such as a grout; however, the ability to ascertain the success of the grouting operation is questionable. Steel is a readily corrodible material and therefore the lifetime of the steel sheet piles is dependent on the corrosive nature of the soil, groundwater, and contaminants with which the steel piles come in contact. A common recommendation is that steel pilings be chemically coated or electrically protected so as to minimize corrosion. Although there are limitations, sheet piling cutoff walls may be used to contain contaminated groundwater, divert a contaminant plume to a treatment facility, and divert groundwater flow around a contaminated area.

Cofferdams: Cofferdams hydraulically isolate a portion of the water body; can be used to isolate contaminated surface water for subsequent pumping to treatment systems, or to isolate uncontaminated surface water for subsequent dredging/sediment removal operations in the surrounding (contaminated) area.

Cofferdams are surface water barriers which are anchored to the soil/sediment at the bottom of a surface water body. They may be constructed of various materials including soil, sheet piling (usually black steel, but galvanized or aluminum coatings are also available), earth–filled sheet pile cells (single–walled or cellular), and sand bags (for short–term structures). Pre–assembled (interlocked) sections of sheet piling are also available. The sheet–piling can be hand–driven using a hand maul or a light pneumatic hammer. Heavy driving equipment such as a drop hammer, pneumatic pile driver, or stem pile driver are also used.

Hydraulic Cage: An engineered system to passively control geohydrological gradients over the long–term in and around either a controlled placement of hazardous materials or an existing contaminated region. The cage is constructed by drilling a series of boreholes around the region of interest and enhancing the hydraulic conductivity of the rock between them. The boreholes can also be used for pre–construction characterization and post–construction monitoring, as well as for dewatering during construction (or underground waste placement), or for groundwater removal during remediation.

The original concept relies on a series of concentric boreholes to create the preferential flow zone. Three enhancements of the concept were considered in a study by C.F. Voss, all based on benefits derived from improving the hydraulic connectivity of the boreholes. The three methods are tunnels, hydraulic fractures, and blast rubblization. Numerical modeling studies were used to evaluate the performance of the alternative hydraulic cage designs in a range of geologic settings. Discrete–fracture modeling was carried out to assess the feasibility of the concept in fractured rock while two–dimensional continuum analyses were used to evaluate the performance in porous media.

The results indicate that the most important factors in determining the efficiency of a hydraulic cage are the contrast between conductivity in the cage and the site. Performance improves as the contrast increases. In the case of cages where the hydraulic connectivity between boreholes was increased using hydraulic fracturing or blasting techniques, the performance improved as the coverage and connectivity of the induced fractures increased. The results demonstrate that hydraulic cages can be effective, even in fractured rock. Flow reductions of at least 70% were observed for cages having a conductivity three orders of magnitude greater than the surrounding medium. Conventional methods are available for constructing the different cages designs at costs commensurate with conventional disposal facilities. The approach has the advantage of greater long-term reliability due to difficulty in maintaining low conductivity barriers in conventional disposal facilities.

Polymer Concrete Barrier: This containment technology uses high strength, impervious polymer concrete to create an in situ barrier. Sealant materials are used that consolidate an earth/sand/gravel matrix into a high strength, impervious polymer concrete useful for the formation of barriers in the earth. These materials have very good chemical resistance and are typically two or three times stronger than structural concrete. This technology is effective for the containment of most contaminated waste. Residual risk from the untreated waste is greatly reduced once contained within a perimeter barrier with a sealant cap over the top (may also be composed of polymer concrete). This containment barrier could be used in conjunction with other in situ technologies.

Membrane and Synthetic Sheet Curtains: Membrane and synthetic sheet curtains can be used in applications similar to grout curtains and sheet piling. The membrane is placed in a trench surrounding or upgradient from the plume of interest, thereby enclosing the contamination or diverting the groundwater flow. Placing a membrane liner in a slurry trench application has also been tried on a limited basis. Attaching the membrane to an impervious layer and having perfect seals between sheets is difficult but necessary in order for membranes and other synthetic sheet curtains to be effective. Impermeable synthetic membranes have also been used on the downgradient side of interceptor trenches to stop the migration of petroleum products for subsequent recovery.

Cryogenic Barrier: This type of barrier is formed by installing freezing pipes around the circumference of a contaminated site. A refrigerant fluid is pumped down the outside pipe and returned through the inner pipe. The double wall design allows the entire volume between walls to freeze, thus containing the site. If necessary, another in situ treatment could then be applied with little risk of contaminant migration.

This technology can be used to isolate or contain all types of contamination and can be used on all media states in which freeze pipes can be installed. It appears to be more cost effective to use this technology for temporary rather than permanent containment because of the high operational costs. Under certain circumstances, containment for a relatively short period of time is sufficient in itself. Cryogenic barriers are compatible with most other in situ technologies.

Plasma Arc Glass Cap: This technology uses a plasma torch to generate a high heat flux in the vicinity of the disposal site surface, thereby vitrifying the surface soil to create an impermeable glass cap. Depending on how the torch is operated, the cap may be anywhere from 1 to 6 inches (2.54 to 15.24 cm) deep.

The mobility of the toxic contaminants will be greatly reduced by placing an impermeable glass cap over the site. Moisture from rain and snow melt will be shielded from the waste, eliminating leaching and downward migration of the contaminants. Contaminants will be constrained from migrating upward. This technology can be used with all contaminants and soils that can be vitrified.

Soil/Cement Wall: The soil/cement wall technology involves fixation, stabilization, and solidification of contaminated soils. Solidification/stabilization agents are blended in situ with the contaminated soils by a multi-axis overlapping hollow stem auger. The product is a monolithic block that extends down to the treatment depth. This technology is effective on soils that are contaminated with metals and semi-volatile organic compounds. This technology has been used on various construction applications, including soil stabilization and cutoff walls.

Vitrified Barriers: In situ vitrification (ISV) is a thermal treatment technology in which a region of soil volume is melted. This process can also be used to produce vitrified barriers. Upon cooling, the resulting product is a glass and crystalline monolith resembling natural obsidian. The process involves creating a barrier by inserting electrodes in the ground and placing a conductive starter path between them. Soil is melted when an electric potential is applied to the electrodes causing the starter path to heat up above the melting point of the soil. Vitrified walls and floors can be joined as need to isolate waste sites from transport mechanisms or to totally contain them, if necessary (e.g., for additional in situ treatment). The vitrified soil barrier is extremely leach resistant and possesses about ten times the strength of unreinforced concrete. It is predicted to be stable over geologic periods of time. It also results in significant volume reduction because no additives are required and the soil is densified in the melting process.

This technology can be used to isolate or contain all contaminant types and can be used on all media states. It can be used to permanently contain a waste site or to temporarily contain a waste site while another method of in situ remediation is applied. There may be a concern in the presence of acids and salts.

Steel Sheet Piling: The joints of conventional sheet piling are designed for mechanical strength but not water tightness. Leakage of water through the unsealed joints is acceptable for most civil engineering applications, but generally not for environmental applications.

A new type of containment wall composed of sealable steel sheet piling has been developed at the University of Waterloo's Institute for Groundwater Research.

With the Waterloo Barrier, the interlocking joints between individual sheet piles incorporate a cavity that is filled with sealant after driving to prevent leakage through the joints. The sealable cavities can be formed in two ways. An internal cavity can be formed as the sheet pile itself is manufactured. Or, an external cavity can be produced adjacent to each joint by attaching a steel 'L' section to conventional sheet piling.

3.3.2 Hydraulic Conveyances

Drainage layer components are designed to collect leachate beneath the waste, or to collect gas above the waste. Both functions require the drainage layer to be significantly more permeable than the adjacent waste or soils. Thus filter systems are required with all drainage systems. Newer geocomposite systems provide both the drainage media and filter layer in a single commercial product.

Passive interceptor systems consist of trenches excavated to a depth below the water table with the possible placement of a collection pipe in the bottom of the trench. These interceptor systems can be used as preventive measures (i.e., leachate collection systems), abatement measures (i.e., interceptor drains), or in product recovery from a groundwater (i.e., oil, gasoline). Interceptor drains are generally used to either lower the water table beneath a contamination source or to collect groundwater from an upgradient source in order to prevent leachate from reaching uncontaminated wells or surface water. Interceptor systems are relatively inexpensive to install and operate and provide a means for leachate collection without impermeable liners. On the other hand, interceptor systems are not well suited to poorly permeable soils and the systems require continuous and careful monitoring to assure adequate leachate collection.

Changing the contour and runoff or runon characteristics of a particular site can be accomplished by several standard engineering techniques. Some of the more common techniques include dikes and berms, ditches, diversions, waterways, terraces, benches, chutes, downpipes, levees, seepage basins, sedimentation basins, and surface grading.

Natural Drains and Collectors: A sand or gravel drainage layer can be used as part of the leak detection component of a waste containment system, as the primary drainage layer above the bottom liner, and as an infiltrating storm water drain in the caping system. Soil drains provide a high permeability medium into which liquids can easily drain to a network of collection pipes. The soil drain usually consists of clean sand or gravel sorted to specific particle sizes by a quarry. In some cases suitable sands or gravels are found on site, but this is unusual.

Drain materials are specified using a particle size distribution. Soil drains typically require only moderate to light compaction and the as–installed drain layer thickness should be verified. Natural drainage layers must be protected by diversion berms or geotextiles from the introduction of waterborne fines from surface erosion of adjacent slopes.

Subsurface Drains: Subsurface drains are used to intercept and/or collect leachate by gravity flow methods. Subsurface drainage systems are generally constructed by excavating a trench and laying tile (tile drain) or piping (pipe drain) end to end along the bottom of the trench. The trench is then backfilled with gravel or other envelope material followed by backfilling the remainder of the trench with soil. The gravel may be lapped with fabric to prevent fine soil from entering the gravel and clogging the drain. If the surrounding soils have a moderately–high to high permeability, an impermeable liner placed downgradient of the drain may be required to prevent intercepted and contaminated water from flowing through to uncontaminated areas.

Waste compatibility tests are required to properly select piping and fill material. Mineral precipitates can cause clogging. Construction of subsurface drains is generally cost prohibitive if substantial hard rock excavation is required. Subsurface drains are not feasible at depths exceeding 40 feet due to the difficulty of shoring and installation. Also, subsurface drains are not suited for areas with very high soil permeability, nor are they applicable to viscous wastes. Subsurface drains are unsuitable for locations adjacent to a recharge area unless used together with an impermeable barrier.

Geosynthetic Drains and Collectors: A geocomposite drain consists of a core material that provides a pathway for drainage and a surrounding geotextile that prevents clogging of the core. Geocomposite drains can be used as part of the leak detection component of a waste containment system, as the primary drainage layer above the bottom liner, and as a lateral drain in the capping system. Similar to the soil drain, a geocomposite drain provides a high transmissivity core through which liquids can easily drain into a network of collection pipes. Geocomposite drains are pre–fabricated and come in rolls or panels up to 300 feet in length.

Overlaps between adjacent roll ends or panels are particularly important with the proper overlap length, orientation, and plastic ties used to bind the overlap. Geosynthetic drainage layers must be protected from damage by vehicle traffic, wind or other disturbances. The layer should also be protected from fines carried by surface erosion during construction.

Plastic Pipes: Both perforated collection and solid transmission pipes are used in the leak detection systems, the primary leachate collection drain, the lateral drain in the cap, and to carry stormwater and leachate away from the waste containment system. Collection and transmission pipes are constructed from a variety of plastics and can be designed for gravity flow lines as well as low or high pressure lines, depending upon the application.

Sumps: Leachate sumps are located in both the primary leachate collection layer and the bottom/lower leak detection layer. Sumps are located at low points in the liner and act as basins in which the leachate can be collected. From the sumps, the leachate is either drained out of the waste containment system by gravity or pumped out. The sumps can be a simple depression in the composite liner or a pre–manufactured plastic basin that is set in the clay liner so that its top is flush with the geomembrane liner. The geomembrane is fusion–welded to a flange on the upper edge of the sump. Pre–manufactured sumps eliminate difficult–to–test field seams or complex grading.

If leachate sumps are fabricated on site, then they will be made from the same materials as the liner system, so certification of the liner materials will already have been provided.

Sumps are the only area of a waste containment system that will continuously receive leachate. As such, any defect in the sump may result in a continuous long–term leak.

Pumps: The leachate can be drained by gravity from the waste containment system or it can be pumped out the system using a submersible pump. The pumps used to move leachate are manufactured for harsh environments and can be powered by electricity or compressed air.

Chutes and Downpipes: Chutes and downpipes are used to convey concentrated flows of surface water from one level of a site to a lower level without erosive damage. Chutes (also referred to as flumes) are open channels that have compacted, smooth linings placed over undisturbed soil or well–compacted fill. Downpipes (also called downdrains or pipe slope drains) consist of rigid piping laid in slope areas. Generally, downpipes extend downslope from earthen embankments (i.e., dikes and berms) and convey water to stabilized waterways or outlets at the base of the slope.

Chutes and downpipes are temporary structures, often used in conjunction with other technologies, that do not require formal design. Chutes and downpipes are useful in emergency situations because they can be quickly constructed.

Dikes and Berms: They prevent excessive erosion of newly–constructed slopes; provide temporary isolation of wastes; and prevent runoff and mixing of incompatible wastes. Dikes and berms are well–compacted earthen ridges or ledges which control surface water runoff from storms or floods by diverting the flow to alternate drainage ways. The terms dikes and berms are generally used interchangeably. Erosion resistant, low permeability, clayey soils or compacted sands and gravel are commonly used in the construction of dikes and berms. Properly constructed earthen dikes are machine compacted. Diversion dikes should have a positive grade toward an outlet. Immediately following construction, diversion dikes should be seeded and mulched or stabilized with, gravel and/or stone rip–rap.

Dikes and berms are only temporary measures until more permanent drainage structures can be installed or until the slope is stabilized with vegetation. They are generally not used for more than a year. Periodic inspection and maintenance are required for dikes/berms.

Ditches, Channels, Swales, and Waterways: Used to intercept runoff and/or reduce slope length; conveys runoff from one area to another.

Ditches and channels are depressions or shallow, excavated areas with V–shaped, trapezoidal, triangular, or parabolic cross–sections, which intercept runoff or reduce slope length. Earthen channels can be used to divert runoff from entering the site. Waterways are channels that have been stabilized with vegetation or stone rip–rap, and are able to collect and transfer diverted water off–site or to an on–site storage/treatment area. A diversion is a modified earthen channel that has a supporting dike or berm along the downhill edge of the channel. Swales are similar to channels except that their side slopes are not as steep, and they have a vegetative cover for erosion control.

Channels and waterways are generally designed to intercept flows from 10 or 25 year storm events, in such a way as to be able to convey these flows at non–erosive velocities. Wider and shallower channel cross–sections have lower flow velocity and thus reduced potential for erosion of channel side slopes. Narrower and deeper channels require stabilization through vegetation or the use of stone rip–rap to line channel bottoms and break up flow. Half–round channels, which are constructed of cut corrugated metal pipe or pre–fabricated asphalt sections, can be placed below grade and have low maintenance and installation costs.

Grading: Grading alters the topography and runoff characteristics of a waste site; optimizes slope and prepares area for surface sealing and/or revegetation. Grading refers to techniques used to reshape the surface of a site in order to manage surface water infiltration and runoff while controlling erosion. Grading techniques include spreading, compaction, sacrification, tracking, and contour furrowing.

Spreading and compaction are used to optimize a slope in such a way that surface runoff is increased while infiltration and ponding are decreased, without increasing erosion. These techniques are used to prevent surface water runoff from contacting waste, and/or to prepare a site for subsequent remediation activities. Sacrification, tracking, and contour furrowing are grading techniques employed to roughen soils in preparation for revegetation. These techniques slow runoff, thereby increasing infiltration and decreasing erosion potential.

Generally, graded slopes should be 3 to 5%; sometimes greater slopes are used to promote more effective drainage, but the maximum slopes usually do not exceed 33%.

Levees and Floodwalls: Levees and floodwalls are flood protection structures in areas subject to inundation from tidal flow or riverine flooding. Levees are earthen embankments that create a barrier to confine floodwaters to a floodway and to protect structures behind the barrier. Levees are constructed of erosion–resistant, low–permeability soils (i.e., clay), or compacted, impervious fill. Floodwalls are similar to levees, except that they are constructed of concrete. Levees generally require a very large base width; therefore, in areas where there is limited space and fill material, concrete floodwalls are preferred.

Levees and floodwalls are generally designed with a height capable of withstanding a 100–year flood (usually 2 ft of freeboard above the 100–yr flood elevation). A 10 ft minimum top width is required for levees to allow access for construction and maintenance equipment. Availability of fill materials on–site reduces construction cost. Drainage structures are often needed to drain the area behind the levee or floodwall. Typically used drainage structures include: diversion ditches, gravel–filled trenches, tile drains, sumps, and/or pressure conduits. If seepage problems occur, it may be necessary to construct a compacted impervious core or sheet–pile cut–off extending below the levee to bedrock. Excess seepage can be collected with gravel–filled trenches or drains along the interior edge of the levee or floodwall. Vegetation or rip–rap can be used to protect levee bank slopes from erosion. Upslope interceptor ditches, diversions, or grassed waterways may be used to prevent backwater flooding from runoff falling on the drainage area behind the levee or floodwall.

Levees and floodwalls are most suitable in flood fringe areas or areas subject to storm tide flooding. They are not suitable for areas with direct open floodways. Federal Emergency management Agency (FEMA) regulations may limit the use or placement of floodwalls and levees. Hydraulic analysis of the impact of the embankment on flooding characteristics of the waterway may be required. Flooding from storm runoff behind a levee and/or floodwall may be a problem; reduced flow storage capacity increases potential for downstream flooding.

Sedimentation Basins/Ponds: These are used to control suspended solids entrained in surface flows (impedes surface runoff carrying solids, allows sufficient time for particulate matter to settle); used in control of diverted surface runoff.

Sedimentation basins remove suspended solids from waterways through gravitational settling. A sedimentation basin is constructed by placing an earthen dam across a waterway or excavated area. It consists of the basin, a principal spillway, and anti–vortex device, and an emergency (overflow) spillway. The principal spillway consists of a vertical pipe (or riser) jointed to a horizontal pipe (barrel) that extends through the dike and outlets beyond the basin. The riser is topped by the anti–vortex device which improves the flow of water into the spillway and prevents floating debris from exiting the basin. Water discharge from the sediment action basin is typically directed toward an existing, stable stream. Additional measures (such as impact basin, rip–rap, excavated plunge pools, and stone facing) may be implemented to protect against scour (erosion).

The size of the sedimentation basin is dependent upon the particle size distribution of the suspended solids, the inflow concentration, the volumetric flow rate, the desired concentration of suspended solids, and the water flow rate to the pond. Given this information, the required area of the sedimentation basin can be calculated.

Seepage/Recharge Basins and Ditches: Intercepts runoff and recharge the water downgradient from the site to minimize groundwater contamination and leachate problems. There are several construction designs for seepage basins and ditches. Typically, a seepage basin consists of an excavated basin, a sediment trap, a bypass for excess flow, and an emergency overflow area. The sidewalls of the basin are constructed of previous material to allow for recharge.

Seepage ditches are usually constructed in parallel with runoff moving through drains set in gravel ditches. Improved percolation occurs when gravel–filled trenches are constructed along the basin floor. Dense turf on the basin sidewalls will prevent erosion while permitting a high infiltration rate.

Terraces and Benches: They control erosion by reducing slope length (terraces); intercept and divert surface water flow (benches). Terraces and benches are embankments, or combinations of embankments, constructed across long or steep slopes. In climates where rainfall is frequent and/or heavy, benches and terraces are typically constructed in association with drainage channels so that concentrated surface flows can be intercepted and transported offsite. Drainage benches may be seeded, mulched, sodded, rip–rapped, chemically stabilized, or lined with concrete or grouted rip–rap (the latter two techniques are more costly alternatives.

Benches are generally designed with sufficient height and width to withstand a 24–hour, 25–year storm. Generally, the spacing between drainage benches should be more frequent for long, steep slopes with erodible soil cover. Structures must be stabilized as soon as possible after grading and compaction.

Terraces and benches are an effective control in areas of high precipitation and can be used for long and steep slopes above, on, or below disposal sites. Terraces and benches should be periodically inspected, especially after heavy rainfall events.

3.3.3 Filters

Filter layers must provide for long–term movement of water through the layer while at the same time limit the movement of waste or soil particles across the layer. Too tight a filter will quickly clog while too loose a filter will result in an excessive loss of solids through the filter. Biological growth can also impact filter layer performance and is currently being studied by the EPA.

Sand/Gravel Filter: A soil filter is used to prevent very fine soil and waste particles from entering into a drain, accumulating, and eventually clogging the drain. Typically, soil filters consist of sand and/or gravel which has been screened to a specified particle size. The sand/gravel filter should have a particle size smaller than the drain particle but larger than the infiltrating particle. The filer may consist of one layer or several successively graded layers depending upon the performance objectives of the designer.

Geotextile Filter: A geotextile filter is used to prevent very fine soil and waste particles from entering into a drain, accumulating and eventually clogging the drain. The geotextile filter is selected by the designer based on its opening size and permittivity. The installation of the geotextile should be inspected to verify that panel overlap or sewn seams meet manufacturer's requirements. Additionally, any folds or wrinkles or damage to the panels must be eliminated and temporary restraint provided if necessary.

3.3.4 Erosion Control

Final cover systems on waste containment systems must be designed to limit the infiltration of surface water while at the same time require only limited maintenance for an extended period of time. Maintenance on such cover systems is significantly influenced by the degree of erosion that is allowed. For example, the EPA suggests that erosion be limited to less than 2 tons of soil per acre per year. The selection of a final cover system may also be influenced by the end use of the cover, e.g., park, or climatic conditions, e.g., lack of rain.

Vegetation and Topsoil: Surface vegetation may be the most economical erosion control system in those regions where rainfall exceeds evapo–transpiration. The vegetation will typically be a native grass tolerant of local climatic conditions. It should also limit spontaneous vegetation by non–desirable plants, germinate rapidly, and be compatible with the cap profile. Vegetation having exceptionally aggressive tap roots should be avoided.

Topsoil is used to support the growth of the vegetation on the cap and other locations that require vegetation. The topsoil is usually obtained on site from a stockpile cut from the construction area or from a nearby borrow area. Project specifications are typically vague regarding topsoil properties, but the organic content of the topsoil should be at leat 3 to 5% to support plant growth.

Revegetation refers to the establishment of a vegetative cover to stabilize the surface of a hazardous waste disposal site. It is frequently preceded by grading and capping, particularly for final cover system designs for waste disposal sites. The process of revegetating a site involves the selection of a suitable plant species, seedbed preparation, seeding/planting, mulching and/or chemical stabilization, and fertilization and maintenance.

Various types of grasses, legumes, shrubs, and trees may be used for vegetation. Generally, grasses provide a quick and lasting ground cover with dense root systems that anchor the soil and enhance infiltration. Legumes are most suited for stabilization and erosion control and enhancing soil fertility (through nitrogen fixation). Shrubs provide a dense surface cover and tend to be more tolerant of acidic soils and other disposal site stresses. Trees provide a long–term protective cover and aid in developing a stable, fertile layer of decaying leaves and branches. Gas migration controls may be required.

Temporary stabilization via straw–bale check dams, mulching, or chemical methods, may be required while vegetation is being established. Also, in cases where revegetation is to be part of a final cover system, it is important to consider the expected root system when selecting the vegetative species, because the roots can interfere with the cover system, e.g., by penetrating liners, etc.

Rip–Rap Cap: Rip–rap consists of natural stones ranging in size from approximately 1 inch to stones that weight hundreds of pounds. Layers of these stones provide a significant impediment to wind and water related erosion. Rip–rap layers are commonly underlain with a geotextile filter to limit potential erosion of underlying fines.

Placement of stone or rip–rap should be monitored to ensure that drop heights are limited. An underlying geotextile should be inspected for damage due to stone placement. When larger stones (>60 pounds) are placed over a geotextile fabric, drop height stone is commonly limited to 18 inches (45 cm). Alternately, a soil layer of 6 inches (15 cm) can be placed over the geotextile to provide a cushion and protect it from damage during rip–rap placement.

3.3.5 Protective Layers

Waste containment systems and covers frequently include layers that are intended to protect functional layers. Outer protective layers include the surface erosion control layers discussed above. This section discusses interior protective layers that function even after they are buried within the system. While some of the protective functions may be short–term or seasonal, e.g., a protective soil layer laced over a liner to protect it from freezing, most protective functions are long–term and are essential to the success of the waste containment system.

Surface sealing is accomplished by covering or capping a waste site with a low permeable material to prevent water from entering the site, thus reducing leachate generation and also controlling vapor or gas produced. Covers or caps can be constructed from native soils, clays, synthetic membranes, soil cement, bituminous concrete, certain waste materials, or asphalt/tar materials. Capping is normally an economical technique, and because the surface is accessible, the cap can be monitored, maintained, and repaired.

Revegetation may be a cost effective method to stabilize the surface of a waste site, especially when preceded by capping and contouring. Vegetation reduces raindrop impact, reduces runoff velocity, and strengthens the soil mass, thereby reducing erosion by wind and water, and improves the site aesthetically.

Biotic Barrier: A Biotic barrier is used in the cap of a waste containment system to prevent small burrowing animals and plant roots from penetrating the drainage layer or the low permeability barrier. The biotic barrier usually consists of a 3 foot (1 m) thick layer of stone or cobbles. Vegetative intrusion can also be limited by herbicide impregnated geotextiles that provide time release protection.

Hardened Layers: Hardened covers provide an alternative to vegetative systems in arid regions that lack sufficient natural moisture. Additionally, hardened systems have been used to provide traffic and parking areas after closure of the waste containment facility. Asphalt and concrete hardened covers are normally limited to slopes less than 10 degrees.

An **asphalt cap** can be used to protect a capping system and provide a potentially usable area over a waste containment system, e.g., a parking lot. The asphalt cap can replace the vegetation, topsoil, drainage layer, and the biotic layer in the cap. In this application, the asphalt layer must provide the erosion resistance of the vegetation/topsoil, the lateral flow capacity of the drainage layer, and the protection of the biotic layer. The porous asphalt layer consists of an asphalt pavement system similar to that used in roadway construction.

During installation, the thickness, temperature, and density of the asphalt should be measured. Additionally, the weather must be monitored to avoid rain or cold temperatures that would hurt asphalt placement.

Like asphalt, a **concrete cap** can protect a capping system and provide a potentially usable area over a waste containment system, e.g., a parking lot. The concrete cap replaces the vegetation, topsoil, drainage layer and the biotic layer in the cap. The concrete layer is similar to that used in roadway construction.

Geotextile Protective Layer: A protective layer over the bottom liner, leachate detection layer and the primary leachate collection layer is required to protect these components from damage during construction and waste placement. The geotextile protective layer is normally a nonwoven material selected according to its unit weight (ounces per square yard. The heavier the nonwoven material, the more cushion it provides.

Geotextiles are either woven or nonwoven fabrics made from polymeric fibers. Woven geotextiles are fabrics made up of webbed fibers that run in perpendicular directions. For filtration, the spaces between the fibers are the most important element. These spaces or voids must be large enough to allow unimpeded liquid flow but be small enough to keep out invading particulates. The geotextiles also must be sufficiently strong to cover and reinforce the apertures, or openings, of the drainage materials they are meant to protect.

In nonwoven geotextiles the fibers are much thinner but far more numerous. The various types are needle–punched, resin–bond, and melt–bond. All contain a labyrinth of randomly oriented fibers that cross one another so that there is no direct line of flow. The fabric must have enough open space to allow liquid to pass through, while simultaneously retaining any upstream movement of particles. The needle–punched nonwoven type is very commonly used as a filter material.

A protective layer of soil or a soil/geotextile layer may be used on the surface of the liner system to protect the underlying geomembrane from construction damage during installation, loads imparted by the waste, weathering, erosion and abrasion, to increase friction, and to dampen potential chemical attack. Above the liquid level of the impoundment, coarser–grained (sometimes rubble) material placed over a geotextile protective layer is often used. The coarse material will generally be more stable on steeper slopes and will dampen wave action and run–up. An even coarser layer (e.g., rip–rap) may be applied over the geomembrane–protective layer to prevent erosion in larger impoundments. Below the liquid level, sand is often used to protect the geomembrane against puncture, and to dampen the effects of strong chemical or high–temperature waste inputs.

Geotextiles may be used at various places in the liner system as protective layers and, where appropriate, as slide–resistant interface materials between soil and geomembranes. On the top, primary, geomembrane liner, a geotextile may be used to shield the geomembrane from any larger, sharper particles in the overlying soil protective layer. A geotextile layer can be used to reduce the thickness of the soil protective layer from 18 inches (45 cm) to perhaps 12 inches (30 cm). Geotextiles may also be used between that geomembrane and the drain layer material, and again at the bottom of the drain layer atop the second geomembrane. Nonwoven geotextiles ordinarily have excellent protection properties, making them ideal for these applications. In general, the greater the mass per unit area, the greater the protection afforded by the geotextile.

Soil Protective Layer: A soil protective layer over the liner, leachate detection layer and the primary leachate collection layer is required to protect these components from damage during waste placement and from the extremes in the weather. The soil should be selected for its resistance to erosion, strength, and stability on the side slopes of the waste containment system. Typically, an on–site soil can be used as the protective layer. Special attention should be given to the installation of a soil protective layer over a geomembrane liner. Such installations require less rigorous compaction specifications for the first lift to avoid damaging the geomembrane during compaction.

Floating Covers: A floating cover is a temporary measure used to prevent overtopping of a waste lagoon prior to final closure; mainly used to cover drinking water supply reservoirs. Also controls volatile air emissions.

A floating cover consists of a synthetic liner placed over an impoundment. The liner is held up by floats, and anchored at the edges of the impoundment. The synthetic liner consists of a 36–mil or 45–mil thick, reinforced Hypalon, chlorinated polyethylene (CPE), or XR–5 material. The material must be tested for compatibility with the waste prior to use. Two basic types of floating cover designs are used. The most commonly used configuration consists of a large center float with several smaller floats attached perpendicularly to the center float. Rainwater is directed to a sump around the perimeter of the floating cover. The rainwater collected in the sump is periodically drained or pumped.

Another type of configuration, directs rainwater through channels in the middle of the cover. The channels consist of sand–filled tubes held at constant depth by floats on either side of the channel. Perforated collection tubes are connected above and parallel to the sand tubes. The collection tubes drain the rainwater off the cover.

3.3.6 Earthworks

Construction or closure of waste containment systems typically requires construction of earthen containment structures, such as dikes and berms, and the development of stable working benches (surfaces) over weak wastes or contaminated soils. Due to cost restrictions, earth work is usually done with either on–site or local soils. In view of the diverse nature of this material, close monitoring during construction is often necessary to achieve design conditions. Weak or soft spots in compacted soil are commonly detected by proof–rolling using a loaded dump truck. Any soil experiencing excessive rutting should be recompacted or excavated and replaced.

Grading: Grading is the general term for techniques used to reshape the surface of covered landfills in order to manage surface water infltration and run–off while controlling erosion. The spreading and compaction steps used in grading are techniques practiced routinely at sanitary landfills. The equipment and methods used in grading are essentially the same for all landfill surfaces, but applications of grading technology will vary by site. Grading is often part of an integrated landfill closure plan.

Grading is considered essential to the continued performance and reliability of a cap. The performance and reliability of a graded surface depends upon effective revegetation. Grading is easy to implement and can generally be performed by local contractors. Grading rarely poses safety hazards to field personnel when performed on a properly capped site. However, when grading is performed to improve site drainage at an uncontrolled site, it may pose a risk to worker safety where drums and explosives are buried near the surface.

Structural Fill: A structural fill is designed to support its own weight and that of any overlying systems without experiencing excessive deformation. Such fills are typically placed to develop a minimum shear strength or compressibility as assumed in design. Laboratory shear strength or compaction tests, made prior to construction, are used to establish acceptable moisture/density requirements for such fills.

Fill soils must be inspected at the site to ensure that they are suitable and have low plasticity (i.e., plasticity index <10) and no large stones (<6 inches). Compacted soils should be tested to verify that they achieve the minimum dry density established by the project specifications.

Soil Bedding Layer: A soil bedding layer is used to level the waste surface immediately below the cap. The bedding layer provides a smooth and stable working surface for the construction of the cap and is usually made from an on–site soil with good strength properties. Other than proof–rolling and measuring the final grade, little field testing is performed on bedding soils.

Geotextile or Geogrid Bedding Layer: A geotextile or geogrid bedding layer is used to level the waste surface and bridge any voids immediately below the cap. Additionally, the bedding layer provides a smooth and stable working surface for the construction of the cap. The geotextile or geogrid must have high tensile strength and be puncture resistant.

The installation is usually monitored to confirm that geotextile or geogrid overlaps meet the manufacturer's specifications. Six (6) inch (15 cm) overlaps are common.

3.3.7 Hydrodynamic Controls

Hydrodynamic controls are employed to isolate a plume of contamination from the normal groundwater flow regime in order to prevent the plume from moving into a well field, another aquifer, or surface water. Isolation of the contaminated plume is accomplished when uncontaminated groundwater is circulated around the plume in the opposite direction of the natural groundwater flow. The circulated zone creates a groundwater (hydrodynamic) barrier around the plume. Groundwater upgradient of the plume will flow around the circulated zone while groundwater downgradient will be essentially unaffected.

Well systems are used for hydrodynamic control of contaminated plumes by manipulating the hydraulic gradient of groundwater through injection and/or withdrawal of water. The three general classes of well systems include (1) well point systems, (2) deep well systems, and (3) pressure ridge systems. All three types of well systems may require the installation of several wells at selected sites.

Well point systems consist of several closely–spaced, shallow wells connected to a main header pipe which is connected to a suction lift pump. Well point systems are used only for shallow aquifers because of the drawdown limitations as determined by the static water level and the limits of the pump. These systems should be designed so that the drawdown of the system completely intercepts the plume of contamination.

Deep wells are similar to well point systems except they are used for greater depths and are normally pumped individually. These wells are used in consolidated formations where the water table is too deep for economical use of suction life systems. Since the maximum depth for suction lift is around 25 feet, deep wells normally employ jet ejector or submersible pumps, or eductor well points.

Pressure ridge systems are produced by injecting uncontaminated water into the subsurface, through a line of injection wells, either up–gradient or downgradient from a plume of contamination. Up–gradient ridges or mounds are used to force up–gradient uncontaminated groundwater to flow around a contaminant plume while the contaminants are being collected by a line of down–gradient pumping wells. The procedure increases the velocity of groundwater into the plume and to the recovery wells, and serves to wash the aquifer. Pressure ridge systems located down–gradient are normally used in combination with up–gradient pumping wells which supply uncontaminated injection water. In either case the injection of fresh water produces an uplift or mound in the original water table which acts as a barrier by forming a ridge which pushes the contaminated plume away from the mound.

3.3.8 Gas Control

For Superfund sites, where a fugitive air emission problem exists, an enclosure structure can be very effective during the excavation and transportation of waste. The enclosure ventilation air will be routed through an emission–control system to prevent the escape of significant air emissions into the area surrounding the excavation zone.

During the trial excavation at the McColl site, a rigid–frame, PVC covered enclosure structure was erected over part of the L–4 Sump prior to the start of excavation. The enclosure proved to be effective in preventing the escape of air emissions during excavation.

The enclosure created a confined work space in which temperatures were approximately 20°F above the outdoor temperature. During the trial excavation, diesel engines were operated on the trackhoe, backhoe/loader, Bobcat, and pug mill. The emissions inside the enclosure resulting from these engines directly contributed to work stoppages due to low visibility, and high THC levels. The exhaust gases from diesel engines add heat, particulate matter, and hydrocarbon species to the enclosure air (SO_2 contributions were no doubt small because of the low sulfur content in diesel fuel).

The high emission levels of SO_2 and THC measured for the tar and char waste materials during the trial excavation caused work stoppages. These were due to health and safety concerns, and interference with equipment steering and braking systems. Since the ventilation air flow rate was fixed, this system was not able to provide enough fresh air to keep pollutant concentrations below design levels.

For landfills, two gas control strategies, passive and active, are available, and may be used at any facility. Passive systems provide corridors to intercept lateral gas migration and channel the gas to a collection point or a vent. These systems use barriers to prevent migration past the interceptors and the perimeter of the landfill. Active systems generate a zone of negative pressure to increase the pressure gradient and, consequently, the flow toward the zone. Active systems also can be used to create a zone of high pressure to prevent gas migration toward the zone.

Passive gas control systems alter subsurface gas flow paths without using mechanical components. Generally, subsurface flow is directed to points of controlled release through the use of high permeability systems, while flow paths to protected areas are blocked through the use of low permeability systems. High permeability systems consist of trenches or wells excavated at the boundary of the landfill and backfilled with a highly permeable material (e.g., coarse, crushed stone). Gas flow is directed to the trench area because its higher permeability is more conducive to gas flow than the surrounding less permeable areas. Low permeability systems, consisting of clay–lined or synthetic–lined trenches are used to block the paths of diffuse gas flow. Gases will then travel through either the ground surface between the barrier and the landfill or through the surface of the landfill. Often, high permeability and low permeability systems are used in combination to control subsurface gas flow.

The maximum recommended depth for the trench is 3 feet. Trench effectiveness is improved by constructing a low permeability system at the perimeter of the high permeability trench to prevent migration past the high permeability trench. Migration underneath the trench can be prevented by extending the trench to bedrock (or impervious strata). Installation of riser pipes and capping of the landfill further facilitates gas movement by enhancing the trench as the path of least resistance. Infiltration of precipitation and/or runoff limits the effectiveness of trench vents. If capping is not employed in conjunction with passive trench vents, then the trenches should not be located in areas of low relief (a slope can be constructed along the trench to control runoff).

Active perimeter gas control systems use mechanical means to alter pressure gradients to redirect the paths of subsurface gas flow. Major components generally include: gas extraction wells, gas collection headers, vacuum blowers or compressors, and gas treatment or utilization systems.

Gas extraction wells can be installed in the landfill or in the soil area surrounding the landfill. They are normally drilled to either the depth of the seasonally low groundwater table or to the base of the landfill. A pipe, which is solid at the top and perforated at the level where the gas is to be collected, is set in crushed gravel (or other permeable material). The area surrounding the pipe at the top of the well is sealed with concrete or clay. The upper portion of the pipe is connected to a gas collection header. The gas collection header is connected to several extraction wells spaced at regular intervals. Vacuum blowers or compressors are used to create a negative pressure area, which causes gases to be drawn up from the extraction well. The gases may subsequently be treated and released to the atmosphere, or recovered for use as fuel.

This design is applicable where site conditions allow drilling through landfilled material to the required depth. Well spacing is a critical factor in the design of the systems. Typically, 100 ft spacing is used. However, appropriate spacing depends upon several factors, including: landfill depth, type of waste, moisture content of waste and surrounding soils, percent compaction of waste, grain-size distribution of surrounding soil, stratigraphy, and soil permeability.

3.3.9 Leachate Collection and Removal Systems (LCRS)

Each leachate collection and removal system, whether above (primary) or between (secondary) the liners, consists of the following components:

1. A low-permeability base which is either a soil liner, composite liner, or flexible membrane liner (FML);
2. A high-permeability drainage layer constructed of either natural granular materials (sand and gravel) or synthetic drainage material (geonet), which is placed directly on the primary and/or secondary liner, or its protective bedding layer;
3. Perforated leachate collection pipes within the high-permeability drainage layer to collect leachate and carry it rapidly to the sump;
4. A protective filter material surrounding the pipes, if necessary, to prevent physical clogging of the pipes or perforations;
5. A leachate collection sump or sumps, where leachate can be removed;
6. A protective filter layer over the high-permeability drainage material which prevents physical clogging of the material; and

7. A final protective layer of material that provides a wearing surface for traffic and landfill operations.

The primary LCRS acts as a leachate collection system to remove leachate from the landfill or waste pile before it can leak through the primary liner component. This system also aids in reducing leakage by removing or reducing the hydraulic head of leachate that exists within the system. The secondary LCRS acts as a leak detection system by collecting and removing any leachate which leaks through the top or primary liner component of the liner system.

The most common method of leachate treatment is pumping to a holding area, and hauling to a waste treatment plant. Due to high costs, on-site treatment facilities are being investigated, using artificial wetlands, or chemical processes.

Another method gaining acceptance is recirculation, where leachate is pumped to the top of the landfill to filter again through the solid waste.

3.3.10 Soil Barrier Alternatives

A 3 ft (0.9 m) thick layer of low-permeability, compacted soil is a required component of secondary liners for hazardous waste landfills and surface impoundments regulated under the Hazardous and Solid Waste Amendments (HSWA to the Resource Conservation and Recovery Act (RCRA) [EPA, 1985]. The recommended designs for cover systems over RCRA hazardous waste landfills and closed surface impoundments include a 60 cm thick layer of low-permeability, compacted soil [EPA, 1989]. Minimum design requirements for liner and cover systems for non-hazardous waste landfills vary from state to state, but many include a layer of low-permeability compacted soil.

Waste disposal facility owners and operators seeking RCRA permits, and responsible parties seeking designs for closure of remedial action sites are requesting approval of commercial materials and soil treatment processes as alternatives to several thickness of compacted soil. Alternative barrier materials possess both advantages and disadvantages in each unique application.

The main function of low-permeability, compacted soil is either to restrict infiltration of water into buried waste (in cover systems) or to limit seepage of leachate from the waste (in liner systems). Other objectives may include enhancement of the efficiency of an overlying drainage system, enhancement of the effectiveness of an overlying geomembrane, adsorption and attenuation of leachate, restriction of gas migration, and others. In the case of a cover system, compacted soil must also be able to withstand subsidence or differential settlement, and must be repairable if damaged by freezing, desiccation, or biologic intruders. For liner systems, the liner must be able to withstand chemical degradation from the liquids to be contained. In addition, low permeability compacted soil must have adequate shear strength to support itself on slopes and to support the weight of overlying materials or equipment.

Fundamental compositional and structural differences between compacted, low-permeability soil and alternative materials create inevitable differences in hydraulic properties, solute attenuation capacity, time of travel of chemical compounds, structural strength, desiccation resistance, freeze/thaw resistance, reaction to settlement, ease of repair, and useful life.

An alternative barrier material, in order to be fully equivalent to a compacted soil layer, must serve the same functions as compacted soil. Due to inherent differences in the composition and construction of compacted soil and alternative materials, the two categories of barrier structures can never be "equivalent" in all possible respects. For example, compacted soil is usually from 2 to 5 ft (0.6 to 1.5 m) thick, whereas the alternative barriers are all typically a fraction of an inch to possibly a few inches (a few mm to a few cm) in total thickness. Due to differences in thickness, an alternative barrier is bound to be more vulnerable to puncture or other damage than a much thicker layer of compacted soil.

Materials such as cements, grouts, and asphalts which are applied as viscous liquids in layers one to four inches (2.5 to 10 cm) in thickness must maintain their integrity after curing. Shrinkage cracks which develop with time must not be allowed. The problem of quality control, for example, assuring consistent thickness of the applied material, has not been addressed in use of these materials in waste management structures. Soil particle binders, such as numerous types of organic polymers, must be proven stable over the expected lifetime of the waste management facility. Although many of these materials, such as polyacrylamides and urethanes, have proven applicability to agricultural soil sealing, their long-term structural and hydraulic performance in waste containment or infiltration prohibitation at hazardous waste sites has not been clearly demonstrated. Barrier materials created by binding mineral particles together are unlikely to possess contaminant sorption properties found in compacted soils.

Materials installed as discrete panels of impervious material suffer from lack of clear demonstration of seam integrity. Mechanical overlapping appears to be adequate with some materials, primarily bentonite blankets installed in cover systems using the "shingle" approach on sloped areas. Joint compounds and installed integrity proposed for rigid panels such as fiberglass, compressed concrete, or other materials must be demonstrated by objective studies before general acceptance can be recognized.

When the potential use of an alternative barrier is evaluated for a specific project, the critical functions of the barrier should be identified. "Equivalency" should be evaluated on the basis of the critical parameters and not necessarily upon all potential areas of comparison. Further, it must be remembered that all liner materials possess inherent advantages and disadvantages—no single type of liner material can be considered optimum for all applications. The site-specific design function of a waste containment liner, a precipitation infiltration barrier, or a groundwater control structure, must be the basis upon which alternative barriers are compared.

Alternative barriers have been claimed to possess several economic and technical advantages over compacted soil:

1. Installation proceeds rapidly and with relative simplicity;
2. A more predictable (than with compacted soil) end-product results where quality of a compacted soil has low assurance;
3. Cost may be as much as one-tenth that of compacted soil;
4. Much less volume is required, providing (1) more landfilling space available, (2) fewer truckloads of materials need for construction, (3) less settlement of underlying wastes because alternative materials may weigh less than thick soil;
5. Lighter construction equipment may be used, resulting in less stress on

underlying geosynthetic components;
6. Retesting of material may be unnecessary after an alternative material or process is initially thoroughly characterized;
7. Unique self–repair or contaminant sorption characteristics may be beneficial where bentonite or similar components are part of the barrier.

Disadvantages include several technical deficiencies in materials and the unknown quantity of future performance due to lack of credible testing data:

1. A general sparsity of objective and independent performance data;
2. Limited field performance experience and data, especially long–term;
3. Vulnerability to damage during construction, due to thin physical nature of materials;
4. Vulnerability of sodium bentonite to adverse chemical reactions with leachate constituents;
5. Unknown tolerance to settlement of underlying waste deposits;
6. Unknown effects of cyclic wetting, drying, freezing, and/or thawing upon bulk shrinkage;
7. Incomplete characterization of hydraulic and structural performance of overlapped seams under field conditions;
8. Potential for instability when installed on slopes common in landfill structures;
9. Unknown performance when overlain by a confining geomembrane which intensifies temperature differentials.

Dynamic Compaction Technology: The goals of dynamic compaction are to increase bearing capacity and decrease total and differential settlement. Dynamic compaction has seen limited use and has only limited documentation. Over the last 10 years, there have been about six examples of applying dynamic compaction to increase the density of wastes prior to capping or final closure of a site. The most recent application was at the Savannah River Plant in South Carolina.

In dynamic compaction, large weights of from 5 to 20 tons free fall from heights of up to 100 feet. The cumulative applied energies of this process typically range from 30 to 150 ft–ton/sq ft, and may succeed in densifying soil or waste down to 50 ft. The spatial distribution and the time sequence of dropping the weights are critical, and have been established by the industry. A few additional factors need to be taken into account, however, such as the effects on structures in the neighborhood, soil conditions, and soil characteristics in transmitting vibration effects. Seismic studies done many years ago with explosives can be applied to determine a safe distance from structures. Other problems requiring the designer's attention include present waste density, the geologic foundation underneath the waste site, impact of local or seasonal weather conditions on equipment operation, proof of effectiveness on a test section, and methods to be used to evaluate degree of success.

The economic advantages of dynamic compaction include reduction of differential settlement and subsidence and increased waste disposal volume. Stabilization of the site against settlement increases the reliability of a cover system in preventing water infiltration. Increases in waste disposal volume have been reported to offset the cost of dynamic compaction operations.

3.3.11 Daily Cover Materials

Daily cover functions to control disease vectors, blowing litter, odors, scavenging, and fires. It should also be effective under various operating conditions, permit controlled management of leachates and gases, and improve aesthetics. Because of its usual availability and traditional use at landfills, soil remains the most commonly employed material for daily cover. However, soil tends to consume landfill capacity, is not always readily and economically available or suitable under various operational conditions, and requires allocation of equipment and personnel. A number of alternative daily cover materials (ADCMs) have been developed.

Foams: Foam ADCMs are usually applied to the landfill working face in 2– to 6–inch– (5– to 15–cm) thick layers by using self–propelled or towed foam generation and application equipment specifically designed for a particular foam. Both hardening and nonhardening foams are available, and they retain their structural integrity from 15 hours to 7 days depending on the specific product and the effect of climatic conditions (particularly rainfall).

Spray–Ons: Slurry or emulsion spray–on ADCMs are applied to the working face using towed or skid–mounted application equipment, similar to hydroseeders but specifically designed for use with a particular product. These products are applied in a 1/16– to 1/2–inch– (0.16– to 1.27–cm) thick layer and allowed to dry to a crust or shell. Spray–ons can retain their matted structure from 1 week to 3 months depending on product and thickness and continuity of coverage.

Geosynthetics: Geosynthetic ADCMs consist of various types of geosynthetic materials that have either been developed or adapted for use as daily landfill cover. Panels fabricated from these materials are placed over the working face at the end of the day and retrieved before the start of the next operating day.

Indigenous Materials: Indigenous ADCMs may consist of various types of locally available waste products, including ash–based materials, shredded automobile components and tires, sludges and sludge–derived products, dredged materials, foundry sand, petroleum–contaminated soils, and shredded green wastes. Many of these same materials are routinely disposed of at landfills. Demonstrating their acceptability may require physical modification, chemical conditioning, or special analysis, since each can vary significantly with respect to physical and chemical characteristics and effectiveness under various operational and climatic conditions.

3.4 STRUCTURAL CONSIDERATIONS

In order for liners and other barriers to properly perform their function, foundations and side slopes of waste containment facilities should be properly prepared.

3.4.1 Foundations

Proper subsoil foundation design of a land disposal system is critical because liner system components, especially leachate collection pipes and sumps, can be easily damaged by stresses caused by foundation movement.

Foundations for hazardous waste land disposal facilities should provide structurally stable subgrades for the overlying facility components. The foundations also should provide satisfactory contact with the overlying liner or other system components. In addition, the foundation should resist settlement, compression and uplift resulting from internal or external pressures, thereby preventing distortion of rupture of overlying facility components. Foundations must also be designed to withstand hydrostatic and gas pressures.

Adequate site investigations are necessary to ensure that the foundation design is developed to accommodate expected site conditions. Site investigations are designed to establish the in-situ subsurface properties, site hydrogeologic characteristics and the area seismic potential, all of which are critical to facility design.

Total settlements of a few inches or less are usually not a problem for oil liner foundations, since most are sufficiently thick and flexible to withstand some differential settlement of the foundation. As long as the topography is fairly uniform and significant subsurface heterogeneities are not present, differential settlement should be minimal. Foundation settlement analyses based on the site's subsurface conditions (determined during site investigation) should be conducted during the design of the facility. These analyses should take into account the loadings of all facility components on the foundations, including footings for pile-type structures such as leachate collection risers, which, if improperly designed, can be forced into or through the liner. Compensated foundation, which implies that the weight of soil extracted from the site balances the weight of fill material, also can be used as part of the design to minimize subgrade settlement. In addition, the expected differential settlement should be compared to the design slope of the leachate collection system to ensure the latter's adequacy is maintained.

Landfill design calculations should include estimates of the expected settlement, even if it is expected to be small. Small amounts of settlement, even a few inches, can cause serious damage to leachate collection piping or sumps. The ability to predict the extent of settlement depends upon the type of process anticipated to cause settlement. There are several settlement processes, each of which should be considered in a land-based unit design, including: (1) primary consolidation, (2) secondary compression, and (3) elastic compression.

Foundations should be designed to control seepage and hydrostatic pressures. Heterogeneities such as large cracks, sand lenses, or sand seams in the foundation soil offer pathways for leachate migration in the event of a release through the liner and could cause piping failures. In addition, soft spots in the foundation soils due to seepage can cause differential settlement possibly causing cracks in the liner above and damaging any leachate collection or detection system installed. Cracks can also be caused by hydrostatic pressure where the latter exceeds the confining pressure of the foundation and liner.

Solutions to these problems include various systems that are available to lower the hydraulic head at the facility. These systems include pumping wells, slurry walls and trenching. Other methods to control foundation seepage include grouting cracks and fissures in the foundation soil with bentonite and designing compacted clay cut-off seals to be emplaced in areas of the foundation where lenses or seams of permeable soil occur.

For waste disposal units, the major issue of concern for foundations is differential settlement. However, for structures such as tank foundations, leachate risers, etc., an additional area of concern is bearing capacity.

The basic criterion for foundation design is that settlement must not exceed some permissible value. This value varies, dependent on the structure and the tolerance for movement without disruption of the unit's integrity. To ensure that the basic criterion is met, the bearing capacity must be established for the foundation soil. The bearing capacity of a soil, often termed its stability, is the ability of the soil to carry a load without failure within the soil mass. The load carrying capacity of soil varies not only with its strength, but often with the magnitude and distribution of the load. After the bearing capacity is determined, the settlement under the expected load conditions should be estimated and compared to the permissible value. The foundation design should be such that the actual bearing stress is less than the bearing capacity by an appropriate factor of safety.

3.4.2 Dike Integrity and Slope Stability

Most landfills and surface impoundments are constructed above natural grade through the use of earthen dikes, excavated below grade slopes constructed around the unit, or some combination of dikes and excavation, depending on site topography. Surface impoundments are often designed to achieve some balance of cut–and–fill, with the excavated soils used to construct the dikes. Landfill cells are excavated below grade in order to provide operating cover materials and to allow for restoration of the site after filling.

These excavated slopes and earthen dikes are vulnerable to stability failures via several mechanisms. Slope and dike failures at hazardous waste management units are potentially very serious; a surface impoundment failure can allow the sudden release of large amounts of hazardous waste to groundwater and surface waters, and a landfill slope failure can seriously damage the liner system, allowing releases of waste and leachate to surrounding soils and groundwater.

For these reasons, earthen dikes must be carefully designed and excavated slopes must be carefully evaluated to assure that they are sufficiently stable to withstand the loading and hydraulic conditions to which they will be subjected during the unit's construction, operation and post–closure periods.

One of the apparent differences between landfill and a surface impoundment unit is that of solid vs. liquid wastes. From the viewpoint of stability, however, there is no real difference; the forces on a slope exerted by liquids are modeled in a manner identical to those of solids. Another issue related to the impoundment of liquids is that of seepage through the dikes, causing piping or hydrostatic uplift pressures; however, this seepage condition is modeled in a manner identical to the condition of groundwater seepage at a cut slope. The failure mechanisms are similar for dikes and excavated slopes.

These three major failure modes are the following:

1. Rotation on a curved slip surface approximated by a circular arc.
2. Translation on a planar surface whose length is large compared to its depth below ground.
3. Displacement of a wedge–shaped mass along one or more planes of weakness in the slope.

In addition to the three major failure modes, dikes and excavated slopes are also vulnerable to failure due to differential settlement, seismic effects including liquefaction, and seepage–inducted piping failure. Safety factors are determined in a manner similar to the three modes.

In order to evaluate an existing, conceptual, or final slope design, the designer or reviewer must consider the following factors:

1. The adequacy of the subsurface exploration program.
2. The stability of the dike slopes and foundation soils.
3. Liquefaction potential of the soils in the dike and the foundation.
4. The expected behavior of the dike when subjected to seismic effects.
5. Potential for seepage induced piping failure.
6. Differential settlements in the dike.

Embankment construction for landfills or surface impoundments involves standard earthwork construction practices. Dike construction activities include fill placement and compaction, drainage system construction, and implementation of erosion control measures.

Compacted fill may be part of the dike core, the dike shell, or may constitute the entire dike. Critical construction activities include emplacement, conditioning, and compaction. To insure that these activities are conducted properly, the following measures must be taken:

1. Placing loose lifts to the thickness established during the test fill program.
2. Removing or reducing clod size material to a maximum size as determined in the test fill.
3. Providing uniform compaction coverage using the type of equipment and number of passes specified in the test fill program.
4. Ensuring uniformity of backfill material.
5. Protecting the surface lifts from desiccation or frost action.
6. Scarifying between compacted lifts.
7. Ensuring adequate connection between lifts.

Geonets are often used on the sidewalls of hazardous waste disposal facilities because of their stability and ease of installation. They should be placed with the top ends in a secure anchor trench and the strongest longitudinal length extending down the slope. They should be installed with a minimum of joints or seams on the slope. The geonets need not be seamed to each other on the slopes, only carefully butted or overlapped and tied. They should be placed in a loose condition, not stretched or in a configuration where they are bearing their own weight in tension.

Geogrids are very strong in transverse and longitudinal directions, making them useful as reinforcing materials for either soil or solid waste. Generally, they are used to steepen the side slopes of interior cells or exterior containment slopes of a facility. Recently they also have been used in the construction of "piggyback" landfills, i.e., landfills built on top of existing landfills, to reinforce the upper landfill against differential settlements within the lower landfill.

3.5 NATURAL UNDERGROUND BARRIERS

3.5.1 Deep Well Injection

Liquid hazardous wastes can be injected underground, and this activity is closely regulated.

U.S. EPA regulations (53 *Federal Register* 28118–28157, July 26, 1988) stipulate that deepwell injection of hazardous wastes is allowed only if either of two no-migration standards is met:

1. Fluid movement conditions are such that the injected fluids will not migrate within 10,000 years: vertically upward out of the injection zone; or laterally within the injection zone to a point of discharge or interface with an Underground Source of Drinking Water (USDW) as defined in 40 CFR Part 146.
2. Before the injected fluids migrate out of the injection zone or to a point of discharge or interface with USDW, the fluid will no longer be hazardous because of attenuation, transformation, or immobilization of hazardous constituents within the injection zone by hydrolysis, chemical interactions, or other means.

There are two important considerations relating to the avoidance of groundwater contamination. Firstly, the well should be dug through a thick layer of impermeable materials such as shale to avoid upward migration of the hazardous material. Secondly, there should be double casing through the saturated zone and some distance into the impermeable zone to avoid leaks into the groundwater.

Two ways to investigate interactions between injected wastes and reservoir materials are (1) direct observation of the injection zone and overlying aquifers using monitoring wells and (2) backflushing the injected waste. In both instances, samples of the fluids in the zone are collected at intervals to characterize the nature of geochemical reactions and to track changes over time.

Monitoring wells have several advantages: time–series sampling of the formation over extended periods is easy and the passage of the waste front can be observed precisely. Disadvantages are cost and the potential for upward migration of wastes if monitoring well casings fail.

The advantages of backflushing are reduced cost compared with that of monitoring wells and reduced sampling time (sampling takes place only during the test period). Disadvantages include less precise time– and distance–of–movement determinations and the need to interrupt injection and to have a large enough area for backflushed fluid storage before reinjection.

Gulf Coast injection wells are typically between 4,000 and 7,000 ft deep with temperatures up to 80°C. A typical injection zone is an arenaceous horizon containing up to 70 wt % detrital quartz, together with 15 wt % of detrital plagioclase and potash feldspars, with the remainder clay minerals with secondary calcite. Confining shale horizons typically consist of about 70 wt % clays with smaller amounts of other detrital minerals and secondary pyrite. A significant although minor amount of organic detritus is present in both shales and sandstones.

3.5.2 Natural Underground Barriers

There are a number of natural subterranean barriers that can act to prevent migration of hazardous waste. A few examples are given here:

(1) Hunter Industrial Facilities, Inc. has received draft permits to construct and operate

a salt dome repository for hazardous wastes near Dayton, Texas. The company plans to stabilize and pulverize all wastes before they are placed in man-made caverns constructed deep within the geologically secure North Dayton salt dome. The repository is designed to hold about 8 million cubic yards of solidified waste, which equates to 20 to 40 times the capacity of a typical hazardous waste landfill. The solidified waste will be permanently isolated from human activities, and groundwater aquifers.

The construction of solution-mined salt caverns is a technique that has been an accepted method of disposal for the past two decades. More than 1,000 salt caverns are currently in use worldwide, storing primarily crude oil, oil products, and natural gas.

Salt makes an excellent material as a geologic repository because it exhibits plastic flow. This property allows the salt to conform to changes in pressure or movement, inhibiting cracking of the dome. Also, it is assumed that since the salt domes have existed for millions of years, their environment is not threatening in any way, and they will continue to be extremely stable for millions of years to come.

In Germany, in the State of Lower Saxony, plans for the siting and permitting of a solution-mined salt cavern for final disposal of hazardous wastes is underway.

(2) Transuronic waste (TRU) is generated in the United States primarily by Department of Energy defense-related activities. This waste is primarily plutonium- and daughter product-contaminated material containing no fission products. There are a wide variety of chemical and physical forms ranging from plastic and paper to contaminated steel, slag, and ash. Most of the wastes are packaged in 55-gal (200-1) drums.

The current plan is to dispose of TRU wastes in the Waste Isolation Pilot Plant (WIPP) in Carlsbad, southern New Mexico. The WIPP facility is a recently completed repository in salt located 665 meters underground. Because it handles only TRU wastes, there is not a significant radioactive decay heat load in the repository. This will allow disposal of wastes in large rooms and simplifies demonstration of regulatory compliance. The WIPP facility is currently completing various performance analyses and tests to meet EPA standards. Assuming that the results are successful, the facility will operate for 5 years as a pilot plant and then proceed to full operation. Some opposition has been experienced.

(3) For disposal of spent nuclear fuel from commercial nuclear power generation, a site at Yucca Mountain, NV, is being evaluated as a deep geological depository. As in the WIPP mentioned above, there is opposition.

(4) The French are investigating the option of storing nuclear wastes in granite (the PACOMA project).

(5) The concept of disposing of high-level waste or spent fuel by burial in suitable geologic media beneath the deep ocean floor is a potential alternative to geological disposal on land. It is based on the same general principle: the main objective is to isolate waste from the biosphere in suitable geological strata for a period of time and in conditions such that any possible subsequent release of radionuclides in the environment will not result in unacceptable radiological risks, even in the long term. Seabed disposal is different from sea dumping which does not involve isolation of low level radioactive waste within a geological strata.

In the seabed concept a multibarrier system would be involved, including a suitable waste form such as glass and the use of corrosion-resistant packages. Deep seabed sediment formations would be chosen in order to contain radionuclides after the waste package fails through corrosion and the radionuclides are released from the waste form by leaching. Such sediments would be made up of very fine-grained particles with the ability to adsorb and impede the movement of most waste radionuclides. Sites in the ocean would have to be chosen on the basis of characteristics of the seabed sediments. They would need to be free from erosion and located away from the edges of tectonic plates where seismic or volcanic movements could disrupt a repository and expose the waste packages. Sites would also be located away from continental margins to avoid areas containing potential mineral and biological resources and away from areas of active pore water movement.

(6) The hydraulic cage concept is currently being considered as part of a pilot study in the German state of Nordheim Westfalen to evaluate the feasibility of using old coal mines for the disposal of hazardous and municipal wastes.

(7) In 1972, Kali and Salz established Untertage-Deponie (UTD) to operate an underground waste disposal facility in the mined-out section of the Kaliwerk Wintershall potash mine, which is located at Herfa-Neurode in the State of Hessen. This area is geologically unique, located in a stable salt formation formed 250 million years ago. The salt deposit is between 200 and 300 meters thick and is at a depth of about 700 to 800 meters. Overburden includes impermeable clay and shale layers and the salt deposit is virtually impervious to water. Storage in the caves is manifested so that special waste can be excavated later on for recycling, although 90 to 95% of the waste is probably permanently stored.

3.6 CONTAMINATED DREDGED MATERIAL

During dredging operations all dredge plants, to differing degrees, disturb bottom sediment, creating a plume of suspended solids around the dredging operation. The suspended solids plume can form relatively low concentrations in the upper water column, high concentrations near the bottom, or both, depending on the type of soil and the amount of energy introduced into the sediment by the dredge. The material suspended in the water column is often referred to as turbidity; the dense near-bottom suspensions are commonly called fluid mud or fluff.

Due to aesthetic and/or biological reasons, it may be generally advantageous to keep resuspension to a minimum. Limitations may be placed on levels of suspended solids when even normal dredging operations occur around public areas or coral reefs or during certain periods in the life cycle of a specific marine species. However, the major problems from suspended solids occur while dredging contaminated sediment. Contaminated sediment may release contaminants into the water column through resuspension of the sediment solids, dispersal of interstitial water, or desorption from the resuspended solids. Once resuspended, fine-grained sediment (clay and silt) tend to remain in the water column longer due to their low settling velocity. These fine-grained sediment fractions also have the highest affinity for several classes of contaminants, such as organics and heavy metals which have made their way into the waterway. For these classes of contaminants, the amounts that are dissolved or desorbed are negligible and basically all contamination transferred to the water column is due to resuspension of solids. Clearly, the control of sediment resuspension during dredging will reduce the potential for release of contaminants and/or their spread to other previously uncontaminated areas.

Most present dredging equipment and techniques evolved as a result of emphasis on economic return as measured by maximum production. Conventional dredges, therefore, are not specifically designed to operate in contaminated sediment and some modification of equipment or operating techniques may be necessary for such use. Fortunately, these modifications have been shown to be minor and generally within reasonable costs. The reason for this can be seen by looking at one definition of resuspension. Resuspension can be viewed as the difference between the amount of sediment loosened or disturbed from the bottom and the amount actually entrained and removed by the dredge. Therefore, the more efficient the dredging process is the less resuspension is likely to occur. Modifications to equipment or techniques that maximize removal without increasing the energy imparted to the sediment in the loosening process will reduce resuspension (and also increase production).

Minimizing sediment resuspension as described above reduces the potential for impact at the dredging site. However, two additional objectives of the dredging process are a result of disposal considerations. It is likely when dealing with contaminated sediment that disposal will be more costly as a result of controls, treatment, handling procedures, and limited disposal areas. In general, the dredging equipment and techniques should seek to minimize the total volume, and therefore, cost, of dredged material that must be handled in the disposal operation. Maximizing the solids concentration decreases the free water that must be treated as effluent and, by reducing the opportunity for extracting soluble contaminants, lowers the levels of contamination in that effluent. A second problem is simply over–dredging the site so that volumes of clean sediment become mixed with the contaminated material, increasing the overall volume that must be disposed of as contaminated. Certainly some deliberate over–depth or over–area is normally designed into a project to ensure complete removal of the contaminants. But, beyond this design, the dredging equipment and technique must provide sufficient control to remove relatively thin layers, e.g., contour dredging, and small hot spots.

In addition to operational factors, there are a number of barrier techniques available as described below.

3.6.1 Stream Diversion and Cofferdams

In some contaminated sediment areas, complete hydraulic isolation of sediments may be desirable so that dewatering followed by dry excavation may be implemented, or so that hydraulic dredging may be conducted in a contained environment. To accomplish stream diversion, all or part of the flow is diverted by cutting off a section of the stream, diverting flow to a pipe or excavated channel, and allowing the flow to re–enter the stream channel at a point further downstream. Stream diversion may be accomplished by placement of cofferdams, and may be temporary or permanent. Generally, permanent diversion is practical only for very small streams.

Cofferdams can be built around a contaminated area in a waterbody to isolate that area from stream flow. The area can then be dredged, dewatered, and excavated, or capped with low permeability material. Cofferdams are most easily constructed for flow containment of shallow ports, streams, and rivers, or waters with low flow velocities. Where flow velocity exceeds 2 feet per second, cofferdam construction is not recommended because of the difficulty of driving sheet piling under these conditions. Cofferdam construction is feasible for some relatively wide and deep rivers (up to about 10 feet), providing that the velocity of flow is not excessive.

Cofferdams may be constructed of many materials, such as soil, sheet piling, earth-filled sheet pile cells, and sand bags (for short duration structures).

3.6.2 Silt Curtains and Booms

One method for physically controlling the dispersion of near-surface turbid water in the vicinity of dredging (and some disposal) operations in quiescent environments involves placing a silt curtain or turbidity barrier either downcurrent from or around the operation. Silt curtains are not recommended for operations in the open ocean, in currents exceeding 1 knot, in areas frequently exposed to high winds and large breaking waves, or around hopper dredges or some cutterhead dredges where frequent curtain movement would be necessary.

Silt curtains are impervious floating barriers that extend vertically from the water surface to a specified water depth. The flexible nylon-reinforced polyvinyl chloride (PVC) fabric, or similar material, forming the barrier is maintained in a vertical position by flotation segments at the top and a ballast chain along the bottom. A tension cable is often built into the curtain immediately above or below the flotation segments (top tension) or some distance below the flotation (center tension) to absorb stress imposed by currents and other hydrodynamic forces. The curtains are manufactured in sections that can be joined together at a particular site to provide a curtain of specified length. Anchored lines hold the curtain in a deployed configuration that is usually U-shaped or circular.

In many cases, especially disposal applications, the concentration of fine-grained suspended solids inside the silt curtain enclosure may be relatively high (i.e., in excess of 1 g/ℓ), or the suspended material may be composed of relatively large rapidly settling flocs. In studying a typical pipeline disposal operation surrounded by a silt curtain, results showed that the vast majority (95 to 99%) of the fine-grained material descended rapidly to the bottom where it formed a low-gradient fluid mud mound. While the curtain provides an enclosure where some of the remaining fine-grained suspended material may flocculate and/or settle, most of the turbid water and fluid mud flow under the curtain. The silt curtain does not indefinitely contain turbid water, but instead diverts its flow under the curtain, thereby minimizing the turbidity in the upper water column outside the silt curtain.

Silt curtain effectiveness, defined as the degree of turbidity reduction outside the curtain relative to the turbidity levels inside, depends on several factors: the nature of the operation; the quantity and type of material in suspension within or upstream of the curtain; the characteristics, construction, and condition of the silt curtain as well as the area and configuration of the curtain enclosure; the method of mooring; and the hydrodynamic conditions (i.e., currents, tides, waves, etc.) present at the site. Because of the high degree of variability in these factors, the effectiveness of different silt curtain operations is highly variable.

Booms are similar to silt curtains and are used to confine contaminants that float, i.e., specific gravity less than 1. Booms tend to decrease advection, dispersion, and photolysis development configurations processes, and may increase volatilization.

3.6.3 Restricted Open–Water Disposal

Restricted open–water disposal as used here simply suggests that some one or more controls beyond those normally applied to conventional projects are required to address either known risks or uncertainties associated with disposal of contaminated sediments. Controls may range from an intensive long–term monitoring program with remedial action plan to a fully engineered and constructed aquatic disposal site. Most positive control measures are based on the concept of isolating the contaminants from the water column or benthic environment. Recently, concepts based on either the separation of contaminants from the dredged material slurry or chemically stabilizing the contaminants in the dredged material have also been proposed.

Dredged material may be transported to and placed at a disposal site in a variety of conditions and by a number of techniques. However, for practical purposes, it may be assumed that either barges/scows, hoper dredges, or a continuous pipeline will be used. Engineered control technologies for discharging include: (1) submerged discharge; (2) submerged diffuser, and (3) gravity–fed downpipe.

Lateral Confinement: An increaseed degree of positive control over the movement of the material placed at a site can be achieved by using lateral barriers to confine the disposed material. Such confinement can be accomplished by using depressions or contour irregularities existing at a site, by excavating such depressions, or by construction of subaqueous dikes, or berms. Lateral confinement addresses the short–term benthic impact by ensuring accurate initial placement and attenuation of the spreading dredged material. It also addresses long–term benthic and water column impacts by providing an inherent degree of isolation from the aquatic environment, reducing the effects of convective currents, and increasing the ease and effectiveness of capping when used.

Capping: Materials, both naturally occurring and man–made, that can be used to cover contaminated dredged material are divided into three categories: inert, chemically active, and sealing agents.

Inert materials include coarse– and fine–grained soils. When soil is used as capping material, the cap should be thick enough to protect the underlying deposit from disturbances caused by storm–generated waves and by propeller wash from navigation traffic and to bury the contaminated sediments out of the reach of benthic organisms.

Capping with chemically active materials involves the placement of a chemical compound over the contaminated dredged material that would react with the contaminants to neutralize or otherwise decrease toxicity. This strategy differs from the use of inert materials in that each contaminated dredged material must be dealt with on a case–by–case basis. Carbon compounds are a common example of chemically active ingredients that can be added to a cap. In the capping of dredged material, the active material should be combined with an inert stabilizer to provide stability to the cap. Another approach would be to cover the active covering layer with an erosion–resistant inert layer. The inert layer would also provide protection for the benthic organisms. While the inert covers have little or no chemically related impact on the organisms, the chemically active covering agents could be harmful to some organisms. Also, greater accuracy would be required for placement of the chemically active materials.

Sealing agents include grout, cements, and polymer films. The unique feature of the grouts and cements is that, when placed on top of contaminated sediment, they will harden and form a crust, preventing erosion and resuspension of the contaminated material. A Japanese firm (Takenaka Komuten) has done work in dredged material stabilization and deep mixing of sediments using grouting compounds. Also, grouting is often used in the offshore oil industry for stabilization of oil–producing facilities. The technology for using grout in the saltwater environment is well developed, and it could be adapted for use in capping contaminated dredged material. However, there are some disadvantages associated with the use of grout in capping dredged material. The thin layer of grout placed over the contaminated material cannot be considered as a permanent cap material. It should be used with a covering of inert material to provide additional stability and habitat for benthic organisms. There could also be problems with the grout cracking as the contaminated dredged material consolidates with time.

Large–mounted deployment systems have been proposed for either hot or cold application of polymer film overlays. The application systems included those for placing coagulable polymers, hot–melt materials, and preformed commercially available films. The application system for the preformed overlay limited its application to water depths of 25 to 30 feet. Concepts for the use of polymer film overlays for cover of contaminated dredged material were developed from early erosion–control efforts related to marine salvage work. The major limitation to these concepts involves the design, construction, and cost of capital equipment required to place them.

Stability of the capping material is a major concern in the design of capping projects. Factors influencing cap erosion include: (1) the particles (size, uniformity, shape, size distribution, texture, etc.; (2) the hydrodynamics of the system; 3) slope of the mound; and (4) the degree of cap material cohesiveness.

A resulting problem with underwater capping is the potential for displacement of the contaminated mass by capping. Depending upon substrate firmness and density of the contaminated mass, the cap material may displace and redistribute the contaminated material, especially if the capping material is of a higher density or coarser size than the contaminated material. Determination of the potential for mass failure and dispersion would require testing with materials physically similar to those which will be placed underwater.

3.6.4 In–Situ Control and Containment

In responding to a situation where bottom sediments are contaminated with hazardous substances, it is sometimes technically infeasible or economically unreasonable to remove all of the contaminated material from its location in the watercourse. If removal is determined to be an unacceptable singular remedial response, in–situ control and containment measures are often considered. These measures are intended to reduce dispersion and leaching of a hazardous substance to other areas in the water body. They may be temporary or permanent response measures.

Inasmuch as in–situ control techniques are similar to those for restricted open–water disposal, discussed earlier, they will be mentioned only briefly.

Retaining Dikes and Berms: Retaining dikes and berms include earthen embankments, earth–filled cellular and double sheet pile walls, water inflated dams and other materials which can be used to minimize transport of contaminated sediments.

Retaining dikes or dams can be constructed perpendicular to the direction of stream flow, or downstream of a dredging operation in order to prevent suspended particulate matter from flowing downstream. This type of dike creates the effect of a holding pond or reservoir, which prevents flow downstream and also promotes the settling of fine particles. The damming creates deeper areas where water velocity is slower and allows more time for small particles to settle. Retaining dikes used for this application are limited to streams with low flow. A water inflated dam constructed from reinforced urethane can also be used for this purpose.

Retaining dikes can also be constructed parallel to a river or stream bank to isolate contaminated deposits from the deeper river channel. When stabilized with vegetation, capping or some other method, these dikes can provide permanent containment.

Cover Techniques: A wide variety of materials can be used to cover contaminated sediments in order to minimize leaching of contaminants and prevent erosive transport of contaminated sediments. Cover materials include inert materials such as silt, clay or sand and active materials or additives which react with contaminants to neutralize or otherwise decrease inherent toxicity. Potentially applicable active cover materials include: limestone and greensand for neutralization; oyster shells or gypsum for metals precipitation; ferric sulfate for both precipitation and base neutralization; and alum for base neutralization.

Cover materials have application for temporary or permanent containment for hazardous waste constituents. Their use is generally limited to protected open waters where bottom currents and flow velocity are generally not sufficient to erode the cap. Some of the active materials can be applied together with inert cover material to treat and contain the sediments.

The active covering strategy differs from the inactive covering strategy because each waste constituent must be evaluated on a case–by–case basis, whereas the performance of the inert materials is not as strongly affected by the waste constituents.

With active cover materials, the criteria for selection are different. If active cover materials are to be used successfully, they must remain in place long enough to react with and treat the contaminants. Limestone and gypsum are pozzolanic in nature and tend to form a thick, cement–like cover that is resistant to erosion. Ferric sulfate, alum, and alumina are very fine–grained and can be expected to behave like clays. Greensand and oyster shells will probably scour most easily, and it may be necessary to mix these materials with a more stable inert cover material.

Surface Sealing: Cement, quicklime, or other grouting materials can be applied to the surface of or mixed with bottom sediments to create a seal which minimizes leaching and erosive transport of contaminated sediments.

Grouts may be applied to the surface of bottom sediments using a number of approaches. These methods can generally be divided into two categories: those which involve stream diversion and those which do not.

Surface sealing methods which involve the use of stream diversion are limited to shallow waters with a low flow velocity, where diversion can be accomplished cost–effectively. The major advantage of this method is that it is unlikely to stir up the sediments and create downstream contamination. Stream diversion also simplifies the application of grouts or sealant materials.

Sealing methods which do not employ diversion are applicable to deep open water, where bottom currents are not sufficient to erode the cap. These methods will provide less resuspension of bottom sediments than in–situ injection methods. Also sealing methods such as concrete pumps can potentially be used in confined areas not accessible to barge–mounted injection systems. However, the grout or sealant may impact the water column during application, application methods would be slow, and it may be difficult to obtain complete coverage.

Following completion of the sealing or stabilizing operation, the sediment bottom can be restored to its natural grade and sediment composition in an effort to restore the habitat for benthic organisms.

In–Situ Grouting: The stabilization of contaminated sediments can be achieved through the injection of grouting materials into sediments. Chemical injection using clay–cement or quicklime has been used widely, particularly by the Japanese, for stabilizing bottom sediments prior to the construction of port and harbor structures. A commonly used Japanese method for grouting with clay–cement is the Deep Cement Mixing Method. The system consists of a number of injection pipes mounted on a barge; the injection pipes are connected to mixing pipes that enter the sediments. Similar equipment is available for deep mixing with quicklime. The process is completed by lowering the operating–mixing apparatus (mixing blades are located within the individual shafts) to the required depth and injecting a cement or lime–based slurry into the sediments. The mixing blades are then reversed and the shafts are removed and relocated (Takenaka Doboku, Co. Ltd.).

A number of other types of grout injection and mixing apparatus are available. Continuous mixing apparatus are available and eliminate the need to continuously raise, relocate and lower the mixing apparatus.

Theoretically, in–situ grouting methods could be used to stabilize sediments to depths of about 80 to 130 feet below sea bottom. However, the feasibility and reliability of these methods for contaminated sediments has not been demonstrated. The use of in–situ methods is restricted by barge orientation, which limits offshore activity to calm waters and periods of good weather. Injection may result in considerable resuspension of sediments.

3.7 SPILL CONTAINMENT

Containment methods prevent the spread of a released hazardous substance by stopping or catching the release or by upstream diversion of a receiving stream. The path of the released hazardous substance may be blocked with barriers or diversions which are either preformed or constructed on site. Containment facilitates subsequent handling of the hazardous substance. Containment methods may include the following: dikes, berms and dams; trenches; booms; barriers in soil; stream diversion; patching and plugging of containers or vessels; portable catch basins; overpacked drums or other forms of containerization; and reorientation of the container.

Dikes, Berms, and Dams: Dikes, berms, and dams may be necessary to contain the spilled or spilling material on land or in water before effective collection can be initiated. Retention dikes and underflow dams may be used to contain floating insoluble materials; however, dike usage is limited to either containment of an entire water body or development of a diversion pathway. Dikes, berms, and dams may consist of earth, sediment, gravel, coarse sand, or polyurethane. Gravel and coarse sand are permeable materials that can be overlaid with reactively impermeable materials, such as clay.

Trenches: Trenches or excavations are also a first step to contain a spilled material prior to treatment. They are effective and relatively inexpensive containing measures which can be used for landspills involving liquids and water spills involving insoluble sinkers. Trenches on land generally require the use of large earth moving equipment such as bulldozers, and usually take advantage of natural conditions and slope to aid in collection of the spilled material. Trenches may also be constructed with small equipment or hand tools in some cases. Trenches are often used to contain the run–off when a knockdown spray is used to reduce vapor hazards associated with volatile liquids. Trenches used to contain volatile liquids will provide very little reduction in the evaporation rate. Dredging equipment such as land–based clamshells, drag lines, and hydraulic and suction dredges are necessary for submerged trench construction to contain insoluble sinkers.

Booms: Booms are used to contain spills of hazardous materials in waterways. There are two general types of booms; surface booms, and sealed booms or curtain barriers.

Surface booms are used to contain water spills involving floating substances, including oil and hazardous materials. A primary concern is compatibility between the spilled substance and the boom material. The manufacturer should be consulted in all cases. If deployed rapidly, surface booms can also be used to contain any undissolved soluble floating substance. They are ineffective for containment of totally dispersed or sinking substances because commercially available surface booms have a draft of only 1 to 2 feet.

Curtain barriers, sealed booms or silt barriers are used to contain hazardous materials that are soluble or sink in water. They are also being used to control turbidity caused by dredging operations. When used with dredging, curtain barriers do not extend all the way to the bottom which makes then unsuitable for containing sinking hazardous substances. Curtain barriers have been designed for bottom to surface coverage. Barriers of flexible reinforced plastics have been developed for depths up to 25 feet. Buoyancy is provided by an air flotation collar.

Barriers in Soil: Barriers in soil are designed to prevent liquid hazardous substances from percolating into the soil and potentially contaminating the groundwater in the area around the spill.

The scenario for use of soil sealants is that they be applied immediately after a spill, preferably in the path of an advancing spill. Sealant/spill compatibility should be verified prior to use. Sealants could also be used in conjunction with other containment measures such as dikes or trenches. Soil sealants and either dikes or trenches would, in conjunction, provide an adequate containment area for spilled material and, at the same time, eliminate most of the potential soil and groundwater contamination problems.

Soil surface sealants are generally grouped into three categories: reactive, nonreactive, and surface–chemical. Nonreactive sealants have previously been polymerized and are dispersed as an aqueous or solvent system. Common nonreactive sealants include bitumastic, rubber, polystyrene and polyvinyl chloride.

Reactive sealants require two or more components to be mixed and reacted at the spill site. One component generally serves as a catalyst for the reaction. Included in this class are epoxy, urea/formaldehyde, and urethane sealants. These sealants are more likely to form a film under adverse weather, but they do have temperature limits. Surface chemical sealants are generally repellents such as silicon and fluorocarbon systems.

Stream Diversion: Stream diversion involves isolating a hazardous material spill by diverting the uncontaminated flow around the spill area. This is not a typical operation but could be accomplished by placing a dam, where possible, upstream of the impacted area, and then diverting the flow around the spill. Stream diversion is a potential countermeasure primarily when insoluble sinking hazardous substances are involved. The spill–impacted area will dry, thus facilitating cleanup of the hazardous materials.

Portable Collection Vessels: Portable collection vessels include any container or storage unit which can be easily transported to the site of a spill, assembled, and used to hold the hazardous substance, contaminated water, soil, etc., either for further treatment or hauling. Typically, spilled liquid materials or contaminated water are pumped into the vessels. It is feasible in some cases to place a collection vessel immediately under a leak or rupture in a liquid storage unit and catch the substance as it leaks out.

Overpacked Drums and Containerization: For spills from relatively small containers which have ruptured it is often possible to place the leaking container inside a larger specially designed container. Overpacking is a general term for this containment procedure.

Overpack drums are usually 85 gallon steel drums, lined with a chemically resistant synthetic material. Leaking containers are placed inside the overpack drum and sorbent material is used to fill in the space between the container and the walls of the overpack. After filling, the overpack drum is sealed. If the leaking containers are small enough, several of them can be placed in one overpack drum in layers. Each container is protected from the one above or below it by a layer or sorbent material. It is absolutely necessary that the contents of the overpack stay securely in place and are not allowed to shift inside the drum.

REFERENCES

1. Barcelona, M., et al, *Contamination of Ground Water, Prevention, Assessment, Restoration,* Noyes Data, 1990.
2. Cullinane, M., et al, *Contaminated Dredged Material, Control, Treatment and Disposal Practices,* Noyes Data, 1990.
3. Daniel, D., et al, *Compilation of Information on Alternative Barriers for Liner and Cover Systems,* EPA/600/S2–91/002, 3/91.
4. Ducharme, A., et al, *Matching Leading DOE Waste Management and Environmental Restoration Needs with Foreign–Based Technologies,* DOE, 10/92.
5. EPA, *Construction Quality Management for Remedial Action and Remedial Design Waste Containment Systems,* EPA/540/R–92/073.
6. EPA, *Demonstration of a Trial Excavation at the McCall Superfund Site,* EPA/540/AR–92/015, 10/92.
7. EPA, *Design and Construction of RCRA/CERCLA Final Covers,* EPA/625/4–91/025, 5/91.
8. EPA, *Ground Water Currents,* (OS–110W), EPA/542/N–93/006, 6/93.
9. EPA, *Guide to Technical Resources for the Design of Land Disposal Facilities,* EPA/625/6–88/018, 12/88.
10. EPA, *How to Meet Requirements for Hazardous Waste Landfill Design, Construction and Closure,* Noyes Data, 1990.
11. EPA, *Innovative Hazardous Waste Treatment Technologies: Domestic and International,* EPA/540/R–92/081, 12/92.
12. EPA, *Landfill Containment and Cover Systems,* EPA/600/A–92/200, 1992.

13. EPA, *Liners and Leak Detection Systems for Hazardous Waste Land Disposal Units,* Final Rulemaking, 57FR3462, 1/29/92.
14. EPA, *Municipal Solid Waste Landfill Regulations,* 56FR50978, 10/9/91.
15. EPA, *RCRA Corrective Action Stabilization Technologies,* EPA/625/R–92/014, 10/92.
16. EPA, *RREL Hazardous Waste Research Symposium (17th),* EPA/600/9–91/002, 4/91.
17. EPA, *Slurry Walls, Engineering Bulletin,* EPA/540/S–92/008.
18. EPA, *The Use of Alternative Materials for Daily Cover at Municipal Solid Waste Landfills,* EPA/600/SR–93/172, 9/93.
19. Goldman, L., et al, *Clay Liners for Waste Management Facilities, Design, Construction, and Evaluation,* Noyes Data, 1990.
20. Grube, W., et al, *Geosynthetic Liners Offer Cover Option,* Environmental Protection, 5/92.
21. Harris, P., *Landfills, A Sudden Leap Into the Limelight,* in Environment Today, 5/92.
22. Hartley, R., *Surface Impoundments, Design, Construction and Operation,* Noyes Data, 1992.
23. Heller, E., *Leachate Demanding Wastewater Solutions,* in Florida Specifier, 2/93.
24. Klos, T., *A Salt Dome Repository for Hazardous Waste,* Proceedings at HMC–South, 2/92, Hazardous Waste Consultant, 7–8/92.
25. Landreth, R., *Resistance of Membranes in Cover Systems to Root Penetration by Grass and Trees,* EPA/600/A–92/202.
26. Noyes, R., *Handbook of Pollution Control Processes,* Noyes Data, 1991.
27. Pettyjohn, W., *Protection of Public Water Supplies from Ground–Water Contamination,* Noyes Data, 1987.
28. Wagner, K., et al, *Remedial Action Technology for Waste Disposal Sites, 2nd ed.,* Noyes Data, 1986.
29. Walker, S., et al, *An Overview of In–Situ Waste Treatment Technologies,* EGG–M–92–342, 1992.

4

Immobilization Technology

Immobilization refers to a broad class of treatment processes that physically or chemically reduce the mobility of hazardous constituents in a waste. There are a number of techniques utilized for immobilization, and their definitions have varied, depending on the source. For the purposes of this chapter, the following definitions will be used.

(1) Stabilization: A process by which a waste is converted to a more chemically stable form. The term entails the use of a chemical to transform the toxic component to a new non–toxic compound or substance.

Chemical fixation, for the most part, has become synonymous with stabilization. However, there is a fine distinction between chemical fixation and stabilization. Chemical fixation is the chemical technology used to detoxify, immobilize, insolubilize or otherwise render a waste component less hazardous when introduced into the environment. It often denotes a chemical reaction between one or more waste components and a solid matrix that is either deliberately introduced or is already in the waste residue.

Almost all chemical fixation processes involve reactions between the waste and some proprietary additives that promote precipitation of soluble metal ions as insoluble hydroxides. These additives can enhance the curing reaction that solidifies the waste and it can increase binding between the waste and the solidifying reagents. The details of a stabilization (fixation) process vary depending upon the nature of the waste being treated. In turn, the physical and chemical properties of the treated products vary according to: (a) characteristics of the waste, (b) types of additives and solidifying reagents, (c) drying conditions, and (d) curing time. Most existing solidification/stabilization technologies were developed to treat inorganic wastes, primarily those that contain dissolved metals. Organic compounds generally interfere with the setting and curing of the solidifying agent.

The preliminary benefit of stabilization techniques is that they limit the solubility or mobility of the contaminants with or without changing or improving the physical characteristics of the waste. Stabilization usually involves adding materials that will ensure that the hazardous constituents are maintained in their least mobile or toxic form.

Alkaline stabilization is a term used by municipalities, for stabilization techniques in sewage sludge management. New forms of chemical stabilization other than lime treatment have been developed by vendors and are being used by municipalities. These technologies add alkaline materials such as cement kiln dust, lime kiln dust, Portland cement, or fly ash for the stabilization of sludge. Most of these technologies are modifications of traditional lime stabilization. The most common modifications include the addition of other chemicals, a higher chemical dose (depending on the chemical type), and supplemental drying. These processes alter the characteristics of the sludge and, depending on the process, increase the stability and physical strength, decrease the odor potential, and/or reduce pathogens.

(2) Solidification: A process in which materials are added to a liquid or semiliquid waste to produce a solid. It may or may not involve a chemical bonding between the toxic contaminant and the additive. They are normally cement, lime, pozzolans, gypsum or silicate based processes.

Solidification implies that the beneficial results of treatment are obtained primarily, but not necessarily exclusively, through the production of a solid block of waste material with high structural integrity—a product often referred to as a "monolith." (In many cases, a monolith is not the end product of the stabilization/solidification processes; however, after placement, the materials may continue to cure into a facsimile of a monolith.) The monolith can encompass the entire waste disposal site—called a "monofill"—or be as small as the contents of a steel drum.

(3) Solidification/Stabilization (S/S): Stabilization processes and solidification processes have different goals. Stabilization systems attempt to reduce the solubility or chemical reactivity of a waste by changing its chemical state or by physical entrapment. Solidification systems attempt to convert the waste into an easily handled solid with reduced hazards from volatilization, leaching, or spillage. The two are discussed together because they have the common purpose of improving the containment of potential pollutants in treated wastes.

Solidification/stabilization technologies have been applied widely and generally have been effective in immobilizing metal and other inorganic contaminants at hazardous waste sites. Solidification/stabilization technologies have been less effective in immobilizing organic contaminants, because solidification alone may not reduce the mobility and toxicity of hydrophobic constituents. Various organic compounds in the waste can interfere with the stabilization chemical reactions and bond formation, thus inhibiting curing of the stabilized material. In addition, treatment of wastes containing volatile organic compounds (VOC) by solidification/stabilization generally has consisted of partitioning VOCs to the air either through aeration (such as materials handling and mixing) or through heat of reaction with treatment reagents.

To constitute treatment under Superfund, a solidification/stabilization technology must demonstrate a significant reduction (90 to 99%) in the concentration of chemical constituents of concern. During the last 10 years, various innovative solidification/stabilization technologies have emerged that are capable of treating wastes containing organic as well as inorganic contaminants. These innovative solidification–stabilization technologies have involved the use of reagents that chemically stabilize contaminants in conjunction with solidification.

Stabilization and solidification waste treatment processes involve the mixing of specialized additives or reagents with waste materials to reduce physically or chemically the solubility or mobility of contaminants in the environmental matrix. The term 'stabilization' is used to describe techniques that chemically modify the contaminant to form a less soluble, mobile, or toxic form without necessarily changing the physical characteristics of the waste. Solidification refers to a technique for changing the physical form of the waste to produce a solid structure in which the contaminant is mechanically trapped. Many stabilization and solidification processes overlap, and the common terminology to describe either or both processes is stabilization/solidification (S/S). Goals of the application of S/S techniques include improving the physical and handling characteristics of liquid or semiliquid contaminated materials, reducing contaminant solubility, and decreasing the rate of transfer of the contaminant. It is important to emphasize that typically S/S does not provide for contaminant destruction and therefore may not be classified as a permanent solution.

Solidification/stabilization is the second most popular technology to treat soil and other wastes at NPL sites. It is selected to remediate metal–containing waste and continues to be the favored technology to treat this material. It can treat most chemical forms of metals, although some compounds are not easily solidified. In some cases, it is selected to treat organic contaminants, primarily semivolatiles (SVOCs). Although solidification/stabilization has several advantages, including low cost, questions remain concerning its effectiveness over time. Consequently, it may require long–term monitoring.

(4) Encapsulation: A process involving the complete coating or enclosure of a toxic particle or waste agglomerate with a new substance, e.g., the additive or binder. Microencapsulation is the encapsulation of individual particles. Macroencapsulation is the encapsulation of an agglomeration of waste particles or microencapsulated materials. These are usually polymer–based processes. Reactive polymer and polymerization techniques are also used.

(5) Sorption: Involves adding a solid to soak up any liquid present, and it may produce a soil–like material. The major use of sorption is to eliminate all free liquid. Nonreactive, nonbiodegradeable materials are most suitable for sorption. Typical examples are activated carbon, anhydrous sodium silicate, various forms of gypsum, celite, clays, expanded mica, and zeolites. Some sorbents are pretreated to increase their activity toward specific contaminants and many are sold as proprietary additives in commercial processes. Sorption is also discussed in Chapter 6.

(6) Vitrification: A high–temperature thermal process that converts sludges or soils into an obsidian–like material and pyrolyzes organic compounds. Also known as glassification. Vitrification technologies are those that involve exposure of hazardous materials to molten glass and related process conditions to affect the destruction, removal, and/or permanent immobilization of hazardous contaminants. Vitrification is defined as conversion of such solids into a glass residual form through the application of heat to the point of fusion. The technologies are applicable to use on solids that are capable of forming a molten, vitreous mass, and of producing a glass–like residual product upon cooling. Typically, the residual product is a solid (super–cooled liquid) containing an amorphous mixture of oxides (primarily silica and alumina) with little or no crystallization present.

4.1 INORGANIC BASED SYSTEMS

The two principal solidification/stabilization (S/S) processes used are cement–based and lime/pozzolan–based processes. In both cement–based and lime/pozzolan–based techniques, the stabilizing process can be modified through the use of additives, such as silicates, that control curing rates, reduce permeability, and enhance the properties of the solid material.

The basic principle of operation for stabilization is that leachable metals in a waste are immobilized following the addition of stabilizing agents and other chemicals. The leachability is reduced by the formation of a lattice structure and/or chemical bonds that chemically bind the metals to the solid matrix and thereby limit the amount of metal constituents that can be leached when water or a mild acid solution comes into contact with the waste material. Stabilization is most effective when the waste metal is in its least soluble state, thereby decreasing the potential for leaching. For example, hexavalent chromium is much more soluble and more difficult to stabilize than trivalent chromium.

Most stabilization/solidification systems being marketed are proprietary processes involving the addition of absorbents and solidifying agents to a waste. Often the marketed process is changed to accommodate specific wastes.

Processing: The stabilization process consists of a weighing device, a mixing unit, and a curing vessel or pad. Commercial concrete mixing and handling equipment is typically used in stabilization processes. Weighing conveyors, metering cement hoppers, and mixers similar to concrete batching plants have been adapted in some operations. When extremely dangerous materials are treated, remote–control and in–drum mixing equipment, such as that used with nuclear waste, is employed.

In most stabilization processes, the waste, stabilizing agent, and other additives, if used, are mixed in a mixing vessel and then transferred to a curing vessel or pad and allowed to cure. The actual operation (equipment requirements and process sequencing) depends on several factors including the nature of the waste, the quantity of the waste, the location of the waste in relation to the disposal site, the particular stabilization formulation used, and the curing rate. Following curing, the stabilized solid formed is recovered from the processing equipment and disposed of.

For the purpose of determining and minimizing organic air emissions, the S/S process can be broken down into three distinct steps:

1. Mixing
2. Curing
3. Storage and landfilling

The mixing step is the basic operation where the waste is placed into a mixer and combined with the binder. The mixer is normally designed for ease of loading of the waste and binder and removal of the mixed material. Once mixed, the material is allowed to cure, at which time chemical reactions between the waste and binder harden the mixture. Curing can take place in the mixing vessel (as when mixing occurs in a drum or other disposal vessel) a temporary storage area, or directly in the landfill where the waste is ultimately placed. It can take as little as a few hours to as many as 30 days or more.

The final step, storage and disposal, is the goal of the S/S process. The material has hardened and if the binders are appropriate for the application, the waste has been stabilized. At this point, the material may be placed in a landfill and covered.

Until relatively recently, the mixing of wastes and binder has been done in open pits, trenches and bunkers. These do not lend themselves to proper collection and control of air emissions. S/S processes can and do produce both particulate and organic air emissions. Particulate control is being required with increasing frequency by states and by federal regulations.

Generally speaking, the type of mixing equipment used will influence how well the organic air emissions are collected and controlled, and the type of binder used will influence the point in the process where they will be released. The mixing equipment's effect on emissions is illustrated by the two extremes of a completely open and a completely closed mixing system. Emissions from a completely open mixing system, an open–pit mixer for example, can be complex, requiring the installation of large enclosures. The enclosures need to be designed so that the waste and binder can readily be added to the mixing vessel, or pit, and the mixture can be removed as well. By comparison, it is relatively easy to duct an enclosed mixer to an air pollution control device. In either case, once the organic air emissions are collected, their removal or destruction can be achieved using readily available equipment.

The type of binder used determines the temperature of the system during mixing or curing. For example, a binder, based on quicklime (CaO), will get very hot when mixed with an aqueous waste. The high temperature will cause a rapid release of the organic constituents. Because the solidified/stabilized product will then have lost most of the volatile constituents during mixing, it will release fewer organic air emissions during curing, storage, and disposal. If the mixing must occur in open equipment, then a binder with a low heat release is desirable. This will result in minimizing the organic air release until the waste can be placed in a sealed and capped landfill.

Inorganic Binders: The binder is typically selected on the basis of cost. As a result, whenever possible, a waste material that reacts with water will be used. Commonly used inorganic binders are:

1. Cement kiln dust
2. Lime kiln dust—typically contains a significant amounts of quicklime
3. Coal fly/bottom ash
4. Mixtures of the above

When these are unavailable or unsuitable, commercial products are use and these include:

1. Natural pozzolans
2. Lime (usually a grade of agricultural lime) mixed with flyash
3. Portland cement, usually mixed with an inert flyash

The solidification of wastes is analogous to the manufacture of concrete, which is a mixture of coarse aggregate (gravel), fine aggregate (sand), and a binder (cement). Water chemically reacts with the binder to form a solid matrix of the components. In solidification, the waste supplies the water and matrix of the components. In solidification, the waste supplies the water and, depending on its composition, a greater or lesser fraction of the aggregate. Clearly, the two mixtures can be formulated differently. When manufacturing concrete, strength is essential. As a result, the binder, water, aggregate and cement are mixed in proportions that optimize this property. The purpose of S/S is to chemically react the water to form a solid, and to immobilize the contaminants. The product only has to achieve a minimal load bearing strength. The hazardous waste regulations only require an unconfined compressibility strength of 50 psi.

It is clearly impossible to discuss all combinations of binders. Regardless of the binder formulations used, the concepts are the same and can be illustrated by the following examples:

1. Aqueous waste solidified/stabilized with portland cement/flyash
2. Aqueous waste solidified/stabilized with lime kiln dust/flyash
3. Aqueous waste solidified/stabilized with agricultural lime/flyash

Flyash from some coal-fired power plants are very reactive and will set-up, much like cement, when mixed with water. Such flyashes are sold in commerce and commonly used as a concrete additive. They can be considered to behave in a similar manner to portland cement with regards to organic air emissions. The type of flyash discussed here is the less valuable variety. When mixed with the lime or lime kiln dust, it participates in the chemical reaction. When mixed with portland cement, however, it is relatively inert. It serves as a source of aggregate and as a bulking agent to absorb the free water.

Concentration of Fine Particulates: For both cement-based and lime/pozzolan-based processes, very fine solid materials (i.e., those that pass through a No. 200 mesh sieve, less than 74 μm particle size) weaken the bonding between waste particles and the cement or lime/pozzolan binder by coating the particles. This coating inhibits chemical bond formation, thereby decreasing the resistance of the material to leaching. If the concentration of fine particulates in an untested waste is significantly higher than that in the tested waste, the system may not achieve the same performance. Pretreatment of the waste may be required to reduce the fine particulate concentration and achieve the same treatment performance, or other, more applicable treatment technologies may need to be considered for treatment of the untested waste.

Concentration of Oil and Grease: Oil and grease in both cement-based and lime/pozzolan-based systems result in the coating of waste particles and the weakening of the bond between the particle and the stabilizing agent, thereby decreasing the resistance of the material to leaching. If the concentration of oil and grease in the untested waste is significantly higher than that in the tested waste, the system may not achieve the same performance. Pretreatment may be required to reduce the oil and grease concentration and achieve the same treatment performance, or other, more applicable treatment technologies may need to be considered for treatment of the untested waste.

Concentration of Organic Compounds: Organic compounds in the waste interfere with the stabilization chemical reactions and bond formation, thus inhibiting curing of the stabilized material. This interference results in a stabilized waste having decreased resistance to leaching. If the total organic carbon (TOC) content of the untested waste is significantly higher than that of the tested waste, the system may not achieve the same performance. Pretreatment may be required to reduce the TOC and achieve the same treatment performance, or other, more applicable treatment technologies may need to be considered for treatment of the untested waste.

Concentration of Sulfate and Chloride Compounds: Sulfate and chloride compounds interfere with the stabilization chemical reactions, weakening bond strength and prolonging setting and curing time. Sulfate and chloride compounds may reduce the dimensional stability of the cured matrix, thereby increasing leachability potential. If the concentration of sulfate and chloride compounds in the untested waste is significantly higher than that in the tested waste, the system may not achieve the same performance. Pretreatment may be required to reduce the sulfate and chloride concentrations and achieve the same treatment performance, or other, more applicable treatment technologies may need to be considered for treatment of the untested waste.

Solubility of the Metal Compound: The metal to be stabilized should be in its least soluble state, or the stabilized waste may exhibit a potential for increased leachability. Pretreatment may be required to chemically reduce or oxidize the metal to a lower solubility state and achieve maximum stabilization performance. For example, hexavalent chromium is much more soluble and more difficult to stabilize than trivalent chromium.

Amount and Type of Stabilizing Agent and Additives: The stabilizing agent and additives used will determine the chemistry and structure of the stabilized material and therefore its leachability. Stabilizing agents and additives must be carefully selected based on the chemical and physical characteristics of the waste to be stabilized. To select the most effective type of stabilizing agent and additives, the waste should be tested in the laboratory with a variety of these materials to determine the best combination.

The amount of stabilizing agent and additives is a critical parameter in that sufficient stabilizing materials are necessary to properly bind the waste constituents of concern, making them less susceptible to leaching. The appropriate weight ratios of stabilizing agent and additives to waste are established empirically by setting up a series of laboratory tests that allow separate leachate testing of different mix ratios. The ratio of water to stabilizing agent (including water in waste) will also impact the strength and leaching characteristics of the stabilized material. Too much water will cause low strength; too little will make mixing difficult and, more important, may not allow the chemical reactions that bind the hazardous constituents to be fully completed.

Degree of Mixing: Mixing is necessary to ensure homogeneous distribution of the waste, stabilizing agent, and additives. Both undermixing and overmixing are undesirable. The first condition results in a nonhomogeneous mixture; consequently, areas will exist within the waste where waste particles are neither chemically bonded to the stabilizing agent nor physically held within the lattice structure. Overmixing, on the other hand, may inhibit gel formation and ion adsorption in some stabilization systems. Optimal mixing conditions are usually determined through laboratory tests. The quantifiable degree of mixing is a complex assessment that includes, among other factors, the amount of energy supplied, the length of time the material is mixed, and the related turbulence effects of the specific size and shape of the mix tank or vessel. The degree of mixing is beyond the scope of simple measurement.

The heart of the stabilization/solidification process is the mixing of the hazardous waste and its stabilization/solidification agent. Unfortunately, this is also the area in which the process may fail to meet expectations.

The objective is to achieve mixing in a practical manner. Although ideally, all hazardous waste should be mixed and reacted with the stabilization/solidification agent, even under ideal laboratory conditions complete mixing is not achieved. Thus, a question of degree of mixing becomes a consideration before one piece of equipment is sent into the field.

Many of the stabilization/solidification activities that have occurred in the United States have used what is known as "area mixing." In area mixing, the stabilization/solidification agent is delivered to the area to be stabilized/solidified and mixed with the waste directly. Often, tracked backhoes are used for the mixing. Typically, enough water is present in the waste to promote reaction with the stabilization/solidification agent and the waste. The waste is mixed until the operator judges the mixing to be complete. The benefits to this type of mixing are (1) the waste usually need not be removed from the site, (2) the operation requires only traditional earth–moving equipment, and (3) it is inexpensive. Stabilization/solidification accomplished by backhoe mixing at Vickery, Ohio has been well documented.

Offsetting these benefits is the probability of incomplete and inadequate mixing of the waste. Successful stabilization/solidification cannot be assured unless it can be demonstrated that proper mixing has occurred. In operations where area mixing has been used, the degree of mixing has been left to the operator's judgment.

Backhoe mixing is unlikely to result in complete mixing that assures that the waste is physically or chemically trapped. Processing equipment, such as pug mills, ribbon blenders, etc., provides more efficient mixing. Due to increasingly stringent volatile and particulate collection requirements, in–vessel mixing is often preferred.

A wide range of mixing equipment is available for use in industries that process solids. Typical mixing equipment suitable for use in stabilization/solidification operations includes pug mills, ribbon blenders, Mueller mixers, extruders, and screw conveyors. The waste and reagent can be metered into this equipment during mixing for quality control purposes.

Residence Time: The residence time or duration of curing ensures that the waste particles have had sufficient time in which to incorporate into lattice structures and/or form stable chemical bonds. The time necessary for complete stabilization depends on the waste and the stabilization process used. The performance of the stabilized waste (i.e., the levels of waste constituents in the leachate) will be highly dependent on whether complete stabilization has occurred. Typical residence times range from 7 to 28 days.

Stabilization Temperature and Humidity: Higher temperatures and lower humidity increase the rate of curing by increasing the rate of evaporation of water from the stabilization mixtures. If temperatures are too high, however, the evaporation rate can be excessive, resulting in too little water being available for completion of the stabilization reaction.

Form of the Metal Compound: Ideally, the waste metal to be stabilized should be in its least soluble state to reduce leaching of the stabilized waste. Pretreatment such as chemical oxidation or chemical reduction may be required to oxidize or reduce the metal to a state of lower solubility. Additionally, the solubility of the metal can be decreased by precipitating it with an appropriate compound. For example, ferric arsenate is less soluble and less leachable than calcium arsenate.

pH Control: It is important that the pH of all cementitious processes be greater than 10.0. This ensures that enough lime is present to initiate and continue the cementitious reactions. At a pH below 10.0, the cementitious reactions virtually cease. Additionally, if there is not substantial excess alkalinity, the initial process mechanisms of calcium alumino–sulfate generation may give a false set because components present in the system will adsorb or react with the lime of hydroxide present. This may reduce the pH to below 10.0, thereby stopping the cementitious reactions. Thus, the practitioner must make certain that enough lime or hydroxide is available to not only react to form the cementitious bonds, but also to act as a buffer to keep the pH greater than 10.0

Waste Type: The types of wastes most likely to be stabilized/solidified are liquids, sludges, and solids. Liquids and sludges are found primarily in lagoons. Solids also may be present in these lagoons, or they may occur as soils that have been contaminated by liquids.

Pretreatment of lagoon materials may be necessary for stabilization/solidification operations. Typical pretreatment operations include neutralization and water removal. Water removal may be a costly and extensive operation, especially if the waste is present beneath the local water table. In this case, dewatering may occur throughout the entire project. The water produced by dewatering must be treated. Several companies offer water treatment operations for remote site operation. These operation can remove solids, dissolved organics, and biological matter.

Access to lagoon materials can be difficult. The distance to the waste will determine whether a backhoe or a dragline is used for removal. For operations that provide for remote removal of the waste, whether the dikes will support the weight of the equipment must be considered. Proper width of the roads/dikes also must be ascertained. Some equipment, e.g., front–end loaders and bulldozers, may actually operate in a lagoon. The feasibility of operating this equipment in the waste must be considered. One consideration would be whether the waste can support the vehicle. If not, the use of wide tracks or low–pressure tires might make this operation feasible. Another consideration is the steepness of the dikes or walls surrounding a lagoon. At the one site, front–end loaders could not be used to carry waste over a dike and into dump trucks for this reason. Instead, an extra operation was necessary, which involved the use of backhoes to transfer the waste from the lagoon bottom to its side.

Soil treatment operations can disturb ongoing operations, and stabilization/solidification operations should be planned to keep this disturbance to a minimum. Dusting of the soil may be a problem, especially if the contaminant is present in high concentrations or if a large receptor population is nearby. Both dust suppression and the stabilization/solidification operation require water. Planners should determine if a source of water is present at the site if it is free of additional contaminants. If the soil contains concrete or large rocks, jackhammers will be needed for their removal. In addition, some soils at a site may not be appropriate candidates for stabilization/solidification even though they are contaminated. For example, an aquitard at the Frontier Hard Chrome Site was contaminated with hexavalent chromium. Removal of this aquitard for stabilization/solidification could have inadvertently introduced contamination into the aquifer.

Land Use and Safety: Stabilization/solidification applications must include an appraisal of future site use. If the waste is very toxic, certain site uses, e.g., residential use, would be precluded after treatment. If the waste will not support structures, this option greatly limits the use of the site. One of the disadvantages of stabilization/solidification is that it increases the waste volume. A significant increase in an already large volume of waste could result in an unexpectedly large waste pile, in which case offsite disposal of a portion of the waste may be necessary.

Site operations require complete health and safety procedures for hazardous waste work. Workers in machinery may require supplied breathing air; others may require air–purifying respirators. Provision also must be made for adequate protective clothing (Level A, B, C, or D). Decontamination areas are essential for hazardous waste stabilization/solidification operations. The proximity of a site to nearby populations must be determined before the stabilization/solidification process is begun. Monitoring for dusts and organics may be necessary, and the presence of these materials in the atmosphere could prevent operations during unfavorable meteorological conditions.

Geography: This topic includes consideration of geology, surface waters, and meteorology. The principal geologic consideration is that relating to site aquifers. Waste that extends to aquifers may have to be dewatered as described earlier. Weather can have an effect on operations or the placement of the waste. For example, stabilization/solidification operations may not be feasible during freezing weather if the stabilized/solidified waste cannot cure. If the waste is subject to failure upon repeated freeze/thaw or wet/dry testing, an engineered solution can be required—i.e., simply placing the waste below the frost line. Finally waste should never be placed in a floodplain.

General: Other considerations relate to the peculiarities of a site. For example, an area may be needed for stockpiling operations, and the absence of such an area could preclude stabilization/solidification as a remedial action. Power and utility sources for treatment operations are also important. If not available locally, electricity can be generated on site and water can be imported. Utilities can have an impact on field operations in another way as well. The presence of overhead power lines should be determined for safe operations at a site. Finally, any debris that is present in the waste may have to be removed. It is not unusual to encounter trees, large rocks, appliances, and other materials within a waste.

Proprietary Processes: The principal difference among the chemical fixation processes offered by various companies is the different additives used to assist in fixing the wastes. Often, the same solidifying agent, for example, cement or lime, may be used in two different processes, while the additives, usually proprietary in nature, are different. The result of such differences is that two processes which may appear similar might be applicable to different wastes. The proprietary additives might be incompatible with certain wastes, breaking down in their presence, and becoming ineffective for bonding. Thus, the type of waste that can be treated (fixed) is process-specific.

The characteristics of the waste that can be treated by the various processes can vary tremendously. For example, one process may only treat wastes containing up to 1% organic compounds, whereas another process may be capable of handling wastes with any organic content. With some exceptions, the vendors do not generally report the treatability limits on metal fixation in detail, nor do they publish data on leaching tests. This lack of information limits the ability to determine the applicability of any of these treatment technologies to a wide spectrum of wastes.

4.1.1 Cement Based

In portland cement systems, the waste is mixed with anhydrous cement powder, water, and frequently, pozzolanic additives. The cement powder consists of a mixture of powdered oxides of calcium, silica, aluminum, and iron produced by kiln burning materials rich in calcium and silica at high temperatures. The major mechanism of [immobilization] in this system is the formation of hydration products from silicate compounds and water. A calcium silicate hydrate gel is formed. This gel swells and forms the cement matrix which is composed of interlocking silicate fibrils. At the same time, constituents present in the waste slurry (e.g., hydroxides of calcium and various heavy metals), form the interstices of the concrete matrix. Metal ions also become incorporated into the crystal structure of the cement matrix itself. A rigid mass results from the interlocking fibrils and other components during setting and curing.

Portland cement is a mixture of powdered oxides of calcium, silica, aluminum, and iron, produced by kiln burning of materials rich in calcium and silica at high temperatures, i.e., 1400° to 1500°C (2552° to 2732°F). When the anhydrous cement powder is mixed with water, hydration occurs and the cement begins to set. The chemistry involved is complex because many different reactions occur depending on the composition of the cement mixture.

As the cement begins to set, a colloidal gel of indefinite composition and structure is formed. Over time, the gel swells and forms a matrix composed of interlacing, thin, densely packed silicate fibrils. Constituents present in the waste slurry, e.g., hydroxides and carbonates of various metals) are incorporated into the interstices of the cement matrix. The high pH of the cement mixture tends to keep metals in the form of insoluble hydroxide and carbonate salts. It has been hypothesized that metal ions may also be incorporated into the crystal structure of the cement matrix, but this hypothesis has not been verified.

In pozzolan–portland cement systems, portland cement and pozzolan materials, i.e., fly ash, are combined to create a high–strength waste and concrete matrix, where solidification/stabilization is achieved through the physical entrapment of waste particles. Fly ash or another pozzolan is often added to the cement to react with free calcium hydroxide and thus improve the strength and chemical resistance of the solidified product. The types of cement used for the solidification can be selected specifically to emphasize a particular cementing reaction, or to enhance cementation, such as sulfate resistance.

Hazardous/toxic waste sites effectively treated by the pozzolan–portland cement process include: (1) heavy metals in metallic or cationic forms, (2) inorganics in anionic form, (3) water–soluble organics, and (4) water–insoluble organics. The wastes that can be treated include aqueous solutions, sludges, and contaminated soils.

To understand the cement–based pozzolan solidification/stabilization process, it is necessary to first understand the binding material. Common (portland) cement is produced by firing a charge of limestone and clay or other silicate mixtures in a kiln at high temperatures. The resulting clinker is ground to a fine powder to produce a cement that consists of approximately 50% tricalcium and 25% dicalcium silicates (also present are about 10% tricalcium aluminate and 10% calcium aluminoferrite). The cementation process is brought about by the addition of water to the anhydrous cement powder. This first produces a colloidal calcium silicate–hydrate gel of indefinite composition and structure. Hardening of the cement is a lengthy process brought about by the interlacing of thin, densely packed, silicate fibrils growing from the individual cement particles. This fibrillar matrix incorporates the added aggregates and/or waste into a monolithic, rock–like mass.

A number of additives, many proprietary, have been developed for use with cement to improve the physical characteristics and to decrease the leaching losses from the solidified mass. Experimental work on the treatment of radioactive waste has shown that nuclear waste retention improvements can be achieved by cement–based stabilization processes with the addition of clay or vermiculite as absorbents. Soluble silicates have been used to bind contaminants in cement solidification processes, but this additive causes an increase in volume to occur during the setting of the cement–waste mixture. A recently proposed modification of this technique involves dissolving the metal–rich waste with fine–grained silica at low pH and then polymerizing the mixture by raising the pH to 7.

The resulting contaminated gel is mixed with cement and hardens within 3 days. The equipment used for the cement–based pozzolan process is similar to that used in a cement batching plant.

Most hazardous waste slurried in water can be mixed directly with cement, and the suspended solids will be incorporated into the rigid matrices of the hardened concrete. This process is especially effective for waste with high levels of toxic metals since at the pH of the cement mixture, most multivalent cations are converted into insoluble hydroxides or carbonates. Metal ions also may be incorporated into the crystalline structure of the cement minerals that form. Materials in the waste, such as sulfides, asbestos, latex and solid plastic wastes, may actually increase the strength and stability of the waste concrete. It is also effective for high–volume, low toxic, radioactive wastes.

The presence of certain inorganic compounds in the hazardous waste and the mixing water can be deleterious to the setting and curing of the waste–containing concrete. Also, impurities such as organic materials, silt, clay or lignite may delay the setting and curing of common portland cement for as long as several days. Dust–like, insoluble materials passing through a No. 200 mesh sieve (74×10^{-6} m particle size) are undesirable, as they may coat the larger particles and weaken the bond between the particles and the cement. Soluble salts of manganese, tin, zinc, copper and lead may cause large variations in setting time and significant reduction in physical strength. In this regard, salts of zinc, copper and lead are the most detrimental. Other compounds such as sodium salts of arsenate, borate, phosphate, iodate and sulfide will retard setting of portland cement even at concentrations as low as a few tenths of a percent of the weight of the cement used. Wastes containing large amounts of sulfate, such as flue–gas cleaning sludges) not only retard the setting of concrete, but, by reacting to form calcium sulfoaluminate hydrate, cause swelling and spalling in the solidified waste–containing concrete. To prevent this reaction, a special low–alumina cement was developed for use in circumstances where wastes containing high sulfate levels are encountered.

Advantages: Cement–based solidification systems are an economically feasible process having the following advantages:

1. The amount of cement used can be varied to produce high–bearing capacities thereby making the waste/concrete material a good subgrade and subfoundation material.
2. Low permeability in the product can also be achieved by varying the amount of cement used.
3. Raw materials are plentiful and inexpensive.
4. The technology and management of cement mixing and handling is well known; the equipment is commonplace and specialized labor is not required.
5. Extensive drying or dewatering of the waste is not required because cement mixtures require water in the hydration process, and thus the amount of cement added can be adjusted to accommodate a wide range of wastewater contents.
6. The system is tolerant of most chemical variations. The natural alkalinity of the cement used can neutralize acids. Cement is not affected by strong oxidizers, such as nitrates or chlorates.
7. Pretreatment is required only for materials that retard or interfere with the

setting action of cement.

8. Leaching characteristics can be improved, where necessary, by coating the resulting product with a sealant.

Disadvantages: Disadvantages include the fact that relatively large amounts of cement are required for most treatment processes (but this may partly be offset by the low cost of material). The weight and volume of the final product is typically about double those of other solidification processes. Uncoated cement–based products may require a well–designed landfill for burial. Experience in radioactive waste disposal indicates that some hazardous constituents are leached from the solidified concrete, especially by mildly acidic leaching solutions. Extensive pretreatment and the use of more expensive cement types or additives may be necessary for waste containing large amounts of impurities, such as borates and sulfates which can effect the setting or curing of the waste–concrete mixture. If ammonia is present in the waste, the alkalinity of cement drives off ammonium ions as ammonia gas. Finally, if energy cost increases dramatically, the cost of cement will likely follow because cement is an energy–intensive material.

Other Considerations: Most hazardous wastes which are aqueous slurries can be mixed directly with cement and the suspended solids will be incorporated into the rigid matrix. Cement solidification is most suitable for treating inorganic wastes, especially those containing metals. When the cement mixture has a high pH, most multivalent cations are converted into insoluble hydroxides or carbonates. Certain materials in the waste, such as latex and solid plastics, may act as reinforcing agents and increase the strength and stability of the waste concrete while other materials like silt, clay, coal or lignite may delay setting and curing of the cement. Other compounds which are especially active as retarders of the setting process include sodium salts of arsenate, borate, phosphate, iodate and sulfide. Products containing large amounts of sulfate not only retard the process, but also react with it to form sulfoaluminate hydrate, which causes swelling and spalling in the solidified waste concrete. Under such circumstances, the special low alumina (Type V) cement is needed to inhibit this reaction.

A number of additives have been developed for use with cement to improve the physical characteristics, for example, strength, and also to decrease the leaching losses from the solidified mass. Many of the additives are proprietary. Very often these additives contain clay or vermiculite, which are used to absorb excessive free liquid, or they contain sodium silicate to increase the binding (through chemical reactions with the cement) between the waste and the cement. Unfortunately, the additives can cause an increase in the final volume of the setting mixture and, hence, increase the cost of ultimate disposal.

Five different types of portland cement are generally used. These are differentiated on the basis of variations in the chemical compositions and the physical properties:

1. Type I is the "normal" cement used for construction purposes and it constitutes over 90% of the cement manufactured in the United States.
2. Type II is used in the presence of moderate sulfate concentrations (150 to 1,500 mg/kg).
3. Type III has a high strength cement and is used when rapid setting is required.
4. Type IV produces a low heat of hydration and is used in large concrete work.
5. Type V is a special low alumina, sulfate–resistant cement used for high

sulfate concentrations (>1,500 mg/kg).

Most waste fixation is done with Type I portland cement, but Types II and V are often used for sulfate or sulfite wastes.

Hydraulic cements have been the primary radioactive waste stabilization agents in the United States since the 1940s. In 1980, Brookhaven National Laboratory was funded by the Department of Energy's Defense Low-level Waste Management Program to test and develop sulfur polymer cement (SPC). It has stabilized routine wastes as well as some troublesome wastes with high waste-to-agent ratios. The Department of Energy's Hazardous Waste Remedial Action Program joined the effort by providing funding for testing and developing sulfur polymer cement as a hazardous-waste stabilization agent. Sulfur polymer cement has passed all the laboratory scale tests required by the U.S. Environmental Protection Agency and U.S. Nuclear Regulatory Commission. Two decades of tests by the U.S. Bureau of Mines and private concrete contractors indicate this agent is likely to exceed other agents in longevity.

4.1.2 Lime/Pozzolan Based

The lime/pozzolan processes use the finely divided, noncrystalline silica in pozzolanic material, e.g., fly ash, and the calcium in lime to produce a concrete-like solid of calcium silicate and alumino hydrates. The waste containment is achieved by entrapping the waste in this pozzolan concrete matrix. In actual operation, the waste, water, and a selected pozzolanic material are mixed to a pasty consistency. Hydrated lime is blended into the mixture and the resulting moist material is packed or compressed into a mold and cured over a sufficient time interval.

Pozzolan, which contains finely divided, noncrystalline silica, e.g., fly ash or components of cement kiln dust, is a material that is not cementitious in itself, but becomes so upon the addition of lime. Metals in the waste are converted to insoluble silicates or hydroxides and are incorporated into the interstices of the binder matrix, thereby inhibiting leaching.

The most common pozzolanic materials are fly ash, blast furnace slag, ground brick, and cement-kiln dust. All of these materials are themselves waste products with little or no value. Therefore, the use of these waste products to consolidate with another waste is often an advantage to the generator, who can dispose two wastestreams at the same time.

For example, by making use of the pozzolanic reaction, power-plant fly ash can be combined with flue-gas-cleaning sludge and lime (along with other additives) to produce an easily handled solid.

The types of additives that are usually used for lime-based chemical fixation include:

1. Certain clays which absorb liquid and bind specific anions or cations.
2. Emulsifiers and surfactants which allow the incorporation of immiscible organic liquids.
3. Proprietary absorbents like carbon, zeolite materials or cellulosic sorbents that selectively bind specific wastes.

Pozzolanic processes are suitable for high-volume, low toxicity wastes containing radioactive materials, inorganics, or heavy metals, with an organic content below 10%.

Certain treatment systems fall in the category of cement-pozzolanic processes and have been in use for some time outside the U.S. In these systems, both cement and lime-siliceous materials are used in combination to give the best and most economical containment for the specific waste being treated. In general, the bulk of the comments, under both classifications above, hold for techniques using a combination of treatment materials.

Advantages: The advantages of lime–based solidification techniques that produce pozzolanic concrete are: solidified material produced has improved handling and permeability characteristics; materials required for the process are often very low in cost and widely available; little specialized equipment is required for processing; the chemistry of lime–pozzolanic reactions are relatively well–known; sulfate content of the waste does not cause spalling or cracking; and extensive dewatering is not necessary because water is required in the setting reaction.

Disadvantages: The lime–based systems have many of the same potential disadvantages as cement–based techniques including: the lime and other additives add to the weight and bulk of the resultant product to be transported and/or landfilled; uncoated lime–treated materials may require specially designed landfills to guarantee that the material does not lose potential pollutants by leaching; the process is temperature sensitive; the waste may require pretreatment; the setting characteristics of the pozzolanic concrete are sensitive to organic content; and the process has a potential for producing fugitive dust emissions.

4.1.3 Silicate Based

The use of silicates in pozzolanic material, in conjunction with lime, has been discussed in the previous section.

Silicate based processes refer to a very broad range of solidification/stabilization methods which use a siliceous material together with lime, cement, gypsum, and other suitable setting agents. Many of the available processes use proprietary additives and claim to stabilize a broad range of compounds from divalent metals to organic solvents. The basic reaction is between the silicate material and polyvalent metal ions. The silicate material which is added in the waste may be fly–ash, blast furnace slag or other readily available pozzolanic materials. Soluble silicates such as sodium silicate or potassium silicate are also used. The polyvalent metal ions which act as initiators of silicate precipitation and/or gelation come either from the waste solution, and added setting agent, or both. The setting agent should have low solubility, and a large reserve capacity of metallic ions so that it controls the reaction rate. Portland cement and lime are most commonly used because of their good availability. However, gypsum, calcium carbonate, and other compounds containing aluminum, iron, magnesium, etc. are also suitable setting agents. The solid which is formed in these processes varies from a moist, clay–like material to a hard–dry solid similar in appearance to concrete.

There is considerable research data to suggest that silicates used together with lime, cement or other setting agents can stabilize a wide range of materials including metals, waste oil and solvent. However, the feasibility of using silicates for any application must be determined on a site–specific basis particularly in view of the large number of additives and different sources of silicates which may be used. Soluble silicates such as sodium and potassium silicate are generally more effective than fly ash, blast furnace slag, etc.

There is some data to suggest that lime–fly ash materials are less durable and stable to leaching than cement fly ash materials. Common problems with lime–fly ash and cement–fly ash materials relate to interference in cementitious reactions that prevent bonding of materials.

Materials such as sodium borate, calcium sulfate, potassium dichromate and carbohydrates can interfere with the formation of bonds between calcium silicate and aluminum hydrates. Oil and grease can also interfere with bonding by coating waste particles. However, several types of oily sludges have been stabilized with silicate based processes.

One of the major limitations with silicate based processes is that a large amount of water which is not chemically bound will remain in the solid after solidification. In open air, the liquid will leach until it comes to some equilibrium moisture content with the surrounding soil. Because of this water loss, the solidified product could require secondary containment.

Silicate–based processes can employ a wide range of materials, from those which are cheap and readily available to highly specialized and costly additives.

4.1.4 Calcination/Self–Cementing/Sintering

Calcination: Calcination is the conversion by thermal decomposition at elevated temperatures of aqueous liquids and sludges into solid materials, without interactions with the gaseous phase (such as air oxidation which occurs during incineration).

Calcination is a well established process with many industrial and waste treatment applications. It is a versatile one–step process for dealing with a variety of simple or complex wastes and can satisfactorily deal with sludges.

The process concentrates the waste, destroys organic components and leaves inorganic components in a more acceptable form for recovery or landfill. For many wastes the results of calcination are predictable although for complex mixtures, some pilot scale work may be desirable in order to determine how effective the process will be. If the waste contains a relatively high proportion of water, it may be desirable to pretreat by filtration, precipitation, etc., to reduce the energy requirements for calcination. Capital and operating costs are substantial, the latter associated principally with fuel requirements. If the waste contains organic components, then some or all of the fuel requirements may be supplied by in–situ combustion of the organic material.

Calcination temperatures normally used are too low to initiate combustion of some types of organic compounds. However, they are high enough to cause volatilization of organics, which have to be removed from process off–gases. For these reasons, the presence of significant levels of organics is undesirable but can be handled with appropriate air pollution controls. Organics, per se, do not interfere with the conversion of oxides to hydroxides or with sintering processes. Afterburners may be required on vitrification units managing high–organic–content wastes to ensure complete combustion of the organics present.

For an aqueous solution, the first reaction that occurs is vaporization of the water, leaving a solid material which can be granular and free flowing, or a compacted solid. A similar process occurs in the initial treatment of a dewatered sludge, i.e., after filtration and centrifugation. In many instances it is possible to proceed further with the calcination to drive off volatile materials from the partially calcined solid, e.g., a salt, to form an oxide that will be more stable or reusable. Typical examples are the calcination of carbonates, hydroxides, sulfites, sulfides, sulfates, and nitrates to the corresponding oxides with evolution of carbon dioxide, water, sulfur dioxide, and nitrogen oxides, respectively. Organic components can also be volatilized from metal organic materials, leaving the metal as a solid residue. The resulting solid produced may be in the form of a dry granular material which is readily handled, or on heating to a higher temperature, the granules may be sintered into a solid mass. On still further heating, certain materials will melt or fuse into a glass-like material. Additives, such as silicates, borax, or phosphates, can be used to decrease the leachability of certain components in the final solid or to assist in glass formation.

Calcination can be a continuous process which generally operates at high temperature and atmospheric pressure. It can be applied to aqueous solutions, slurries, sludges, and tars with the objective of producing a dry powder or solid material. It can also be used for solids and powders to produce a more acceptable form of waste, i.e., one which is less soluble and therefore does not represent a leachability problem after landfill. Calcination of mixed organic/inorganic wastes can be advantageous because combustion of the organic portion will provide some or all of the heat necessary to sustain the process.

The process results in a substantial volume reduction of about 90% in the case of liquids, and 50 to 75% in the case of inorganic sludges. Only a minor reduction in volume occurs with the sintering or calcination of solids to drive off volatile components. In all cases, a solid material is obtained after calcination which is generally much more suitable for storage or landfill than the original uncalcined material. However, the calcine may still be toxic unless the toxic component was destroyed or was removed as a volatile material during the calcination. In the latter case, additional treatment would be required on the air or water effluents.

A major advantage of calcination is that several operations can often be carried out in a single step, i.e., concentration, destruction, and detoxification. A potential disadvantage is the fuel requirement if the waste does not contain any combustible material. Calcination temperatures are generally in the range of 650° to 1100°C although temperatures up to 1400°C are feasible. Temperatures of 500° to 900°C are more common. Above 1300°C the choice of refractory lining becomes more limited and costs increase sharply.

The calcination temperature selected is generally a temperature above which metal hydroxides present will decompose to the corresponding oxides. The temperature chosen is normally high enough to cause extensive sintering (surface area loss) of the oxides formed, while at the same time not volatilizing these materials.

Calcination temperatures are normally selected based on the temperatures at which hydroxides are thermally decomposed to the corresponding oxides and water vapor. To select an optimum operating temperature, one should know the approximate composition of the waste. A few toxic metal oxides have fairly low volatilization temperatures. Arsenic oxide, selenium dioxide, and mercuric oxide all volatilize below 500°C. High-temperature calcination should not be used for wastes that contain these volatile constituents unless the wastes are blended with materials such as lime, which will react with the constituents before they can vaporize. Nonvolatile arsenic compounds such as ferric and calcium arsenates can be calcined without concern for vaporization of material.

The high-temperature treatment also significantly reduces the surface areas of the oxides formed by sintering, thereby reducing the reactivity of the material. After the waste material has been calcined at an elevated temperature, it is withdrawn from the oven or kiln, cooled, and either land disposed or forwarded to another process, such as stabilization, for further treatment.

In some instances, the waste may be blended with lime prior to heating. In those cases, chemical reaction may occur during the calcining process. Water present as either free water or water of hydration is evaporated, and hydroxides present are thermally decomposed to the corresponding oxides and water vapor. At the higher temperatures, the surface area of the dehydrated material is decreased by thermal sintering. Conversion of hydroxides to oxides and substantial losses of surface area render the material less reactive in the environment and lower the leachability of characteristic toxic metals present. In general, the higher the calcination temperatures used, the more complete the loss of water and the greater the accompanying loss of surface area, resulting in lower leachability potential.

Calcination is generally a batch process, and sufficient time must be allowed for samples to be brought to the operating temperature. Residence times of several hours are normally used to minimize the effects of heat-up time. Residence time is a function of the time needed to bring the calcination furnace or kiln to the desired temperature and the time needed to complete the dehydration and sintering processes at the selected temperature.

Calcination invariably produces particulates and gaseous products in the exit gas stream which cannot usually be emitted to the atmosphere. Some effluents may require an extensive air pollution control system. One exception is the decomposition of carbonates to carbon dioxide, which can be safely emitted to the atmosphere. However, in a number of industrial applications, it is economical to recover at least part of this carbon dioxide for recycle to the process. Particulates can be removed by cyclones, filters, or by electrostatic precipitators. Water-soluble vapor components can be removed by aqueous wet scrubbers (but the spent scrubbing liquid may require further treatment or recycle to the process to avoid potential water pollution problems). Remaining gaseous components can be removed by adsorption on carbon, alumina, silica gel, or alumino-silicates.

Equipment utilized can be rotary kiln, hearth, or fluidized bed furnaces. The fluidized bed furnaces would be applicable to liquids or slurries.

Self-Cementing: Some industrial wastes such as flue-gas-cleaning sludges contain large amounts of calcium sulfate and calcium sulfite. A technology has been developed to treat these types of wastes so that they become self-cementing. Usually a small portion (8 to 10% by weight) of the dewatered waste sulfate/sulfite sludge is calcined under carefully controlled conditions to produce a partially dehydrated cementitious calcium sulfate or sulfite. This calcined waste is then reintroduced into the waste sludge along with other proprietary additives. Fly ash is often added to adjust the moisture content. The finished product is a hard, plaster-like material with good handling characteristics and low permeability.

Self-cementing processes require large amounts of calcium sulfate and calcium sulfite and are appropriate for immobilizing heavy metals.

The primary advantage for using a self-cementing process is the material produced is stable, nonflammable and nonbiodegradable. There are reports of effective heavy metal retention, which is perhaps related to chemical bonding of potential pollutants. Other advantages are: (1) no major additives have to be manufactured and shipped to the processing site; (2) the process is reported to produce a faster setting time and more rapid curing than comparable lime–based systems; and (3) these systems do not require completely dry waste because the hydration reaction uses up water.

The disadvantages associated with self–cementing processes include: the self–cemented sludges have much the same leaching characteristics as cement– and lime–based systems; only high calcium sulfate or sulfite sludges can be used; and additional energy is required to produce the calcined cementitious material. The process also requires skilled labor and expensive machinery for calcining waste and for mixing the calcined wastes with the bulk waste and the proprietary additives.

Sintering: Sintering can be defined as a limited form of calcination in which the physical structure, but not the chemical nature, of the solid is changed. For instance, dry powders may be heated to sinter them into a solid mass, usually with some reduction in volume. Additives such as silicates, which also sinter readily, can be added to improve this process.

4.1.5 Sorption

Sorption is the addition of solid sorbents to soak up and prevent the loss of drainable liquids through the mechanisms of capillary action, surface wetting, and chemical reaction. To prevent undesirable reactions, the sorbent material must be matched to the waste. Zeolite, kaolite, vermiculite, calcite, amorphous entonites, silicates, acidic and basic fly ash, and kiln dust are all typical sorbents. There are also synthetic sorbents available. Sorbents can be spiked with scavengers to bind trade metals, flocculating agents, and agents to improve subsequent solidification (cementing) processes.

Sorbents are widely used to remove free liquid and improve waste handling. Some sorbents have been used to limit the escape of volatile organic compounds. They may also be useful in waste containment when they modify the chemical environment and maintain the pH and redox potential to limit the solubility of wastes. Although sorbents prevent drainage of free water, they do not necessarily prevent leaching of waste constituents and secondary containment could be required.

The quantity of sorbent material necessary for removing free liquid varies widely depending on the nature of the liquid phase, the solids content of the waste, the moisture level in the sorbent, and the availability of any chemical reactions that take up liquids during reaction. It is generally necessary to determine the quantity of sorbent needed on a case–specific basis.

This process results in high concentrations of contaminants at the surface of the material, and contaminants may leach. The treated material is permeable.

Advantages to this technology include plentiful raw materials, known mixing technology, improved handling, inexpensive additives, minimum pretreatment, and adequate bearing strength for landfill. The disadvantages include a large volume of additives, poor leachate control, placement sensitivity, limited bearing strength, temperature sensitivity, and free water may be released under high pressure.

4.1.6 In Situ Methods

In situ techniques can be used to immobilize heavy metals or organics in soil.

One remedial action option available to mitigate the leaching potential of contaminant metals into groundwater and their subsequent transport through underground aquifers is in situ immobilization. In situ immobilization can be carried out by introducing treatment chemicals into the ground by various means. If soluble chemicals are used, they can be applied by saturating the soil with the chemical in solution. This fluid application may be carried out at a high rate by surface flooding the site or more gradually by spraying and allowing the solution to drain freely into the soil. The variation in application rate will affect the period of soil exposure to the treatment material, the degree of void filling accomplished, and the amount of air present in the soil during the treatment period. A complementary confinement or pumping system may be appropriate if the soluble treatment chemical has undesirable environmental effects or is worth recycling due to high chemical costs.

Insoluble chemicals can be introduced into the ground by spreading, filling, forced injection, suspension transport, or by placing it in a low permeability encapsulation barrier. Spreading may suffice as a means of treating metals if the soil has a high moisture content and the metal contaminants lie very close to the surface. This may be most applicable to soils with high organic content. Tilling is the most common method of introducing a soil treatment chemical into the ground. Routine tilling can mix dry chemical additives into the soil to a depth of one to two feet. Special deep tilling equipment is available which can reach as deep as five feet into the ground. Fine insoluble chemicals can be transported short distances through soil voids by placing them in suspension in water or in a weak solvent or acid. The suspended material is then injected in a fashion similar to chemical grouting or through nozzles in close spaced probes. Typically, fine material can be transported several feet from the nozzle in this fashion. The particle size can be correlated to soil grain size using traditional grouting guidelines. In formations with high permeability and low organic content, where metals have migrated to depths greater than 10 feet or more, mixing insoluble treatment materials into the soil may be impractical. Under these circumstances, the treatment chemical can be placed into a barrier material, such as bentonite soil or asphalt emulsions used for slurry wall construction, jet grouting or block displacement.

In situ immobilization of heavy metals in contaminated soils can be accomplished by adding natural or synthetic chemical additives to the soil. These additives must have certain desirable properties to successively immobilize heavy metals. Treatment additives fall into two classes of chemicals, strongly adsorbing and weakly adsorbing.

By their nature, once strongly–adsorbing insoluble chemical additives are added and distributed throughout the soil, they will not migrate down through the soil to groundwater. The heavy metals must be adsorbed, complexed and/or chelated on the additive and must not hydrolyze nor be desorbed under exposure to varying conditions in the soils, such as a low pH or a varying Eh which tends to solubilize the metals. The chemical additives must be resistant to chemical and microbial degradation in the soil environment so that metals are not released from the additives over long periods of time, say, for at least a few years. Finally, the chemicals additives, themselves, must not leach any deleterious organic or inorganic substances that could contaminate groundwater.

For weakly adsorbing chemical additives to be effective in immobilizing heavy metals, they must either cause the metals to precipitate or complex and/or chelate the metals and then attach themselves to the soil structure. In either case, the metals will not migrate down through the soil. Complexation and/or chelation to the weakly adsorbing additive is not sufficient since there is the distinct possibility the complexed metal could migrate to and be transported in groundwater. In any event, after the metals are precipitated, they must not be resolubilized under varying soil conditions, such as over a range of pH and Eh. Finally, as with the strongly adsorbing additives, the weakly adsorbing additives must be resistant to chemical and microbial degradation in the soil and must not leach out any deleterious organic or inorganic substances to the soil water.

Today, many chemical additives are used in the treatment of wastewaters to reduce heavy metals concentrations. Many of these chemical additives can reduce the metal concentrations to levels below NPDES effluent discharge limits, and in some cases to below drinking water standards. Candidate treatment chemical additives include:

1. Standard cation ion exchange resin
2. Chelate ion exchange resins
3. Devoc–Holbein metal scavenging molecules
4. Natural materials—clays, molecular sieves, and greensand

Various techniques include:

Injection Method: Solidifying/stabilizing agents can be injected into the waste material in liquid or slurry form. Injection can be achieved by flow of the solidifying/stabilizing reagent inside a porous tube to the required depth. Variations include permeation grouting, and jet grouting.

Surface Application: When the waste material is sufficiently shallow and permeable, stabilizing agents can be applied in a solid or liquid form onto the surfaces and allowed to penetrate. This application technique is especially appropriate for rendering a specific waste component less toxic. Shallow soil mixing has been used extensively in Japan, and has been introduced in the United States.

4.2 ORGANIC ENCAPSULATION SYSTEMS

One common technique for stabilizing organic contaminants is blending them into resins and then solidifying the mixtures. Plastic solidifying agents fall into two main categories, thermoplastics and thermosets. Thermoplastics are materials that become fluid upon heating and include asphalt, polyethylene, polypropylene, and nylon. Thermoplastic techniques generally call for the waste to be dried, heated, dispersed through the heated plastic matrix, and then cooled (solidified) and placed in containers. Thermosets include urea formaldehyde, polyester, and phenolic and melamine resins. Thermoset techniques call for the waste to be mixed with the thermoset prior to reaction of the mixture to form a solid matrix through crosslinking reactions. This matrix will remain solid throughout subsequent heatings. Containers may or may not be needed with thermosets.

In early work, asphalt and bitumen were the most widely applied materials for solidifying organics. These fixative materials are chemically stable and lost in cost. At low waste–to–fixative loadings, these materials were generally found to exhibit acceptable solidification properties (e.g., good solid product formation and dimensional stability remained upon immersion in water). However, for high contaminant loadings, above about

30% by weight, or, in general, for most organics of lower molecular weight, high vapor pressure or hygroscopic nature, these materials often yielded unacceptable products. More recently, these products have been replaced by thermoplastic or thermosetting resins; e.g., linear polyethylene has been employed as a stabilizer for certain organics.

Soil stabilization chemicals are also available that react with moisture in the soil or an aqueous catalyst to form a hydrophobic crosslinked polymer–based gel. The semi–solid gel forms in situ coats and binds the soil particles together. The chemical and water (or catalyst) mixture is sprayed on cultivated or loosened soil to react with the upper 3 to 4 inches of soil. The resulting gel–soil mixture then becomes a barrier to water infiltration.

Commonly offered grouts include organic polymers based on acrylamides, polyurethanes, urea, and phenolics. The advantages some of the chemical grouts offer are that they are easy to mix, they penetrate soil much like water (since they are polar and have a viscosity similar to water), they can be applied by spraying, and they are generally nontoxic when handled properly. The grouts form highly stable compounds of extended but unknown life. However, grouts are sensitive to freeze–thaw and wet–dry conditions, and some grouts will deteriorate under ultraviolet light and other degradative mechanisms.

EPA believes that encapsulation technologies are applicable primarily to wastes containing hazardous metal constituents. Encapsulation may immobilize hazardous organics as well as metals; however, incineration is more applicable to organics since incineration destroys organics completely, whereas encapsulation can only immobilize them.

There are a number of very similar encapsulation processes that differ from each other only in the encapsulating agent used. In all of the processes, the waste is first dried to remove moisture. The waste is then usually reheated and mixed with hot asphalt or thermoplastic material such as polyethylene. The mixture is then cooled to solidify the mass. The ratio of matrix (fixative or encapsulating agent) to waste is generally high, i.e., 1:1 or 1:2 fixative to waste on a dry basis. The matrix, once solidified, coats the waste to minimize leaching.

Encapsulation processes can take the form of macroencapsulation, microencapsulation, or both. Microencapsulation is the containment of individual waste particles in the polymer or asphalt matrix. Macroencapsulation is the encasement of a mass of waste in a thick polymer coating. The waste mass may have been microencapsulated prior to macroencapsulation.

4.2.1 Thermoplastic Microencapsulation

The use of thermoplastic solidification systems in radioactive waste disposal has led to the development of waste containment systems that can be adapted to industrial waste. In processing radioactive waste with bitumen or other thermoplastic material (such as paraffin or polyethylene), the waste is dried, heated and dispersed through a heated, plastic matrix. The mixture is then cooled to solidify the mass.

The process requires some specialized (expensive) equipment to heat and mix the waste and plastic matrices, but equipment for mixing and extruding waste plastic are commercially available. The plastic in the dry waste must be mixed at temperatures ranging from 130° to 230°C, depending on the melting characteristics of the material and type of equipment used.

A variation of this process uses an emulsified bitumen product that is miscible with a wet sludge. In this process, the mixing can be performed at any convenient temperature below the boiling point of the mixture. The overall mass must still be heated and dried before it is suitable for disposal. Ratios of emulsions to waste of 1:1 to 1:1.5 are necessary for adequate incorporation.

Thermoplastic microencapsulation is commonly used for high–toxicity, low–volume wastes and is suitable for inorganic and most organic waste. In many cases, the waste type can rule out the use of an organic–based treatment system. Organic chemicals that are solvents for the matrix obviously cannot be used directly in the treatment system. Strong oxidizing salts, such as nitrates, chlorates or perchlorates, will react with the organic matrix materials and cause slow deterioration. At the elevated temperatures necessary for processing, the matrix–oxidizer mixtures are extremely flammable.

Leach or extraction testing undertaken on anhydrous salts embedded in bitumen as a matrix indicates that rehydration of the embedded compound can occur. When the sample is soaked in water, the asphalt or bitumen can swell and split apart, thereby greatly increasing the surface area and rate of waste loss.

Some salts, such as sodium sulfate, will naturally dehydrate at the temperatures required to make the bitumen plastic; thus, these easily dehydrated compounds must be avoided in thermoplastic stabilization.

Some of the major advantages of using a thermoplastic matrix are:

1. Leaching rates of the contaminants from the treated mixture are significantly lower than those from the cement–based or lime–based processes.
2. Overall volume of the waste may be reduced, since the waste needs to be dewatered before using the thermoplastic technique.
3. Thermoplastics usually adhere well to the materials being encapsulated.
4. End–product is fairly resistant to attack by aqueous solution. Microbial degradation is minimal.
5. Materials embedded in the thermoplastic matrix can be reclaimed if needed.
6. End–product will tend to be lighter than if a cement–based system is used since the weight of the thermoplastic matrix is less. This low density would reduce the transportation costs on a per weight of treated stream basis.

The disadvantages of using a thermoplastic matrix are:

1. Expensive equipment and skilled labor are necessary for processing.
2. They cannot be used with materials that decompose at high temperatures, especially citrates and certain types of plastics.
3. Thermoplastic materials are flammable. There are workplace hazards associated with working with organic materials such as bitumen at elevated temperatures.
4. During heating, some mixtures that contain volatile organics can produce objectionable oils and odors, causing secondary air pollution.
5. The waste materials must be dried before they can be mixed with the thermoplastic materials. This requires a large amount of energy. Incorporating wet wastes greatly increases losses through leaching and

poses the possibility of considerable gas (steam) evolution during processing. This steam likely carries with it dissolved contaminants which pose both workplace and general air quality hazards.

6. Strong oxidizers are usually not suitable for thermoplastic processes due to the possible reaction between the oxidizers and the binding material.
7. Dehydrated salts within the thermoplastic matrix slowly rehydrate if the mixture is soaked in water. This causes the waste block to fragment, thus increasing the rate of leaching.

4.2.2 Surface Encapsulation (Macroencapsulation):

Many waste treatment systems depend on binding particles of waste material together. To the extent to which the binder coats the waste particles, the wastes are encapsulated. However, the systems addressed by surface encapsulation processes are those in which the waste has been pressed or bonded together and then is enclosed in a coating jacket of inert material. A number of systems for coating solidified industrial waste have been examined. In most cases, coated materials have suffered from lack of adhesion between coatings and bound wastes, and lack of long–term integrity in the coating materials.

Surface encapsulation (macroencapsulation) is appropriate for both organic and inorganic wastes.

Surface coating of concrete–waste composites has been examined extensively. The major problems encountered have been poor adhesion of the coating onto the waste or lack of strength in the concrete material containing the waste. Surface coating materials that have been investigated include asphalt, asphalt emulsion and vinyl. However, no surface coating system for cement–solidified material is currently being advertised.

Advantages: Major advantages of an encapsulation process involve the fact that waste materials never come into contact with water, therefore, soluble materials, such as sodium chloride, can be successfully surface–encapsulated. The impervious jacket also eliminates all leaching into contacting waters as long as the jacket remains intact.

Disadvantages: A major disadvantage of surface encapsulation is that the resins required for encapsulating are expensive. The process requires large expenditures of energy in drying, fusing the binder and forming the jacket. Polyethylene is combustible, with a flash point of 350°C, making fires a potential hazard. The system requires extensive capital investment and equipment. Skilled labor is required to operate the molding and fusing equipment.

4.2.3 Reactive Polymers (Thermosetting):

In contrast with the thermoplastic techniques in which a polymerized material is heated and mixed with the substance to be solidified, reactive polymer processes usually are carried out at ambient temperature. They involve the mixing of monomers, such as urea–formaldehyde, with a catalyst to form a polymer. The polymer is formed in a batch reactor where the wet or dry waste is blended with a prepolymer using a specially designed mixer. When the two components are thoroughly mixed, a catalyst is added and mixing is continued until the catalyst is completely dispersed. Mixing is terminated before the polymer forms and the resin–waste mixture is transferred to a waste container, if necessary. The polymerized material does not chemically combine with the waste; it forms a sponge–like mass that traps solid particles but leaves liquid wastes alone. The polymer mass is usually dried in order to increase the binding between the polymer and the waste prior to land disposal. Reactive polymer processes generally require less solidifying agent per weight of waste than do other solidification systems, and they produce a less dense material for disposal. The degree of binding (cross–linking) between the waste and the polymer is influenced by parameters such as pH, water content, and ionic constituents in the feed stream.

Several alternative polymers have been used for the reactive polymer technique including: ureas, phenolics, epoxides, polyesters and vinyls.

The major advantages of the reactive polymer process are:

1. Less fixative material is required for solidifying the same amount of waste than using cement or lime–based techniques.
2. Waste material is usually dewatered, but not necessarily completely dried. The finished, solidified polymer, however, must be dried before ultimate disposal, with the resulting reduction in the amount of waste to be disposed.
3. End–product has a lower density (specific gravity about 1.3) than cement. The low density reduces the transportation costs for the fixed product.
4. Solidified resin is non–flammable and high temperature is not required to form the resin.

The major disadvantages of the reactive polymer process include:

1. No chemical reactions occur in the solidification process that chemically bind the potential pollutants. The particles of the waste material are simply entrapped in an organic matrix.
2. Catalysts used in the urea–formaldehyde process are strongly acidic. Most metals are fairly soluble at low pH and can escape in water not trapped in the mass during the polymerization process. The catalyst may be highly corrosive and require special mixing equipment, reactors, containers, etc.
3. Some of the reactive polymers are biodegradable.
4. Secondary containment in steel drums is required before disposal, raising costs in processing and transportation.

4.2.4 Polymerization

Polymerization uses catalysts to convert a monomer or a low–order polymer of a particular compound to a larger chemical multiple of itself. Often, such large polymers have greater chemical, physical and biological stability than the monomers (or dimers or trimers) of the same chemical.

This technology treats organics including aromatics, aliphatics, and oxygenated monomers such as styrene, vinyl chloride, isoprene, and acrylonitrile. It has application to spills of these compounds.

4.3 VITRIFICATION

Vitrification technologies are those that involve exposure of hazardous materials to molten glass and related process conditions to affect the destruction, removal, and/or permanent immobilization of hazardous contaminants. Vitrification is defined as conversion of such solids into a glass residual form through the application of heat to the point of fusion. The technologies are applicable to use on solids that are capable of forming a molten, vitreous mass, and of producing a glass-like residual product upon cooling. Typically, the residual product is a solid (super-cooled liquid) containing an amorphous mixture of oxides (primarily silica and alumina) with little or no crystallization present.

Exposure of contaminants to vitrification processing results in several desirable results: (1) destruction of hazardous organics by pyrolytic decomposition and/or oxidation, (2) removal (partial or full) of low-solubility, high-volatility, high-solubility inorganics in the residual glass product through chemical incorporation and/or encapsulation. Thus, the vitrification processes may be considered as both thermal treatment (destruction) and immobilization-processes.

The various vitrification processes similarly produce a glassy residual product resembling natural obsidian in physical and chemical characteristics. The residual product may be made in granular form, cast into containers, or in multi-thousand ton monoliths. Typically the product has excellent structural, weathering, and biotoxicity characteristics, making it suitable for long-term environmental exposure. The residual typically is able to surpass EPA leach testing requirements (e.g., EP-Tox and TCLP), making it a candidate for delisting as a hazardous waste.

Vitrification of wastes involves combining the wastes with molten glass at a temperature of 1350°C or greater. However, the encapsulation might be done at temperatures significantly below 1350°C (a simple glass polymer such as boric acid can be poured at 850°C). This melt is then cooled into a stable, noncrystalline solid.

This process is quite costly and so has been restricted to radioactive or very highly toxic wastes. To be considered for vitrification, the wastes should be either stable or totally destroyed at the process temperature. Classification of wastes is an extremely energy intensive operation and requires sophisticated machinery and high trained personnel.

Of all the common solidification methods, vitrification offers the greatest degree of containment. Most resultant solids have an extremely low leach rate. Some glasses, such as borate-based glasses, have high leach rates and exhibit some water solubility. The high energy demand and requirements for specialized equipment and trained personnel greatly limit the use of this method.

Classification of Vitrification Processes

	Examples
1. Electric Process Heating	
A. Joule Heating	
1. ex situ	Ceramic Melter
2. in situ	In Situ Vitrication
B. Plasma Heating	Plasma Furnace
C. Microwave Heating	Microwave Melter
D. Miscellaneous Electric Heating	Resistance Heating, Induction Heating, Electric Arc Heating
2. Thermal Process Heating	Rotary Kiln Incinerator (operated in slagging mode)

Some interesting new applications include vitrification of asbestos by Tokyo Electric Power Co., Georgia Tech, and Vitrifix of North America, as well as vitrification of municipal incinerator fly ash by Corning, Inc., and Battelle Pacific Northwest Laboratories.

Commercial processes include:

1. Pyro–Converter™ (Penberthy Electromelt)
2. High Temperature Fluid–Wall Reactor (Thagard, Huber, Vulcan)
3. Electric Pyrolyzer (Westinghouse)
4. Pyro–Disintegrator™ (Pyrogenics)
5. In–Situ (DOE, Battelle)
6. Cyclone Vitrification (Babcock & Wilcox)
7. Cyclone Melting System (Vortec)

High–level nuclear waste from past commercial reprocessing operations, defense programs, and any future reprocessing must be solidified before it can be transported to a geologic repository. The final waste form must meet a number of different requirements at various stages of the waste disposal process, including processing that is safe and practical at acceptable cost and unaffected by small variations in waste composition and process conditions; a final form that can withstand handling, short–term corrosion, and coolant loss or sabotage without dispersing its contents; and a final form that can resist transportation accidents, such as impacts and fires. In addition, the final form must meet requirements for emplacement in a repository; the requirements include structural integrity, resistance to surface contamination and fire, dimensions, weight, retrievability, low leachability under both static and flowing water conditions, compatibility with the host rock, and resistance to dispersal after accidents or deliberate intrusion.

To date, borosilicate glass has been the most–studied waste form; alternative forms are also being evaluated. Waste can be fired to form a mixture of oxides (calcine) at 300° to 700°C. Waste can be solidified by mixing it with clay to absorb water; the mixture can also be fired to form a ceramic. Waste can be mixed with concrete; the mixture can be hot–pressed to eliminate excess water. Calcine can be agglomerated with additives to reduce water solubility, eliminate excess water. Calcine can be agglomerated with additives to reduce water solubility, or treated to form supercalcines, which are highly stable, leach–resistant, silicate minerals.

Titanate and zirconate minerals similar to natural minerals are known to have been stable in a wide range of geologic and geochemical environments for billions of years. Vitrified wastes can be converted to a more stable crystalline form (partial devitrification); high–temperature glasses are also being studied. Pellets of glass, supercalcine, or other waste forms can be incorporated into a metal binder (matrix); a similar alternative is to form small waste particles in situ in the metal matrix this is known as cermet). Waste can also be coated with carbon, aluminum oxide, or other impervious materials before encapsulation in metal to form multiple barriers.

Current program activities are focused on the development of HLW immobilization technologies at three sites: (1) the Savannah River Site (SRS); (2) the Hanford Reservation (HANF); and (3) the Idaho National Engineering Laboratory (INEL).

A comprehensive evaluation of a number of alternative HLW forms was performed by each of these three sites, as well as an independent Alternative Waste Form Peer Review Panel, to determine their relative scientific merits and engineering practicality.

Processes to produce selected waste forms were also subject to a quantitative evaluation procedure based on criteria in reliability/complexity, resource requirements, personnel safety, and quality control. Seven selected waste form processes have been evaluated. The process for vitrification of HLW in the form of borosilicate glass using a Joule–heated continuous ceramic melter was finally selected for further development, while a process to convert HLW into a ceramic form was chosen as the best alternative to glass.

The EG&G Rocky Flats plant is testing the use of microwaves for reducing and solidifying radioactive waste. The process reduced waste volume and weight by 87% in several earlier experiments. The Rocky Flats method uses microwaves to melt sludge–type waste at temperatures of up to 2800°F. The result is vitrification of the waste into a glass matrix that is denser and more leach–resistant than the usual sludge by–product.

4.3.1 Ex–Situ Processing Considerations

Vitrification technologies include glass and slag vitrification. Vitrification processes involve dissolving the waste at high temperatures into glass or a glass–like matrix.

High–temperature vitrification is applicable to nonwastewaters containing arsenic or other characteristic toxic metal constituents that are relatively nonvolatile at the temperatures at which the process is operated. This technology is also applicable to many wastes containing organometallic compounds, where the organic portion of the compound can be completely oxidized at process–operating conditions. Afterburners may be required to convert unburned organics to carbon dioxide.

The process is not generally applicable to volatile metallic compounds or to wastes containing high levels of constituents that will interfere with the vitrification process. High levels of chlorides and other halogen salts should be avoided in the wastes being processes because they interfere with glassmaking processes and cause corrosion problems.

The basic principles of operation for vitrification technologies depend on the technology used. In glass and slag vitrification processes, the waste constituents become chemically bonded inside a glass–like matrix in many cases. In all instances, the waste becomes surrounded by a glass matrix that immobilizes the waste constituents and retards or prevents their reintroduction into the environment. Arsenates are converted to silicoarsenates, and other metals are converted to silicates.

In the **glass vitrification process,** the waste and normal glassmaking constituents are first blended together and then fed to a glassmaking furnace, where the mixed feed materials are introduced into a pool of molten glass. The feed materials then react with each other to form additional molten glass, in which particles of the waste material become dissolved or suspended. The molten glass is subsequently cooled. As it cools, it solidifies into a solid mass that contains the dissolved and/or suspended waste constituents. Entrapment and chemical bonding within the glass matrix render the waste constituents unavailable for reaction.

Soda ash, lime, silica, boron oxide, and other glassmaking constituents are first blended with the waste to be treated. The amount of waste added to the blend is dependent on the waste composition. Different metal oxides have differing solubility limits in glass matrices. The blended waste and glass raw material mixture is then fed to a conventional, heated glass electric furnace.

The introduced material typically is added through a port at the top of the furnace and falls into a pool of molten glass. The glass constituents dissolve in the molten glass and form additional glass. Molten glass is periodically withdrawn from the bottom of the furnace and cooled. This material then solidifies on cooling into solid blocks of glass–like material. Organics present in the feed mixture undergo combustion at the normal operating temperatures of 1100° to 1400°C and are fully oxidized to carbon dioxide and water vapor.

The top of the furnace is normally cooled so that volatile materials, such as arsenic oxides, that are present in the feed mixtures can condense on the cooled surface and fall back into the melt, where they can undergo chemical reaction to form silicoarsenates involved in the glassmaking process. Most of the arsenic used in making glass by this method is present as salts such as calcium arsenate. This approach was introduced into the glass industry to minimize fugitive arsenic losses.

Gases, such as carbon dioxide, that are liberated during the glassmaking process exit the furnace through the top and are generally wet–scrubbed prior to reentering the atmosphere.

Slag vitrification differs from glass vitrification in that finely ground slag from metal–refining processes and waste are premixed and fed to the same type of furnace as that used for glassmaking. The slag liquifies at the process temperature (1100° to 1200°C), and the waste constituents either dissolve or become suspended in the molten slag. Subsequent cooling of the slag causes it to solidify, trapping the waste inside a glass–like matrix and rendering it unavailable for chemical reaction or migration into the environment.

The slag vitrification process is basically similar to glass vitrification except that granulated slag, instead of the normal glassmaking constituents, is blended with the waste for feed to the system. A pool of liquid slag is present in the furnace, and the blended raw material mix typically is introduced at the top of the furnace and falls into this molten slag. The granulated slag–waste mixture liquifies to form additional slag. Slag is periodically withdrawn from the slag pool and cooled into blocks.

The type of furnace used for glass vitrification can also be used for slag vitrification. The operating parameters are similar.

Organic Content: At process operating temperatures (1100° to 1400°C), organics are combusted to carbon dioxide, water, and other gaseous products. The combustion process liberates heat, reducing the external energy requirements for the process.

The amount of heat liberated by combustion is a function of the Btu value of the waste. The Btu content merely changes the energy input needs for the process and does not affect waste treatment performance. The amount of material that may not oxidize completely is a function of the organic halogen content of the waste. The presence of these halogenated organics does impact process performance because sodium chloride has a low solubility in glass. The presence of high chlorides results in a porous glass that is undesirable. If the halogenated organic content of an untested waste is the same as or less than that present in an already tested waste, the system should achieve the same performance for organic destruction.

Concentrations of Specific Metal Ions: Most metal oxides have solubility limits in glass matrices. Hence, their concentration determines the amount of glass–forming materials or slag with which they must be reacted in this process to generate a nonleaching slag or glass. Oxides for which extensive solubility information is available are alumina, antimony oxide, arsenic oxides, barium oxide, cadmium oxide, chromium oxides, copper oxides, cobalt oxides, iron oxides, lead oxides, manganese oxides, nickel oxides, selenium oxides, tin oxides, and zinc oxides. If the concentrations of specific metals in an untested waste are less than those in a tested waste, then the same ratio of slag or glass raw materials to waste may be used for vitrification purposes. If, however, the concentration of metal is greater than that in the tested waste, a different formulation must be used.

Concentrations of Deleterious Materials: Some waste constituents, such as chlorides, fluorides, and sulfates, interfere with the vitrification process if they are present at high levels. These salts have limited solubilities in glass; therefore, when they are present, additional glass–forming raw materials must be added to compensate for their presence. The solubility limits of various salts in glasses are discussed in references on glass production such as the *Handbook of Glass Manufacture.* Generally, if the concentrations of such materials in an untested waste are lower than those in a tested waste, then the same ratio of glass–forming constituents to waste may be used. Reducing agents such as carbon or ferrous salts reduce arsenates and selenates to lower valence compounds that are more volatile. These compounds should not be present in significant quantities in arsenic– or selenium–containing wastes to be vitrified.

Moisture Content: Materials fed to the vitrification process should be reasonably dry (i.e., contain less than 5% free moisture). If a waste has excess moisture above this level, it should be thermally dried before it is blended with glass–forming materials; otherwise, it may react violently when introduced to the molten glass or slag pool.

Composition of the Vitrifying Agent: Slag and various glassmaking formulations are used as vitrifying agents. The choice of the vitrifying agent is determined by the solubility of the waste constituents to be vitrified. Different inorganic oxides have differing solubilities in various glass matrices.

For slags, the presence of carbon or other reducing agents is undesirable when arsenic–bearing or selenium–bearing wastes are vitrified. Carbon or ferrous salts in the slag reduce arsenates in the waste to arsenic trioxide, which has a low volatilization temperature. In a similar manner, these same reducing agents reduce selenates to elemental selenium, which also has a low volatilization temperature.

In glass vitrification, various glassmaking formulations can be used. EPA examines the proposed formulations to ensure that the toxic metal ion concentrations of the final product do not exceed solubility limits. Hence, EPA examines the material balances based on waste composition and glassmaking additives and the published solubility limits for metal oxides in various glasses to ensure that the vitrified product is indeed a glass containing the solubilized toxic waste constituents.

Operating Temperature: Vitrification furnaces are normally operated in the 1100° to 1400°C range. The exact operating temperature is usually selected based on the desired composition of the final product. Furnaces are normally equipped with automatic temperature control systems.

Residence Time: Sufficient time must be allowed for the materials added to glass furnaces to reach operating temperatures and then undergo the chemical reactions needed to produce glasses. Residence times are normally on the order of 1 to 2 hours for processes operated at 1100° to 1200°C. For glasses or slags requiring slightly higher temperatures, slightly longer residence times are usually selected.

Vitrification Furnace Design: Vitrification furnaces normally incorporate the following design features:

1. Withdrawal of the product in liquid form from the base of the furnace.
2. Maintenance of a liquid pool of product in the furnace.
3. Addition of product constituent mix at the top of the furnace.
4. Design of the top area of the furnace in a manner that allows for cooling of this area. (This is important because volatile constituents of the input feed may vaporize from the melt. The cool top area allows these constituents to condense and fall back into the melt.)
5. Presence and proper operation of an air emissions control afterburner and scrubbing system to manage vent gas emissions from the system such as volatilized noncombusted organics and hydrogen chloride vapors from combustion of any chlorinated organics present.

4.3.2 Ex–Situ Methods

Molten Glass Furnace: This technology uses a pool of molten glass as the heat transfer mechanism to destroy organics and to capture ash and inorganics. The emissions include acid gas and any particulates while all residues are contained in the glass. The advantages include significant volume reduction, most wastes are treatable and the residual is stabilized, nonbreaking glass. The process is based on existing glass making technology.

The electric furnace/melter class includes processes that utilize a ceramic–lined, steel–shelled melter to contain the molten glass and waste materials to be melted. Some of these processes utilize equipment quite similar to electric glass furnaces that have widespread use for the manufacture of glass products, e.g., bottles, plate products. Such melters involve placement of waste materials and glass batch chemicals directly on the surface of a molten glass bath. The majority of melting occurs at the waste/molten glass interface as heat is transferred from the molten glass. As such waste is heated, organics and inorganic volatiles are evolved and either pyrolyzed or oxidized prior to off–gas treatment to ensure safe air emissions.

Another class of melters involve feeding mechanisms that introduce the waste materials below the molten glass surface. Such method of introduction results in pyrolysis of organic contaminants within the molten glass, followed by evolution of pyrolyzed off–gases to the space above the glass surface and thence to the off–gas treatment system. Both classes of melters result in the incorporation of nonvaporizable inorganics into the molten glass.

Periodically, the electric melters must be tapped to remove the accumulated glass product. The molten glass may be cast directly into containers. Another alternate utilizes a water bath to produce a granular residual product. The containerized or loose residual product must then be disposed.

The molten glass furnace is a tunnel–shaped reactor, lined with refractory brick, in which a pool of glass is maintained in a molten state by electric current passing through the glass between submerged electrodes. Such furnaces are used extensively in the glass manufacturing industry. The unit is designed to withstand temperatures as high as 1260°C (2300°F), and corrosion by acidic gases. They are equipped with heat recovery and air pollution control systems, and can be combined with a preconditioning heater or primary incineration unit.

In the absence of a primary incineration unit, wastes can be fed directly to the furnace chamber, above the pool of molten glass. Solids, slurries, and highly viscous liquids are usually charged via a screw feeder. Liquids may also be sprayed into the chamber through nozzles located at the top of the unit. Combustion air is fed to the system from two locations, one near the top, and the other nearer to the surface of the pool on the opposite side, in order to maximize the turbulence within the reaction space. The temperature within the chamber is maintained at 2300°F. Residence time of gases within the chamber is about 2 seconds although this can be increased if desired by reducing load. Residence time of solids within the glass will be appreciably longer, and is measured in terms of hours.

During operation, volatile waste materials mix with air, ignite, and react in the space above, and at the surface of, the pool of molten glass. The solid products of combustion, dirt, and other noncombustible materials, e.g., heavy metal contaminants or the solid waste being treated, will be incorporated into the glass bed. Gaseous products flow out of the chamber, through a series of ceramic fiber filters, which catch most of the particulate matter. The hot gases, consisting primarily of CO_2, water vapor, and HCl (if chlorinated organics are incinerated) then pass through a heat exchanger for heat recovery (heat is used to warm the combustion air). The exhaust gases flow next to a series of water spray–type scrubbers. The first spray chamber is designed to use a slightly alkaline scrubbing liquor, to capture acidic vapors. Water is used in the other spray chamber (or chambers), to remove remaining particulates and other scrubbable vapors. The gases are then reheated above the dew point, and passed through charcoal and HEPA filters before being vented out the stack. The entire system is maintained under negative pressure by means of the exhaust blower.

After a period of usage, the molten glass bed, with the solid waste materials incorporated, is tapped out of the chamber into metal canisters, and, after cooling, is sent to a disposal facility. The ceramic filters, which eventually become loaded with particulate matter, can be disposed of by dissolving them in the molten glass bed. The glass bed can also be used to encapsulate the sludge from the spray chambers, and the spent charcoal and HEPA filters.

Advanced Electric Reactor (High Temperature Fluid Wall) (HTFW): Advanced electric reactors use electrically heated fluid walls to pyrolyze waste contaminants. The resulting thermal radiation causes pyrolysis of the organic constituents in the waste feed. At these high temperatures inorganic compounds melt and are fused into vitreous solids. Most metal salts are soluble in these molten glasses and can thus become bound in a solid matrix (vitrified beads). Following pyrolysis in the reactor, granular solids and gaseous reactor emissions are directed to a post reactor zone, where radiant cooling occurs.

This process is used to treat organics or inorganics, in solid, liquid or gaseous form (solid or liquid may require pretreatment) and for PCB or dioxin contaminated soils. It is limited to treating solids less than 35 U.S. mesh and liquids atomized to less than 1,500 micron droplets. A post treatment process may be needed in order to remove products of incomplete combustion from the emissions.

The process can be made available in a mobile version. Capital and operating costs are high.

The advantages are: transportability, extremely high treatment efficiencies (because of the high process temperatures and long gas residence time), intrinsic safety features (such as the activated carbon beds, fail safe design, electrical driven solids feeder, and large amount of thermal inertia in the reactor), essentially no stack or fugitive emissions, and the ability to detoxify wastes in a pyrolytic atmosphere thereby avoiding products of oxidation such as dioxins and furans.

Cyclone Furnaces: In a cyclone combustor power plant, coal is burned in a separate chamber outside the furnace cavity. The hot combustion gases then pass into the boiler where the actual heat exchange takes place.

The advantage of a cyclone combustor is that the ash is kept out of the furnace cavity where it could collect on boiler tubes and lower heat transfer efficiency. To keep ash from being blown into the furnace, the combustion temperature is kept so hot that mineral impurities melt and form slag, hence the name *"slagging combustor."* A vortex of air (the "cyclone") forces the slag to the outer walls of the combustor where waste can be removed. This concept has been modified to vitrify metal–contaminated soil.

In the **Babcock & Wilcox** process, a 6 million Btu/hr pilot–scale cyclone furnace was demonstrated using a synthetic soil matrix (SSM). This non–mobile cyclone furnace is a scaled–down version of B&W's commercial cyclone boiler and is capable of firing natural gas, oil, or coal. The cyclone furnace is water cooled and simulates the geometry of B&W's single cyclone, front–wall–fired cyclone boiler. The furnace has a horizontal cylinder (barrel) lined with a refractory layer suitable for operation at high temperatures. This unit is designed to achieve very high release rates, temperatures, and turbulence

The SSM was contained in 55 gal drums. A drum tumbler was used to mix each drum before it was transferred into the feeder tank. The feed SSM was introduced at a nominal feed rate of 170 lb/hr via a soil disperser (atomizer) at the center of the cyclone. The cyclone furnace was fired with natural gas during the demonstration and preheated combustion air (nominal 800°F) entered the furnace tangentially. Particulate matter from the feed soil is retained along the walls of the furnace by the swirling action of the combustion air and is incorporated into a molten slag layer. Organic material in the soil is incinerated in the molten slag or in the gas phase. The slag exits the furnace from a tap at the cyclone throat at a temperature of approximately 2400°F, then drops into a water–filled quench tank, where it cools and solidifies.

The gas residence time in the furnace is approximately two seconds. The gas exits the cyclone barrel at a temperature of over 3000°F and exits the furnace at a temperature of over 2000°F. A heat exchanger cools stack gases to approximately 200°F before they enter the pulse–jet baghouse. A small portion of the soil exits as fly ash in the flue gas and is collected in the baghouse. The cyclone facility is also equipped with a scrubber (a lime spray dryer) to control any acid gases that may be generated. The scrubber and baghouse are followed by an induced draft (ID) fan, which draws flue gases into a process stack for release to the ambient air.

The **Vortec CMS** developed by Vortec Corp., which is the primary thermal processing system, consists of three major assemblies: a precombustor, an in–flight suspension preheater, and a cyclone melter chamber. Contaminated soil (waste in slurry or dry form) is introduced into the precombustor as the first step in the process, where heating and oxidation of the waste materials are initiated. The precombustor is a vertical vortex combustor designed to provide sufficient residence time to vaporize water and to initiate oxidation of organics in the waste materials before melting of the material. The suspension preheater is a counter rotating vortex (CRV) combustor designed to provide suspension preheating of the materials, as well as secondary combustion of volatiles emitting from the precombustor and combustion of auxiliary fuel introduced directly into the CRV combustor. The average temperature of materials leaving the CRV combustion chamber is typically between 2200° and 2700°F.

The preheated solid materials exiting the CRV combustor enter the cyclone melter, where they are separated to the chamber walls to form a molten glass layer. The vitrified glass product and the exhaust gases exit the cyclone melter through a tangential exit channel and enter a glass and gas separation chamber assembly. The exhaust gases then enter an air preheater for waste heat recovery and are subsequently delivered to an air pollution control subsystem for particulate and acid gas cleanup. The molten glass product exits the glass and gas separation chamber through a slag tap and is delivered to a water quench assembly for subsequent disposal.

The **Institute of Gas Technology (IGT)** has developed a two-stage, fluidized-bed cyclonic agglomerating incinerator based on a combination of technologies developed at IGT over many years.

The first stage of the incinerator is an agglomerating fluidized-bed reactor, which can operate either under substoichiometric conditions or with excess air. The system can operate over a wide range of conditions, from low temperature (desorption) to high temperature (agglomeration), including the gasification of high British thermal units (Btu) wastes (such as natural gas). With a unique distribution of fuel and air, the bulk of the fluidized-bed is maintained at 1500° to 2000°F, while the central spout temperature can be varied between 2000° and 3000°.

When the contaminated soils and sludges are fed into the fluidized-bed, the combustible fraction of the waste undergoes a rapid gasification and combustion, producing gaseous components. The solid fraction, containing metal contaminants, undergoes a chemical transformation in the hot zone and is agglomerated into glassy pellets.

The product gas from the fluidized-bed is fed into the second stage of the incinerator, where it is further combusted at a temperature of 1600° to 2200°F. The second stage is a cyclonic combustor and separator that provides sufficient residence time (2½ seconds) to oxidize carbon monoxide and organic compounds to carbon dioxide and water vapor, with a combined destruction removal efficiency greater than 99.99%. Volatilized metals are collected downstream in the flue gas scrubber condensate.

Entrained Bed Gasification: The Texaco entrained-bed gasification process is a noncatalytic partial oxidation process in which carbonaceous substances react at elevated temperatures to produce a gas containing mainly carbon monoxide and hydrogen. This product, called synthesis gas, can be used (1) to produce other chemicals or (2) to be burned as fuel. Ash in the feed melts and is removed as a glass-like slag.

The process treats waste material at pressures above 20 atmospheres and temperatures between 2200° and 2800°F.

Wastes are pumped in a slurry form to a specially designed burner mounted at the top of a refractory-lined pressure vessel. The waste feed, along with oxygen and an auxiliary fuel such as coal, flow downward through the gasifier to a quench chamber that collects the slag for removal through a lock hopper. The synthesis gas is then further cooled and cleaned by a waste scrubbing system; a sulfur recovery system may be added. Fine particulate matter removed by the scrubber may be recycled back to the gasifier. Metals and other ash constituents become part of the inert slag.

This process can treat contaminated soils, sludges, and sediments containing both organic and inorganic constituents, such as used motor oils and lubricants, chemical wastes, and petroleum residues. Solids in the feed must be ground and pumped in a slurry form containing 40 to 70% solids by weight and 60 to 30% liquid, usually water.

Flame Reactor Process: The flame reactor process (Horsehead Resource Development Co., Inc. is a patented, hydrocarbon–fueled, flash smelting system that treats residues and wastes containing metals. The reactor processes wastes with a hot (>2000°C) reducing gas produced by the combustion of solid or gaseous hydrocarbon fuels in oxygen–enriched air. In a compact, low–capital cost reactor, the feed materials react rapidly, allowing a high waste throughput. The end products are a nonleachable slag (a glass–like solid when cooled), and a recyclable, heavy metal–enriched oxide, and a metal alloy.

The achieved volume reduction (of waste to slag plus oxide) depends on the chemical and physical properties of the waste. The volatile metals are fumed and captured in a product dust collection system; nonvolatile metals condense as a molten alloy. The remaining trace levels of metals are encapsulated in the slag. At the elevated temperature of the flame reactor technology, organic compounds are destroyed. In general, the process requires that wastes be dry enough (up to 5% total moisture) to be pneumatically–fed, and fine enough (less than 200 mesh) to react rapidly. Larger particles (up to 20 mesh) can be processed; however, the efficiency of metals recovery is decreased.

The flame reactor technology can be applied to granular solids, soil, flue dusts, slags, and sludges containing heavy metals. Electric arc furnace dust, lead blast furnace slag, iron residues, zinc plant leach residues and purification residues, and brass mill dusts and fumes have been successfully tested. Metal–bearing wastes previously treated contained zinc (up to 40%), lead (up to 10%), chromium (up to 4%), cadmium (up to 3%), arsenic (up to 1%), copper, cobalt, and nickel.

High Temperature Metals Recovery (HTMR): Several types of high–temperature metals recovery (HTMR) processes are currently available or under development for the recovery of metals from sludges generated either directly by industrial processes or from the treatment of industrial wastewaters. These HTMR processes may involve plasma–based or high–temperature fluid–wall reactor systems (which use electricity as the energy source) or coal/natural gas–based technologies. The HTMR processes have several potential advantages: (1) maximum volume reduction, which reduces the ultimate disposal requirements of any residual materials; (2) potential for destruction of other toxic organic constituents in the wastes; and (3) the potential for energy recovery through the combustion of waste products. Disadvantages include (1) high capital and operating costs; (2) high maintenance requirements because of high–temperature operations; (3) need for highly skilled and experienced operators; and (4) the potential for adverse environmental impacts, primarily from atmospheric discharges. Because of differences in the design and configuration of HTMR processes, no unique process description is applicable to all HTMR processes. Pretreatment and posttreatment requirements also vary by the type of process.

Processes include:

1. Flame reactor technology: Horsehead Resource Development Co., Inc.

2. Sirosmelt furnace technology: Ausmelt Pty. Ltd. (Australia).
3. Silico/soda ash technology: Rostocker, Inc.
4. INMETCO process

High-temperature metals recovery (HTMR) processes are applicable only for the processing of sludges, not for wastewaters. One significant advantage of the HTMR processes is that other toxic constituents in the wastes, such as complexed cyanides/organics, would also be destroyed at the high temperatures (>1100°C) prevailing in the furnaces.

Important waste characteristics affecting the performance of HTMR processes include the following: (1) concentrations of undesirable volatile metals, (2) boiling points of the metal constituents, and (3) thermal conductivity of the waste. Depending on their relative volatility, metals are partitioned between a crude oxide dust (usually recovered from the reactor gases in a fabric filter) and a slag (which may be sold for use in road beds, parking lots, and other fill or ballast-type applications). The processes are generally not sensitive to variations in the composition of the sludges; however, the economics are dictated by the metals content. INMETCO has the following waste feed limitations for their process: Cu, <2 wt %; Co, <2 wt %; Mo, <5 wt %; P, <0.1 wt %; SiO_2, <15 wt %; and S, <1 wt %. Personnel responsible for operating the 20,000 tons/yr Horsehead Flame Reactor in Monaca, PA, indicated that high levels of mercury or antimony in the wastes may result in contamination of the crude oxide product. Chlorides or other halides in the wastes may also result in reduced metallic oxide recoveries from the process. Plasma-based technologies are often very sensitive to the moisture content of the feed materials; hence they may be unable to treat sludges unless they are sufficiently dried prior to high-temperature treatment.

HTMR is generally not used for mercury-containing wastes even though mercury will volatilize readily at process temperatures present in the high temperature units. The retorting process is normally used for mercury recovery because mercury is very volatile and lower operating temperatures can be used.

Several types of plasma arc systems are being investigated by different vendors. The heart of all these systems is the plasma torch, which consists of a closely spaced pair of electrodes installed in a furnace to produce an electrical arc. A process gas is injected into the space between the electrodes. This gas can be inert, oxidizing, or reducing in nature. Wastes are introduced into the reactive zone of the furnace, where the molecular bonds of the waste material are destroyed and the waste materials are converted to basic elements (e.g., metals, carbon, oxygen) or simple molecules (e.g., carbon monoxide, water). The product collection system can consist of a condenser or a combination condenser and baghouse.

A high-temperature fluid-wall reactor uses radiative heat to pyrolyze waste components and causes them to form elements or simple compounds. The wastes are gravity-fed into a cylindrical graphite core that is electrically heated to about 2200°C. The waste materials are heated by radiation in the absence of oxygen. Volatile metals (e.g., lead, zinc) are separated as a fume and can be collected in a fabric filter system, whereas the nonvolatile metals can be collected from the bottom of the furnace, either in the form of a metallic mixture of a slag.

Joule Heated Glass Melter: In the glass industry, joule heating refers to utilizing the material being heated as the resistance element in an electrical circuit, thus avoiding the necessity of transferring heat from a metallic resistance element.

It is thought that the glass melter would be highly adaptable to the destruction and fixation of industrial contaminated soils. The process would accept contaminated soils directly, eliminating any pretreatment steps. In a single step, the organic constituents would be destroyed and/or combusted at temperatures of 1200°C (2200°F), and the inorganic constituents and soil would react with a relatively small amount of glass formers to create a glass matrix. It is a promising technology for incorporating high-level nuclear waste into a stable glass.

With the special feeding technique employed in the melter, a more rapid processing rate is expected with the joule-heated glass melter than with conventional melters. The melter could be scaled to meet the requirements of a waste treatment operation designed to process numerous tons per day. The melter would be expected to have a 10 year lifetime, requiring replacement of the ceramic linings and some of the other high-temperature components once ever 2 to 3 years.

Plasma Arc Systems: Plasma is highly ionized gas that contains equal numbers of positively and negatively charged particles. Plasmas can be created by passing gas through an electrical discharge and thereby ionizing it.

The plasma centrifugal reactor (Retech, Inc.) varies significantly from the melter/furnace concept. In the plasma reactor, prepared waste materials are fed into a rotating reactor well in which a transferred-arc plasma torch is operating. The plasma torch, which is capable of temperatures exceeding 10,000°C heats the waste material beyond the point of melting to about (typically) 1600°C. The melted material is allowed to fall into a slag chamber where it is collected in a container.

Organics and other volatiles emitted during the plasma heating are passed to a secondary combustion chamber into which an oxidizing gas is added. The resulting off-gases are then transferred to an off-gas treatment system to ensure safe air emissions. The containerized slag must eventually be disposed.

Contaminated soils enter the sealed furnace through the bulk feeder. The reactor well rotates during waste processing. Centrifugal force created by this rotation prevents material from falling out of the bottom and helps to evenly transfer heat and electrical energy throughout the molten phase. Periodically, a fraction of the molten slag is tapped, falling into the slag chamber to solidify.

Off-gas travels through a secondary combustion chamber where it remains at 2000° to 2500°F for more than 2 seconds. This allows the complete destruction of any organics in the gas. After passing through the secondary combustion chamber, the gases pass through a series of air pollution control devices designed to remove particulates and acid gases. In the event of a process upset, a surge tank allows retention for reprocessing.

Residuals from the cleanup system can sometimes be fed back to the furnace. Salts resulting from neutralizing chlorides must eventually be discarded. In some circumstances, metals can be recovered from the scrubber sludge.

Westinghouse and Pyrolysis Systems, Inc. (PSI) have developed a mobile plasma reactor (Pyroplasma System). Westinghouse has also developed a plasma cupola process for metal recovery. Processes are also being developed in Japan, France, and Germany. An in-situ process has been developed at Georgia Tech, for vitrification of landfills, as well as vitrification of asbestos at demolition sites.

Advantages:

1. Since radiative heat transfer proceeds as a function of the fourth power of temperature, a plasma system has very intense radiative power and therefore is capable of transferring its heat much faster than a conventional flame.
2. Organic chlorides are known to dehydrogenate when excited by the ultraviolet radiation which is abundant from thermal plasmas.
3. Because the plasma arc for waste destruction is a pyrolytic process, it virtually does not need oxygen at all. Compared to conventional incinerators which normally require about 150% excess air to ensure proper combustion, the plasma arc will save the energy required to heat the excess air to the combustion temperatures and will thereby produce significantly less oxygenated by-products that would otherwise need to be treated in downstream equipment.
4. The process has a very short on/off cycle.
5. Because of its compactness, a plasma arc system has potential for use in a mobile trailer for movement of the system from site to site.

Disadvantages:

1. Because the temperatures are so high (about 18,000°F at the arc's centerline), the durability of the arc and the refractory materials could be a potential problem.
2. Because the arc is very sensitive to many factors such as sudden drops in voltage, the operation of the system requires highly-trained professionals.

Slagging Incinerators: Rotary kiln incinerators can operate in the ashing mode, or in the slagging mode. A standard incinerator is normally operated in the ashing mode to keep slagging to a minimum. Slagging incinerators run at higher temperatures (1100° to 1300°C), and can be utilized to vitrify heavy metals in the slag.

In order to optimize the properties of slag to pass EPA's leaching test, efficient furnace operation is required; refractory corrosion, and erosion should be minimized. The four primary slag control parameters are fusion temperature, molar basicity, apparent viscosity, and oxidation state.

PSI Technology Company's metals immobilization and decontamination of aggregate solids (**MeIDAS**) process is a modified incineration process for the destruction of organics and treatment of metals in contaminated soils. In this process, the contaminated soil is treated in a typical incinerator, in conjunction with sorbents. The high temperatures present during the processing destroys the organics, while simultaneously encapsulating the metals in the soil into a form that is nonleachable (as determined by toxicity characteristic leaching procedure analysis). With a one-to-one soil to sorbent ratio, the metals present in the soil can be effectively treated.

The **Pyrakiln Thermal Encapsulation Process** developed by Allis Mineral Systems is a rotary kiln process that traps metals in a controlled melting process operating in the temperature range between slagging and nonslagging modes, producing nodules of ash that are ¼ to ¾ inch in diameter.

The use of fluxing reagents is a key element in this technology. These will be introduced into the kiln in the proper amount and type to lower the softening temperature of the ash. Proper kiln design is required to allow the outlet of the kiln to function as an ash agglomerator. Good temperature control is required to keep the agglomerates at the correct particle size, yielding the desired ¼ to ¾ inch size nodules. The production of nodules, rather than a molten slag, potentially avoids a multitude of operating problems, such as ash quenching, overheating, and premature failure of refractory. It also simplifies cooling, handling, and conveying of the ash.

The controlled nodulizing process should immobilize metals with high boiling points. Lead, zinc, and other metals with lower vaporization temperatures tend to leave the kiln as a fine fume and can be removed in the air pollution control system. Reagents can be injected into the kiln, the air pollution control devices, or a final solids mixer for stabilizing fines collected from the gas stream.

Thermally–Driven Chemical Bonding: Ceramic Bonding, Inc. (CBI) has developed a process and specialized equipment (patents pending) to render metal bearing sludges harmless, and create a tailored product which can be used as a construction raw material. The technology involves high temperature physical–chemical bonding of metals into naturally occurring alumina–silicates, to produce a non–soluble, non–leachable ceramic aggregate.

The CBI process is an endothermic, physical–chemical bonding of metal ions to alumina–silicate crystals in a high temperature environment. The process is accomplished in two stages: proportioning and mixing of the additive material with the metal sludge; and thermal processing.

Proportioning and mixing is accomplished in the CBI system by commercially available feeders and blenders. Once mixed, the material is nodulized to a size and shape which is mechanically and thermally appropriate for further processing. The mixing and nodulizing are performed continuously, and are monitored by a supervisory control system. Adequate bonding will occur for a wide range of sludge/additive stoichiometries, however, it is economically desirable to maintain mixture ratios and moisture contents within pre–determined limits.

In the CBI thermal process, the nodules are first dried and heated to the required chemical bonding temperature. Upon reaching the necessary temperature, the metal ions are incorporated into the lattice structure of the alumina–silicate additive. Concurrently, the physical characteristics of the additive are altered by the high temperatures and bonding between the heavy metal cations and the additive crystals is enhanced. This process is a type of sintering, wherein the pore volume of the material decreases, and a thermodynamic phase change takes place. After sintering, the material is cooled, and discharged. Appropriate residence times for each of the above processes are assured, by careful design of the thermal processor, and through the automatic control system, which monitors and governs the material throughput rate, and the temperature profile.

4.3.3 In Situ Vitrification

In situ vitrification (ISV) involves the electric melting of contaminated soils in place. ISV uses an electrical network typically consisting of four electrodes placed in a square pattern and at the desired depth, to electrically heat and melt contaminated soils and solids at temperatures of 2900° to 3600°F (1600° to 2000°C). ISV destroys organic pollutants by pyrolysis. Inorganic pollutants are immobilized within the vitrified mass, which has properties of glass. Both the organic and inorganic airborne pyrolysis by-products are captured in a hood, which draws the contaminants into an off-gas treatment system that removes particulates and other pollutants of concern.

ISV is effective on aqueous media, organic liquids, sediments, soils, and sludges contaminated with halogenated volatiles, halogenated nonvolatiles, nonhalogenated volatiles, nonhalogenated nonvolatiles, pesticides, dioxins/furans, organic cyanides, organic corrosives, volatile metals, nonvolatile metals, and PCBs.

On saturated soils or sludges, the initial application of the electric current is needed to reduce the moisture content before the vitrification process can begin. This increases energy consumption and associated costs. Also, sludges must contain a sufficient amount of glass-forming material (nonvolatile, nondestructible solids) to produce a molten mass that will destroy or remove organic and immobilize inorganic pollutants. The ISV process, however, has the following limitations: (a) individual void volumes in excess of 150 ft^3 (4.25 m^3); (b) buried metals in excess of 5% of the melt weight or continuous metal occupying 90% of the distance between two electrodes; (c) rubble in excess of 10% by weight; and (d) the amount and concentration of combustible organics in the soil or sludge. These limitations must be addressed for each site.

Acids and salts in the soil can also be a concern when using this technology. Acids and salts can cause the soil to have an abnormally high electrical conductivity (hence, a low electrical resistance), which is generally more pronounced as the moisture content of the soil increases. This low resistance will require the application of more electrical energy to the treatment area in order to achieve a vitrified melt. This may also result in a much higher melt temperature.

This technology is particularly applicable to nuclear contaminated soils, including mixed wastes.

In Situ Vitrification (ISV) is an innovative, mobile, onsite remediation technology for contaminated solids. ISV has been under development for the U.S. Department of Energy (DOE) since 1980 by Battelle Memorial Institute's Pacific Northwest Laboratories. DOE has licensed the technology to Battelle, who has in turn exclusively sublicensed it to Geosafe Corporation for commercial application purposes.

The ISV process involves in situ electric melting of contaminated solids at very high temperatures, typically in the range of 1600° to 2000°C for most soils. An array of four electrodes is either placed to the desired treatment depth in the volume to be treated prior to treatment (fixed electrodes), or the electrodes are lowered into the treatment volume as the melt progresses (moveable electrodes). A conductive mixture of graphite and glass frit is placed on the surface between the electrodes to serve as an initial conductive starter) path. As electric potential is applied between the electrodes, current flows through the starter path, heating it and the adjacent solids to the solids melting point. Upon melting, typical soils become electrically conductive; thus the molten mass becomes the primary electrical conductor and heat transfer medium allowing the process to continue beyond startup. The molten mass grows downward and outward as long as electric power is applied.

An off-gas collection hood gathers gases that evolve from the treatment zone during processing. Water vapor is usually the predominant evolved gas present in the hood, since most soils contain 15 to 30% moisture above the saturated zone. Secondarily, organic contaminant pyrolysis products and soil decomposition products will evolve to the surface under the collection hood. A large amount of ambient air is allowed to enter the hood, where it supplies oxygen for the combustion of flammable pyrolysis products and for purposes of cooling the hood. The air and other gases are then drawn through an off–gas treatment system to ensure their acceptability for release.

Significant volume reduction (25 to 45 vol % for most soils) occurs as solids particles melt and interstitial void volume is removed. Volume reduction results in a subsidence of the melt surface below the starting grade. When power is terminated to the melt, it cools to a monolithic, vitrified (glassy with microcrystallinity) residual product which resembles natural obsidian (natural volcanic glass) for most soil applications. Single melts as large as 1,000 tons can be produced by existing large–scale equipment capable of processing 120 tons per day. Adjacent melts fuse together to produce a single impermeable monolithic structure. Completion of each melt setting involves placement of clean backfill to the desired depth in the subsidence volume.

Since ISV is a batch or setting type process, its time–operated efficiency increases with depth of processing. The process is most economical when dealing with large quantities (e.g., 300 to 1,000 tons treated per setting of electrodes). Processing depths greater than 10 feet are ideal, but not necessary.

ISV applications may be categorized relative to the primary location and/or condition of the waste. Such categorization includes: (1) contaminated soil, (2) buried waste, and (3) underground structures. Most ISV development work has focused on contaminated soil applications wherein the contaminated media is primarily soil. The soil has typically become contaminated in such cases through exposure to contaminated liquids. In many cases the contaminated liquid is water that has percolated through impounded or buried waste that may or may not have been removed prior to addressing remediation of the contaminated soil. Contaminated soil applications are relatively straightforward compared to other types of applications; and the ISV technology is considered to be developed and demonstrated for many such applications.

Buried waste applications address wastes that have been covered by soil such as backfilled impoundments and landfills. Substantial amounts of test work has been performed on a variety of process sludges, ash, and containerized waste. The ISV technology is not considered generically ready for such applications; at this time a specific test and demonstration plan is necessary for each one.

Buried waste applications involving wastes which were highly heterogeneous at time of burial typically pose a problem of site characterization. It is necessary to know worst case conditions within the treatment zone to allow appropriate remedial design for the site. Homogeneous wastes, such as some settled lagoon and impoundment sludges and sediments, pose less of a characterization problem. However, the chemical composition of such wastes must be analyzed relative to the soil in the treatment zone to allow prediction and evaluation of melt behavior when the sludge/sediment zones are encountered. In some cases it may be necessary to intermix the soil and waste layers to allow proper treatment. The effect of the wastes on overall residual product chemistry and properties must also be evaluated.

Containerized wastes such as buried drums, crates, and cartons pose additional problems. Whereas the ISV process conditions may be adequate for treating such materials, the site characterization challenge becomes even more severe. The ISV technology is not considered ready for application to such sites at this time except on a test and demonstration basis. ISV is being developed for such applications within the DOE community because of the high cost of alternative technologies. Treatment of such sites may require use of equipment with larger than normal off–gas treatment capacity and/or the use of secondary off–gas containment to protect against unforseen high gas generation events.

The primary qualification regarding type of soil that may be treated by ISV is whether or not the soil will form and support a melt. ISV test results have indicated that most natural soils may be processed by ISV without modification. Various sludges, sediments, and process tailings have also been successfully tested. For proper application, it is necessary that the soil and/or other solids contain sufficient inorganic material that will remain in the molten state during treatment. It is the molten mass that serves as the electrical conductor during ISV; and the flow of electricity through the melt results in the generation of heat which is then passed into adjacent soil by thermal conduction.

Molten soil must possess sufficient electrical conductivity to allow the process to be performed economically. Electrical conductivity within a soil melt is typically provided by the monovalent alkali earth cations, e.g., sodium, potassium. It is desirable that such cations be present in the 2 to 5 wt % range, which is common for most soils. In the event a soil possesses insufficient molten conductivity, it is possible to obtain the needed conductivity through addition of other materials, e.g., materials that provide Na_2O and/or CaO, such as suitable soil, soda ash, and lime.

The chemical (oxide) composition of the soil is important in determining the quality of residual product produced. Soil is the result of weathering of rocks, and rocks are made up of many minerals (complex metal oxides). Upon melting, minerals decompose to a melt mixture of major oxides, in which silica is predominant for most soils. Silicate melts typically produce a residual product of excellent properties relative to environmental exposure. Other low–silica soils, e.g., limestone/dolomite, have also been treated by ISV to produce a high quality residual product. It is possible to determine the applicability of ISV to various soils by performing and evaluating oxide composition analyses and small–scale melt tests.

As the high temperature ISV melt moves slowly downward and outward through the contaminated solids, a very steep thermal gradient (150° to 250°C per inch) precedes the melt. At appropriate temperature regimes within this gradient, or within the melt itself, the solids and contaminants undergo change of physical state and decomposition reactions. The possible dispositions of particular contaminants include: (1) chemical and/or thermal destruction, (2) removal from the treatment volume to the off–gas treatment system, and (3) chemical and/or physical incorporation within the residual product. Many site– and application–specific variables affect the disposition of specific contaminants. The primary variables include: (1) contaminant physical and chemical properties, (2) melt chemistry, (3) melt be removed during processing, (4) depth of processing, and (5) analytical chemistry requirements associated with process control and permit compliance.

Advantages: ISV advantages relative to alternative technologies include its capability to: (1) simultaneously process mixed waste types (organic, heavy metal, radioactive); (2) achieve destruction, removal, and immobilization performance beyond regulatory criteria; (3) be performed onsite and in situ; (4) accept significant quantities of rubble and debris in the treatment zone; (5) achieve a significant volume reduction (25 to 45% for most soils); and (6) produce an unequalled residual product with a geologic time life expectancy (thousands to millions of years). ISV also possesses significant differences from other vitrification technologies, e.g., plasma, joule–melter, slagging kiln, many of which are considered advantages.

Limitations: The primary limitations on ISV applications relate to: (1) total organic concentration, (2) water recharge rate, (3) depth of processing, and (4) presence of inclusions. Since organics become gaseous pyrolysis products during ISV, the concentration of organics must be limited in relation to the off–gas collection and treatment equipment capacity. The average allowable concentration for most organics falls in the range of 5 to 10 wt %.

Fully saturated soils may be processes, however, it is economically advantageous to minimize soil moisture content and water recharge rate. These factors influence cost through consumption of energy and impacting of processing rate. Processing rate may be limited by the amount of energy going into water removal, or by operating at less than full power to maintain acceptable water vapor generation and removal rates peculiar to a specific application.

The maximum depth processed by ISV to date is 19 to 20 feet. Greater depths will be attempted in the continuing ISV development program.

The ISV process is capable of accommodating significant inclusions within the treatment zone, e.g., rocks, roots, drum remnants and other metal scrap, concrete, asphalt, construction debris, etc. However, the concentration of these must be limited so as to not interfere with proper formation and advancement of the melt. All of the above limitations are subject to consideration during applicability analyses, treatability testing, and project remedial design.

Development is underway at Georgia Tech on a plasma–arc torch process that is essentially an in–situ process for vitrifying wastes in landfills. The goal is to bore as much as 100 feet into an existing landfill and melt the accumulated waste from the bottom up. The vitrified waste could be left in place, or possibly used as fill in construction projects.

REFERENCES

1. Arienti, M., et al, *Dioxin–Containing Wastes,* Noyes Data, 1988.
2. Barth, E., et al, *Stabilization and Solidification of Hazardous Wastes,* Noyes Data, 1990.
3. Berkowitz, J., *Unit Operations for Treatment of Hazardous Industrial Wastes,* Noyes Data, 1978.
4. Breton, M., et al, *Treatment Technologies for Solvent Containing Wastes,* Noyes Data, 1988.
5. Burton, D., et al, *Treatment of Hazardous Petrochemical and Petroleum Wastes,* Noyes Publications, 1989.
6. Chambers, C., et al, *In Situ Treatment of Hazardous Waste Contaminated Soils,* Noyes Data, 1991.
7. EPA, *Approaches for Remediation of Uncontrolled Wood Preserving Sites,* EPA/625/7–90/011, 11/90.
8. EPA, *Babcock & Wilcox Cyclone Furnace Vitrification,* EPA/540/SR–92/017, 9/92; and EPA/540/SR–93/507, 5/93.
9. EPA, *A Compendium of Technologies Used in the Treatment of Hazardous Wastes,* EPA/625/8–87/014, 9/87.
10. EPA, *Corrective Action: Technologies and Applications,* EPA/625/4–89/020, 9/89.
11. EPA, *Eighteenth Annual Risk Reduction Engineering Laboratory Research Symposium,* EPA/600/R–92/028, 4/92.
12. EPA, *Forum on Innovative Hazardous Waste Treatment Technologies,* EPA/540/2–89/056, 9/89.
13. EPA, *Forum on Innovative Hazardous Waste Treatment Technologies (Third),* EPA/540/2–91/015, 9/91.
14. EPA, *Handbook—Stabilization Technologies for RCRA Corrective Actions,* EPA/625/6–91/026, 8/91.

15. EPA, *Handbook-Vitrification Technologies for Treatment of Hazardous and Radioactive Waste,* EPA/625/R-92,002, 5/92.
16. EPA, *Horsehead Resource Development Co., Inc., Flame Reactor Technology,* EPA/540/A5-91/005, 5/92.
17. EPA, *Innovative Treatment Technologies, Overview and Guide to Information Sources,* EPA/540/9-91/002, 10/91.
18. EPA, *Proceedings of the U.S. EPA Municipal Wastewater Treatment Technology Forum,* EPA 430/09-91-020, 9/91.
19. EPA, *Remedial Action, Treatment, and Disposal of Hazardous Waste (17th),* EPA/600/9-90/006, 2/90.
20. EPA, *Remedial Action, Treatment, and Disposal of Hazardous Waste (19th),* EPA/600/9-91/002, 4/91.
21. EPA, *Site Program Demonstration Test: Plasma Centrifugal Furnace,* EPA/540/S5-91/007, 8/92.
22. EPA, *Summary of Treatment Technology Effectiveness for Contaminated Soils,* EPA, 9355.4-06, 6/90.
23. EPA, *Superfund Innovative Technology Evaluation Program (Fourth),* EPA/540/S-91/008, 11/91.
24. EPA, *Superfund Innovative Technology Evaluation Program (Fifth),* EPA/540/R-92/076, 10/92.
25. EPA, *Synopses of Federal Demonstrations of Innovative Site Remediation Technologies,* EPA/542/B-92/003, 8/92.
26. EPA, *Treatment Technology Background Documents,* OSW, EPA, 1/91.
27. Freeman, H., *Innovative Thermal Hazardous Waste Treatment Processes,* Noyes Publications, 1985.
28. Holden, T., et al, *How to Select Hazardous Waste Treatment Technologies for Soils and Sludges,* Noyes Data, 1989.
29. Jackman, A., et al, *Hazardous Waste Treatment Technologies,* Noyes Publications, 1991.
30. Krishnan, E., *Recovery of Metals from Sludges and Wastewaters,* Noyes Data, 1993.
31. Meltzer, M., et al, *Metal-Bearing Waste Streams,* Noyes Data, 1990.
32. Nickelson, D., et al, *ISV Technology Development Plan for Buried Waste,* DOE, EGG-WTD-10325, 7/92.
33. Noyes, R., *Handbook of Pollution Control Processes,* Noyes Publications, 1990.
34. Nunno, T., et al, *International Technologies for Hazardous Waste Site Cleanup,* Noyes Data, 1990.
35. Quenau, P., et al, *Slag Control in Rotary Kiln Incinerators,* Poll. Eng., 1/5/92.
36. Roberts, R., et al, *Hazardous Waste Minimization Initiation Decision Report,* Naval Civil Engineering Laboratory, 71-021A, 6/88.
37. Surprenant, N., et al, *Halogenated-Organic Containing Wastes,* Noyes Data, 1988.
38. Wagner, K., et al, *Remedial Action Technology for Waste Disposal Sites,* Noyes Data, 1986.
39. Walker, S., *An Overview of In-Situ Waste Treatment Technologies,* EGG-M-92-342, DOE, 1992.

5

Membrane Technology

In recent years, membranes and membrane separation techniques have grown from a simple laboratory tool to an industrial process with considerable technical and commercial impact. Today, membranes are used on a large scale to produce potable water from the sea by reverse osmosis, to clean industrial effluents and recover valuable constituents by electrodialysis, to fractionate macromolecular solutions in the food and drug industry by ultrafiltration, to remove urea and other toxins from the blood stream by dialysis in an artificial kidney, and to release drugs such as scopolamin, nitroglycerin, etc. at a predetermined rate in medical treatment. Although membrane processes may be very different in their mode of operation, in the structures used as separating barriers, and in the driving forces used for the transport of the different chemical components, they have several features in common which make them attractive as a separation tool. In many cases, membrane processes are faster, more efficient and more economical than conventional separation techniques. With membranes, the separation is usually performed at ambient temperature, thus allowing temperature–sensitive solutions to be treated without the constituents being damaged or chemically altered. This is important in the food and drug industry and in biotechnology where temperature–sensitive products have to be processed. Membranes can also be "tailor–made" so that their properties can be adjusted to a specific separation task.

The interest in membrane technology in pollution control applications is increasing rapidly. Membrane processes are particularly advantageous for dilute feedstreams.

Industrial separation processes consume a significant portion of the energy used in the United States. A 1986 survey by the Office of Industrial Programs estimated that about 4.2 quads of energy are expended annually on distillation, drying and evaporation operations. This survey also concluded that over 0.8 quads of energy could be saved in the chemical, petroleum and food industries alone if these industries adopted membrane separation systems more widely.

Membrane separation systems offer significant advantages over existing separation processes. In addition to consuming less energy than conventional processes, membrane systems are compact and modular, enabling easy retrofit to existing industrial processes.

The design of an energy–efficient, cost–effective membrane always involves a tradeoff between two factors: selectivity and permeability. Selectivity is a measure of the separation efficiency of the membrane, and determines the amount of energy required to accomplish the separation. In general, the higher the selectivity, the lower the amount of energy consumed.

Permeability is a measure of the flux of a given gas species across the membrane, and determines the amount of membrane material required. In general, the higher the permeability, the smaller the surface area of membrane needed (and the less the membrane capital cost).

To increase permeability, membrane designers could decrease the thickness of the membrane. However, as the thickness is decreased, defects or open passages in the membrane will often allow some feed gas to pass through directly; this reduces selectivity. On the other hand, an increase in thickness—as a way to boost selectivity—often reduces permeability. In general, any modifications made during the design of a given polymer to improve one of these two attributes will result in a sacrifice in the other.

Surface modifications of membranes is assuming increased importance. Surface modification techniques include enzyme attachment, affinity, plasma, grafting, and high-frequency discharge treatments.

For an efficient process, the chemical and physical properties of the semi-permeable membrane must be compatible with the waste streams chemical and physical characteristics. Some membranes may be dissolved by some wastes. Suspended solids and some organics will clog the membrane material. Low-solubility salts may precipitate onto the membrane surface. Temperature considerations are important.

Three processes, microfiltration, ultrafiltration, and reverse osmosis, are related filtration techniques, in which a solution containing dissolved or suspended solutes is forced through a membrane filter. The solvent passes through the membrane; the solutes are retained. These processes differ principally in the size of the particles separated by the membrane.

Microfiltration is considered to refer to membranes that have pore diameters from 0.1 μm (1,000 Å) to 10 μm. Microfiltration membranes are used to filter suspended particulates, bacteria or large colloids from solutions. Ultrafiltration refers to membranes having pore diameters in the range 20 to 1,000 Å. Ultrafiltration membranes can be used to filter dissolved macromolecules, such as proteins, from solution. In the case of reverse osmosis, the membrane pores are so small, in the range of 5 to 20 Å in diameter, that they are within the range of the thermal motion of the polymer chains. Reverse osmosis membranes are used to separate dissolved microsolutes, such as salt, from water. The principal application of reverse osmosis is the production of drinking water from brackish groundwater, or the sea.

The fourth fully developed membrane process is electrodialysis, in which charged membranes are used to separate ions from aqueous solutions under the driving force of an electrical potential difference. The process utilizes an electrodialysis stack, built on the filter-press principle, and containing several hundred individual cells formed by a pair of anion and cation exchange membranes. The principal application of electrodialysis is the desalting of brackish groundwater.

Gas separation with membranes is the more mature of the two developing technologies. In gas separation, a mixed gas feed at an elevated pressure is passed across the surface of a membrane that is selectively permeable to one component of the feed. The membrane separation process produces a permeate enriched in the more permeable species and a residue enriched in the less permeable species. Current applications are the separation of hydrogen from argon, nitrogen and methane in ammonia plants, the production of nitrogen from air and the separation of carbon dioxide from methane in natural gas operations.

Pervaporation is a relatively new process that has elements in common with reverse osmosis and gas separation. In pervaporation, a liquid mixture is placed in contact with one side of a membrane and the permeate is removed as a vapor from the other. The mass flux is brought about by maintaining the vapor pressure on the permeate side of the membrane lower than the potential pressure of the feed liquid. Currently, the only industrial application of pervaporation is the dehydration of organic solvents. However, pervaporation processes are being developed for the removal of dissolved organics from water and the separation of organic solvent mixture.

Other membrane processes include facilitated and coupled transport. Facilitated transport usually employs liquid membranes containing a complexing or carrier agent. The carrier agent reacts with one permeating component on the feed side of the membrane and then diffuses across the membrane to release the permeant on the product side of the membrane. The carrier agent is then reformed and diffuses back to the feed side of the membrane. The carrier agent thus acts as a shuttle to selectively transport one component from the feed to the product side of the membrane.

Other processes discussed include nanofiltration, dialysis, donnan dialysis, and electrolytic water dissociation. Electrophoresis is discussed in Chapter 7.

Membrane–based separations of solids from liquids and liquids from liquids have enjoyed increasing popularity over the last decade. This popularity has been generated by the unique advantages offered by membranes over competing separations techniques. These advantages include: perfect separation and crystal clear filtrate, controlled size exclusion through selection of pore sizes extending from microns to molecular dimensions, and excellent materials compatibility due to a wide variety of available membranes.

On the other hand, membrane separations have had many limitations which restricted their use. For many applications, membrane separations are largely confined to the processing of low solids and low volume streams. This is due to the relatively high capital cost and low throughput capacity of these systems. In addition, membrane separators suffer from fouling (a long–term loss in throughput capacity due to membrane degradation). Traditional crossflow membrane separators also cannot concentrate feed slurries to high solids concentrations due to the rheological requirement that the feed material remain watery in consistency. The highly variable characteristics of many waste streams, unlike product streams, increased the problems of fouling, plugging , and degradation.

New Logic has developed a system which may overcome the limitations of traditional membrane systems. This system is called Vibratory Shear Enhanced Processing (VSEP). In a VSEP system, the feed slurry remains nearly stationary, moving in a meandering flow between parallel membrane leaf elements. The leaf elements move in a vigorous vibratory motion tangent to the face of the membranes. The shear waves produced by the membrane vibrations cause solids and foulants to be repelled and liquid to flow to the membrane pores unhindered.

A batch reactor incorporating a membrane–biofilm module for removing polyaromatic hydrocarbons (PAHs) from groundwater has been developed at the Technical University of Hamburg–Harburg (TUHH; Germany) and the GKSS Forschungzentrum Geesthacht

GmbH (Teltow, Germany).

5.1 DIALYSIS

Dialysis is one of several membrane separation procedures, which consist in essence of a barrier which will preferentially pass certain components of a fluid mixture or solution, and a driving force to cause such transfer to take place. Dialysis uses s semi-permeable membrane capable of passing small solute molecules (such as salts and small organic species) while retaining colloids and solutes of higher molecular weight. The driving force for this transfer is the concentration gradient—the difference in chemical activity of the constituents on either side of the membrane. The transfer through the membrane is by diffusion, that is, the progress of individual molecules, rather than by the hydrodynamic flow that would occur through a porous medium.

Requirements for a membrane are good transfer rates, suitable mechanical strength and durability, and resistance to chemical degradation, and low cost. Thermal stability may also be desired, since there are advantages to carrying out dialysis at somewhat elevated temperatures if possible. The dialysis membranes in most extensive use today are those based upon cellulosic or vinyl materials in either tubular, flat sheet, or recently hollow fiber form.

Although dialysis will occur under "static" conditions, concentration gradients develop on either side of the membrane, and this retards the transfer. To prevent this, it is common practice to provide stirring or high flow rates at the membrane surfaces which minimize stagnant film thickness. Commercial dialysis usually involves the flow of the solute-containing stream across one face of the membrane and the flow of the second higher volume wash stream on the other side.

Because concentration gradient is the sole driving force, dialysis is generally suited only to concentrated streams (5 to 20%) in which large concentration gradients may be achieved. Also, the concentration of solute at the feed stream outlet (known as the dialysate) is greater than that of the exiting wash stream (known as the diffusate), except under exceptional circumstances.

A water-permeable membrane may experience significant water transfer from the wash stream to the waste stream (an osmotic effect); this tends to diminish the effective concentration gradient. This transfer may be minimized by proper choices of membrane and, when necessary, adding "acceptable" solutes to the wash stream in order to reduce the chemical activity of the water.

Applications: Current applications of dialysis include (1) hemodialysis, (2) the separation of caustic soda from hemicellulose wastes in the rayon industry, (3) separating nickel sulfate in the electrolytic copper refining industry, and (4) purification techniques in pharmaceutical and biochemical laboratories.

The potential applicability of dialysis to hazardous wastes is limited to liquids containing high concentrations of low-molecular weight dissolved species. (In theory, it could be useful in nonaqueous systems, assuming a compatible membrane material, but this does not seem to have been investigated.) Caustics and the mineral acids dialyze readily, although oxidizing materials, such as nitric or chromic acid, or spent etchants, may cause severe membrane degradation unless they are properly chosen. Because of their small size, one would expect cyanides to permeate the membrane fairly well. Particulate matter which could plug the flow system or cause membrane erosion should be removed by a pretreatment, as should any constituents which might tend to form sludges or films on the membrane surfaces.

Wherever membrane filtration is employed, users choose reverse osmosis, ultrafiltration, electrodialysis, or some other similar process. The disadvantages of dialysis are its low rate, its unsuitability to dilute feed solutions (less than 0.1%), and the fact that both output streams are more dilute than the feed, and thus the material to be disposed of would require additional concentration.

5.2 DONNAN DIALYSIS

Donnan dialysis uses an anion– or cation– selective membrane, which functions similarly to ion exchange resins. For an anion exchange membrane, cations in both solutions (on each side of the membrane) are prevented from diffusing across the membrane, but anions will redistribute themselves across the membrane until equilibrium is reached, i.e., ratios of all similarly charged anions are equal. With a cation–exchange membrane, cations will diffuse across the membrane and movement of anions will be restricted. The driving force for these exchange reactions is the potential created by the displacement of the system from the equilibrium ratios which can be controlled by adjusting solution concentrations.

The major difference between Donnan dialysis and coupled transport processes is the type of membrane used and the transport mechanisms involved. The coupled transport membrane is highly selective and therefore has more specific process applications, whereas the Donnan dialysis membrane has application to a wider variety of solution constituents. Greater purity, however, can be achieved with the coupled transport membrane.

Applications: An example of anion Donnan dialysis is the sweetening of citrus juices. In this process, hydroxyl ions furnished by a caustic solution replace the citrate ions in the juice.

Donnan dialysis appears to have some application for the treatment of metal–containing aqueous wastes. However, problems of membrane stability have limited development. DuPont presently markets "Nafion," a perfluorosulfonic acid membrane that is being evaluated as a Donnan dialysis membrane for removal of nickel in electroplating wash water. Anion exchange membranes for the removal of copper, cadmium, and zinc cyanide complexes are also being evaluated. Studies are being conducted using quaternized polyvinyl pyridine and polyvinyl benzylchloride films grafted on a polyethylene base. Ion transport rates were reportedly proportional to ion exchange capacity.

The Southwest Research Institute is conducting research on Donnan dialysis. The Donnan dialysis process could prove to be a cost–effective alternative to conventional treatment practices because of its minimal operating requirements.

Future applications for Donnan dialysis could include:

1. Water softening
2. Heavy metal recovery

3. Acid recovery from etching baths
4. Recovery of acids and bases from salts
5. Recovery of organic acids from salts
6. pH control w/o adding acid or base

5.3 ELECTRODIALYSIS / ELECTROLYTIC WATER DISSOCIATION

5.3.1 Electrodialysis

Electrodialysis is a process by which electrically charged membranes are used to separate ions from an aqueous solution under the driving force of an electrical potential difference.

The arrangement consists of a series of anion– and cation–exchange membranes arranged in an alternating pattern between an anode and a cathode, to form individual cells. If an ionic solution, such as an aqueous salt solution, is pumped through these cells and an electrical potential established between anode and cathode, the positively charge cations in the solution migrate toward the cathode and the negatively charged anions migrate toward the anode. The cations pass easily through the negatively charged cation–exchange membrane but are retained by the positively charge anion–exchange membrane. Likewise, the negatively charged anions pass through the anion–exchange membrane and are retained by the cation–exchange membrane. The overall result is an ion concentration increase in alternate compartments, while the other compartments simultaneously become depleted of ions. The depleted solution is generally referred to as the diluate and the concentrated solution as the brine, or concentrate.

The efficiency of electrodialysis as a separation process is mainly determined by the ion–exchange membranes used in the system. The operating costs are dominated by the energy consumption and investment costs for a plant of a desired capacity, which are a function of the membranes used in the process and various design parameters such as cell dimensions, feed flow velocity and pressure drop of the feed solution in the cell.

The energy required in an electrodialysis process is an additive of two terms: one, the electrical energy to transfer the ionic components from one solution through membranes into another solution, and two, the energy required to pump the solutions through the electrodialysis unit. Depending on various process parameters, particularly the feed solution concentration and the current utilization, either one of the terms may dominate. At high feed solution ion concentration, the energy needed for the transfer of ions is generally the dominating factor. The capital cost of an electrodialysis plant is proportional to the membrane area required for a certain plant capacity, which is determined mainly by the feed solution concentration and the limiting current density.

With the development of ion–exchange membranes with low electrical resistance directly after the second world war, multicompartment electrodialysis became commercially available for demineralizing or concentrating various electrolyte solutions. During the 1960s the United States Office of Saline Water directly supported research and development of electrodialysis for the production of potable water from brackish water. Similar smaller programs in Europe, Israel and South Africa stimulated the development of new membranes and improved processes. At the same time, several Japanese companies developed electrodialysis as a means of concentrating seawater for use as a brine for the chlor–alkali industry and for producing table salt. These early electrodialysis systems were all operated unidirectionally, i.e., the polarity of the two electrodes, and hence the position of the dilute and concentrated cells, was permanently fixed in an electrodialysis stack. This mode of operation often led to scale formation and membrane fouling caused by the precipitation of low solubility salts on the membrane surface. Scaling affects the efficiency of electrodialysis significantly and the materials precipitated at the membrane surface have to be removed by flushing with cleaning solutions, the frequency depending on the concentration of such materials in the feed solution. The control of scaling and membrane fouling generally leads to an increase in the capital and operating costs. Extreme cases of membrane scaling and fouling can make the process economically unattractive. A significant advance in scaling control was the introduction of a special operating mode referred to as electrodialysis reversal (EDR). EDR was introduced by Ionics Inc. to continuously produce demineralized water without constant chemical addition.

In EDR systems, the polarity of the electrodes is periodically reversed. This reverses the direction of ion movement within the membrane stack, thus controlling membrane fouling and scale formation. Typically, reversal occurs approximately every 15 minutes and is accomplished automatically. Upon reversal the streams that formerly occupied concentrate compartments become demineralized streams. Automatic valves switch the inlet and outlet streams, so that the incoming feed water flows into the new demineralizing compartments and any concentrate stream remaining in the stack must now be desalted. This creates a brief period of time in which the demineralized stream (product water) salinity is higher than the specified level.

Because of reversal, no flow compartment in the stack is exposed to high solution concentrations for more than 15 to 20 minutes at a time. Any build–up of precipitated salts is quickly dissolved and carried away when the cycle reverses. EDR effectively eliminates the major problems encountered in unidirectional systems.

In summary, the development of electrodialysis as an efficient demineralization and ion exchange process is characterized by three major innovations:

1. The use of highly selective cation– and anion–exchange membranes
2. The use of a multi–compartment stack design
3. The polarity–reversal operating mode

As with reverse osmosis systems, electrodialysis systems are available as packaged units equipped with electrical components, pumps, motors, pretreatment features, recycle, temperature control, cleaning, and other features. These can be arranged in parallel or series as required by the application and its process streams. Properly designed and operated, electrodialysis units have proven to be effective and reliable.

By packaging several cell pairs of membranes (typically 50 to 300 cell pairs) between electrodes and manifolding the streams, a concentrated stream and a depleted stream, from which 45 to 55% of the ions have been removed, are generated. Further ion reduction of the depleted stream can be accomplished in additional stages. However, electrodialysis cannot process highly deionized water because of the poor electrical conductivity of such waters. Excessive feed pressures must be avoided to prevent leakage. However, the flow rate in any stage must be sufficient to create adequate turbulence to keep concentration polarization below scaling limits.

There are two general types of commercially available ion–exchange membranes: heterogeneous and homogeneous. Both types usually contain a reinforcing fabric to increase tensile strength and improve dimensional stability. Heterogeneous membranes have two distinct polymer phases. They are rather easily made by grinding up ion–exchange resins and dispersing the particles in a film–forming polymer. Most commercially available membranes are of the homogeneous type with a continuous polymer phase containing ionic groups attached to the polymer chains. The most common methods of membrane preparation are listed below.

Heterogeneous. Ion exchange resin particles dispersed in polymer film.

1. Calender or press mixture of resin and polymer.
2. Cast film from dispersion of resin in polymer solution.
3. Disperse resin in prepolymer, cast film, and polymerize.

Homogeneous. Ionic groups attached to polymer.

1. Polymerize ionic monomers.
2. Crosslink polyelectrolyte.
3. Graft onto preformed films.
4. Cast solution of polyelectrolyte in film–forming polymer.
5. Imbibe graftable monomers into film, polymerize, and graft.

A potential problem with all applications is the possibility of reaching excessive current densities because of the high concentration of ions at the membrane interfaces. Possible consequences of this are the precipitation of metals such as calcium and magnesium and the electrolysis of water to hydrogen and hydroxide ions. Undesirable effects leading to membrane fouling and local overheating of membranes can result.

Pretreatment or system design features can avoid problems resulting from electrolysis. For example, introducing turbulence or reducing the total ionic content of the concentrate stream have been successful in reducing fouling and electrolysis. To avoid fouling tendencies, almost all manufacturers recommend periodic reversal of the applied voltage while simultaneously re–routing the feed and concentrate.

There are no fundamental limits, other than solubility, on the maximum concentration level obtained in the concentrate. However, power consumption is directly proportional to the ion content of the feed. This contrasts with reverse osmosis, in which separation costs are less strongly influenced by concentration. Consequently, electrodialysis operating costs are favorable for low feed ion concentration and become less so as concentration increases. In addition, electrodialysis is generally used to produce a concentrated solution, such that evaporation units are not required. Where a valuable concentrate is being provided, salts may be concentrated to 20% or more, significantly beyond that feasible for reverse osmosis systems.

Applications (Non–Environmental): Current applications include:

1. Desalination of rainwater and brackish water.
2. Production of table salt from rainwater.
3. Boiler feedwater and industrial process water demineralization.
4. Ultrapure water production
5. Removal of organic acids from wine and fruit juices.
6. Recovery of amino acids from fermentation.
7. Separation of proteins
8. Desalting of protein solutions, whey, milk and molasses.
9. Washing of photographic emulsions.
10. Solution enrichment or depletion techniques in the chemical industry.
11. Regeneration of ion exchange resins

Applications (Environmental): Metal finishing processes offer numerous applications for electrodialysis in pollution control and material recovery. The rinse streams from such processes pose particularly troublesome pollution problems. They are usually too dilute for direct metal recovery and too concentrated for disposal.

The principal area of application of electrodialysis appears to be the recovery of metals from electroplating bath rinsewaters. Electrodialysis and reverse osmosis are competitive processes for these applications. Electrodialysis would appear to have the advantage when concentration levels are low (operating costs are low), or when recovery values justify the expense of achieving concentration levels higher than those possible with reverse osmosis. Over 100 electrodialysis systems are now employed commercially for the recovery of metals from electroplating rinsewaters.

Complete recycling of the water and the metal ions is achieved. Compared to reverse osmosis, electrodialysis has the advantage of being able to utilize thermally and chemically stable membranes, so the process can be run at elevated temperatures and in solutions of very low or high pH values. Furthermore, the concentrations which can be achieved in the brine can be significantly higher. The disadvantage of electrodialysis is that only ionic components can be removed and additives usually present in a galvanic bath cannot be recovered.

An application which has been studied in a pilot–plant stage is the regeneration of chemical copper plating baths. In the production of printed circuits, a chemical process is often used for copper plating. The components which are to be plated are immersed into a bath containing, besides the copper ions, a strong complexing agent, for example, ethylene diamine tetra acetic acid (EDTA), and a reducing agent such as formaldehyde. Since all constituents are used in relatively low concentrations, the copper content of the bath is soon exhausted and copper sulfate has to be added. During the plating process, formaldehyde is oxidized to formate. After prolonged use, the bath becomes enriched with sodium sulfate and formate and consequently loses useful properties. By applying electrodialysis in a continuous mode, the sodium sulfate and formate can be selectively removed from the solution, without affecting the concentrations of formaldehyde and the EDTA complex. Hereby, the useful life of the plating solution is significantly extended. Several other successful applications of electrodialysis in wastewater treatment systems that have been studied on a laboratory scale are reported in the literature. Large, commercially operated plants are at present, however, rare. Ionics and Tokuyama Soda are the market leaders. However, because the plant capacities needed for this application are smaller than in desalination, there are good opportunities for smaller companies to compete.

Electrodialysis processes using bipolar membranes may become increasingly important because of their low energy consumption compared with alternative technologies. Electrodialytic regeneration of ion-exchange membranes may become increasingly attractive as stricter environmental discharge regulations are enforced.

Current and future applications of electrodialysis in the environmental area include:

1. Electroplating rinse water (heavy metals removal).
2. Etch bath rinse water.
3. Pickling wastewater.
4. Desalination of special effluent (e.g., glycerine).
5. Radioactive wastewater treatment.

6. Regeneration of ion–exchange resins.
7. Paper mill waste treatment.
8. Recovery of organic acids from corresponding salts.
9. Concentration of reverse osmosis brines.

5.3.2 Electrolytic Water Dissociation

A process referred to as electrodialytic water dissociation or water splitting to produce acids and bases from salts, has been known for a number of years.

A cell system consisting of an anion–, a bipolar–, and a cation–exchange membrane as a repeating unit is placed between two electrodes. Sodium sulfate or other salt solution is placed in the outside phase between the cation– and anion–exchange membranes. When a direct current is applied, water will dissociate in the bipolar membrane to form an equivalent amount of hydrogen and hydroxyl ions. The hydrogen ions will permeate the cation–exchange side of the bipolar membrane and form sulfuric acid with the sulfate ions provided by the sodium sulfate from the adjacent cell. The hydroxyl ions will permeate the anion–exchange side of the bipolar membrane and form sodium hydroxide with the sodium ions permeating into the cell from the sodium sulfate solution through the adjacent cation–exchange membrane.

Applications: The production of sulfuric acid and sodium hydroxide at significantly lower cost than by conventional techniques can be achieved. Laboratory tests have demonstrated that production costs for caustic soda by utilizing bipolar membranes are only one–third to one–half the costs of the conventional electrolysis process. The process has recently been commercialized by Allied Corporation's Aquatech Systems Division. The process is affected by limited alkaline and temperature stability of the anion–exchange part of the bipolar membrane.

No specific pollution control applications are known to this author.

5.4 GAS SEPARATION

The study of gas transport in membranes has been actively pursued for over 100 years. This extensive research resulted in the development of good theories on single gas transport in polymers and other membranes. The practical use of membranes to separate gas mixtures is, however, much more recent. One well–known application has been the separation of uranium isotopes for nuclear weapon production. With few exceptions, no new, large scale applications were introduced until the late 1970's when polymer membranes were developed of sufficient permeability and selectivity to enable their economical industrial use. Since this development is so recent, gas separations by membranes are still less well–known and their use less widespread than other membrane applications such as reverse osmosis, ultrafiltration and microfiltration.

Membranes for the separation of gas mixtures are of two very different kinds: one a microporous membrane, the other nonporous. Microporous membranes were the first to be studied and the basic law governing their selectivity was discovered by Graham. When pore size of a microporous membrane is small compared to the mean–free path of the gas molecules, permeate will be enriched in the gas of the lower molecular weight. Since molecular weight ratios of most gases are not very large and since the selectivity is proportional to the square root of this ratio, not only practical but theoretical enrichments achievable by this method necessarily will be small. In order to have an efficient separation of a gas mixture, many separation stages are required. On the other hand, since this method of separation is based strictly on mass ratios and not chemical differences, it is the only membrane based method capable of separating isotopes of a given compound. This is the reason it was chosen as a method to enrich uranium in the fissionable isotope of mass 235 in the development of the atomic bomb in World War II. This separation method inherently is very expensive—it requires a large amount of hardware for a given amount of processed gas, the membrane specifications are stringent (high porosity, small pore size), and the energy requirements are high. The large scale utilization of such a process has been feasible only because economic considerations are not of prime importance in this application.

The other membrane–based gas separation method utilizes non–porous membranes. In permeating through the membrane, the gases are separated due to differences in their diffusivity and solubility in the membrane matrix (normally an organic polymer). Molecular size will play a role in such separations but so will the chemical nature of the gas. Thus, conceptually very efficient separations should be possible this way. As polymer science developed, many polymers were tested for gas permeabilities and indeed some with very good selectivities were found.

Membrane Technology and Research, Inc., (MTR) has developed organic selective composite membranes that have been successfully demonstrated for the separation and recovery of volatile organic compounds (VOCs). The vapor separation process developed at MTR typically consists of two steps: compression–condensation and membrane vapor separation. The compression–condensation step is conventional. The membrane step is new and is based on high performance composite membranes that are 10 to 100 times more permeable to organic compounds than to air. The membrane separation step enhances the recovery possible with compression and condensation alone, allowing the process to operate at much higher recovery rates, or allowing the temperature and pressure conditions to be relaxed.

Most gas–separation work to date has been performed with polymeric membranes. However, membranes made from palladium–silver alloys have been the subject of substantial research and limited commercial development. These membranes have extremely high hydrogen/methane selectivities, and can be operated at elevated pressures and temperatures above 400°C. These systems are still available today, but have not been a large scale commercial success.

Ceramic, glass, and carbon/molecular–sieve membranes have also been investigated. Molten salt membranes have recently been developed.

Membrane durability is of prime importance when considering polymeric membranes for industrial gas separations. Membranes that can operate at higher temperatures and pressures are required if membranes are to become a significant unit operation in industry. Research is needed to answer questions about aging, temperature, and pressure effects on membrane permeability, selectivity, and mechanical strength. Studying the effect of post–treating membranes (to reduce surface porosity), as well as the performance of polymer blends may be helpful in developing new, more durable membranes.

Research into how the molecular structure of a polymer chain affects the performance of the membrane would yield better insights as to how to tailor membranes to specific separations. The molecular geometry of the polymer chain and the radicals and molecules at the membrane surface will undoubtedly have a marked effect on the permeation of a given compound across the membrane. If these characteristics were to be studied in detail in parametric fashion, the criteria for selecting a membrane with a particular molecular structure to perform the desired separation could be easily set. Finally, as the materials science of membranes becomes better understood, new membrane formulations and configurations can be developed to improve the selectivity, permeability, and durability of membranes under increasingly hostile applications.

Although cryogenic methods of gas separation predominate, gas separation membrane techniques are now overtaking the other noncryogenic technique—Pressure Swing Adsorption (PSA).

Applications (Non-Environmental): Current or potential non-environmental applications include:

1. Recovery of hydrogen from purge streams such as encountered in ammonia plants, other chemical streams, from hydroprocessors in petroleum refineries, and from methane.
2. CO_2 removal from well-head gas in enhanced oil recovery, as well as H_2S removal from sour gas.
3. Separation of oxygen and nitrogen in air.
4. Removal of water from natural gas, and water/alcohol mixtures.
5. Adjust the hydrogen/carbon ratio in synthesis gas.
6. Separation of olefins from alkanes.
7. Helium separation.

Applications (Environmental): There are some current and potential applications for gas separation with membranes in the environmental field.

1. CO_2 recovery from bio and landfill gases has been successful.
2. Removal of chlorinated hydrocarbons from air, with solvent recovery has been successful. Permeate tends to become oxygen enriched (flammability concern) and vacuum system design is exotic.
3. Acid gas separations, such as flue-gas desulfurization, industrial gas treating, and organic sulfur removal.
4. Separating hydrogen from coal gas in coal gasification.
5. Cleaning solvent vapor separation in semiconductor manufacturing.
6. Separation of H_2S from coal gasification process streams.

5.5 LIQUID MEMBRANES / COUPLED AND FACILITATED TRANSPORT

5.5.1 Liquid Membranes

Liquid membranes have gained increasing significance in recent years in combination with the so-called facilitated transport which utilizes selective "carriers" transporting certain components such as metal-ions selectively and at a relatively high rate across the liquid membrane interphase. It is relatively easy to form a thin fluid film. it is difficult, however, to maintain and control this film and its properties during a mass separation process. In order to avoid a break-up of the film, some type of reinforcement is necessary to support such a weak membrane structure. Two different techniques are used today for the preparation of liquid membranes. In the first case, the selective liquid barrier material is stabilized as a thin film by a surfactant in an emulsion-type mixture. In the second technique for making liquid membranes, a microporous polymer structure is filled with the liquid membrane phase. In this configuration, the microporous structure provides the mechanical strength and the liquid–filled pores the selective separation barrier. Both types of membranes are used today on a pilot–plant stage for the selective removal of heavy metal–ions or certain organic solvents from industrial waste streams. They have also been used rather effectively for the separation of oxygen and nitrogen.

Supported Liquid Membranes: The preparation of supported liquid membranes is extremely simple, when certain requirements concerning the selective barrier and the microporous support material are fulfilled. The liquid membrane material should have a low viscosity and low vapor pressure, i.e., high boiling point and, when used in aqueous solutions, a low water solubility. Otherwise, the useful lifetime of the membrane is rather limited. The microporous substructure should have a high porosity, a pore size small enough to support the liquid membrane phase sufficiently under hydrostatic pressure and the polymer of the substructure should be hydrophobic in nature for most liquid membranes used in contact with aqueous feed solutions. In practice, liquid membranes are prepared by soaking a hydrophobic microporous membrane, such as a Goretex (Gore Corp.) or Cellgard (Celanese Corp.) typestretched polytetrafluoroethylene or polyethylene membrane, in the hydrophobic liquid which may consist of a selective carrier such as certain oximes or tertiary or quaternary amines dissolved in kerosene. The disadvantage of supported membranes is their thickness which is determined by the thickness of the microporous support structure, which is in the range of 10 to 50 μm, and therefore about 100 times the thickness of the selective barrier of an asymmetric polymer membrane. Thus, the fluxes of supported liquid membranes can be low even when their permeabilities are high.

Unsupported Liquid Membranes: Very thin unsupported liquid membranes may be obtained, when the selective membrane material is stabilized by an appropriate surfactant in an aqueous emulsion. A hydrophobic membrane phase is transformed into an emulsion by stirring with an aqueous phase. Ideally, droplets form in this process, in which the aqueous phase is surrounded by a relatively thin hydrophobic membrane forming phase which is surrounded by a second aqueous phase. The mass exchange occurs between the inner and outer aqueous phases through the liquid membrane interphase. In reality, the hydrophobic membrane phase and the surrounding aqueous phases are more fractionated and the diffusion pathways become longer as a result. With another aqueous solution, the component to be eliminated is supplied to the original emulsion and passes through the membrane into the internal solution. The emulsion membrane technique was popularized and fully developed by Li and his co–workers at Exxon.

The National Fertilizer and Environmental Research Center, Muscle Shoals, AL is assessing the feasibility of using liquid membranes to extract pesticides from rinsewaters typical of those generated by fertilizer/agrichemical dealers.

5.5.2 Facilitated Transport

Facilitated transport is a form of extraction carried out in a membrane. As such, it is different from most of the other membrane processes described in this book. Many of these, for example ultrafiltration, are alternative forms of filtration. Others, especially gas separations and pervaporation, depend on diffusion and solubility in thin polymer films.

In contrast, facilitated transport involves specific chemical reactions like those in extraction.

Facilitated transport usually has four characteristics that make it different from other membrane separations:

1. It is highly selective.
2. It reaches a maximum flux at high concentration differences.
3. It can often concentrate as well as separate a given solute.
4. It is easily poisoned.

Characteristics (2) and 3 are the most powerful evidence that facilitated transport is occurring.

The advantage of facilitated transport membrane processes lies not in their energy consumption, but in their speed. Facilitated membrane processes are much faster than their conventional counterparts. Used in the form of membrane contactors, facilitated transport processes are about eighty times faster than absorption towers of equal volume, and six hundred times faster than extraction columns of equal volume. This greater speed is primarily the result of the very large surface area per volume possible in membrane devices. This surface area can be achieved at a wide range of flows, independent of the constraints of loading and flooding which can compromise conventional unit operations. Although facilitated transport membranes are likely to be ten times more expensive per volume than existing equipment, their speed and space efficiency may not provide sufficient motivation to displace fully depreciated equipment in a stable chemical process industry. It could, however, have substantial impact in other smaller-scale applications, especially pharmaceuticals.

Facilitated membrane diffusion can be much more selective than other forms of membrane transport, but the membranes used in the process are usually unstable. This instability is a tremendous disadvantage, and is the reason why this method is not usually commercially practiced.

In addition to liquids, facilitated transport can include molten salt membranes, gels, glasses, and microporous ceramics.

Coupled and facilitated transport are very similar, with only a subtle distinction between them.

5.5.3 Coupled Transport

Coupled transport is a membrane process for concentrating ions and separating ions from aqueous solutions. The membrane used in this process consists of a water-insoluble liquid containing an ion-complexing agent that is specific for the ion of interest. The desired ion is complexed at one interface of the membrane, forming a neutral-ion complex. The neutral ion-complex then diffuses across the membrane to the opposite interface, where the reaction is reversed by making appropriate changes in the external solution conditions. The reformed complexing agent then diffuses back across the membrane, where it picks up more of the desired ion. Thus, the complexing agent acts as a shuttle to carry ions across the membrane. Bend Research Corporation has done most of the development work in coupled transport technology.

Applications: In contrast to fundamental efforts, commercial successes are sparse. The most serious, publicly detailed effort has been by Rolf Marr and his associates at the

University of Graz, in Austria. These workers developed a large-scale emulsion liquid membrane process for the selective extraction of zinc from textile waste. They used liquid membranes of diamines dissolved in kerosene. The product solution is first emulsified in the carrier medium. This emulsion is then re-emulsified with the feed solution. In the permeation step the zinc migrates into the interior product solution. The emulsion is then passed to a settler and the feed solution, now stripped of zinc, is removed as a raffinate. In a final step, the carrier-product emulsion phase is passed to a de-emulsifier where the emulsion is broken to produce a carrier solution, which is recycled, and a product solution containing the separated zinc. The zinc process was installed in Lenzing, Austria, where it processed 1,000 m^3/hr for at least six months.

Another commercial application of liquid membranes is in oil production as a well control fluid.

Other efforts to demonstrate industrial processes have been stillborn. General Electric explored the possibility of acid gas removal using an aqueous carbonate liquid membrane supported by cellulose acetate. The project was abandoned at the pilot stage. Bend Research and General Mills Chemicals (now a division of Henkel) both studied copper removal using an oxime solution. They apparently abandoned the work because they found no compelling advantages over conventional extraction. Recovery of uranium from ore leach solutions was also developed by Bend Research to the pilot-plant stage, using amines immobilized in hollow-fiber modules. This work was also abandoned because of stability problems. In by far the largest effort, Exxon studied the recovery of metals and of drugs using emulsion liquid membranes, a geometry on which they hold broad patents, now beginning to expire. Both uranium and copper recovery were investigated, but neither process became commercial. One additional effort, a collaboration with Mitsui Chemical for recovering dissolved mercury, reached the pilot stage, but has since been abandoned.

Many Chinese and Eastern European commercial efforts on facilitated transport apparently continue, for reasons that are not clear. Pilot units for processing chromium and other metals are reported to be operating in China. The future prospects of these efforts are unknown.

Cobalt and nickel recovery has been investigated. Coupled transport might be applied to electroplating rinse waters for a chrome-plating operation. Similar schemes have been proposed for other metallic rinse waters. Hydrocarbon separations have also been investigated.

One of the most promising applications of coupled transport is the renovation of circuit board etchant solutions that contain copper. The printed circuit board industry produces more than ten million gallons per year of spent etchant solutions in which copper, other salts and etchant chemicals are concentrated. Coupled transport permits continuous on-site removal of copper from the etchant solutions and simultaneous regeneration of the etchant solution. This represents a considerable savings in the costs of manufacturing circuit boards. A relatively small unit is able to process a large volume of solution.

Finally, the closely related passive and facilitated-transport processes for phenol and ammonia recovery should be mentioned. In these processes, dilute phenol or ammonia feed solutions are contacted with a liquid membrane in which they are freely soluble. They dissolve in the membranc, diffusing to the product side where they are removed by neutralizing with a base (in the case of phenol) or an acid (in the case of ammonia).

Although the transport mechanism does not involve a carrier and these are, therefore, passive transport processes, the actual process is quite similar, and Li et al published the details of these separations using emulsion membrane techniques.

Stevens Institute of Technology has developed a liquid membrane hollow fiber process that appears to be suited to industrial gas separation strategies.

The Tennessee Valley Authority investigated the use of liquid membrane emulsions for removal of pesticides from aqueous solutions, and it appears that the process is feasible.

5.6 MICROFILTRATION

Microfiltration (MF) is a process for separating material of colloidal size and larger from true solutions. It is usually practiced using membranes. In this section only microfiltration accomplished by membranes is covered. Microfilters are typically rated by pore size, and by convention have pore diameters in the range 0.1 to 10 μm. A microfiltration membrane is generally porous enough to pass molecules which are in true solution even if they are very large. Thus, microfilters can be used to sterilize solutions, because they may be prepared with pores smaller than 0.3 μm, the diameter of the smallest bacterium *Pseudomonas diminuta.*

There are several key characteristics necessary for efficient microfiltration membranes. These are (1) pore size uniformity, (2) pore density, and (3) the thinness of the active layer or the layer in which the pores are at their minimum diameter.

All current MF membranes may be classified as either *"tortuous-pore"* or *"capillary-pore"* membranes. The *"capillary-pore"* structure is distinguished by its straight-through cylindrical capillaries, whereas the *"tortuous-pore"* structure resembles a sponge with a network of interconnecting tortuous pores.

The *"tortuous-pore"* membranes are the most common and include typical cellulosic membranes and virtually all other polymers. The *"capillary-pore"* membranes are currently manufactured commercially by Nuclepore Corp. and Poretics Corp. They are available as polycarbonate or polyester membranes.

Polymer membranes are by far the market leaders in microfiltration. The major firms in the worldwide microfiltration business all sell polymer membranes in overwhelmingly greater quantities than the more trendy and more discussed inorganics. Nylon, polysulfone, and polyvinylidene fluoride are the major polymers used, in addition to the old workhorse cellulosics. Polypropylene is widely used in process microfiltration.

Some of the newer microfiltration membranes are ceramic membranes based on alumina (Alcoa-Ceraver), membranes formed during the anodizing of aluminum (Anotec), and carbon membranes (GFT). Glass is also used as a membrane material. Zirconium oxide is also deposited on a porous carbon tube. Sintered metal membranes are fabricated from stainless steel, silver, gold, platinum, and nickel, in disks and tubes.

Microfiltration modules began as flat-sheet filters for use in the laboratory, and were soon incorporated into plate-and-frame devices. A growing diversity of applications has led to the development of numerous devices such as the spiral-wound module, copied from other membrane applications, the pleated cartridge, referred to above, the stack-filter module, designed for certain pharmaceutical applications, and capillary modules. Continued development of module types is likely, but the low manufacturing cost of spirals and capillaries makes them the product to beat for volume applications in crossflow. Pleated cartridges enjoy a similar advantage in dead-end applications.

Some of the more innovative membranes developed recently are amenable to compact, economical fabrication. Carbon membranes may be pyrolized from fibers, and are described as being formed already sealed to an end plate. Most ceramics are available in tubular form, but one firm has pioneered a low-cost monolith, and another makes ceramic capillary tubules which can be wound up into a cartridge.

Microfiltration, a relatively mature industry, has had its most profitable growth in relatively small filters operating in high-value applications. The industry trend is to build on this base, but to expand into lower value, higher volume applications. Most of this growth will be with membrane devices operating in a manner that handles retained material efficiently, meaning either crossflow or backwash.

Applications (Non-Environmental): The heart of the microfiltration field is sterile filtration, using microfilters with pores so small that microorganisms cannot pass through them. These disposable filters, typically in the form of pleated cartridges, are sold to a variety of users, but the major customer is the pharmaceutical industry.

A great many of the drugs and solutions produced by the pharmaceutical industry or made up in the hospital pharmacy have to be both sterile and relatively free of particulate matter, especially if the product is to be injected into the bloodstream. For drugs and other products that will not withstand heat, sterilizing filtration is the only alternative. Tissue culture media, parenteral solutions, vaccines, human plasma fractions, antibiotics, diagnostic injectables are all sterilized by membrane filters.

A second major application for microfilters is in the electronics industry for the fabrication of semiconductors. As semiconductor devices shrink in size, the conductive paths on their surfaces get closer together. Dirt particles represent potential short circuits in the semiconductor device. Therefore, filtration of various streams throughout the manufacturing process is a vital concern. For microfiltration companies schooled in the sterile filtration discipline, the electronics applications seemed made-to-order. A particularly attractive application in this industry is final filtration of the water used to rinse semiconductors during fabrication.

Other major applications are in the food industry and include dextrose clarification, milk fat removal, haze removal from gelatin, wine clarification, juice filtration, and beer sterilization and bottoms recovery. Microfiltration displaces diatomaceous earth if economics are favorable. The Japanese are using MF units to remove corrosion products from the coolant loop in their nuclear reactors.

Applications (Environmental): There are numerous interesting possibilities for MF in the environmental area as follows:

(1) Water treatment: The largest emerging opportunity for microfiltration is for the treatment of municipal water, permitting it to be sterilized without chlorine. This would take microfiltration back to its World War II roots. There is no doubt that microfiltration membranes have the ability to remove bacteria from water. A recent Australian study showed that microfiltration membranes can also remove viruses from contaminated surface water. Since viruses are much smaller than the pores in a microfiltration membrane, the finding has been attributed to the viruses being adsorbed on clay particles, which are large enough to be caught by a microfilter. There are, however, concerns about subsequent contamination in the water distribution system, since chlorine has a residual effect that protects water against contamination after treatment. Recent Federal regulations probably

mean that chlorination will continue to be required for drinking water supplies.

(2) Sewage treatment: The other potentially very large market for which microfiltration might be a candidate is the treatment of municipal sewage. A scheme proposed by Memtec (Australia) would shift the treatment of sewage to distributed processing, a plan that envisions many small sewage treatment facilities, centrally monitored. Should this plan ever become reality, the market for microfiltration membranes would be immense.

(3) Heavy metal removal: MF may be used to remove heavy metals from waste streams provided pretreatment chemicals are added to precipitate the metals to particles of filterable size. The chemical pretreatment step is crucial since it will affect the performance of the membrane and the resultant sludge volume as well as the contaminant removal efficiency. Reduction/oxidation, absorption/oxidation, and/or catalytic reactions are utilized along with pH adjustment to provide the optimum precipitation. Although conventional methods of wastewater treatment may use a similar pretreatment chemistry, the final solid/liquid separation by gravity settling is usually not as effective as membrane filtration.

(4) Organics removal from waste streams: Could be applicable to groundwaters and process waters with COD levels between 100 and 500 mg/L where the molecular weight of contaminants being concentrated are over 200. Based on a site demonstration, waste streams rich in polyaromatic hydrocarbons (PAHs) would probably be suitable, while those with a goal of concentrating phenols would probably not be appropriate. Crossflow membrane filtration may be applicable to waste streams containing high molecular weight or non–polar organic contaminants, such as polychlorinated biphenyls. The process may also be useful for separating other emulsified or dispersed organics which do not lend themselves to simple physical phase separation.

(5) Industrial laundry wastewater.

(6) Fuel oriented hydrocarbon separations.

(7) Particulate removal from stack gas.

5.7 PERVAPORATION

Pervaporation is a membrane process used to separate mixtures of dissolved solvents. A liquid mixture contacts one side of a membrane; the permeate is removed as a vapor from the other side. Transport through the membrane is induced by the difference in partial pressure between the liquid feed solution and the permeate vapor. This partial–pressure difference can be maintained in several ways. In the laboratory a vacuum pump is usually used to draw a vacuum on the permeate side of the system. Industrially, the permeate vacuum is most economically generated by cooling the permeate vapor, causing it to condense. The components of the feed solution permeate the membrane at rates determined by their feed solution vapor pressures, that is, their relative volatilities and their intrinsic permeabilities through the membrane. Pervaporation has elements in common with air and steam stripping, in that the more volatile contaminants are usually, although not necessarily, preferentially concentrated in the permeate. However, during pervaporation no air is entrained with the permeating organic, and the permeate solution is many times more concentrated than the feed solution, so that its subsequent treatment is straightforward.

Results have shown that pervaporation membranes can be used to separate azeotropic mixtures efficiently, a result that is not achievable with simple distillation.

Application (Non-Environmental): The most important application is the separation of water from concentrated alcohol solutions. GFT of Hamburg, West Germany, the leader in this field, installed their first major plant in 1982. Currently, more than 100 plants have been installed by GFT for this application.

Although most of the installed solvent dehydration systems have been for ethanol dehydration, dehydration of other solvents including isopropanol, glycols, acetone and methylene chloride, has been considered.

No commercial systems have yet been developed for the separation of the more industrially significant organic/organic mixtures. However, current technology now makes development of pervaporation for these applications possible and the process is being actively researched in a number of laboratories. The first pilot-plant results for an organic-organic application, the separation of methanol from methyltertbutyl ether/isobutene mixtures, was recently reported by Air Products. Texaco is also working on organic-organic separations. This is a particularly favorable application and currently available cellulose acetate membranes give good separation. It can only be a matter of time, however, before much more commercially significant organic-organic separations are attempted using pervaporation.

The major field of application for pervaporation will be either the separation of organic components with almost identical boiling characteristics or azeotropic mixtures. In most cases, the desired product quality cannot be achieved in a single step—either a combination of different processes or a "multi-stage" process will be necessary. In this case, a reflux-cascade has to be designed.

Applications (Environmental): The most important environmental application is the separation of small amounts of organic solvents from contaminated waters, for which the technology has been developed by MTR. Both of the current commercial processes concentrate on the separation of organics from water. This separation is relatively easy, because organic solvents and water, due to their difference in polarity, exhibit distinct membrane permeation properties. The separation is also amenable to membrane pervaporation because the feed solutions are relatively non-aggressive and do not chemically degrade the membrane.

A number of applications exist for pervaporation in the removal or recovery of organic solvents from water. If the aqueous stream is very dilute, pollution control is the principal economic driving force. However, if the stream contains more than 1 to 2% solvent, recovery for eventual reuse can significantly enhance the process economics.

Wastewater Technology Center (Canada) has developed a cross-flow pervaporation system. The performance of the cross-flow pervaporation system increases with temperature, with an equipment limitation of 35°C. Permeable membranes that preferentially adsorb VOCs are used to partition VOCs from the contaminated water. The VOCs diffuse from the membrane and water interface through the membrane and are drawn off by a vacuum pump. Upstream of the vacuum pump, a condenser traps and contains the permeating vapors, condensing all the vapor, therefore, allowing no discharge to the atmosphere.

Calculations suggest that pervaporation can be economically applied to a wide range of aqueous industrial streams. The process is still very new and its ultimate competitive position compared to more conventional techniques is uncertain. When the feedstream contains less than 100 ppm solvent, carbon adsorption beds become small and this is probably the preferred technique. Similarly, when the stream contains more than 5 to 10% solvent, distillation, steam stripping or incineration will probably be less expensive than pervaporation. However, in the intermediate range of between 100 ppm and 5% solvent, pervaporation will find a number of important applications.

5.8 REVERSE OSMOSIS

Reverse osmosis (RO), the first membrane–based separation process to be widely commercialized, is a liquid/liquid separation process that uses a dense semipermeable membrane, highly permeable to water and highly impermeable to microorganisms, colloids, dissolved salts and organics.

A pressurized feed solution is passed over one surface of the membrane. As long as the applied pressure is greater than the osmotic pressure of the feed solution, "pure" water will flow from the more concentrated solution to the more dilute through the membrane. If other variables are kept constant, the water flow rate is proportional to the net pressure.

In normal osmotic processes, solvent will flow across a semi–permeable membrane from a dilute concentration to a more concentrated solution until equilibrium is reached. The application of high pressure to the concentrated side will cause this process to reverse. This results in solvent flow away from the concentrated solution, leaving an even higher concentration of solute. The semi–permeable membrane can be flat or tubular, but regardless of its shape it acts like a filter due to the pressure driving force. In application the waste stream flows past the membrane while the solvent, such as water, is pulled through the membrane's pores and the remaining solutes such as organic or inorganic components do not pass through, but become more and more concentrated on the influent side of the membrane.

An industrial reverse osmosis plant usually will consist of three separate sections. The first section is the pretreatment section in which the feedwater is treated to meet the requirements of reverse osmosis element manufacturers and the dictates of good engineering practice. Following pretreatment, the feedwater is introduced into the reverse osmosis section where the feedwater is pressurized and routed to the reverse osmosis elements which are in pressure vessels. The feedwater flows across the membrane surface where product water permeates through the membrane and a predetermined amount remains behind as reject. The reject is discharged to waste while the product water is routed to the post–treatment section. The third or post–treatment section treats the product water to remove carbon dioxide and adds chemicals as required for industrial use of the product water.

The factor which has the greatest influence on reverse osmosis membrane system design is fouling, caused by particulate and colloidal matter that become concentrated at the membrane surface. Pretreatment is used to remove particulate matter from the feed water, but is seldom completely effective. The concentration of foulants at the membrane surface increases with increasing permeate flux and product recovery rate. A system designed to operate at a high permeate flux is likely to experience high fouling rates and will require frequent chemical cleaning.

Since the development of the crosslinked, fully aromatic polyamide thin–film composite membrane by Cadotte in 1977, emphasis has focused on this membrane type throughout the industry.

In most plants that use reverse osmosis in the preparation of process water, the reject stream is routed directly to waste discharge without any additional post–treatment. In industrial plants that use reverse osmosis to treat industrial wastes, the reject stream may contain valuable materials and this stream would be sent back to the process. In other applications of industrial waste treatment by reverse osmosis, the reject stream may require additional treatment prior to ultimate discharge. In this case, the reverse osmosis unit will have provided a large volume of water that is disposable or can be reused (the product) and a smaller volume of reject which can be treated more economically.

The product water from a reverse osmosis unit will have a low pH and most probably a high concentration of carbon dioxide. The carbon dioxide can be removed and the pH of the product increased by use of a decarbonator. A decarbonator is a packed column in which product water is introduced at the top while either forced or induced air is introduced at the bottom. The air and water flow countercurrently over and around the column packing. The carbon dioxide is stripped from the water and exits from the decarbonator at the top in the air stream. In a well–designed decarbonator, the carbon dioxide content can be reduced to about 5 mg/ℓ in the water effluent.

There are six basic types of reverse osmosis membranes: asymmetric microporous membranes, Loeb–Sourirajan thin–skinned anisotropic membranes, composite membranes, dynamically–formed membranes, liquid membranes, and plasma–polymerized membranes. Only three of these membrane types—Loeb–Sourirajan, composite, and dynamically–formed membranes—are currently available for commercial application.

The major advantage of reverse osmosis for handling process effluents is its ability to concentrate dilute solutions for recovery of salts and chemicals with low power requirements. No latent heat of vaporization or fusion is required for effecting separations; the main energy requirement is for a high pressure pump. It requires relatively little floor space for compact, high capacity units, and it exhibits good recovery and rejection rates for a number of typical process solutions.

A limitation of the reverse osmosis process for treatment of process effluents is its limited temperature range for satisfactory operation. For cellulose acetate systems, the preferred limits are 18° to 30°C (65° to 85°F); higher temperatures will increase the rate of membrane hydrolysis and reduce system life, while lower temperatures will result in decreased fluxes with no damage to the membrane. Another limitation is inability to handle certain solutions. Strong oxidizing agents, strongly acidic or basic solutions, solvents, and other organic compounds can cause dissolution of the membrane. Poor rejection of some compounds such as borates and low molecular weight organics is another problem. Fouling of membranes by slightly soluble components in solution or colloids has caused failures, and fouling of membranes by feedwaters with high levels of suspended solids can be a problem. A final limitation is inability to treat or achieve high concentration with some solutions. Some concentrated solutions may have initial osmotic pressures which are so high that they either exceed available operating pressures or are uneconomical to treat.

Although the main applications for reverse osmosis have related to the processing of water, there is considerable interest in water/organic liquid separations, and organic liquid mixtures separations.

Extensive experimental studies by Battacharyya and Williams (University of Kentucky) showed that thin–film, composite membranes can be used effectively for the separation of selected hazardous organic compounds. This waste treatment technique offers definite advantages in terms of high solute separations at low pressures (<2 MPa) and broad pH operating range, and the use of charged membranes would allow the selective separation of some organics from feeds containing high salt concentrations. In addition, feed pre–ozonation of selected organics provided significant improvement in flux and rejection characteristics for both charged and uncharged membranes because of the formation of ionizable organic acid intermediates during the ozonation that did not interact as strongly with the membrane. The overall ozonation/membrane process effectively produced permeate water of high quality while it minimized the volume of waste that must be further treated.

Applications (Non–Environmental):

1. The largest single application area at present is desalination of seawater and brackish waters, which accounts for about 50% of total sales. Much of this application is the production of potable water for municipal water supplies.
2. Home "point–of–use" units.
3. Ultrapure water for semiconductor manufacturing, pharmaceuticals, and medical uses.
4. Boiler feedwater and cooling tower blowdown recycle for utilities and power generation.
5. Food processing applications such as: dairy processing, sweeteners concentration, juice and beverage processing, and production of light beer and wine.
6. Upgrade waters for agricultural purposes.

Applications (Environmental): Current and emerging applications include:

1. Chemical process industries: process water production and reuse, effluent disposal and water reuse, water/organic liquid separation, and organic liquid mixtures separation.
2. Metals and metal finishing: mining effluent treatment, and plating rinse water reuse and recovery of metals.
3. Textiles: dyeing and finishing, chemical recovery, and water reuse.
4. Pulp and paper: effluent disposal and water reuse.
5. Hazardous substance removal: removal of environmental pollutants from surface and groundwaters.
6. Municipal wastewater reclamation.

5.9 ULTRAFILTRATION / NANOFILTRATION

5.9.1 Ultrafiltration

Ultrafiltration is a membrane process with the ability to separate molecules in solution on the basis of size. An ultrafiltration membrane acts as a selective barrier. It retains species with molecular weights higher than a few thousand Daltons (macrosolutes), while freely passing small molecules (microsolutes and solvents). The separation is achieved by concentrating the large molecules present in the feed on one side of the membrane, while the solvent and microsolutes are depleted as they pass through the membrane.

For example, an ultrafiltration membrane process will separate a protein (macrosolute) from an aqueous saline solution. As the water and salts pass through the membrane, the protein is held back. The protein concentration increases and the salts, whose concentration relative to the solvent is unchanged, are depleted relative to the protein. The protein is, therefore, both concentrated and purified by the ultrafiltration.

Ultrafiltration may be distinguished from two related processes, reverse osmosis and microfiltration. Given the same example of a solution of protein, water and salt, the reverse osmosis membrane will pass only the water, concentrating both salt and protein. The protein is concentrated, but not purified. The microfiltration membrane will pass water, salt and protein. In this case, the protein will be neither concentrated nor purified, unless there is another larger component present, such as a bacterium.

Ultrafiltration membranes are typically rated by molecular weight cutoff, a convenient but fictitious value giving the molecular weight of a hypothetical macrosolute that the membrane will just retain. Microfiltration membranes, however, are rated by pore size, specifically the pore diameter. In practice, the distinction between the two membrane processes is blurred. There is some overlap in size between the largest polymer molecules and the smallest colloids. Microfiltration membranes with the smallest pore sizes sometimes retain large macrosolutes. An ultrafiltration membrane is sometimes used for what appears to be a microfiltration application because in that use it has a greater throughput.

Most ultrafiltration processes operate in crossflow mode, although a few laboratory devices and industrial applications operate in dead–end flow. The use of check filters or guard filters at the point–of–use in ultrapure water applications is an example of dead–end ultrafiltration.

Most ultrafiltration membranes are made from polymers, by the phase inversion process. Ultrafiltration membranes may also be formed dynamically, for example, by the deposition of hydrous zirconium oxide on a porous support under controlled conditions of flow and pressure. A few ultrafiltration membranes prepared from ceramic materials are available.

Ultrafiltration membranes may be characterized in terms of pore size and porosity, even though there is little direct evidence for the kinds of pores that the terminology suggests. A frequently used model characterizes the membrane as a flat film with conical pores originating at its surface. The surface pores are large enough to permit passage of solvent and microsolute molecules, but are too small for effective penetration of the larger microsolute. The conical shape is desirable, in that any entity that makes it through the opening at the membrane surface can continue unimpeded; there is no danger of pore–plugging.

Membranes may be made with differing pore sizes, pore densities, and pore–size distributions. These attributes are determined by measuring the flux of pure water through the membrane. The membrane is tested with dilute solutions of well characterized macromolecules, such as proteins, polysaccharides, and surfactants of known molecular weight and size, to determine the molecular weight cutoff. The above description implies the possibility of separating large macrosolutes from small macrosolutes by using the appropriate pore size. In practice, however, due to concentration polarization, only limited changes in the relative concentration of macrosolutes of different size are feasible.

There are three primary ultrafiltration device configurations: tubular, spiral wound, and hollow fiber. The tubular device is often used for small–flow, high value applications. Because of the tubular construction, mechanical cleaning can be done easily. The tubular device is the unit of choice for cases involving severe fouling. The spiral–wound design is often used for high volume applications. The spiral–wound configuration is more vulnerable to fouling and mechanical cleaning is difficult to perform. The hollow fiber design consists of a membrane wound into a hollow cylinder with the inside diameter varying from 500 to 1,100 microns. The choice of diameter size depends on whether the application is high–fouling or low–fouling. Capital costs for ultrafiltration systems range from medium to high. The operating/maintenance costs are medium.

Applications (Non–Environmental): Some of the applications also have environmental implications.

1. Ultrapure water for the semiconductor industry.
2. Pyrogen removal in the pharmaceutical industry.
3. Removal of metals in crude oil to prevent contamination of catalysts.
4. Size (PVA) recovery in the textile industry.
5. Dyestuff recovery and purification.
6. Concentration and recovery of latex emulsions.
7. Recovery of lignin compounds, and removal of color bodies in the pulp and paper industry.
8. Concentration to increase cheese yield.
9. Concentration of egg white, and gelatin.
10. Wine clarification.
11. Treatment of tank sediment, and alcohol reduction in breweries.
12. Biotechnology separations.
13. Enzyme separations.
14. Fruit juice clarification.

Applications (Environmental):

1. Ultrafiltration of electropaint wastewater for paint recovery.
2. Concentration of oily emulsions for pollution abatement.
3. Concentration of grey water.
4. Reclamation of waste lubricating oil.
5. Removal of heavy metals from metal plating wastes.
6. Waste treatment in the pulp and paper industry.
7. Reduction of high COD levels in corn starch plants.
8. Addition of a UF membrane to an activated sludge reactor (Dorr–Oliver process).
9. Treatment of sanitary waste in isolated locations.
10. Cheese whey protein recovery.
11. Selective removal of dissolved toxic metals from groundwater in combination with chemical treatment.

5.9.2 Nanofiltration

A nanofiltration membrane could be thought of as a tight ultrafiltration membrane. Nanofiltration membranes can be used to produce drinking water. Because they have good molecular rejection characteristics for divalent cations such as calcium and magnesium, nanofiltration membranes can be used effectively to remove hardness in place of conventional chemical softening. These membranes also reject higher molecular weight organics that contribute to taste and odor or that can react with chlorine to form trihalomethanes or other byproducts. A third benefit is that nanofiltration membranes reject viruses and other particles, which improves the efficiency of downstream disinfection processes.

Based on the above properties, nanofiltration could have interesting pollution control potential. In Flagler Beach, Florida, a study was conducted to evaluate the rejection of six synthetic organic compounds (SOCs) from a potable water source by a nanofiltration membrane process.

5.10 FORMED–IN–PLACE TECHNOLOGY

SBP Technologies, Inc. has developed a formed–in–place membrane technique that was demonstrated at the American Creosote works site in Pensacola, Florida.

SBP uses a proprietary formed–in–place membrane technology. The membrane is formed on porous sintered stainless steel tubes by depositing microscopic layers of inorganic and polymeric chemicals. The properties of the formed–in–place membrane can be varied by controlling the type of membrane chemicals used, their thickness, and the number of layers. This important feature allows for customization of the membrane system to a wide variety of waste characteristics and clean–up criteria. The formed–in–place membrane can be quickly and economically reformulated in the field to accommodate changes in waste characteristics or treatment requirements.

The filtration unit consists of porous sintered stainless steel tubes arranged in a modular, shell–and–tube configuration. Multi–layered inorganic and polymeric "formed–in–place" membranes are coated at microscopic thickness on the inside diameter of the stainless steel tubing by the recirculation of an aqueous slurry of membrane formation chemicals. This "formed–in–place" membrane functionally acts as a hyperfilter, rejecting species with molecular weights as low as 200. In addition, surface chemistry interactions between the membrane matrix and the components in the feed play a role in the separation process. A relatively clean steam, called the "permeate," passes through the membrane while a smaller portion of the feedwater, retaining those species that do not pass through the membrane, is retained in a stream called the "concentrate" or "reject."

The claim that the system can be operated to recover 80% of the feedwater volume as permeate was achieved.

REFERENCES

1. Baker, R., et al, *Membrane Separation Systems—Recent Developments and Future Directions,* Noyes Data, 1991.
2. Berkowitz, J., et al, *Unit Operations for Treatment of Hazardous Industrial Wastes,* Noyes Data, 1978.
3. Bhattacharyya, D., et al, *Separation of Hazardous Organics by Low Pressure Reverse Osmosis Membranes,* EPA/600/S2–91/045, 1/92.
4. Bravo, J., et al, *Fluid Mixture Separation Technologies for Cost Reduction and Process Improvement,* Noyes Data, 1986.
5. Cheremisinoff, N., *Reverse Osmosis Offers Advanced Waste Treatment,* Nat. Env. Jnl., 1–2/93.

6. EPA, *Innovative Hazardous Waste Treatment Technologies: Domestic and Foreign (4th Forum),* EPA/540/R-92/081, 12/92.
7. EPA, *Membrane Treatment of Wood Preserving Site Groundwater by SBP Technologies, Inc.,* EPA/540/AR-92/014, 8/93.
8. EPA, *Remedial Action, Treatment and Disposal of Hazardous Waste, (15th),* EPA/650/9-90/006, 2/90.
9. Hardenburger, T., Chem. Engr., 10/92.
10. Huling, S., *Facilitated Transport,* EPA/540/4-89/003, 8/89.
11. Kobylinski, E., et al, "Using Membranes for Material Reuse," *Environmental Protection,* 9/92.
12. Krishnan, K., et al, *Recovery of Metals from Sludges and Wastewaters,* Noyes Data, 1993.
13. McCoy & Assoc., *Commercialization of Innovative Treatment Technologies,* Haz. Waste Cons., 5-6/93.
14. Norwood, V., "Removal of Pesticides from Aqueous Solutions Using Liquid Membranes," in EPA/600/9-91/047, 1/92.
15. Noyes, R., *Handbook of Pollution Control Processes,* Noyes Data, 1991.
16. Palmer, S., et al, *Metal/Cyanide Containing Wastes—Treatment Technologies,* Noyes Data, 1988.
17. Porter, M., ed., *Handbook of Industrial Membrane Technology,* Noyes Publications, 1990.
18. Strathmann, H., et al, *Handbook of Industrial Membrane Technology,* M. Porter, ed., Noyes Publications, 1990.
19. Taylor, J., et al, *Synthetic Organic Compound Rejection by Nanofiltration,* EPA/600/S2-89/023, 1/90.
20. Wilk, L., et al, *Corrosive-Containing Wastes—Treatment Technologies,* Noyes Data, 1988.

6

Physical Technology

Physical processes are those that remove the hazardous material, without any chemical changes taking place. These physical processes can be included as one portion of a treatment train.

Typically, at NPL sites, treatment trains are used to address media and wastes containing both metals and organics. Some of the most frequently–selected treatment trains for these wastes using innovative technologies include soil washing or thermal desorption followed by solidification/stabilization. Solvent extraction is another technology potentially applicable to mixed organic and metal waste; however, it has not gained the same level of acceptance as thermal desorption. In a few cases, innovative technologies have been selected to treat both metals and organics simultaneously. In situ flushing is being used for both metals and organics at three sites.

6.1 ABSORPTION

Absorption can be considered as either a chemical or a physical process. Chemical reactant absorption is covered in Chapter 2. One form of physical absorption involves transferring a component from a gas stream to a liquid, utilizing a solvent. Another form is the sorption of a liquid by a solid sorbent, such as utilized in spill clean up.

6.1.1 Gas Stream Absorption

Although often used to recover products or raw materials, absorption can also serve as an emission control device. In this capacity, absorption has been used to control alcohols, acids, chlorinated and fluorinated compounds, aromatics, esters, aldehydes and other organics. Absorption devices can be used separately or in conjunction with other air pollution control equipment, e.g., to provide additional pollutant removal after incineration or after condensation. Liquids are used as the absorbent; therefore, a media transfer of toxic pollutants can occur. In general, more soluble compounds are removed with greater efficiency. Liquid–to–gas ratios, liquid temperature, and column height are also important parameters affecting efficiency.

Water may be used for the absorption of organic compounds that have relatively high water solubilities (e.g., most alcohols, organic acids, aldehydes, ketones, amines, and glycols). For organic compounds that have low water solubilities, other solvents (usually organic liquids with low vapor pressures) are used. Although absorption will be attractive for some process vents, it cannot be used to control all process vents. Because its use depends on the economics of recovery, absorption can be better classified as a product recovery device rather than an organic control device. Absorption is attractive if a suitable solvent is available, a significant amount of organics can be recovered, and the recovered organics can be reused. It is usually not considered when the organic concentration is below 200 to 300 ppmv. Generally, vent gas streams will consist of low-concentration organics. The control of low-concentration organics by absorption, however, usually requires long contact times and large quantities of absorbent for adequate emissions control.

The mechanism of absorption consists of the selective transfer of one or more components of a gas mixture into a solvent liquid. The transfer consists of solute diffusion and dissolution into a solvent. For any given solvent, solute, and set of operating conditions, there exists an equilibrium ratio of solute concentration in the gas mixture to solute concentration in the solvent. The driving force for mass transfer at a given point in an operating absorption tower is related to the difference between the actual concentration ratio and the equilibrium ratio. The absorbing liquids (solvents) used are chosen for high solute (VOC) solubility and include liquids such as water, mineral oils, and nonvolatile hydrocarbon oils.

Devices based on absorption principles include spray towers, venturi scrubbers, packed columns, and plate columns. Spray towers require high atomization pressure to obtain droplets ranging in size from 500 to 1,000 μm in order to present a sufficiently large surface contact area. Spray towers generally have the least effective mass transfer capability and, thus, are restricted to particulate removal and control of high-solubility gases such as sulfur dioxide and ammonia. Venturi scrubbers have a high degree of gas-liquid mixing and high particulate removal efficiency but also require high-energy input and have relatively short contact times. Therefore, their use is also restricted to high-solubility gases. As a result, VOC control by gas absorption is generally accomplished in packed or plate columns. Packed columns are mostly used for handling corrosive materials, for liquids with foaming or plugging tendencies, or where excessive pressure drops would result from use of plate columns. They are less expensive than plate columns for small-scale operations where the column diameter is less than 0.6 m (2 ft). Plate columns are preferred for large-scale operations, where internal cooling is desired or where low liquid flowrates would inadequately wet the packing.

In a packed tower, the gas to be absorbed is introduced at the bottom of the tower and allowed to rise through the packing material. Solvent flows from the top of the column, countercurrent to the vapors, absorbing the solute from the gas phase and carrying the dissolved solute out of the tower. Cleaned gas exits at the top for release to the atmosphere or for further treatment as necessary. The saturated liquid is generally sent to a stripping unit where the absorbed VOC is recovered. Following the stripping operation the absorbing solution is either recycled back to the absorber or sent to a treatment facility for disposal.

The major tower design parameters to be determined for absorbing any substance are column diameter and height, system pressure drop, and liquid flowrate required. These parameters are derived from considering the total surface area provided by the tower packing material, the solubility and concentrations of the components, and the quantity of gases to be treated.

The VOC removal efficiency of an absorption device is dependent on the solvent selected and on design and operating conditions. For a given solvent and solute, an increase in absorber size or a decrease in the operating temperature can increase the VOC removal efficiency as well as by a change in the absorbent.

Systems that utilize organic liquids as solvents usually include stripping and recycle of the solvent to the absorber. In this case the VOC removal efficiency of the absorber is also dependent on the solvent stripping efficiency.

Absorption is attractive if a significant amount of VOC can be recovered for reuse. Absorption is not usually considered for use when the VOC concentration in a process vent stream is below 200 to 300 ppmv. Furthermore, the use of absorption is subject to the availability of the appropriate solvent for a particular VOC.

A number of chemical processes use absorption systems as an integral part of the production scheme. A typical acetic anhydride manufacturing facility is an example of one such production scheme. Acetic anhydride is produced via the pyrolysis of acetic acid to form ketene. The ketene produced in the pyrolysis furnaces contains by-products and other impurities. Ketene is separated from these by-products and impurities by contacting the product stream with glacial acetic acid in a ketene absorber. Ketene is absorbed from the product stream and routed to further processing and eventual acetic anhydride purification.

Systems can also include wet scrubbers of the high energy venturi type, jet venturi (ejector scrubber) type, and wet cyclones.

The jet-venturi fume scrubber works on the principle of an ejector, entraining and scrubbing large volumes of gas without baffles or moving parts. Motivating fluid enters the scrubber from a nozzle in a hollow cone spray, creating a draft that draws the gases and vapors into the moving stream where they are continuously scrubbed or absorbed.

The trend is to hybrid systems using venturis or cyclones plus scrubbing towers and mist eliminators, or a combination of various types of venturis or towers. This combination provides both rough and fine scrubbing to achieve highest efficiency, and is effective for mixtures of gases or gases and particulate.

The efficiency of absorption as an organic control technique depends on several factors: the solubility of the organic in the solvent; the concentration of the organic in the gas stream; temperature; the L/G ratio; and the contact surface area. Higher gas solubilities and inlet concentrations provide a larger driving force for more efficient absorption. Since lower temperatures correspond to higher gas solubilities, absorption is also enhanced at reduced temperatures.

6.1.2 Absorption of Liquids by Solids

Sorbents are used for cleanup or control of spilled and other liquids.

The evaluation and selection of a sorbent that is effective and, when cost is a factor, economical, can be a difficult task. One source of difficulty is that many factors influence the performance and cost of using sorbents for spill cleanup operations. Factors that require consideration include safety, sorption capacity, desorption characteristics, sorbent cost, availability and ease of application, collection, regeneration, and disposal. The difficulties in evaluating these factors and judging their relative effects on overall performance are compounded by the lack of information, data, and guidance on sorbent use for liquid hazardous substance control, and the dispersion and nonuniformity of much of the data that are available.

Sorbents may be applied to a spill of a hazardous substance either manually or with mechanical aid. The choice of which general method to employ, either manual or mechanical, depends on the size of the spill and on the medium into which the spill has occurred. For example, the application of a sorbent to a small spill of 500 gallons or less of a hazardous substance onto land is generally most easily accomplished manually, using a shovel in the case of a particulate sorbent, or by throwing on a pillow, pad, or mat sorbent. On the other hand, the same size spill into a large body of water (assuming the substance floats) would probably require using a boat as a platform from which to manually apply either a pillow, pad, mat, or boom type sorbent.

Sorbents can be quite effective for the cleanup of spills onto land, particularly when the size of the spill is small, for example, less than 500 gallons. In fact, using sorbents for small spills is often the method of choice because other amelioration techniques are too expensive on a small scale or are other wise impractical. Unconfirmed analysis of spill size data indicates that most spills are 100 pounds or less and that most spills are onto land.

Spills of hazardous substances into water are also common occurrences, but as with spills onto land, spills into water are usually small and probably occur primarily during bulk cargo transfer operations. Small spills of floating hazardous substances into water are generally most easily cleaned up with sorbents. Other methods of cleanup are ineffective on this type of spill because they generally collect large amounts of water with the recovered hazardous substance, thus requiring another separation step.

Care must be exercised when choosing a sorbent to clean up a floating hazardous substance because there are two liquid phases involved, not a single liquid phase as with a land spill. The sorbent must selectively sorb the hazardous substance instead of water and must not sink. In other words, only hydrophobic floating sorbents are considered for this scenario. Also, particulate sorbents that float are not considered feasible since their collection would be difficult. In a non–flowing or slow moving water body, sorbent booms and pillow, pad, and mat sorbents can be effective. In a flowing water body, sorbent booms may be effective depending on the water flow and the size of the spill. Also, depending on the nature of the flowing water body and the degree of containment rendered by the sorbent boom, pillow, pad, or mat sorbents may be used effectively in a complementary role.

Landfills that accept containerized hazardous waste for disposal are required to comply with EPA regulations regarding the disposal of those containers. Containers holding liquids or free liquids must not be placed in a landfill unless, before placement, the liquids or free liquids are treated or stabilized chemically or physically (e.g., by mixing with a solid sorbent), so that free liquids are no longer present. Free liquids can thus be retained by a sorbent or otherwise immobilized.

For spills, drips, and leaks in an industrial setting either granular sorbents (clay, sawdust, rice hulls, corn cobs) or fiber sorbents (vegetable fibers, polypropylene, polyester) could be used. Fibers have considerably higher absorptive capacity than granular products, and could have overall lower costs due to labor savings, compressibility for removing oils or other contaminants resulting in lower disposal costs.

Particulate Sorbents: Manual application of particulate sorbents is limited to small spills of hazardous substances on land. Because of the difficulty in collection, particulate sorbents are not considered for use on spills into water. Also, manual application of particulate sorbents is infeasible for a large land spill because of the labor intensive nature of such an operation.

Because particulate sorbents are powders, granules, or other loose forms, the only practical method for manually applying them is shoveling. The type of shovel that is most satisfactory is a coal type scoop shovel, or other similar type with a flat bottom and turned up sides. This configuration will produce a maximum rate of application. Also, spark proof tools should be used around flammable hazardous substances.

Mechanized application of particulate sorbents to a land spill may be appropriate when the spill is of medium size (500 to 5,000 gallons). A mechanized application technique allows for a more rapid cleanup of the spill, and makes application of the sorbent onto the center of the pool easier and more complete. When using mechanical means to apply sorbent materials to a spill, the goal should be to apply an even, uniform layer of sorbent on top of the spill from the upwind side of the spill, working from the edges towards the middle of the spill. Sorbent should be applied until the sorbent is no longer wet by the spilled hazardous substance. If the spill is into water and the hazardous liquid floats, then pillow, pad, or boom type sorbents should be used since their collection is easier.

The best mechanical techniques for applying particulate sorbents will depend to a large extent on the bulk density of the sorbent. For sorbents of low to moderate bulk density (e.g., cellulose–based sorbents, expanded minerals), devices that entrain the solid in a high velocity stream of air are judged to probably be the most effective. The sorbent is propelled by repeated collisions with the high–velocity air molecules. The distance the sorbent can be dispersed is limited, albeit greater than can be achieved manually, due to the comparatively low momentum which results from low density particulates.

For higher density particulate sorbents (e.g., clay minerals), centrifugal devices would probably be more effective than the air–entrainment devices described above. These devices impart momentum to the sorbent particles via a rotating impeller.

Pillow, Pad, and Mat Sorbents: Pillows, pads, and mats may be applied to spills on land or to floating spills on non–flowing water. The only practical method of application is to manually toss the sorbent onto the spill by hand, regardless of spill size or the size of the water body.

For small spills on land, pillow, pad, or mat sorbents can be applied by throwing the sorbent onto the spill either underhand or in a frisbee–throwing type of movement. This operation should be performed from the upwind side of the spill. The entire spill area should be covered with the sorbent.

For large spills on land it may be difficult to throw the sorbent onto the middle of the spill. Use a pillow type sorbent for this purpose since these are heavier than pads or mats and can therefore be thrown the farthest and are least affected by the wind. If the spill is so large that the middle of the spill cannot be reached by throwing sorbent from the periphery of the spill, then proceed by working from the perimeter inwards. Place sorbent around the perimeter of the spill, then collect it and slowly work inwards toward the center of the spill.

Pillows, pads, or mats can be thrown onto a floating spill from the shore when the water body is small enough. When the water body is large, a boat will be required so that complete sorbent application can be achieved. However, this procedure must be very carefully evaluated with respect to the nature of the spilled substance and the hazards it presents since the personnel in the boat cannot easily escape if a problem arises. A worst case scenario would be the accidental ignition of a flammable hazardous substance while the boat and personnel are near the spill, or the failure of a breathing apparatus in a spill of a hazardous vapor that is highly toxic. The only time such an operation should be attempted is if the safety of the personnel involved can be assured.

Sorbent Booms: Sorbent booms (rather than pads, mats, or pillows) may be deployed against floating spills of hazardous substances in flowing watercourses when the current is below 2 knots and spill thickness is less than 1 mm. If the spill is ongoing, a sorbent boom can be deployed just downstream of the point of entry of the hazardous substance into the watercourse, or as close as practical. Additional sorbent booms could be deployed at suitable intervals downstream to collect any of the hazardous substance that has already spilled or that might subsequently escape the first sorbent boom deployed. Although sorbent/hazardous liquid compatibility was evaluated in developing the Sorbent Selection and Use Guides, the compatibility of the outer sorbent boom containment fabric with specific hazardous liquids must be verified before use.

If the spill has been halted, a sorbent boom should be placed far enough downstream so that it is completely deployed prior to the arrival of the spill slug. Backup sorbent booms should also be placed downstream to catch any hazardous substance which escapes the first-line sorbent boom.

If the spill is into a narrow stream, the sorbent boom can be deployed by throwing or casting an attached line to a person standing on the other side. The sorbent boom is then pulled across the stream. Alternatively, the sorbent boom can be hand carried across the stream, but only if it is safe to do so.

When the spill is into a wide river, the only practical way of deploying a sorbent boom is to use a boat to carry it out from shore a sufficient distance to intercept the floating spill. Anchors may be required to secure the sorbent boom to the bottom to prevent it from being carried downstream. Alternatively, the sorbent boom can be secured to the shoreline. The "no-oil-loss tow speed" for three sorbent booms tested was 1 ft/sec in calm water conditions. Sorbent booms used for hazardous substance pickup, therefore, would be generally effective unless current flow appreciably exceeds 1 ft/sec.

The people involved in the deployment of sorbent booms must necessarily position themselves downstream of the spill. There may be times when downstream is also downwind. When such a condition exists, extra caution must be exercised so that individuals are not exposed to toxic vapors in excess of safe levels. Such precaution might include the use of fully encapsulating suits that contain SCBA (self-contained breathing apparatus). The user should refer to sources that will provide information regarding the proper protective equipment. In addition, monitoring of spill vapor concentrations may be necessary.

6.2 ADSORPTION

Adsorption is a surface phenomenon which separates a hazardous liquid from a bulk, nonhazardous liquid like water by contact with a solid adsorbent. It is also important for removing volatile organic compounds (VOCs) from gas streams. Other minor uses include removal of metals from waste streams, use with biological treatment processes, adsorptive filtration, and as an in situ soil treatment.

Adsorbent materials include:

1. Activated Carbon
2. Synthetic Resins
3. Activated Alumina
4. Natural Adsorbents

6.2.1 Activated Carbon for Organics Removal

The chemistry of carbon is such that most organic compounds and many inorganics will readily attach themselves to carbon atoms. The strength of that attachment (and thus, the energy required for subsequent desorption) depends on the bond formed, which in turn, depends on the specific compound being adsorbed. Carbon to be used for adsorption is usually treated to produce a product with large surface–to–volume ratio, thus, exposing a practical maximum number of carbon atoms to be active adsorbers. Carbon so treated is said to be "activated" for adsorption. Activated carbon which has adsorbed so much contaminant that its adsorptive capacity is severely depleted is said to be "spent." Spent carbon can be regenerated, but for strongly adsorbed contaminants, the cost of such regeneration can be higher than simple replacement with new carbon.

This process is used to treat single–phase aqueous organic wastes with high molecular weight and boiling point and low solubility and polarity, chlorinated hydrocarbons such as tetrachloroethylene, and aromatics such as phenol. It is also used to capture volatile organics in gaseous mixtures. Limitations are usually economic and related to the rapidity with which the carbon becomes spent. Rule of thumb guidelines are that contaminant concentrations should be less than 10,000 ppm, suspended solids less than 50 ppm, dissolved inorganics and oil and grease less than 10 ppm.

The basic principle of operation for carbon adsorption is the mass transfer and adsorption of a molecule from a liquid or gas onto a solid surface. Activated carbon is manufactured in such a way as to produce extremely porous carbon particles whose internal surface area is very large (500 to 1,400 square meters per gram of carbon). This porous structure attracts and holds (adsorbs) organic molecules as well as certain metal and inorganic molecules.

Adsorption occurs because (a) the contaminant has a low solubility in the waste, (b) the contaminant has a greater affinity for the carbon than for the waste, or (c) a combination of the two. The amount of contaminants that can be adsorbed by activated carbon ranges from 0.10 to 0.15 gram per gram of carbon.

The two most common carbon adsorption processes are the granular activated carbon (GAC), which is used in packed beds, and the powdered activated carbon (PAC).

The activated carbon adsorption process is one of the most frequently applied technologies for the removal of trace organic compounds from an aqueous solution. Adsorption is a surface phenomenon in which soluble molecules from a solution are bonded onto a particular substrate. Therefore, one of the most desirable properties of an adsorbent is a high surface to volume ratio. Activated carbon (with a surface to volume ratio ranging from 500 to 1,400 m^2/g is a good adsorbent for effective removal of organic compounds. Activated carbon will adsorb most organic compounds to some degree. Factors that affect the adsorption process include: (1) carbon pore structure, (2) carbon contact time, (3) temperature, and (4) pH.

Mixtures of organics may cause significantly reduced adsorption capacity for certain compounds due to the preferential adsorption of other compounds by the carbon. Competitive adsorption of organic compounds is extremely complicated and difficult to predict. Therefore, it is recommended that pilot treatability tests be performed on the waste in question. The effectiveness of activated carbon adsorption is limited by the following waste characteristics: (1) low molecular weights, (2) high polarities, and (3) high solubility.

Carbon adsorption can be applied to aqueous and gaseous wastes containing a wide range of organic compounds. The following are compounds that can be successfully removed from waste streams:

1. Organic liquids with metals and halogens,
2. Organic nitrogen compounds,
3. Chelated heavy metals, and
4. Volatile organics.

The following is a list of applications for which the activated carbon adsorption process is not recommended:

1. High solids content (greater than 500 mg/).
2. Unassociated metals, and
3. High humidity gas streams.

The exhausted carbon will contain all of the waste constituents removed from the waste streams. The carbon must be either regenerated (on– or off–site) or disposed of in a secure landfill (carbon with PCBs or dioxin are not currently regenerated by the vendors). Thermal regeneration of the used carbon is the most common method currently used. Other methods of regeneration employed are solvent and steam regeneration.

Periodic backwashing of the carbon will require holding tanks for the backwash. Often the backwash is allowed to settle and the liquid portion is sent back through the carbon. The small amount of sludge generated during settling contains high concentration of organics and requires disposal. There are both granular and powdered systems.

The principal advantages of carbon adsorption technology is its ability to achieve low effluent concentration levels for a large number of compounds, including many halogenated organics. The technology appears particularly applicable to high molecular weight compounds such as the chlorinated pesticides. It is also applicable to the treatment of many compounds which are normally toxic and resistant to biological treatment. Material recovery may also be possible if regeneration methods other than thermal regeneration can be used.

Limitations are largely associated with high capital and operating costs, particularly when thermal reactivation must be used. Thermal reactivation, the most effective means of regeneration, is cost effective only for relatively large installations (i.e., greater than 1,000 lb/day) and for wastes with relatively low (less than 1%) organic concentrations. The technology is sensitive to other impurities such as suspended solids and oil and grease, thus, some degree of pretreatment is usually required to ensure effective performance. The adsorption process is also not effective for many low molecular weight and highly water soluble organics.

The use of activated carbon for metals removal is discussed in Section 5.2.

Removal of Contaminants from the Liquid Phase: Adsorption is a widely–used process for the removal of organic contaminants from liquid waste streams. Activated carbon is the most commonly used adsorbent. Largely nonpolar, carbon is particularly effective for the removal of hydrophobic, high molecular weight organic compounds from aqueous streams. Thus, it is a good adsorbent for many of the halogenated organic compounds. Activated carbon adsorption must be considered a potentially viable treatment technology for many halogenated organic–bearing wastewater streams, either as a primary treatment for moderately high (up to 0.5%) concentrations of organic compounds or as a secondary polishing type treatment for much lower levels of contamination. The cost effectiveness of adsorption is dependent on flow rates and concentrations of the organic contaminants and on the adsorptive capacity of the carbon for the contaminants. Adsorption should be cost effective for concentrations of organic compounds up to about 1,000 mg/ , and could be cost effective for concentrations up to 5,000 mg/ . For concentrations above 5,000 mg/ , other unit processes are generally more cost effective, unless nondestructive chemical regeneration can be used to recover the adsorbed materials.

Activated carbon is available in powder (PAC) or granular (GAC) form. GAC is more commonly used because its larger size is more amenable to handling in the equipment used to achieve contact and regeneration. Both types of carbon adsorbent have large contact surface areas, far in excess of their nominal external surface areas. Surface areas, resulting from a network of internal pores 20 to 100 Å in diameter, are of the order of 500 to 1,500 m^2/g. Porosities can be as large as 80%. The adsorption capacity of an activated carbon for a contaminant is a function of the surface area and the surface binding process and can approach 1 g/g of carbon.

Adsorbent binding forces result from the interaction of the contaminant surface molecules with the carbon surface atoms. The attractive forces are generally weaker and less specific than those of chemical bonds and, hence, the term physical adsorption is used to describe the binding mechanism.

A typical continuous adsorption system consists of multiple columns filled with activated carbon and arranged in either parallel or series. Total carbon depth of the system must accommodate the "adsorption wavefront;" i.e., the carbon depth must be sufficient to purify a solution to required specifications after equilibrium has been established. Bed depths of 8 to 40 ft are common. Minimum recommended height–to–diameter ratio of a column is 2:1. Ratios greater than 2:1 will improve removal efficiency, but result in increased pressure drop for the same flow rate. Optimum flow rate must be determined in the laboratory for the specific design and carbon used. For most applications, 0.5 to 5 gpm/ft^2 of carbon is common.

Various configurations are available for GAC adsorption applications. Based on influent characteristics, flow rate, size and type of carbon, effluent criteria and economics, each design is unique in its mode of operation.

The adsorption beds of both series and parallel design can be operated in either an upflow or downflow direction. A downflow mode of operation must be used where the GAC is relied upon to perform the dual role of adsorption and filtration. Although lower capital costs can be realized by eliminating pretreatment filters, frequent backwashing of the adsorbers is required. Application rates of 2 to 10 gpm/ft^2 are employed, and backwash rates of 12 to 20 gpm/ft^2 are required to achieve bed expansions of 20 to 50%.

The use of supplemental air increases efficiency of the backwashing.

Prefiltration is normally required to prevent blinding upflow–expanded beds with solids. In this configuration, smaller particle sizes of GAC can be employed to increase adsorption rate and decrease adsorber size. Application rates can be increased even to the extent that the adsorbent may be in an expanded condition.

The design arrangements are:

1. Adsorbers in Parallel
2. Adsorbers in Series
3. Moving Bed
4. Upflow–Expanded

The above systems are not generally used with powdered activated carbons. The PAC systems now used involve mixing the PAC with the waste stream to form a slurry which usually can be separated later by methods such as filtration or sedimentation. PAC is generally used simultaneously with biological treatment to enhance organic removal by biological processes.

The economic success of an adsorption system usually depends on the regenerability of the adsorbent. The exception is where there are very long adsorption or loading cycles due to very low concentrations of halogenated organic constituents in the inlet feed. This type of system usually operates on a "throw away" basis. If very large quantities of adsorbent are involved, then regeneration and reuse are required for economical operation. The regeneration techniques employed in industry are thermal regeneration, steam regeneration, and acid or base regeneration. Solvent washing or biological treatment are other methods that are occasionally used for regeneration. Solvent recovery, if possible, can lead to adsorbent recovery with attendant cost benefits. Thermal regeneration is the most commonly applied technique for GAC systems, since this is the only method that can generally ensure effective regeneration.

Air and water discharges from carbon adsorption systems employing carbon regeneration can be relatively innocuous. Under proper design and operation conditions, the treated water will generally be suitable for discharge to surface waters. Other aqueous streams such as backwash, carbon wash and transport waters are recycled or sent to a settling basin. Emissions will result from thermal reactivation, but when afterburners and scrubbers are used, the controlled emissions are essentially non–polluting. In some installations, particulates must be removed from the air stream (e.g., via a cyclone and baghouse) resulting in a solid waste.

There are several advantages of using Granulated Activated Carbon (GAC) systems for **radon removal** in small community water supplies: (1) GAC units can be installed in–line so no additional pumping is required (systems are already under pressure), (2) radon is contained, not released into the air, and (3) GAC has lower maintenance requirements than aeration systems and had no moving parts. The disadvantages of GAC are: (1) the potential for substantial gamma emissions from radon progeny, (2) the potential for accumulation of long–lived radionuclides (i.e., uranium 238 and 235, radium–226 and lead–210), (3) the long empty bed contact time (EBCT) required, and (4) the potential for fouling of the GAC by oxidized metals, organics, particulates, and/or microorganisms, requiring frequent backwashing.

Many wastewater streams contain dilute concentrations of organic pollutants that are not treated effectively by conventional activated sludge processes. These pollutants, however, can often be treated effectively anaerobically. If the pollutants were treated anaerobically, pass–through of the pollutants to the receiving stream and stripping of volatile compounds during aeration could be minimized. To treat the entire wastewater stream in an anaerobic digester would not be economical. However, if the bulk liquid stream could first be passed through a sorbent bed such as granular activated carbon (GAC) prior to aeration, only the sorbent material, a much smaller volume, would require anaerobic stabilization at elevated temperatures.

Powdered activated carbon (PAC) has a particle size typically less than 100 micrometers in diameter and can be added at several locations in the conventional drinking water treatment process train plant to remove trace compounds that cause undesirable taste and odor.

The PAC process is based on the same principles of adsorption as GAC. However, PAC is not a granular filter medium; rather it is a powder added directly to the water at one or more points during the treatment process. Typically, PAC can be added during coagulation, flocculation, sedimentation, and filtration. The addition of PAC:

1. Improves the organic removal effectiveness of conventional treatment processes.
2. Addresses short–term and emergency problems with conventional treatment systems.
3. Acts as a coagulant aid.
4. Removes taste and odor.

PAC is also an attractive treatment technology because it is less expensive than GAC in addressing seasonal problems, is easily started and stopped, creates no headloss, does not encourage microbial growth, and has relatively small capital requirements.

The chief disadvantage of this process is that some contaminants require large dosages of PAC for removal. Another disadvantage is that PAC is suitable only for conventional treatment systems. PAC also requires specific system hydraulics, space, and sludge–handling practices. PAC has proven ineffective in removing natural organic matter, due to the competition from other contaminants for surface adsorption and the limited contact time between the water and PAC. In addition, PAC adsorption is not amenable to basin–mixed flow reactions (as opposed to column–mixed flow reactions).

Two potential techniques for improving PAC's effectiveness for organic contaminant removal are the Roberts–Haberer process and the fluidized–bed PAC adsorber.

Removal of Contaminants from the Gaseous Phase: Activated carbon is effective in capturing certain organic vapors by the physical adsorption mechanism. However, activated carbon has a finite adsorption capacity. When the carbon becomes saturated (i.e., all of the carbon surface is covered with organic material), there is no further organic removal; all vapors pass through the carbon bed. At this point (referred to as "breakthrough"), the organic compounds must be removed from the carbon before adsorption can resume. This process is called desorption or regeneration.

The two basic configurations for carbon adsorption systems are regenerative and nonregenerative systems. Regenerative systems can be categorized as fixed, moving, or fluidized. The most common adsorption system for controlling air pollutants is the fixed carbon bed. Fixed–bed carbon adsorbers are used for controlling continuous, organic gas streams with flow rates ranging from 30 to 3,000 m^3/min (1,000 to over 100,000 ft^3/min). The organic concentration can be as low as several parts per billion by volume (ppbv) or as high as 25% of the lower explosive limit of the vapor stream constituents. Fixed–bed carbon adsorbers may be operated in either intermittent or continuous modes. For intermittent operation, the adsorber removes organics only during a specific time period. Intermittent mode of operation allows a single carbon bed to be used because it can be regenerated during the off–line periods. For continuous operation, the unit is equipped with two or more carbon beds so that at least one bed is always available for adsorption while other beds are being regenerated. The process vent gases are filtered and cooled before entering the carbon bed. The inlet gases to an adsorption unit are filtered to prevent bed contamination. The gases are cooled to maintain the bed at optimum operating temperature and to prevent fires or polymerization of the hydrocarbons. Vapors entering the adsorber stage of the system are passed through the porous activated carbon bed.

Adsorption of inlet vapors occurs in the bed until the activated carbon is saturated with organics. The dynamics of the process may be illustrated by viewing the carbon bed as a series of three layers or mass transfer zones (MTZ). Gases entering the bed are highly adsorbed first in the upper zone. Because most of the organic is adsorbed in the upper zone, very little adsorption takes place in the middle and lower zones. Adsorption in the middle zone increases as the upper zone becomes saturated with organics and proceeds through the lower zone. When the bed is completely saturated (breakthrough), the incoming organic–laden vent gases are routed to an alternate bed while the saturated carbon bed is regenerated. Typically, the duration of the adsorption cycle varies considerably depending on the solvent being reclaimed and its regeneration characteristics.

Regeneration of the carbon bed is accomplished by heating the bed or applying vacuum to draw off the adsorbed gases. Low–pressure steam is frequently used as a heat source to strip the adsorbent of organic vapor. The steam–laden vapors from regeneration are then sent to a condenser, and the condensate typically is sent on to some type of solvent recovery system. The regenerated bed is put back into active service while the saturated bed is purged of organics. (Note: Organic emissions resulting from regeneration should also be controlled and accounted for in the efficiency determination of the overall system.) The regeneration process may be repeated many times, but eventually the carbon must be replaced.

The life span of activated carbon depends on the nature of the pollutants being controlled. For clean organics, a carbon life of 10 to 20 years can be expected; for a stream containing trace amounts of high–boiling–point materials, 5 to 10 years is reasonable; but the presence of polymerized organics may require carbon reactivation every 1 to 3 years.

Nonregenerative systems (e.g., carbon canisters) are applicable for controlling organic emissions that are expected to vary in types of organics and concentrations and to occur at relatively low total mass rates. Nonregenerated systems are carbon canisters typically consisting of a 0.21 m^3 (55 gal) drum with inlet and outlet pipe fittings. Use of carbon canisters is limited to controlling low–volume gas streams with flow rates less than 3 m^3/min (100 ft^3/min). Carbon cannot be regenerated directly in the canister. Once the activated carbon in the canister becomes saturated by the organic vapors, the carbon canister must be removed and replaced with a fresh carbon canister. The spent carbon canister is then recycled or discarded depending on site–specific factors.

The design of a carbon adsorption system depends on the chemical characteristics of the organic compound being recovered (the adsorbate), the physical properties of the vent gas stream (temperature, pressure, humidity, and volumetric flow rate), and the physical properties of the adsorbent. The adsorbent concentration and type are key factors in the design of a carbon adsorption system. The adsorption characteristics of each compound are assessed based on their physical properties data, for example, polarity, refractive index, boiling point, molecular weight, and solubility in water. Nonpolar compounds and compounds with high refractive indices tend to be adsorbed more readily than polar compounds such as water. High vapor pressure/low boiling point adsorbates and low molecular weight compounds adsorb less readily. Compounds with molecular weights greater than 142 adsorb readily but are difficult to desorb.

Although carbon adsorption is an excellent method for recovering some valuable process chemicals, it cannot be used as a universal control method for process vents. The conditions under which carbon adsorption is not recommended may exist in some process vents. These include streams with very high or low molecular weight compounds, and mixtures of high- and low-boiling-point organic compounds. The range of organic concentration to which carbon adsorption can be applied is from only a few parts per million to concentrations of several percent. Adsorbing process vent streams with high organic concentration may result in excessive temperature rise in the carbon bed due to the accumulated heat of adsorption of the organic loading. However, high organic concentrations can be diluted to make a workable adsorption system. The molecular weight of the compounds to be adsorbed should be in the range of 45 to 130 g/g–mol for effective adsorption. Carbon adsorption may not be the most effective emission control technique for compounds with low molecular weights (below 45 g/g–mol) owing to their smaller attractive forces or for high molecular weight components (>130 g/g–mol) that attach so strongly to the carbon bed that they are not easily removed.

Properly operated adsorption systems can be very effective for homogeneous offgas streams but can have problems with a multicomponent system containing a mixture of light and heavy hydrocarbons. The process vent gas streams addressed in this document are likely to be a mixture of organics, with one or two major constituents and one or more minor constituents. Two or more organics in the vent gas streams, as a general rule, will have the following effects:

1. The adsorption of organic compounds having higher molecular weights will tend to displace those having lower molecular weights. Lighter compounds will tend to be separated or partitioned from the heavier compounds and will pass through the bed at a faster rate. This will increase the MTZ and may require additional carbon bed depth or shorter operating cycles.
2. Carbon retentivity may be reduced.
3. Efficiencies of any given system will tend to be lower on a multiple organic application.
4. The lower explosive limit (LEL) of the mixture will vary directly with the LEL of the individual components. Safety considerations may dictate more or less dilution air to reduce flammability potential.

Considerable progress has been made in the science and technology of adsorption for separation and purification of gas mixtures during the last thirty years. Pressure Swing Adsorption (PSA) processes have become the state of the art method for many bulk gas separations in the chemical and petrochemical industries. Novel Thermal Swing

Adsorption (TSA) processes provide the most energy efficient methods for many trace or dilute impurity removal applications. New and more efficient adsorptive separation processes as well as new adsorbent compositions are emerging frequently and the scope and size for the application of this technology is expanding. This technology is being used to produce ultra pure gases, more than one product from multicomponent feed gas mixtures, and adjusted product compositions containing extraneous components not present in the original feed gas mixture.

Virgin carbon is used for organics removal. However, if impregnated carbon is used, the impregnating agent is sodium or potassium hydroxide.

6.2.2 Resin Adsorption

Resin adsorption is an alternative treatment technology for the removal of organic contaminants from aqueous waste streams. The underlying principle of operation is similar to that for carbon adsorption; organic molecules contacting the resin surface are held on the surface by physical forces and subsequently removed during the resin regeneration cycle. Resin adsorbents can be made from a variety of monomeric compounds which differ in their polarity and thus, their affinity for different types of compounds. The choice of resin type can lead to an adsorbent tailored specifically for effective removal of special classes of compounds. For example, hydrophobic resins such as those prepared from styrene–divinyl benzene monomers, are most effective for nonpolar organics and bonding is largely the result of Van der Waal's forces; acrylic based resins on the other hand are most polar and dipole–dipole interactions may play the major role in the binding of polar molecules to the resin surface. The general concept is that like molecules attract. Polar resins will attract polar organics; nonpolar compounds will be attracted by the more hydrophobic or nonpolar resins.

A significant aspect of resin adsorption is that the bonding forces are usually weaker than those encountered in granulated activated carbon (GAC) adsorption. Regeneration can be accomplished by simple, nondestructive means such as solvent washing, thus providing the potential for solute recovery. Thermal regeneration (generally not possible with resin adsorbents because of their temperature sensitivity) is usually required for carbon adsorbents, eliminating the possibility of solute recovery. The resins differ in many other respects from activated carbon adsorbents. In addition to differences in the ease and usual methods of regeneration associated with the chemical nature of the two adsorbents, there are significant differences in shape, size, porosity and surface area. Resin adsorbents are generally spherical in shape rather than granular, and are smaller in size and lower in porosity and surface area than GAC adsorbents. Surface areas for resins are generally in the range of 100 to 700 m^2/g, as opposed to 800 to 1,200 m^2/g for activated carbon. Adsorptive capacities are thus less for the resin adsorbents, although the chemical nature and the pore structure of the resin can be tailored to enhance the selectivity of the resin and, therefore, its adsorption capacity for specific organic components. Other notable properties of resin adsorbents include their nondusting characteristics, their low ash content, and their resistance to bacterial growth. The last characteristic is primarily a result of the fine pore structure which inhibits bacterial intrusion.

Another significant difference between resin and carbon adsorbents is their cost. Resin adsorbents are much more expensive. They generally will not be competitive with carbon for the treatment of waste streams containing a number of contaminants with no recovery value. However, resin adsorption should be considered if material recovery is practical, selectivity is possible, and for cases where carbon regeneration is not effective. Like carbon adsorption systems, resin adsorption can produce an effluent with low levels of contaminant concentrations, particularly in cases where contaminants are well characterized and few in number. Resin adsorption combined with carbon adsorption may be effective for certain waste streams containing a number of contaminants.

Resin adsorption systems are designed and operated in similar fashion to GAC systems. A principal difference will be in the regeneration step; regeneration of the resin is usually performed in situ with aqueous solutions or solvents. Solute recovery from the regeneration liquor will also be required, with distillation the most likely method.

Because of their expense, resins are not commonly used full–scale to remove organics from wastewaters. There is also little publicly available information on current or proposed industrial applications. Information of a general nature does report that resins are being used for color removal from dyestuff and paper mill waste streams, for phenol removal, and for polishing of high purity waters.

The following applicants have been identified as being particularly attractive for resin adsorption technology.

1. Treatment of highly colored wastes where color is associated with organic compounds.
2. Material recovery where solvents of commercial value are present in high enough concentration to warrant material recovery since it is relatively easy to recover solutes from resin adsorbents.
3. Where selective adsorption is an advantage and resins can be tailored to meet selectivity needs.
4. Where low leakage rates are required; resins exhibit low leakage apparently as a result of rapid adsorption kinetics.
5. Where carbon regenerations is not practical, e.g., in cases when thermal regeneration is not safe.
6. Where the waste stream contains high levels of inorganic dissolved solids which drastically lowers carbon activity; resins activity can usually be retained, although prerinses may be required.

Resin adsorption appears to offer advantages in certain situations; e.g., (1) for treatment of highly colored wastes, (2) for material recovery, (3) where low leakage is required, and (4) in instances where carbon adsorption is not practical. The advantages of resin adsorption are a result of their potential for selectivity, rapid adsorption kinetics, and ease of chemical regeneration.

Major limitations of resin adsorbents result from: (1) the generally lower surface area and usually lower adsorption capacities than those found in activated carbon; (2) possible susceptibility to fouling due to poisoning by materials that are not removed by the regenerant; and (3) their relatively high cost. The high cost of the resin may be balanced by its ease of regeneration and their predicted long lifetimes in situations where carbon must be thermally regenerated and carbon losses become appreciable (up to 10%).

6.2.3 Metals Removal

Activated Carbon: Studies in the field of metallurgy have indicated that carbon adsorption of many metallic compounds can be successfully achieved and has found commercial application for certain aqueous waste streams. However, adsorption efficiency varies considerably between different compounds.

Activated carbon has a fixed adsorption capacity for each type of metallic compound. Once this capacity is saturated, contaminants will no longer be adsorbed and the activated carbon must be regenerated or replaced. The carbon can be reactivated by using a strong acid or base to remove metal particles and bring them back into the solution.

The adsorption characteristics of activated carbon for metals removal are more complex than those for organic compounds because the charged nature of the metals affects their rate of removal from the solution. In general, the specific surface area, pore structure, and surface chemistry of the activated carbon significantly affect its adsorption characteristics for removal of contaminants.

Other parameters that influence the metals removal efficiency of activated carbon are pH, temperature, presence of chelating agents, ionic strength, carbon dose, and metal concentration. The pH of the solution affects contaminant removal by influencing the surface charge of the activated carbon and affecting the distribution of the metal ions in the solution. As the pH decreases, the solubility of metal ions generally increases. Complexing the metal ions in the solution by using chelating agents considerably increases the adsorption of metallic compounds onto the activated carbon. Chelating agents such as ethylenediaminetetraacetic acid (EDTA) and nitrilotriacetic acid (NTA) significantly increase the removal of mercury and cadmium by carbon material.

The process of metals removal by activated carbon often involves the use of multiple columns or tanks filled with carbon and operated in series or parallel configurations. The carbon bed depth should be high enough to remove all the metals from the solution to the required concentration. Carbon columns are operated in two basic modes: fixed–bed or moving/pulse modes. In the fixed–bed mode of operation, all of the activated carbon in the column is replaced with fresh or regenerated material when breakthrough of the contaminants occur. In the moving or pulsed mode of operation, only that portion of the carbon that has been exhausted is removed and replaced.

Activated carbon has been widely used to remove organic materials from aqueous wastes; however, its application for removal of metals from wastewaters has been limited. Areas in which activated carbon is applied to remove metals include material technology, in which valuable inorganics such as gold and silver are extracted from solution by activated carbon; analytical chemistry, in which activated carbon is used to enrich specific metal ions for quantitative analysis; and in water and wastewater treatment, in which carbon is used to remove metallic compounds.

A few bench– and pilot–scale studies have been conducted on the treatment of waste streams by activated carbon adsorption. Also, a few full–scale carbon–adsorption systems are commercially available for treatment of wastewaters containing chromium and mercury.

Activated carbon can also be used to remove metals from gases.

There are two methods used to adsorb toxic metals, particularly mercury with carbon. The first method is to inject very small quantities of activated carbon powder into the flue gas. This process is similar to DSI, and when both processes are used, the carbon and the sorbent can be injected together through the same nozzle or port in a mixture of typically 95 to 98% sorbent and 2 to 5% carbon. The carbon is then carried in the flue gas with the sorbent and is collected in the particulate removal device, typically a baghouse. (Because of the corna discharges in an ESP and the combustible nature of the carbon, an ESP is not the best method for activated carbon collection).

The second method used to adsorb toxic metals onto carbon is to flow the flue gas through one or more beds of activated carbon. To prevent the beds from plugging, they are normally placed downstream of the particulate removal device. It is also beneficial if the gas is not saturated so that moisture condensation is not a problem. Carbon beds are available in a variety of different housings, with one of the simplest being a filter frame that will fit into a normal HEPA filter housing.

Occasionally, two beds in series are used with the first bed using a lower-grade carbon and the second bed using a higher grade carbon. With this concept, low-cost carbon filters can do the majority of the work and, therefore, must be replaced more often. The higher grade, more expensive carbon filters are then used only as a final element and do not have to be replaced as often.

Adsorptive Filtration: This technology uses adsorptive filtration to remove inorganic contaminants (metals) from the liquid phase. An adsorbent, ferrihydrite, is applied to the surface of an inert substrate, such as sand, which is then placed in a vertical column. The column containing the coated sand acts as a filter and adsorbent. Once the adsorptive capacity of the column is reached, the metals are removed and concentrated for subsequent recovery using a pH-induced desorption process.

In a test at a Superfund site simultaneous sorption and filtration of Cu, Cd, and Pb were determined to be feasible using iron oxide-coated sand under reasonable engineering conditions. Soluble effluent concentrations of a few tens of µg/ or less are achievable. The media can remove particulate metals simultaneously from the water, probably with an efficiency comparable to that achievable with conventional sand filtration. The media can be regenerated by exposure to an acid solution, yielding regenerant solutions containing metal concentrations a few hundred times as concentrated as the influent. In the tests, filtration limited process performance more so than sorption, although this outcome is not generalizable. The limiting factor would certainly depend on the specific chemical composition of the influent solution.

The iron oxide surface has a strong pH-reversible affinity for the metals. The fact that the oxide is physically attached to sand grains rather than a suspended material makes it much easier to work with, and allows the filtration of particulate metal contaminants to proceed simultaneously with adsorption from the soluble phase. Finally, the process is applicable to anionic metals such as chromate, selenite, and arsenate, which are difficult to precipitate efficiently.

In the University of Washington process, sand can be coated using a few different procedures. All involve using an iron nitrate or iron chloride salt (as the source of the iron), sand, heat, and, in some cases, base (sodium hydroxide). The resulting ferrihydrite-coated sand is insoluble at pH above about 1, thus acidic solutions can be used in the regeneration step to ensure complete metal recovery. There has been no apparent loss of treatment efficiency after numerous regeneration cycles. Anionic metals (such as arsenate, chromate, and selenite) can be removed from the solution by treating it at a pH near 4 and regenerating it at a high pH.

The advantages this technology has over conventional treatment technologies are that it (1) acts as a filter to remove both dissolved and suspended contaminant from the waste stream, (2) removes a variety of complex metals, (3) works in the presence of high concentrations of background ions, and (4) removes anions.

Devoe–Holbein Technology: Devoe–Holbein Technology uses coordinating compounds covalently bonded to the surface of an inert carrier material to capture metal ions. In waste treatment applications, the reactants are used in equipment similar to that employed for ion exchange resins.

The technology was originally developed by DeVoe–Holbein as an adaptation of biological mechanisms in which living cells selectively extract a variety of metal nutrients (e.g., Na, K, Mg, Ca, Cu, Zn, Co, Fe, Se, and Mn) from their environment. Cells can acquire target metals by means of specialized molecular sites on their surfaces that recognize and bind only that species. Examples of such selective reactants are the nonprotein iron–binding molecules, collectively known as siderophores.

DeVoe–Holbein has synthesized a series of metal–capturing compositions with catechol, or substituted catechols, as the active component. Such compositions have similar properties to those of *Enterobactin.* Catechol was covalently bound to solid surfaces with bifunctional linking agents of defined lengths. Highly porous glass is the solid substrate which has been found to be most practical for the composition synthesis.

According to DeVoe–Holbein, the resulting compositions proved to be highly efficient, typically achieving 99% or higher removal rates, and are selective for individual or groups of metals. The rapid adsorption kinetics minimizes required contact time and the compounds are mechanically and chemically stable. In addition, the compositions are regenerable, requiring only small volumes of regenerant.

The synthesized compounds were employed for waste treatment applications in a manner similar to classical ion exchange. The media is contained in a fixed bed, and the metal–laden solution is passed through the bed during the service cycle. Following saturation of the media with metals, the bed is backwashed and the bound metal is displaced by an appropriate regenerant. Co–current and counter–current fixed bed systems have been developed. The basic modular system which is now commercially available, can be expanded or realigned to correspond to end users' varying throughput requirements and spatial limitations.

DeVoe–Holbein adsorption units are only able to treat contaminants in solution. Similar to ion exchange, high concentrations of suspended solids which can foul the adsorbent bed are typically pretreated through some form of filtration. Waste streams from the adsorption process include: contaminated regenerant and filtered solids from the pretreatment system. Filtrate from the pre–filtering system can generally be land disposed without further treatment. The regenerant may require treatment (e.g., neutralization, precipitation, dewatering) and disposal if not amenable to recycling.

One of the reported advantages of the DeVoe–Holbein system is that it is capable of yielding a more highly concentrated regenerant than ion exchange. Several options for downstream utilization of the concentrated metal regenerant are therefore possible. When it is compatible with the parent solution bath and metal concentrations are sufficiently high, the regenerant stream may be reused directly. If higher metal concentrations are required, an intermediate recovery step can be employed. For example, metal may be recovered from the regenerant electrolytically, recycling the regenerant to the adsorption process and selling the metal as scrap.

6.2.4 Biologically Activated Systems

Powdered Activated Carbon Treatment (PACT™): Powdered activated carbon treatment is a relatively new process in which powdered activated carbon is added to the aeration basin of an activated-sludge facility. The physical adsorption properties of the powdered activated carbon (ranging in concentration from 1,000 to 10,000 mg/ in the aeration basin) enhances removal of hazardous wastes. Full-scale PACT systems have been reported in the literature.

A number of advantages have been reported for this combined physical/biological process. These include the removal of non-biodegradable organics, reduced emission of organics to the air, particularly during the period of acclimation, better settling properties of the biomass/powdered activated carbon sludge, and protection of the microbial population from toxic shocks. In addition, the powdered activated carbon helps reduce effluent concentrations of organics during the acclimation period. Some evidence also exists for the ability of microorganisms to bioregenerate the powdered activated carbon during periods of low organic loading. The major disadvantage of the system is cost, which may be greater than those of other bioremediation processes.

Just as for the conventional activated-sludge process, emissions of VOCs to the air are likely to occur from the aeration basin, and possibly from the settling tank. Data have not yet been located on the magnitude of those emissions. Qualitatively, it can be assumed to be smaller than for the activated-sludge process.

Fluidized Beds (Expanded Beds): In fluidized-bed reactors, solid material, which is colonized by microorganisms is suspended by water flowing upward through the tank. The solid material is either inert (e.g., sand, coal, or plastic) or active (i.e., granular activated carbon). Both aerobic and anaerobic types of fluidized beds are in use or under investigation. In aerobic systems, air is diffused from the bottom through the bed.

There are several advantages that fluidized beds have over packed beds. Because gas bubbles can pass through the bed easily, smaller particles can be used. The use of smaller particles results in a larger biofilm surface area, which can handle higher organic loading. Also, the beds expand rather than clog as the biofilm grows. Growth can easily be controlled by removing particles from the top of the bed, washing them, and returning the cleaned particles to the reactor bottom.

Fluidized beds which use granular activated carbon as the solid material are referred to as biological activated-carbon systems. In these systems, adsorption and biodegradation of organics occurs simultaneously. This has advantages over a design in which the mechanisms occur in series. The granular activated carbon protects the system from organic shock loads, extends the retention time of the less readily biodegraded organics and adsorbs refractory compounds, just as it does in the powdered activated-carbon process. The microorganisms not only degrade the organics in the waste liquid, but also have been shown to regenerate the granular activated carbon. A major drawback to the use of biological activated-carbon systems is the large capital investment in granular carbon required.

No information has been located on emissions of organic vapors from fluidized-bed reactors. For anaerobic systems, they are not assumed to be significant. For aerobic reactors, using diffused–air systems, air emissions are likely to be important.

Soil Filter Beds (Biofilters): Soil filter beds rely on the mechanisms of adsorption, mechanical filtration, and aerobic biodegradation to treat air streams passing through the soil. The process has been used since the 1950s to control odors, such as from wastewater treatment plants or stock farms. Recently, particularly in Europe, they are being used to remove toxic vapors from industrial gas streams or from contaminant recovery systems used in environmental cleanup. The use of soil filter beds in the treatment of hazardous wastes is a new technology.

A typical soil filter bed has five major components: (1) an air blower to push the waste air stream through the bed, (2) a system to distribute the air uniformly through the bed, (3) a system to maintain proper moisture content, which is crucial for removal efficiency, (4) a medium, either soil or peat compost, for microbial growth, and (5) an underdrain system to remove excess water. The soil bed is normally three to ten feet deep. Sewage sludge can also be incorporated into the soil or compost medium as an inoculum.

Work done thus far with soil filter beds indicates that they are simple to use, inexpensive, reliable and self–regenerating. They can be used intermittently. Microorganisms have been reported to survive two weeks between loadings. In addition, the low cost and small space requirements of biofilters make them a promising bioremediation technology, particularly for use by small industries.

Post–Ozonized Wastewater: An interesting development in carbon technology is its use after the wastewater is ozonized. This combination (known as Bacteriologically Activate Carbon or BAC) has proved effective in treating otherwise biologically inactive organic compounds. The process involves chemical modification of the organics by the ozone. Maintenance of an aerobic region on the carbon allows a biologically activated film to develop and the modified organics are further treated by a mixed process of biological oxidation and carbon adsorption. The system has the advantage of being a potential add–on to existing BPT systems, and should be cost effective since it has been found that the carbon only needs regeneration at infrequent intervals.

Expanded–Bed Bioreactor: The objective of an EPA–funded study was to examine the effectiveness of the anaerobic Granulated Activated Carbon (GAC) expanded–bed bioreactor as a pretreatment unit for the detoxification of a simulated high strength industrial wastewater containing several volatile RCRA compounds present in backgrounds consisting of non–RCRA organic compounds. As a pretreatment unit, the goal was not to maximize COD destruction but to reduce the VOC concentrations to acceptable levels. This goal was achieved very satisfactorily. The reactor demonstrated excellent treatment; removals of greater than 97% were achieved for all the VOCs. Chloroform was found to be inhibitory to the system at effluent concentrations of about 100 μg/ . It was found to inhibit the degradation of acetate and acetone, two of the three base flow organic compounds. Chloroform itself, however, was removed to greater than 97%. The only limiting factor in this treatment study was the high effluent COD experienced during the inhibitory phase, which was composed almost entirely of acetate and acetone and, as such, should easily be removed by any of several treatment options. The amount of stripping occurring was negligible compared to the amount of stripping anticipated to occur in an aerobic biological process. The anaerobic GAC expanded–bed bioreactor represents an excellent pretreatment unit for the treatment of wastes containing VOCs.

6.2.5 Activated Alumina

Activated alumina adsorbs arsenic and fluorides. Alumina's removal efficiency depends on the wastewater characteristics. High concentrations of alkalinity or chloride and high pH reduce activated alumina's capacity to adsorb. This reduction in adsorptive capacity is due to the alkalinity causing (e.g., hydroxides, carbonates, etc.), and chlorine anions competing with arsenic and fluoride ions for removal sites on the alumina.

While chemical precipitation can reduce fluoride to less than 14 mg/ by formation of calcium fluoride, activated alumina can reduce fluoride levels to below 1.0 mg/ on a long–term basis. An initial concentration of 30 mg/ of fluoride can be reduced by as much as 85 to 99+%. Influent arsenic concentrations of 0.3 to 10 mg/ can be reduced by 85 to 99+%. However some complex forms of fluoride are not removed by activated alumina. Caustic, sulfuric acid, hydrochloric acid, and alum are used to chemically regenerate activated alumina.

Activated alumina has been used at potable water treatment plants for many years. Furthermore, the equipment is similar to that found in ion–exchange water softening plants which are commonly used in industry to prepare boiler water.

High capital and operation costs generally limit the wide application of this process in industrial applications.

Activated alumina can be used for odor control. Several commercial products are available which consist of dry pellets of activated alumina impregnated with potassium permanganate. Odorous compounds are adsorbed into the surface of the pellets and are subsequently oxidized by the potassium permanganate.

The NOXSO SO_2/NO_x removal process is a dry, regenerable system capable of removing both SO_2 and NO_x in flue gas from coal–fired utility boilers. In the basic process, the flue gas passes through a fluidized–bed adsorber located downstream of the precipitator; the SO_2 and NO_x are adsorbed by the sorbent. The sorbent consists of spherical beads of high–surface–area alumina impregnated with sodium carbonate. The cleaned flue gas then passes to the stack.

6.2.6 Peat Adsorption

Peat moss is a complex natural organic material containing lignin and cellulose as major constituents. These constituents, particularly lignin, bear polar functional groups, such as alcohols, aldehydes, ketones, acids, phenolic hydroxides, and ethers, that can be involved in chemical bonding. Because of the polar nature of the material, its adsorption of dissolved solids such as transition metals and polar organic molecules is quite high. These properties have led to the use of peat as an agent for the purification of industrial wastewater.

Peat adsorption is a "polishing" process which can achieve very low effluent concentrations for several pollutants. If the concentrations of pollutants are above 10 mg/, then peat adsorption must be preceded by pH adjustment for metals precipitation and subsequent clarification. Pretreatment is also required for chromium wastes using ferric chloride and sodium sulfide. The wastewater is then pumped into a large metal chamber called a kier which contains a layer of peat through which the waste stream passes. The water flows to a second kier for further adsorption. The wastewater is then ready for discharge. This system may be automated or manually operated.

Peat adsorption can be used in nonferrous metals manufacturing for removal of residual dissolved metals from clarifier effluent. Peat moss may be used to treat wastewaters containing heavy metals such as mercury, cadmium, zinc, copper, iron, nickel, chromium, and lead, as well as organic matter such as oil, detergents, and dyes. Peat adsorption is currently used commercially at a textile plant, a newsprint facility, and a metal reclamation operation.

The major advantages of the system include its ability to yield low pollutant concentrations, its broad scope in terms of the pollutants eliminated, and its capacity to accept wide variations of wastewater composition.

Limitations include the cost of purchasing, storing, and disposing of the peat moss; the necessity for regular replacement of the peat may lead to high operation and maintenance costs. Also, the pH adjustment must be altered according to the composition of the waste stream.

6.2.7 Permeable Treatment Beds

A trench, excavated down to a confining layer, is filled with adsorbent or chemical treatment material, such as activated carbon, diatomaceous earth, fly ash, zeolites, lime or sodium carbonate (to raise pH). Contaminated groundwater is treated as it percolates through the beds. Beds must be sufficiently permeable to allow passage of groundwater. Bed pores may clog up, beds require renovation or replacement.

6.3 AIR SPARGING

Air sparging, in conjunction with soil vapor extraction, can effectively remove volatile organic compounds (VOCs) from the saturated and unsaturated (vadose) zones.

Air sparging wells are used to force air into soil or strata below the water table. When contaminants dissolved in the groundwater and adsorbed onto soil come into contact with the air, they volatilize into the air phase. The stripped contaminants are then drawn upward through the vadose zone and are ultimately collected by a vapor extraction system. Soil vapor extraction is discussed in Section 6.13.

Hydrocarbon contamination can exist in five forms: (1) free phase product floating on the water table/ (2) dissolved phase in the groundwater; (3) vapor phase in the soil above the water table; (4) sorbed phase on the soils above the water table; and (5) sorbed phase on the soils below the water table.

The technologies most commonly used, pump–and–treat, and vapor extraction, fail to adequately address the fifth phase, contaminated soils below the water table. Air sparging can remediate this fifth phase.

An air compressor forces the air into the ground, and to enhance vaporization, the air may be heated prior to injection.

Air sparging, also called "in situ air stripping" and "in situ volatilization," is a technology utilized to remove VOCs from the subsurface saturated zone. It introduces contaminant–free air into an impacted acquifer system, forcing contaminants to transfer from subsurface soil and groundwater into sparged air bubbles. The air bubbles are then transported into soil pore spaces in the unsaturated zone where they can be removed by soil vapor extraction (SVE).

Air sparging systems must operate in tandem with SVE systems that capture volatile contaminants stripped from the saturated zone. Using air sparging without accompanying SVE could create a net–positive, subsurface pressure extending contaminant migration to as–yet–unaffected areas. Thus, the treatment could increase the overall zone of contamination. Without SVE, uncontrolled contaminated soil vapor could also flow into buildings (i.e., basements) or utility conduits (i.e., sewers), creating potential explosion or health hazards.

6.3.1 Remediation Mechanisms

The SVE system alone may affect the rate of volatilization of VOCs from the saturated zone. However, transport of immiscible contaminants from the saturated zone to the vadose zone necessitates channeling them to the air/water interface for removal by an SVE system. Thus, the rate of contaminant transport from groundwater to soil vapor phase has increased with the addition of air sparging to an SVE system.

The effectiveness of combined SVE/air sparging systems results from two major mechanisms: contaminant mass transport and biodegradation. Depending on the system configuration, the operating parameters, and contaminant types found on–site, one mechanism usually predominates. In both remediation mechanisms, oxygen transport in the saturated and unsaturated zones plays a key role. Although the exact nature of the saturated zone vapor phase is not completely understood, sparging seems to create air bubbles, which move through the groundwater to the unsaturated soil, like bubbles in an aeration basin. Other theories trace the movement of air through irregular pathways in the saturated zone and, ultimately, to the surface of the water table. These theories suggest that the air would move as pockets through soil pathways, rather than forming bubbles, because groundwater travels in a porous medium.

The nature of air transport affects mass transfer to and from the groundwater regime. Bubbles exhibit higher surface area for transfer of oxygen to the groundwater and for volatile migration to the unsaturated zone, than the area provided by continuous, irregular air–flow pathways.

Mass Transfer: Mass transfer employs several mechanisms that move contaminants from saturated zone groundwater to unsaturated soil vapors, including (a) dissolving soil–sorbed contaminants from the saturated zone to groundwater; (b) displacing water in soil pore spaces by introducing air; (c) causing soil contaminants to desorb; (d) volatilizing them, and (e) enabling them to enter the saturated zone vapor phase. Due to the density difference between air and water, the sparged air migrates upwards in the aquifer. The pressure gradient resulting from the creation of a vacuum in the unsaturated zone pulls the contaminant vapors toward and into the SVE wells.

The action of the air passing through the saturated zone in response to sparging leads to turbulence and mixing of the groundwater. This in turn increases the rate at which contaminants adsorbed to the saturated zone soils dissolve into the groundwater. Light non–aqueous phase liquids (LNAPLs) floating on the water table are also subject to increased rate of transfer to the unsaturated zone because they are volatilized by the air sparging process.

In summary, air sparging increases the speed at which the following occur:

1. Volatilization of contaminants from the groundwater to the vadose zone;
2. Desorption and dissolution of adsorbed contaminants from the soil into the groundwater; and
3. Dissolution of NAPLs due to mechanical mixing.

The mass transfer of contaminants may be further enhanced by heating the air prior to sparging. The increase in air temperature will increase the rate of volatilization of contaminants.

Biodegradation Mechanism: Aerobic biodegradation of contaminants by indigenous microorganisms requires the presence of a carbon source, nutrients, and oxygen. Air sparging increases the oxygen content of the groundwater thus enhancing aerobic biodegradation of contaminants in the subsurface. Certain organic contaminants, such as petroleum constituents, serve as a carbon source for microorganisms under naturally occurring conditions. The rate of biodegradation can be enhanced by optimizing nutrient status of the system.

Remediation of an aquifer via the biodegradation mechanism has distinct advantages since a portion of the contaminants will be biologically degraded to carbon dioxide, water, and biomass—yielding a lower level of VOCs in the extracted air. This in turn can substantially reduce vapor treatment costs. The possibility of off–site contaminant vapor migration is also reduced when sparged vapors entering the vadose zone contain lower levels of contaminants.

Certain contaminants, such as chlorinated solvents, can undergo biodegradation under anaerobic conditions. Air sparging, in these instances, could adversely affect this biodegradation process.

6.3.2 Technology Applicability

Although air sparging is a relatively new technology for contaminated subsurface soil remediation, it has been applied at hundreds of sites in the United States and Europe since 1985. However, the design of these systems has been, for the most part, empirically based. The effectiveness of air sparging depends on various site conditions.

Depth to Groundwater: Air sparging has been effective in an aquifer 150 ft below surface. There appears to be no depth limit at which air sparging would not be effective, but significant cost implications may accompany the installation of an air sparging system in a very deep aquifer. However, a water table located at a shallow depth (<5 ft), may increase the difficulty of recovering vapors with SVE. It could release VOC emissions to the atmosphere. Capping such a site with pavement or other impervious material might reduce atmospheric emissions.

Volatility of Contaminants: Enhancing mass transfer of contaminants from the soil and groundwater into the vapor phase, a key mechanism of the air sparging process, requires highly volatile contaminants. Volatility is directly related to the Henry's Law Constant of a compound and its vapor pressure—the higher the Henry's Law constant, the higher the volatility. In general, compounds which are effectively removed from contaminated water by air stripping are sufficiently volatile for adequate air sparging treatment. Compounds with Henry's Law Constants of 10^5 atm–m^3/mol or greater can be air stripped or sparged. Due to their high volatility, petroleum compounds (e.g., benzene and toluene), and solvents (e.g., trichloroethylene) are very amenable to air sparging technology.

Solubility of Contaminants: The solubility of a contaminant in water determines its ability to be stripped by air sparging. In general, the more soluble a contaminant is in water, the greater the difficulty there is in using air sparging.

Biodegradability of Contaminants: Since biodegradation is enhanced by air sparging, compounds that are readily aerobically degraded are amenable to remediation by air sparging. Biodegradation of petroleum hydrocarbons, such as those found in gasoline and diesel leaks from USTs, has been significantly increased with air sparging. Prior to designing an air sparging system for bioremediation, electrolytic respirometry should be used to analyze samples of the soils and groundwater. This will make it possible to gauge the effectiveness of the indigenous microorganisms and their energy sources to metabolize the petroleum hydrocarbons.

Soil Permeability: Soil permeability, which measures the ease of fluid flow through the soil column, is a critical parameter in the design of air sparging systems. Injected air must flow freely throughout the aquifer to achieve adequate removal rates. In most aquifers, horizontal permeability is greater, by a factor of ten, than vertical permeability. Successful sparging systems require air flow in both horizontal and vertical directions. Vertical flow is particularly important since the contaminant must migrate to the vadose zone for removal by SVE.

If the geology restricts the vertical flow, contaminants may migrate laterally into previously uncontaminated areas. Hydraulic conductivity of 0.001 cm/sec or greater is required to obtain sufficient subsurface air flow. Bench-scale experiments have shown coarse sand (d_{50} = 0.8 mm) forming the dividing line between soils, which permits injected air to rise by hydraulic uplift alone from soil that required additional pressure to inject air and through which air escaped at only a few points.

Due to the heterogeneity of soils at all sites, it may be necessary to concentrate wells in areas with lower permeability. The spacing of the wells depends on the radius of influence. In general, highly permeable soils will have larger radii of influence and higher air flow rates than lower permeable soils. Screen placement requires a good understanding of the stratigraphy of a site. Well layout should overlap the radii of influence. This will ensure the treatment of all soil areas.

Clogging of the injection well screen or the aquifer in the vicinity of the sparging wells could reduce permeability and, therefore, decrease the effectiveness of the method. Clogging may result from enhanced bacterial growth under increased oxygen levels. In addition, oxidation at sites with high iron and manganese levels could cause further clogging. Some applications have injected nitrogen instead of ambient air to minimize problems associated with fouling. However, the use of nitrogen also prevents the enhancement of aerobic biodegradation.

Confining Layers: Some air sparging proponents point out that it can only achieve success at sites with water table (i.e., unconfined) aquifers. Confined aquifers, where a low permeability layer lies above the water-bearing zone, would inhibit the flow of air upward from the saturated zone to the vadose zone. The injected air in these situations would flow radially away from the injection point; the vapor extraction system would not recover it. Such a situation could build up pressure in the aquifer.

For unconfined aquifers, stratigraphic layers with different permeabilities will also affect air and water flow patterns as well as influence the air sparging system. In such situations, optimal air flow will occur in the more permeable zones. Air flow may travel horizontally away from the injection point and create a wider zone of influence than would otherwise be expected.

Soil Characteristics Air sparging systems are most applicable for sites with sandy soil, due to its permeability. Soil containing a large organic carbon fraction may impede the desorption of volatile organic contaminants, thus reducing air sparging effectiveness. In extraction wells, the presence of a large amount of monomers in the soil may cause clogging of well screens possibly due to polymerization.

Presence of LNAPL: Low-density (or light) nonaqueous phase liquids (LNAPL) floating on the water table presents a particular problem during air sparging. The air sparging action creates a mounding effect in the proximity of the sparge well. In sites with steep hydraulic gradients, this mounding effect may be sufficient to move a plume of LNAPL, possibly contaminating clean areas. While it is possible to prevent the plume movement by modulating the sparged air pressure, it is more important to recover the mobile portion of the LNAPL to a residual saturation phase.

Contamination in Bedrock Aquifer: The effectiveness of air sparging hinges on the mass transfer of air to the groundwater and movement of the contaminants vapor through the saturated zone upward into the unsaturated zone where they can be extracted. Unless the rock formation is highly fractured, with fractures vertically oriented, this technology will not provide sufficient mass transfer to effectively remediate a bedrock aquifer.

Metals in Groundwater: In addition to the possibilities of clogged well screens resulting from oxidation of metals in groundwater and the growth of bacteria, precipitation of metals can also be an inhibiting factor. Since ambient air contains carbon dioxide, calcium carbonate precipitation may occur in some aquifers during air sparging. This may also reduce the air flow through the system.

Contaminant Location: Air sparging targets contaminants in the saturated zone and the capillary fringe. For compounds with a density less than water such as many petroleum constituents, much of the contamination may lie in the capillary fringe and just below the water table, depending on such factors as water table fluctuations, the amount of product released, contaminant density, and contaminant solubility. Dense non–aqueous phase liquids (DNAPL), such as trichloroethylene, often migrate through the aquifer to a lower confining unit and to greater depths. For dissolved contaminants in the aqueous phase, groundwater flow and direction will control the distribution of contaminants throughout the site. Depending on soil characteristics, air sparging would remediate DNAPL–contaminated soil as well.

Combination with Other Technologies: Air sparging is always used in conjunction with SVE. The implementation of SVE addresses the vadose zone contamination, and incorporates air sparging wells to treat saturated zones.

Groundwater extraction at air sparging sites may serve as a hydraulic control. Injected air may mobilize contaminants adsorbed to soil, either by displacement from the soil matrix or through increased dissolution of the adsorbed contaminant into the groundwater during mixing caused by air injection. If this occurs and the rate of volatilization is insufficient, downgradient groundwater concentrations could actually increase. Air sparging may have fallen into disfavor in Germany due to increased downgradient dissolved contamination. To prevent this situation, a groundwater pumping system could hydraulically contain the site groundwater flow.

6.4 CONDENSATION

Condensation is used as a control technique for some organic compounds. It cools the gas stream and transforms the gaseous compound to a liquid. Like absorption, condensation is one of the primary techniques used for product recovery; however, it is also used as an air–pollution–control device. Control of storage and process emissions is a common application. Condensers are frequently used in series with other control equipment, including absorbers, incinerators, and adsorbers.

Two common methods of condensation are:

1. Lowering the temperature of the air stream to below the contaminants' dew point is the most common method. Either refrigeration or cryogenic processes can be used to lower the temperature of the gas stream.
2. Increasing the pressure of the gas stream while holding the temperature constant, thereby forcing the contaminants to a liquid state, is the second method of condensation.

Condensation is a process of converting all or part of the condensible components of a vapor phase into a liquid phase. This is achieved by the transfer of heat from the vapor phase to a cooling medium, If only a part of the vapor phase is condensed, the newly formed liquid phase and the remaining vapor phase will be in equilibrium. In this case, equilibrium relationships at the operating temperatures must be considered. The heat removed from the vapor phase should be sufficient to lower the vapor phase temperature to at or below its dewpoint temperature (i.e., the temperature at which the first drop of liquid is formed).

Condensation devices are of two types: surface condensers and contact condensers. Surface condensers are shell–and–tube type heat exchangers. The coolant and the vapor phases are separated by the tube wall and they never come in direct contact with each other. Surface condensers require more auxiliary equipment for operation but can recover valuable VOC without contamination by the coolant, minimizing waste disposal problems. Only surface condensers are considered in the discussion of control efficiency and applicability since they are used more frequently in industry.

The major equipment components used in a typical surface condenser system for VOC removal includes: (1) shell and tube dehumidification equipment, (2) shell–and–tube heat exchanger, (3) refrigeration unit, and (4) VOC storage tanks and operating pumps. Most surface condensers use a shell–and–tube type heat exchanger to remove heat from the vapor. As the coolant passes through the tubes, the VOC vapors condense outside the tubes and are recovered. The coolant used depends upon the saturation temperature of the VOC stream. Chilled water can be used down to 7°C (45°F), brines to –34°C (–30°F), and chlorofluorocarbons below –34°C (–30°F). Temperatures as low as –62°C (–80°F) may be necessary to condense some VOC streams.

VOC removal efficiency of a condenser is dependent upon the type of vapor stream entering the condenser, and on condenser operating parameters (flowrate and temperature of the cooling medium). High VOC removal efficiencies are achievable for condensers, but the design and operation of condensers for large heat removals from dilute VOC streams may be costly. Efficiencies of condensers in actual operation usually vary from

50 to 95%.

In one study a primary condenser system was used in 19 out of 66 units with vent streams (about 29%). In some cases, additional (secondary) condensers are used to recover more VOC from the vent stream exiting the primary condenser. Condensers are sometimes present as accessories to vacuum generating devices (e.g., barometric condensers).

Condenser systems are not well suited for vent streams containing VOC with low boiling points. In addition, condensers are not well suited for vent streams with low concentrations of VOC, such as streams containing large quantities of inerts such as carbon dioxide, air, or nitrogen. Low boiling point VOC and inerts contribute significantly to the heat load that must be removed from the vent stream, resulting in costly design specifications and/or operating costs. In addition, some low boiling point VOC cannot be condensed at normal operating temperatures. For example, process units producing chlorinated methanes have vent streams with substantial amounts of methane, methyl chloride, and methylene chloride. These compounds are not readily condensed and, as a result, are usually destroyed in a combustion device.

A primary condenser system is usually an integral part of distillation operations. These condensers are needed to provide reflux in fractionating columns and to recover distilled products. At times, additional (secondary) condensers are used to recover more organics from the vent stream exiting the primary condenser. Condensers are sometimes present as accessories to vacuum-generating devices (e.g., barometric condensers).

The use of a condenser to control organic emissions may not be applicable to some process vent streams. Secondary condensers used as supplemental product recovery devices are not well suited for vent streams containing organics with low boiling points or for vent streams containing large quantities of inerts such as carbon dioxide, air, and nitrogen. Low boilers and inerts cannot be condensed at normal operating temperatures, and they usually carry over some organics. For example, condensation is not generally considered effective for process vents on air stripping units and other streams that contain less than 10,000 ppm organics.

6.5 DISTILLATION

Distillation is simply the process of evaporation followed by condensation whereby separation of volatile materials can be optimized by controlling both the evaporation-stage temperature (and pressure) and the condenser temperature. Distillation separates miscible organic liquids for solvent reclamation and for waste volume reduction. The resulting residuals are still-bottoms (often containing toxic metals from ink and paint pigments) and intermediate distillate cuts. Two major types of distillation processes are batch distillation and continuous fractional distillation.

The distillation process is used to recover solvents and chemicals from industrial wastes that otherwise would be destroyed by waste treatment. Although effective in recovering solvents and other organic compounds for recycle and reuse, distillation is not a widespread wastewater treatment technology because effluent levels that are acceptable for recycle and reuse are generally too high for wastewater discharge.

Distillation is a thermal treatment technology applicable to the treatment of wastes containing organics that are volatile enough to be removed by the application of heat.

Constituents that are not volatilized may be reused or incinerated as appropriate. The four most common distillation processes are batch distillation, fractionation, steam stripping, and thin film evaporation. Steam stripping is discussed later in this chapter.

Batch distillation can be used to treat wastes having a relatively high percentage of volatile organics. In general, batch distillation is applied to spent solvent wastes where the wastes are highly concentrated in the solvent and yield significant amounts of recoverable materials upon separation. Batch distillation is particularly applicable for wastes that have both very volatile and very nonvolatile components since the separation of that combination of components is amenable to the relatively unsophisticated batch distillation equipment.

Fractionation is typically applied to wastes containing greater than about 7% organics. It is designed to achieve the highest degree of distillate purity of the separated components. Fractionation can be operated to produce multiple product streams for recovery of more than one organic constituent from a waste while generating relatively small amounts of residue to be land disposed. In general, this technology is used where recovery of multiple constituents is desired and where the waste contains minimal amounts of suspended solids.

Steam stripping is a form of distillation applicable to the treatment of wastewaters containing organics that are volatile enough to be removed by the application of heat using steam as the heat source. Typically, steam stripping is applied where the waste contains less than 1% volatile organics. This process is discussed in a later section.

Thin film evaporation is typically applied to wastes containing greater than 40% organics. However, the feed stream to the thin film evaporator must contain low concentrations of suspended solids.

6.5.1 Principles of Operation

The basic principle of operation for distillation processes, i.e., batch distillation, fractionation, thin film evaporation, and steam stripping, is the volatilization of hazardous components through the application of heat. The components that are volatilized are then condensed and typically either reused or further treated, usually by liquid injection incineration. In thermal drying, the basic principle of operation for drying is the removal of a liquid from a solid waste by evaporation. This is similar to distillation in that volatilization of organic constituents also occurs. However, the primary purpose of thermal drying is to volatilize water. Liquid constituents will vaporize as a result of applied heat. In thermal drying, the rate at which liquid evaporation occurs depends on the thermal conductivity of the solid waste to be dried and the boiling points of the volatile liquid constituents to be evaporated.

An integral part of the theory of the distillation process is the principle of vapor–liquid equilibrium. When a liquid mixture of two or more components is heated, the vapor phase present above the liquid phase becomes more concentrated in the more volatile constituents (those having higher vapor pressures). The vapor phase above the liquid phase is then removed and cooled to yield a condensate that is also more concentrated in the more volatile components. The degree of separation of components depends on the relative differences in the vapor pressures of the constituents; the larger the difference in the vapor pressures, the more easily the separation can be accomplished.

If the difference between the vapor pressures is extremely large, a single separation cycle or a single "equilibrium stage" of vaporization and condensation may achieve a significant separation of the constituents. In such cases, batch distillation or thin film evaporation would be used. Typically, batch distillation units and thin film evaporation units contain only one equilibrium stage and are thus limited in the degree of separation by the relative volatilities of the constituents.

If the difference between the vapor pressures of volatile components is small, then multiple equilibrium stages are needed to achieve effective separation. In practice, the multiple equilibrium stages are obtained by stacking "trays" or placing "packing" into a column. Essentially, each tray represents one equilibrium stage. In a packed steam stripping column or fractionation column, the individual equilibrium stages are not discernible, but the number of equivalent trays can be calculated from mathematical relationships. The vapor phase from a tray rises to the tray above it, where it condenses; the liquid phase falls to the tray below it, where it is again heated and separated.

The vapor–liquid equilibrium of the waste components can be expressed as relative volatility, which is the ratio of the vapor–to–liquid concentrations of a constituent divided by the ratio of the vapor–to–liquid concentrations of another constituent. The relative volatility is a direct indicator of the ease of separation. If the numerical value is 1, then separation is impossible because the constituents have the same concentrations in the vapor and liquid phases. When the relative volatility is 1, the liquid mixture is called an azeotrope. Separation becomes increasingly easier as the value of the relative volatility becomes increasingly different from unity.

6.5.2 Batch Distillation

A batch distillation unit usually consists of a steam–jacketed vessel, a condenser, and a product receiver. As the name implies, it is a batch process, not a continuous process. The steam jacket provides the heat required to vaporize the volatile constituents in the liquid fraction of the waste. The rising vapor is collected in the condenser, cooled, and condensed. The liquid product stream is then routed to the product receiver.

It is important to note that this technology treats wastes by vaporizing constituents, not destroying them. Accordingly, an integral part of this technology is a condensation system to collect the organics, as well as an air emission control system to collect or destroy those organics that are not condensed. The cooling load of the condenser is calculated in the design to ensure that the product recovery rate is maximized and emissions from condenser venting are minimized. The "bottoms," which are the least volatile constituents of the waste, are withdrawn from the bottom of the batch still. Because batch distillation is used to remove the volatile organics from wastes, the bottoms are reduced in volatile organic content. However, prior to disposal, the bottoms generally require additional treatment, such as incineration, for residual, less volatile organics.

6.5.3 Fractionation

Fractionation is a continuous process conducted in a unit that consists of a reboiler, a column containing stripping and rectification sections, a condenser, and a "reflux" system. The reboiler is a device that provides the heat required to vaporize the liquid fraction of the waste. It supplies enough heat to maintain the liquid in the column at its boiling point. The stripping and rectifying sections are composed of a set of trays in a vertical column. The discrete trays may be replaced by loose packing consisting of plastic, metal, or ceramic geometric shapes that provide surface area for the continuous boiling/condensing that takes place in the column. In the stripping section, vapor rising from the boiler is contacted with the downflowing liquid feed. Through this contact, the constituents with lower boiling points (i.e., those that are more volatile) are concentrated in the vapor. In the rectification section, the vapor rising above the feed tray is contacted with downflowing condensed liquid product (reflux). Through this contact, the vapor is further enriched in the constituents with lower boiling points (i.e., the more volatile constituents). The rising vapor is collected at the top of the column and condensed in a condenser.

The liquid product stream, except for the portion returned to the column as reflux, is then routed to a product receiver. The "bottoms," which are the least volatile components (i.e., those with the highest boiling points), are continuously withdrawn from the reboiler. Because the liquid composition varies slightly from one equilibrium stage to the next, it is also possible to withdraw streams of differing quality (sometimes called "fractions") from different locations throughout the column. This is typically done in refining petroleum, resulting in different grades or "cuts" of petroleum products.

6.5.4 Thin Film Evaporation

Thin film evaporation is a continuous process conducted in a unit that typically consists of a steam–jacketed cylindrical vessel and a condenser. The steam–heated surface of the cylindrical vessel provides the heat required to vaporize the volatile constituents in the waste. The evaporator walls are heated from the outside as the feed trickles down the inside walls in a thin film. Unique to this form of distillation is the distribution device that spreads the thin film over the heated surface. The feed rate of waste is controlled to allow the more volatile material adequate time to vaporize. The heat transfer from the heating medium (steam) to the waste is determined by their relative temperatures, the heat transfer rate of the vessel materials, and the thermal properties of the waste stream forming the film. The rising vapor is collected at the top of the column, cooled, and condensed in a condenser. The condensed liquid product stream is then routed to a product receiver. The "bottoms," which are the least volatile components of the waste, are continuously withdrawn from the bottom of the thin film evaporator. Because thin film evaporation is used to remove the volatile organics from wastes, the bottoms are reduced in volatile organic content. However, prior to disposal, the bottoms usually require additional treatment, such as incineration, for residual, less volatile organics.

6.5.5 Metal Finishing Applications

Evaporation/distillation in the metal finishing industry is one of the oldest recovery techniques, and is widely used in industry. Over 600 units are currently in operation in the United States. They are most commonly used in metal finishing and electroplating industries to recover plating solutions, chromic acid and other concentrated acids, and metal cyanides. In addition, water recovered from the evaporation process is of high purity and can be reused in process waters. These systems are most effective in recovering acids, bases, and metals from rinsewaters. Systems can be designed cost–effectively with capacities ranging from 20 to 300 gph. These systems are cost–competitive with conventional neutralization and disposal technologies. Greater cost savings are realized with larger operations.

The use of evaporation/distillation systems to recover concentrated streams directly from the spent solution is limited. Pilot–scale evaporation/distillation systems for recovery of nitric/hydrofluoric acid pickling liquors have been tested at facilities in Europe. However, cost–effective systems for direct recovery of spent solutions via evaporation–distillation have not been developed at the commercial–scale for application in the United States. Other technologies, such as thermal decomposition appear to be more cost–effective for this purpose.

6.5.6 Vacuum Distillation

Gas processors are frequently faced with problems associated with accumulated impurities in gas–treating chemicals which are used to remove water vapor and acid gases such as hydrogen sulfide or carbon dioxide. The EPS (Canada) developed and demonstrated a new technology that offers the petroleum industry an opportunity to economically regenerate their waste gas treating chemicals, reduce the total waste volume, and save energy required to manufacture and dispose of process chemicals. The process employs high vacuum distillation under conditions of flow and temperature such that no risk of further degradation is allowed and the treatment chemicals can be used as makeup or fresh chemical in the gas–treating process.

Spent acid streams that contain heavy metals are one of industry's most persistent waste treatment problems.

A cost–saving method for reducing the volume and toxicity of this type of acid waste has been developed at the Department of Energy's Pacific Northwest Laboratory (PNL), operated by Battelle, and recently licensed to a private firm, Viatec/Recovery Systems Inc. This process includes vacuum distillation as part of the treatment sequence.

A typical application would involve a spent acid stream containing sulfuric, phosphoric, or nitric acids along with copper, zirconium, or nickel. In the first step, the spent acid mixture is pumped through a steam–heated boiler where it is partially vaporized. Then, in the flash separator, it is separated into liquid, which contains the heavy metals, and vapor, which contains the acid and water. The liquid is collected in a tank, where heavy metals can be removed by precipitation. The vapor is further separated in a vacuum distillation column, yielding clean acid that is suitable for recycling.

6.6 EQUALIZATION

Equalization involves the process of dampening flow and pollutant concentration variation of wastewater before subsequent downstream treatment. By reducing the variability of the raw waste loading, equalization can significantly improve the performance of downstream treatment processes that are more efficient if operated at or near uniform hydraulic, organic, and solids loading rates and that reduce effluent variability associated with slug raw waste loadings. Equalization is accomplished in a holding tank manufactured from steel or concrete, or in an unlined or lined pond. The retention time of the tank or pond should be sufficiently long to dilute the effects of any highly concentrated continuous flow or batch discharges on treatment plant performance.

Provisions should be made for mixing of the leachate to prevent the deposition of solids. Discharge pumping and flow control are also required.

Equalization tanks or basins may be designed as either in–line or side–line units. With in–line equalization, the entire daily flow passes through the basin, and leachate is discharged to the treatment plant at an essentially constant rate. With side–line equalization, only the flow above the average daily flow rate is diverted into the basin; when the flow rate falls below the daily average, leachate from the equalization basin is discharged to the treatment plant to bring the flow rate up to the average level. In–line equalization is the preferred arrangement for leachate treatment applications because contaminant concentrations as well as flow rates are equalized in this approach. Side–line systems typically only provide flow–rate equalization.

Because the composition and volume of leachate vary greatly, equalization is required to achieve optimum performance of the treatment system. When placed ahead of chemical operations in the treatment process train, equalization improves chemical feed control and process reliability. When placed ahead of biological operations, equalization minimizes shock loadings, dilutes inhibitory substances, stabilizes pH, and improves secondary settling. In plants that operate on an intermittent schedule, equalization tanks/basins double as influent storage tanks.

Few disadvantages are associated with equalization. Solids and oil and grease present in the leachate may tend to accumulate on the basin walls, but these materials can be removed by spraying the walls with water. Mixing of the tank/basin contents may strip highly volatile components from the leachate. The need for control of volatilized compounds must be determined on a site–specific basis.

Equalization is reliable from both equipment and process standpoints, and is used to increase the reliability of the flow–sensitive treatment processes that follow by reducing the variability of flow and pollutant concentrations.

6.7 EXTRACTION

Extraction technologies (i.e., solvent extraction and critical fluid extraction) are used to treat wastes containing a variety of organic constituents and a broad range of total organic content. Extraction technologies have been demonstrated for treatment of API separator sludges and other hydrocarbon–bearing wastes generated by the petroleum and petrochemicals industries. These technologies are also applicable to wastes of similar composition generated by other industries, such as the organic chemicals industry. Solvent extraction is also used in the hydrometallurgical field for the recovery of metals, and for the recovery of spent acid pickling liquors in the steel industry.

In solvent extraction, the selection of an extraction fluid (solvent) is dependent on the solubility of the organic waste constituents in the extraction fluid.

Critical fluid extraction is applicable to wastes containing organics that are soluble in pressurized fluids such as carbon dioxide, propane, butane, or pentane (extraction fluids). Compounds that have been successfully extracted from wastes by this process include aliphatic hydrocarbons, alkenes, simple aromatics such as benzene and toluene, polynuclear aromatics, and phenols.

The basic principle of operation in extraction technologies is that constituents are removed from a waste by mixing the waste with an extraction fluid (solvent) that will preferentially dissolve the waste constituents of concern from the waste. In the simplest extraction systems, two components are mixed: (a) the waste stream to be extracted, and (b) extraction fluid. The extraction fluid and waste stream are mixed to allow mass transfer of the constituent (the solute) from the waste stream to the extraction fluid.

Except for the waste constituents that are to be extracted, the waste must be immiscible in the extracted fluid, so that after mixing, the two immiscible phases can physically separate by gravity. Separation of the extraction fluid phase and the waste stream phase occurs under quiescent conditions, relying on the density differences between the two phases.

The extraction fluid, which now contains the extracted contaminants, is called the extract; the extracted waste stream from which the contaminants have been removed is called the raffinate. The extract can be either the heavy (more dense) phase or the light (less dense) phase. In theory, the maximum degree of separation that can be achieved is provided by the selectivity value, which is the ratio of the equilibrium concentration of the contaminants in the waste. The solvent extraction process can be either batch or continuous.

In critical fluid extraction, the extraction fluids used are compounds that are usually gases at ambient conditions. Since the extraction fluid being used is a gas, it is first pressurized, which converts it to a liquid. As a liquid, it leaches (dissolves) the organic constituents out of the complex waste with which it is mixed. The enhanced solubilities of various organic compounds in hydrocarbons and other extraction fluids at high pressure aid in their removal from a waste. The process is usually carried out at or near the extraction fluid's "critical pressure," which is the pressure above which the liquid form of the extraction fluid cannot be gasified, no matter how much the fluid is heated. (In fact, above the critical pressure liquid and gas phases become physically indistinguishable, in that two phases do not really exist.)

6.7.1 Solvent Extraction

Solvent extraction, also referred to as liquid–liquid extraction, involves the separation of the constituents of a liquid solution by contact with another immiscible liquid for which the impurities have a high affinity. The separation is based on physical differences that affect differential solubility between solvents and may be enhanced by adding reagents to cause a definite chemical reaction.

The end result of solvent extraction is to separate the original solution into two streams—a treated stream and a recovered solute stream (which may contain small amounts of water and solvent). Solvent extraction may thus be considered a recovery process because the solute chemicals are generally recovered for reuse of further treatment and disposal. The process for extracting a solute from solution will typically include three basic steps:

1. Mixing of solvent with waste stream;
2. Extraction and separation; and
3. Recovery of solvent from the treated stream, either by distillation or steam stripping.

Solvent extraction generates a treated wastewater residual, which is discharged, and an extract, which in some cases may be recycled and reused. Often, the process function and wastewater treatment function of solvent extraction are integrated as water contaminants are returned with the solvent to the process; in these cases, the facility often does not consider the extraction to be a treatment process, although the net result is to reduce total loading of pollutants discharged from the process. Solvent extraction is most effectively applied to segregated process streams where the potential for collecting specific residuals for reuse is greatest.

Complete separation is rarely achieved so that some form of post–treatment is required for each separated stream. To effectively recover solvent and solute from the process, other treatments are needed such as distillation or stripping.

Though not as popular as stripping or carbon adsorption processes due to higher costs, extraction is widely used to recover valuable solvents. Extraction can achieve 98% recovery. Extraction is a proven method for removing phenol, acetic acid, salicylic (and other hydroxy aromatic) acids, and petroleum oils from aqueous solutions. It is also used to recover methylene chloride from isopropyl alcohol, Freon from organic wastestreams containing oil and alcohol, and a mix of chlorinated hydrocarbon from alcohol or acetone. Many valuable organics present in moderate to high concentrations in aqueous or organic solutions can potentially be extracted (e.g., phenols, acids, alcohols, amines, glycols, tetrahydrofuran, and dimethylformamide). The major limitation to solvent extraction is the difficulty in finding a suitable solvent low in cost, high in extraction efficiency, and easily separable from the extracted substance.

Extraction Processes: The simplest, least effective solvent extraction unit is a single–stage system (mixer–settler system). The solvent and the liquid waste stream are mixed together; the raffinate and extract are separated by settling without further extraction.

The more effective multistage contact extraction is basically a series of batch mixer–settler units. The waste stream is contacted with solvent in a series of successive steps or stages. Raffinate from the first extraction stage is contacted with fresh solvent in a second stage, and so on.

In countercurrent, multistage extraction columns, fresh solvent and the waste stream continuously enter at opposite ends of a column consisting of a series of extraction stages. Extract and raffinate layers pass continuously and countercurrently from stage to stage through the system.

Several types of extraction systems are used for contact and separation. Three of the most common systems—mixer–settler systems, extraction columns, and centrifugal contactors—are discussed below.

Single–stage mixer–settler systems are extraction systems composed of a mixing chamber for phase dispersion, followed by a settling chamber for phase separation. Mixer–settler systems are typically used to treat solids or highly viscous wastes and can handle difficult–to–mix components. The vessels may be either vertical or horizontal. Dispersion in the mixing chamber occurs by pump circulation, nonmechanical in–line mixing, air agitation, or mechanical stirring. In a two–stage mixer–settler system (a simple multistage contact extractor), the extract from the first stage is sent to a recovery unit to separate the solvent from the remaining extract containing the organic constituents of concern. The recovered solvent is used, and the remaining extract is either reused or further treated in an incineration unit. The raffinate from the first stage is sent to the second–stage unit for additional extraction. Recycled solvent from recovery of the first–stage extract and/or fresh solvent makeup is added to the first–stage raffinate before it is mixed and sent to the second stage. The extract from the second stage, containing mainly solvent, is recycled to the first–stage unit as the solvent stream. The resulting raffinate from the second stage may require filtering to remove solids before it is sent to further treatment (if required). If solids collected during filtration contain treatable levels of hazardous constituents, they will require further treatment, such as stabilization (for metals) and/or incineration (for organics) prior to disposal.

Parameters such as the density or specific constituent concentrations in the extract may be monitored to determine when the second–stage extract must be sent to solvent recovery and when fresh or recycled solvent must be added to the first–stage unit.

Extraction columns are continuous flow, countercurrent, multistage contact systems. Two types of extraction columns are packed extractors and sieve–tray extractors. A packed extractor contains plastic or ceramic materials in various geometric shapes or structured packings of wire gauze or mesh. Mass transfer of the contaminants from the waste to the extract is promoted because of breakup and distortion of both phases as they contact the packing, resulting in the intimate mixing of the waste and the solvent.

The sieve–tray extractor is similar to the sieve–tray column used in fractionation distillation. Tray perforations cause the formation of liquid droplets that aid the mass transfer process by allowing for more intimate contact between the solute and the solvent.

Centrifugal contactors are based on the application of centrifugal force to increase rates of countercurrent flow and enhance separation of the phases. Centrifugal units are used when short contact times are required, such as when unstable materials are being processed. One type of centrifugal contactor consists of a drum that rotates around a shaft equipped with annular passages at each end for feed and raffinate. The light phase is injected under pressure through a shaft and is then routed to the periphery of the drum through perforations. The heavy phase is also charged through the shaft, but it is channeled to the center of the drum through perforations. Centrifugal force acting on the phase–density difference promotes dispersion as the phases are forced through the perforations. Centrifugal extractors provide short contact time, have minimal space requirements, and easily handle emulsified materials and fluids with small density differences.

Application: Integration of equipment into an overall system for successful treatment of waste streams will require considerable analysis of the waste stream of interest and the candidate processes. Liquid–liquid extractions are most useful when separations involve materials that are not easily separated by distillation or other treatment processes. Generally, liquid–liquid extractions of water streams are conducted to remove materials which have high water solubility and therefore almost invariably a low Henry's Law constant. Air or stream stripping do not appear to be viable options for wastes of this type. Liquid–liquid extractions may be particularly applicable when the relative volatilities of solute/solvent compounds make separation by distillation difficult or when high waste concentrations make carbon adsorption uneconomical. Extraction of the soluble compounds could be viable. Solvent extraction of emulsified material might also be considered provided mass transfer considerations are acceptable. This would have to be determined experimentally.

Solvent choice and design parameter options are many and varied. Although design and operation of a liquid–liquid extraction system to achieve acceptable effluent levels is theoretically possible, existing experimental and field data for halogenated and other organics indicate that most units, as presently designed and operated, fall short of this goal.

Properly designed and operated, the liquid–liquid extraction process does not appear to pose significant environmental problems. Both process exit streams contain potential contaminants that must be addressed as part of the process. The solvent will contain solute (contaminant in the feed) that must be removed if the solvent is to function adequately in recycle. The treated waste stream (assuming all significant traces of contaminant have been transferred to the solvent) could contain dissolved solvent which may or may not be significant and warrant additional treatment. Since these potential conditions are recognized and must be dealt with by system designers, the environmental impacts of a viable liquid–liquid extraction system should be minimal.

Advantages and limitations of liquid–liquid extraction processes are:

Advantages:

1. Recovery of costly materials can be accomplished with little threat of thermal decomposition or chemical interaction.
2. Recovery (separation) of materials which have similar relative volatilities or adsorption isotherms can be achieved.

Limitations:

1. Some residuals will generally be present in both the raffinate and extract streams, thus, some provision must be made for their removal and subsequent disposal.
2. Economics may not be favorable.
3. Deviations that limit the extent of removal may occur upon scale–up.

In Situ Extraction: Solvent extraction differs from soil washing in that it employs organic solvents rather than aqueous solutions to extract contaminants from the soil. Like soil washing, it is a separation process that does not destroy the contaminants. The contaminants will have greater solubility in the solvent than in the soil. The equilibrium concentration gradient drives the mass transport process such that the contaminant transfers from the soil to the solvent. When the soil is separated from the solvent, the soil contaminant concentrations are presumably lower than before contact with the solvent. Soil washing treats organic compounds much more effectively than inorganic compounds and metals. It can be used in conjunction with other processes to reduce remediation costs.

Sediments, sludges, and soils contaminated with volatile organic compounds (VOCs), polychlorinated biphenyls (PCBs), halogenated solvents, and petroleum wastes can be effectively treated with solvent extraction. The removal of inorganic compounds such as acids, bases, salts, and heavy metals is limited, but the compounds usually do not hinder the remediation process. Metals may undergo a chemical change to a less toxic or leachable form but their presence in the waste streams may also restrict disposal and recycle options.

The remediation process begins with excavating the contaminated soil and feeding it through a screen to remove large objects. In some cases, solvent or water is added to the waste in order to pump it to the extraction unit. In the extractor, solvent is added and mixed with the waste to promote dissolving of the contaminants into the solvent. Laboratory testing can determine which solvent adequately separates the contaminants from the soil. Generally, the solvent has a higher vapor pressure than the contaminants so that with an appropriate pressure or temperature change, the solvent may be separated from the contaminants, compressed, and recycled to the extractor.

Up to five waste streams may result from the solvent extraction process:

1. Concentrated contaminants;
2. Solids;
3. Wastewater;
4. Oversized rejects; and
5. Treated air emissions.

With the contaminants in a more concentrated form, they may be analyzed and subsequently designated for further treatment, recycle, or reuse before disposal. Solvent extraction has presumably improved the condition of the solids but often the solids need dewatering, treatment for residual organic compounds, additional separation, stabilization, or other treatment. The water from the dewatering process, the solids, and the water from the extractor need analysis to choose the most appropriate treatment and disposal. Typically solvent extraction units are designed to produce negligible air emissions, but significant levels of emissions may occur during waste preparation. The units are a closed–loop design and the solvent is recycled and reused.

The primary advantage of solvent extraction is the treatability of a wide variety of media. This is in contrast to soil washing, the success of which is heavily dependent on the particle size distribution.

Some disadvantages of the process are that solvent extraction:

1. Does not destroy the contaminants;
2. May not be appropriate for contaminants with high vapor pressures because these compounds may be removed with the solvent in the separation process instead of remaining with the concentrated contaminant stream;
3. Is compromised by the presence of detergents and emulsifiers which compete with the solvent in dissolving the contaminants;
4. May leave residual solvent and contaminant concentrations in the treated waste;
5. Is not effective for high molecular weight or hydrophilic compounds; and
6. May use flammable or mildly toxic solvents.

Commercial Processes: There are a number of processes available including the following:

The **CF Systems Corporation** technology uses liquified gases as solvent to extract organics from sludges, contaminated soils, and wastewater. Propane is the solvent typically used for sludges and contaminated soils, while carbon dioxide is used for wastewater streams. The system is available as either a continuous flow unit for pumpable wastes or a batch system for non–pumpable soils and sludges.

The **Carver–Greenfield Process for Extraction of Oily Waste** (Dehydro–Tech Corp.) is designed to separate materials into their constituent solid, oil (including oil–soluble substances), and water phases. The process is intended mainly for soils and sludges contaminated with oil–soluble hazardous compounds. The technology uses a food–grade carrier oil to extract the oil–soluble contaminants. Pretreatment is necessary to achieve particle sizes of less than 0.25 inch.

The **BEST** process (Resources Conservation Company) is a mobile solvent extraction system that uses one or more secondary or tertiary amines [usually triethylamine (TEA)] to separate organics from soils and sludges. TEA is hydrophobic above 20°C and hydrophilic below 20°C. This property allows the process to extract both aqueous and nonaqueous compounds by simply changing the temperature.

Extraksol™ (CET Environmental Services—Sanivan Group) is a solvent extraction technology on a modular transportable system. This batch process extracts organic contaminants from soil using proprietary, nonchlorinated organic solvents. The solvents are regenerated by distillation, and the contaminants are concentrated in the distillation residues.

6.7.2 Dissolution

Dissolution may be defined as the complete or partial transfer of one or more components from a solid phase into a liquid phase in contact with the solid. The liquid phase may consist of an active consumable reagent in a carrier solvent or a simple solvent alone. The carrier solvent used most frequently in dissolution of inorganic species is water because it aids in holding the dissolved species in solution (i.e., it "solvates" the dissolved species), and is by far the most common and inexpensive solvent. Dissolution reactions have been employed purposefully in hydrometallurgy, an area related to hazardous waste treatment, since the 1800's and in the synthetic chemicals industry since its inception.

Dissolution is a preliminary or first treatment step that can be used to remove either major or minor constituents from solids. The major output from the process is a liquid stream containing the hazardous components that must be subjected to further processing for recovery or destruction. The process is applicable to any solids or mixture of solids that can be wetted by the suitable liquid. Emphasis should be placed on more complete characterization of solids to aid in the specification and testing of possible dissolution processes for these solids. Furthermore, in view of the sensitivity to chemical costs, efforts should be made to utilize lower quality or waste reagents wherever possible.

6.7.3 Supercritical Fluid Extraction

At a certain combination of temperature and pressure, fluids reach their critical point, beyond which their solvent properties are greatly altered. These properties make extraction more rapid and efficient than processes using distillation and conventional solvent extraction methods. Presently, the EPA is investigating the use of supercritical carbon dioxide to extract hazardous organics from aqueous streams. The National Bureau of Standards is investigating the potential of various fluids to serve as supercritical extractive solvents.

A critical fluid extraction system consists of a blending tank, one or more extraction vessels, one or more decanters, one or more filters, and an evaporation unit. Organic wastes such as oil refinery sludges are first combined in a blending tank and mixed well to yield a homogeneous, pumpable mixture. This mixture is then pumped under pressure to an extraction vessel filled with a liquified gas such as carbon dioxide, propane, butane, or pentane and is mixed under pressure to extract (dissolve) hydrocarbon components from the waste into the liquified gas extraction fluid. After the hydrocarbon components of the waste dissolve in the pressurized liquid, the resulting solution is gravity-separated in a decanter into a wastewater (or waste solids) stream and an extraction fluid-rich stream. The wastewater stream, containing inorganic solids, is sometimes filtered under pressure to remove the insoluble components. The extraction fluid–rich stream is fed to a pressurized evaporation unit. The extraction fluid (carbon dioxide, propane, butane, or pentane) is evaporated by dropping the pressure and is subsequently recovered, repressurized, and recondensed for reuse. The residue from the evaporation, consisting of a liquid hydrocarbon mixture, is then reprocessed or reused, blended with fuels for heat recovery, or incinerated. If the extracted waste stream still exceeds treatment requirements, it may be extracted again with fresh extraction fluid at the high pressure conditions.

If inorganic residuals (or waste solids) filtered from the waste/solvent mixture contain treatable levels of hazardous constituents such as certain metals (e.g., chromium, lead) or organics, they will require further treatment such as stabilization or incineration prior to disposal.

Critical fluid extraction processes are designed to extract hydrocarbon components from mixed oily and organic waste liquids and sludges. For critical fluid extraction to be economically applied, the waste should contain at least a few percent by weight of extractable hydrocarbons. This process has been demonstrated on wastes containing from 5 to 34% hydrocarbons. For lower concentrations, batch distillation or conventional solvent extraction may be more economical. Also, critical fluid extraction is not economical for wastes containing high concentrations of extractable hydrocarbons (higher than 95%). These wastes are more amenable to fractionation treatment.

The constituents in the waste feed that are to be extracted determine the type of extraction fluid that is best suited for the process. For example, polar organic molecules (e.g., phenol) in an organic feed can be extracted with an aqueous solvent. Organic constituents in aqueous feeds can be extracted with various organic solvents. Metal–containing wastes can be extracted with acids (e.g., trialkylphosphoric and carboxylic acids) or amine solvents. If the solubility of the waste constituents of concern in the extraction fluid in the untested waste is significantly lower than that in the tested waste, the untested system may not achieve adequate performance. Use of another extraction fluid may be required to increase the solubility of the waste constituents of concern and achieve the same treatment performance, or other, more applicable treatment technologies may need to be considered for treatment of the untested waste.

In critical fluid extraction, when carbon dioxide is used as the extraction fluid, high alkalinity will interfere with the process because carbon dioxide will react to form carbonates and bicarbonates. This will result in excessive carbon dioxide consumption. For wastes having high alkalinity levels, an extraction fluid other than carbon dioxide should be used in the critical fluid extraction process. The same problem does not arise if a hydrocarbon fluid (propane, butane, or pentane) is used. If carbon dioxide was the extraction fluid used on a tested waste and the alkalinity of the untested waste is significantly higher than that of the tested waste, the system may not achieve the same performance. Use of another extraction fluid may be required to achieve the same treatment performance, or other, more applicable treatment technologies may need to be considered for treatment of the untested waste. Alkalinity of the waste has not been found to affect the performance of conventional solvent extraction processes.

Processes are also being commercialized to clean products and equipment with supercritical carbon dioxide techniques. The benefits include (1) elimination of solvents, (2) faster processing times, and (3) will work on a wide variety of materials.

Advantages of SFE over distillation and liquid extraction:

1. The ability of a supercritical fluid to dissolve nonvolatile compounds at moderate temperatures and the strong variation of solubility with pressure combine to reduce the energy requirements of SFE compared to distillation and liquid extraction/stripping.
2. In SFE, many options are available for achieving and controlling the desired selectivity, which is extremely sensitive to variations in pressure, temperature, and choice of solvent.
3. The extract is virtually free of residual solvent. The solvent can be recovered with minimal losses by either isobaric heating or isothermal decompression to the gaseous state.
4. Supercritical fluids can be used to vaporize thermally labile nonvolatile substances at moderate temperatures, at which they are nondistillable. At these conditions, the selectivity may be too low for liquid extraction.
5. Nontoxic, nonhazardous supercritical CO_2 can be used in the food and pharmaceutical industries without contaminating the product. Liquid extraction leaves residual toxic organic solvent in the product even after energy–intensive distillation or vacuum–processing stages.
6. While the density, and thus the solvent capacity, of a supercritical fluid is nearly comparable to that of a liquid, the lower viscosity and higher diffusivity provide advantages in transport rates. For example, the settling rate is higher for precipitates, and the mass transfer rate is improved for solvent diffusion through solid phases.
7. By reducing the density of the solvent over a continuum, the extract may be fractionated into numerous components, even if they have similar volatilities.
8. Additional components, usually intermediate in volatility between the solute and the supercritical solvent, can be used for further manipulation of the phase behavior, for example, volatility enhancement of the solute.

6.8 FREEZING PROCESSES

6.8.1 Ground Freezing

Ground freezing has potential for remediation because contaminants apparently can be concentrated ahead of freezing fronts. By concentrating contaminants, artificial ground freezing can be used to reduce the volume of contaminated soil at a site and thereby facilitate remediation.

Laboratory studies have shown artificial ground freezing to be a potentially effective method of driving volatile organic contaminants from soil matrices. Contaminants can be removed from soils by using the differences in the physical and chemical properties of the water and contaminants within the soil. When the temperature of soil is gradually lowered below 0°C, ice nucleation starts in the soil water and ions are rejected. As the dissolved chemicals in the soil solution are excluded from the ice, they become more concentrated in the remaining liquid. As the temperature drops further, the amount of ice becomes larger and the amount of unfrozen water decreases. The concentrated solution is pushed ahead of the ice lense in a desired direction, as determined by the location of the cooling system. This large bulb of concentrated contaminated soil may then be treated separately without treating the entire soil mass.

If a situation requires the contaminants to be immobilized, the temperature should be dropped rapidly rather than gradually. In this case, the ions and salts are entrapped between the ice particles and the separation process will not be complete.

Ground freezing has been used at the INEL Idaho Falls nuclear waste site, in order to avoid the creation of hazardous dusts while excavating.

6.8.2 Freeze Crystallization

Freeze crystallization operates on the principal that when water freezes, the ice crystal structure that forms naturally excludes all contaminants from the water molecule matrix. Thus, when the ice crystals are recovered and washed with pure water to remove any adhering brine contaminants, that which remains is very pure water.

A version of freeze crystallization that directly injects the refrigerant into the waste has been created by Freeze Technologies Corp. In this process, the mixed waste liquid enters through the feed heat exchanger where it is cooled to within a few degrees of its freezing temperature. The cooled feed then enters the crystallizer where it is mixed with boiling refrigerant. Water is crystallized in the stirred solution, and is maintained at a uniform concentration or ice fraction by continuous removal of a slurry stream (liquid + ice) that flows to the eutectic separator–growth column. The eutectic separator is used in the system when dissolved materials in the feed are themselves crystallized because of high water recoveries. The growth column is a zone where water crystals are increased in size to better accommodate subsequent washing.

Ice slurry from the crystal separator is pumped to the wash column where it forms a porous pack. The slurry liquid is removed from the column via screened openings, and is then either returned to the eutectic separator or is removed from the system for recycle/disposal. The ice is separated from the liquid in the wash column by filtering screens that allow passage of liquid concentrate but not the ice crystals. Hydraulic forces generated (by the flow of liquid to the screens in the middle of the ice pack) provide the mechanism for propelling the ice pack upward in the column. Ice is washed with melt water and scraped from the top of the pack into a reslurry chamber in the wash column. Within the wash column, melted product is used to transport the ice to a shell and tube heat exchanger, where the slurry is heated on the tube side and hot refrigerant gas is condensed on the shell side.

This technology will remove both organic and inorganic, ionic and non–ionic species, from contaminated aqueous streams. It works on both surface waters and groundwaters as well as directly on process wastes.

The process is applicable to free liquids, whether the solvent is water or an organic. It can also be used in conjunction with other processes to treat other media. For example, contaminated soils can be washed to transfer the contaminant into a liquid (i.e., water) medium. This has not been particularly attractive because of the low concentrations in the washing medium. Freezing can concentrate this to allow by–product recovery or more economical final destruction.

6.8.3 Other Processes

Cryogenic Cooling: A technique for reducing the evaporation of spilled hazardous materials, diminishing both the toxicity and flammability hazards. Deployment of the technique consists of distributing a cryogen (usually ice, liquid nitrogen or dry ice) over the surface of the spill. Alternatively, leaking containers can be packed in the cooling medium. In addition to the primary effect of reducing the vapor pressure of the hazardous substance, cryogenic cooling can, in many cases, immobilize a liquid spill by freezing it. Also, the use of dry ice creates an inert gas blanket which prevents the ignition of flammable substances. The efficiency of cryogenic cooling to ameliorate spills of hazardous chemicals is controlled by four factors: delivery, physical properties of the spilled chemical, the freezing point of water, and environmental factors.

Electrostatic Concentration: A process for concentrating aqueous mixtures by freezing, pulverizing, and then separating the liberated particles electrostatically. It could be an economical method for separating aqueous mixtures.

Suspension Freezing: In a Japanese process, power plant sludge, with water contents of 80 to 85%, is frozen at −25°C for 40 hours, using LNG at −165°C as the cooling source. The sludge is then thawed in a concrete tank. This freezing–and–thawing process allows the majority of the solid particles to grow from about one micrometer to 20 μm in size, resulting in separation of the solid and liquid components. Overall treatment costs are half those of incineration.

Surface Soil Cooling: A technique for decreasing soil temperatures so as to reduce the vapor pressure of volatile constituents and thus their volatilization rates. One way to lower soil temperature is to apply cooling agents to the soil surface. Applicable cooling agents include solid carbon dioxide, liquid carbon dioxide, and liquid nitrogen.

VOC Removal: Airco Gases is commercializing a liquid nitrogen (LN_2) system to remove volatile organic compounds (VOC) from process emissions. The LN_2 cools emissions to condense VOCs that can then be recycled.

The system is efficient for processes blanketed in a nitrogen atmosphere. The nitrogen passes through and cools the condenser, moves to the process vessel, where it picks up emissions, then loops back into the condenser, where fresh nitrogen chills out the VOCs.

6.9 OIL/WATER SEPARATION

Oil/water separators are another class of devices that utilize density to achieve separation. Oil/water separators have wide application at petroleum refineries, shipboard bilge water processing, and metal processing where oil and water emulsions are used as lubricants.

The API separator has been the standard gravity device, in use for many years in the petroleum industry. A newer device is the CPI (Corrugated Plant Interceptor), a process similar to lamellar classification for solids removal.

Oil/water separation is also an important environmental procedure used as a pretreatment step ahead of easily fouled operations such as ion exchange or membrane separation technologies.

Oil separation techniques are used to remove oils and grease from wastewater. Oil may exist as free or emulsified oil. The separation of free oils and grease is accomplished by gravity, and normally involves retaining the oily waste in a holding tank and allowing oils and other materials less dense than water to float to the surface. This oily top layer is skimmed off the wastewater surface by a mechanism such as a rotating drum–type or a belt-type skimmer. Emulsified oil, after it has gone through a "breaking" step involving chemical or thermal processes to generate free oil, can also be separated using a skimming system.

There are two types of emulsions—mechanical, and chemical. Mechanical emulsions are produced by pumping, mixing, and related processes and can be destabilized by coalescence, chemical coagulation, or agglomeration. Chemical emulsions are usually purposely induced to stabilize the emulsion for an industrial use. Chemical emulsions can be broken by the use of coagulants, acids, pH adjustment, heat, centrifuging, or high-potential alternating current.

6.9.1 Gravity Separation

Gravity separation, the primary and most common treatment, is based on the specific gravity difference between water and immiscible oil globules, and is used to move free oil to the surface of a water body for subsequent skimming and removal.

The effectiveness of a gravity separator depends on the proper hydraulic design and the period of wastewater detention for a given rise velocity. Longer retention times generally increase separation efficiency.

6.9.2 Skimming

Pollutants with a specific gravity less than water will often float unassisted to the surface of the wastewater. Skimming removes these floating wastes. Skimming normally takes place in a tank designed to allow the floating debris to rise and remain on the surface, while the liquid flows to an outlet located below the floating layer. Skimming devices are therefore suited to the removal of non-emulsified oils from raw waste streams. Common skimming mechanisms include the rotating drum type, which picks up oil from the surface of the water as it rotates. A doctor blade scrapes oil from the drum and collects it in a trough for disposal or reuse. The water portion is allowed to flow under the rotating drum. Occasionally, an underflow baffle is installed after the drum; this has the advantage of retaining any floating oil which escapes the drum skimmer. The belt type skimmer is pulled vertically through the water, collecting oil which is scraped off from the surface and collected in a drum.

Gravity separators, such as the API type, utilize overflow and underflow baffles to skim a floating oil layer from the surface of the wastewater. An overflow-underflow baffle allows a small amount of wastewater (the oil portion) to flow over into a trough for deposition or reuse while the majority of the water flows underneath the baffle. This is followed by an overflow baffle, which is set at a height relative to the first baffle such that only the oil bearing portion will flow over the first baffle during normal plant operation. A diffusion device, such as a vertical slot baffle, aids in creating a uniform flow through the system and in increasing oil removal efficiency.

The removal efficiency of a skimmer is partly a function of the retention time of the water in the tank. Larger, more buoyant particles require less retention time than smaller particles. Thus, the efficiency also depends on the composition of the waste stream. The retention time required to allow phase separation and subsequent skimming varies from 1 to 5 minutes, depending on the wastewater characteristics.

API or other gravity-type separators tend to be more suitable for use where the amount of surface oil flowing through the system is consistently significant. Drum and belt type skimmers are applicable to waste streams which evidence smaller amounts of floating oil and where surges of floating oil are not a problem. Using an API separator system in conjunction with a drum type skimmer would be a very effective method of removing floating contaminants from non–emulsified oily waste streams.

There is a very high level of oil and grease removals attainable in a simple two–step oil removal system. Based on the performance of installations in a variety of manufacturing plants and permit requirements that are consistently achieved, it is determined that effluent oil levels may be reliably reduced below 10 mg/ℓ with moderate influent concentrations. Very high concentrations of oil may require two–step treatment to achieve this level.

Skimming which removes oil may also be used to remove base levels of organics. Plant sampling data show that many organic compounds tend to be removed in standard wastewater treatment equipment. Oil separation not only removes oil but also organics that are more soluble in oil than in water. Clarification removes organic solids directly and probably removes dissolved organics by adsorption on inorganic solids.

Skimming as a pretreatment is effective in removing naturally floating waste material. It also improves the performance of subsequent downstream treatments. Many pollutants, particularly dispersed or emulsified oil, will not float "naturally" but require additional treatments. Therefore, skimming alone may not remove all the pollutants capable of being removed by air flotation or other more sophisticated technologies.

6.9.3 Coalescing

The basic principle of coalescence involves the preferential wetting of a coalescing medium by oil droplets which accumulate on the medium and then rise to the surface of the solution as they combine to form larger particles. The most important requirements for coalescing media are wettability for oil and large surface area. Monofilament line is sometimes used as a coalescing medium.

Coalescing stages may be integrated with a wide variety of gravity oil separation devices, and some systems may incorporate several coalescing stages. In general, a preliminary oil skimming step is desirable to avoid overloading the coalescer.

One commercially marketed system for oily waste treatment combines coalescing with inclined plate separation and filtration. In this system, the oily wastes flow into an inclined plate settler. This unit consists of a stack of inclined baffle plates in a cylindrical container with an oil collection chamber at the top. The oil droplets rise and impinge upon the undersides of the plates. They then migrate upward to a guide rib which directs the oil to the oil collection chamber, from which oil is discharged for reuse or disposal.

The oily water continues on through another cylinder containing replaceable filter cartridges, which remove suspended particles from the waste. From there the wastewater enters a final cylinder in which the coalescing material is housed. As the oily water passes through the many small, irregular, continuous passages in the coalescing material, the oil droplets coalesce and rise to an oil collection chamber.

Coalescing is used to treat oily wastes which do not separate readily in simple gravity systems. The three–stage system described above has achieved effluent concentrations of 10 to 15 mg/ℓ oil and grease from raw waste concentrations of 1,000 mg/ℓ or more.

Coalescing allows removal of oil droplets too finely dispersed for conventional gravity separation–skimming technology. It also can significantly reduce the residence times (and therefore separator volumes) required to achieve separation of oil from some wastes. Because of its simplicity, coalescing provides generally high reliability and low capital and operating costs. Coalescing is not generally effective in removing soluble or chemically stabilized emulsified oils. To avoid plugging, coalescers must be protected by pretreatment from very high concentrations of free oil and grease and suspended solids. Frequent replacement of prefilters may be necessary when raw waste oil concentrations are high.

6.9.4 Removal from Aquifers

Contained recovery of oily wastes (CROW) is a process that displaces oil wastes with steam and hot water. The contaminated oils and groundwater seep up into a more permeable area and are pumped out of the aquifer. A process has been developed by Western Research Institute.

6.9.5 Decantation

Decantation is a gravity separation technique used to separate immiscible liquids of different densities. The unstable mixture of immiscible liquids is slowly fed into a decant tank where it is continuously separated, coalesced, and withdrawn. Dust and dirt particles can interfere with the process of coalesence. They are sometimes filtered out before decantation. Depending on composition, these particles must be treated and disposed of. Decantation is often used to remove insoluble oils from spent solvents in the dry cleaning and petroleum refining industries. Because the nonaqueous layer removed by decantation is often saturated with water, further processing to dry the recovered liquid is often necessary.

Decant tanks are simple in design and relatively compact. The main factors in designing a decanter are the droplet size of the discontinuous phase, and the volume fraction of the discontinuous fluid. More complex units, used mainly by the petroleum refining industry to separate oil–water mixtures, are the API and tilted–plate separators.

6.9.6 Air Flotation

Induced (IAF) or dissolved (DAF) air flotation is utilized where gravity separation is not particularly effective, although gravity separation usually precedes flotation. Air bubbles can be introduced at atmospheric or higher pressures, which attach to oil particles and float them to the surface. More rapid oil removal can be accomplished than in gravity operations, with smaller treatment units. Additional introduction relating to flotation is discussed in a later section devoted to suspended solids treatment.

6.9.7 Other Techniques

Other oil/water separation methods include:

1. Centrifugal separation.
2. Emulsion breaking (chemical, electrical, or physical methods).
3. Biological techniques.

4. Carbon adsorption.
5. Membrane processes.

6.10 PARTICULATE REMOVAL

6.10.1 Dry Particulate Removal

Fabric Collectors: A baghouse is a large filter housing filled with numerous long filter bags. Typically, the bags are cylindrical and made of fabric, although a flat bag or a pleated filter can also be used, and ceramic and sintered metal bags are available. In operation, a cake of solids is built up on the fabric surface, and it is this porous cake that actually does the particulate filtering. If the cake did not build up, the fine particulate present in the flue gas would penetrate into the fabric pores and quickly plug or blind the filter bag. With the cake, the blinding process is substantially slowed, and the bags may last from weeks to years, depending on the bag and particulate characteristics.

There are three types of baghouses: reverse air, mechanical shaker, and pulse jet. For high–temperature, continuous duty, a pulse jet baghouse is normally used. The pulse jet housing will consist of one or more bag chambers with a hopper beneath each chamber. The bags normally hang vertically in the bag chambers, suspended from a tubesheet that separates the dirty bag chamber from the clean air plenum.

Dust–laden gases enter the baghouse and pass through fabric bags that act as filters. The bags can be of woven or felted cotton, synthetic, or glass–fiber material in either a tube or envelope shape.

The high efficiency of these collectors is due to the dust cake formed on the surfaces of the bags. The fabric primarily provides a surface on which dust particulates collect through the following four mechanisms:

1. Inertial Collection: Dust particles strike the fibers placed perpendicular to the gas–flow direction instead of changing direction with the gas stream.
2. Interception: Particles that do not cross the fluid streamlines come in contact with fibers because of the fiber size.
3. Brownian Movement: Submicron particles are diffused, increasing the probability of contact between the particles and collecting surfaces.
4. Electrostatic Forces: The presence of an electrostatic charge on the particles and the filter can increase dust capture.

In mechanical–shaker baghouses, tubular filter bags are fastened onto a cell plate at the bottom of the baghouse and suspended from horizontal beams at the top. Dirty gas enters the bottom of the baghouse and passes through the filter, and the dust collects on the inside surface of the bags.

In reverse–air baghouses, the bags are fastened onto a cell plate at the bottom of the baghouse and suspended from an adjustable hanger frame at the top. Dirty gas flow normally enters the baghouse and passes through the bag from the inside, and the dust collects on the inside of the bags.

In reverse–jet baghouses, individual bags are supported by a metal cage, which is fastened onto a cell plate at the top of the baghouse. Dirty gas enters from the bottom of the baghouse and flows from outside to inside the bags. The metal cage prevents collapse of the bag.

<u>Advantages:</u> The advantages of the baghouse make it one of the best particulate

removal devices available. The best advantage of the baghouse is the high fractional removal efficiency. When properly designed, baghouses remove essentially all particulate greater than 2 μm and can remove greater than 97% of all particulate larger than 0.5 μm. With extra care in design and operation, even higher submicron particulate removal efficiencies can be obtained. Baghouses are very flexible with essentially an infinite turndown ratio, and are capable of handling a great variety of dust loads. In addition, as long as the fabric is intact, the bags are a positive barrier that will not be penetrated by the particulate even if other problems arise.

Baghouses are frequently chosen when dry or semidry and acid–gas scrubbing is employed. Unlike wet particulate collectors, a baghouse keeps the acid–gas scrubbing residue dry. Baghouses can also improve acid conversion as the large volume in the bag chamber provides additional time for alkali to come in contact with and react with acid gases. In this service, the cake that builds up on a bag surface also contains alkali reagent. Because all gases must pass through this cake, additional acid gas neutralization will occur.

Disadvantages: Although baghouses have proven to be useful tools, they are not without certain disadvantages. Baghouses cannot tolerate moisture, which may cause bag blinding and severe corrosion. This implies that the baghouse must be operated above the dew point. Bag blinding will also result from extremely fine particulate penetrating into the fabric media and from volatile particulate (such as heavy metals and PICs) condensing on the fabric. Due to physical stress, the bags tend to get holes and tears that release particulate matter. Fine dust is easily re–entrained while pulsing, and so the chamber may need to be removed from service to perform the required cleaning.

Ceramic Candles: Ceramic candles are very similar in concept to a baghouse. The ceramic candles are housed in a chamber much like a baghouse, with a hopper beneath the candles and a tube sheet to support the candles and separate the dirty section from the clean section. The candle filters can be made from a variety of ceramic materials using either ceramic granules or fibers, and may have a soft or hard form.

Ceramic candle filters have been used in industry for several years. Approximately 100 of these filters are in service throughout the world. Of this number, approximately 10 to 15 units are in service in the United States. Most of these units are relatively small, and only a few are used on thermal treatment systems such as incinerators. The low number of incinerator installations may be due in part to some common problems that were associated with the original filters; however, recent improvements have corrected many of these problems. For example, the original ceramic candle filters were easily plugged and required frequent replacement. Current filters use an outer layer that has very fine pores and an inner layer that has larger pores. The outer layer protects the filter from penetration of fine particulate and therefore minimizes blinding. With this improvement, incinerator operators with ceramic candle filtration are reporting filter life of 6,000 to 8,000 hours of service. Some ceramic candle filters now being used in France have disposable outer layers that can be removed and discarded. New outer layers can then be placed on the original base layers. As a result, the filters last longer and disposal volume is reduced.

Many of the improvements in ceramic candle filters are a result of the development of soft filters. Soft ceramic filters have a maximum operating temperature of 2800°F (as compared with 1900°F for hard ceramic filters). The soft filters are approximately seven times lighter in weight and cost slightly less than half as much. In addition, the soft candles are not subject to thermal shock and result in less waste volume when the candles are replaced.

The ceramic candles are similar to baghouses in operation, with only a few differences. Dirty flue gas enters near the bottom of the filter housing just above the hopper. The gases pass through the filters, depositing particulate on the outside of the ceramic filters. After a layer of particulate matter builds up on the surface, the filters are cleaned by pulsing with compressed air. For a typical application, the initial pressure drop is approximately 5 inches of water. The maximum pressure drop is 5 psi; however, custom designs have specified a 15 psi pressure drop.

With ceramic candles, the pulsing air is not used to expand the filter like it does in a baghouse; rather, the pulsing air creates a reverse flow through the filter. As with baghouses, only a portion of the filters, approximately 10% are pulsed at a time so that the other filters remain in service. The gas volume used in pulsing is not controlled by valves like it is in a baghouse; rather, a pulse measuring tank is used, providing more precise control of the amount of air that is used.

The equivalent of an air–to–cloth ratio in the ceramic candle filters is referred to as face velocity. Ceramic candles are normally designed for face velocities of 10 to 15 ft/min. For high dust loadings a face velocity of 2 to 4 ft/min is used. Alternatively, the high dust loading can be partially reduced with a cyclone so that the higher face velocities can be used. A typical can velocity to match a 15 ft/min face velocity is approximately 30 ft/sec. Another important parameter related to ceramic filters is the ratio of the largest pore to the mean pore. A value of two is desirable for this ratio.

<u>Advantages:</u> Ceramic candles have many of the same advantages as fabric filters. They can be turned down to minimal or no flow with no problems, and may perform better at the lower flow rate. They act as a positive barrier to particulate emissions should there be a failure in the system. Ceramic filters are also reported to have high particulate fractional removal efficiencies; over 99.5% of the particulate 0.3 μm or larger are collected with ceramic candles. This is one of the highest removal efficiencies available.

Another advantage of ceramic filters is their ability to operate at very high temperatures. As previously discussed, hard ceramic filters can operate at 1900°F, and soft ceramic filters can operate at 2800°F. At these temperatures, combustion is still taking place, and the candles can be used to collect and completely burn the soot particles on the candle surface.

<u>Disadvantages:</u> As with a baghouse, the filters should be kept dry. If moisture collects on the surface, it can normally be baked off by raising the temperature, but this temperature increase will result in a slight increase in pressure drop. Therefore, the filters should be operated above the dew point. Re–entrainment of ultrafine particulate during cleaning may also be a problem when using ceramic candles.

Ceramic candle filters also have a few unique problems. At high temperatures, the structural strength of the tube sheet is lowered and thermal expansion may cause some difficulties. To resolve these issues, the tube sheet is cooled with water or steam. In addition, the housing must be refractory lined to prevent thermal damage.

Because ceramic candles are made from the same types of materials as refractories, they are subject to the same types of failures. Frequent thermal cycles cause some wear. Likewise, the ceramic can be attacked by alkali metals, forming eutectics that degrade or ruin the filters. To mitigate this problem, "getters" can be used to precoat the filters. The getters interact with the alkali metals and help protect the ceramic material. In addition, the particle size of the getters can be controlled so that a porous cake forms on the filter surface. As with a baghouse, such a cake provides better filtration at a lower pressure drop over long time periods.

Electrostatic Precipitation: An electrostatic precipitator (ESP) is a large chamber filled with long, vertical electrodes referred to as discharge electrodes and collection electrodes. Particulate removal in an ESP involves two key steps that are brought about by the discharge electrodes and the collection electrodes. In the first step, particulate is given an electrical charge by applying a large voltage, as high as 100,000 V, to the discharge electrodes. This high voltage causes a partial breakdown of the flue gas, generating gaseous ions that are easily attached to particulate matter. In this manner, the particles become charged.

The second step involves the particulate migration or precipitation resulting from the electrostatic forces developed once the particles are charged. Because the particles pick up the same charge as the discharge electrodes, the particles are repulsed away from the discharge electrodes. Further particulate migration and ultimate collection is then brought about by the collection electrodes which have a charge that is the opposite of the particulate charge. As such, the particulate is attracted to the collection electrodes. For proper precipitation to occur, the drag force on the particles from the gas flow must be lower than the electrostatic force, and the residence time in the ESP must be sufficient for the particles to reach a collection electrode. Gas velocity in an ESP typically ranges from 2 to 5 ft/sec, and gas residence time can be as high as 15 sec.

The discharge electrodes can be either positive or negative. Positive discharge electrodes have the advantage of generating less ozone, and negative discharge electrodes have the advantage of creating greater particulate charges and, therefore, better collection. In practice, it is possible to have both positively and negatively charged particles generated in the same ESP. To take advantage of this, ESPs with both positive and negative collection electrodes are available commercially.

There are two arrangements of electrodes used in electrostatic precipitation. In the single-stage arrangement, the discharge electrodes and the collecting electrodes are located in the same chamber. The collection electrodes are plates or tubes, and the discharge electrodes are typically made from wires or rods. When plates are used for the collection electrodes, rows of discharge electrodes are placed between the rows of collection plates. When tubular collection electrodes are used, one wire discharge electrode is suspended down through the center of each of the tubes. In this manner, the collection electrodes are in close proximity to the discharge electrodes.

In the two-stage arrangement, the discharge electrodes are placed in the first chamber, where ionization and particulate charging are achieved. The collection electrodes are placed in the second chamber, where removal is achieved. Regardless of how the precipitator is arranged, charges on the particles are neutralized when they contact the collection electrodes. When this happens, some particles fall into the collection hopper, and some of the particles adhere to the collection electrode. The collection electrodes are periodically cleaned to limit the amount of buildup by rapping with hammers that are normally automatically controlled.

<u>Advantages:</u> ESPs have numerous advantages that have made them a popular

particulate removal device. ESPs have a high removal efficiency. The typical removal efficiency in an incinerator application is greater than 95%; however, by increasing the size of the ESP and adding more electrodes, removal efficiencies greater than 99.9% can be obtained. Because the collector does not rely on impaction or filtration, ESPs have proven to be an excellent device for submicron particulate removal. Furthermore, ESPs have an infinite turndown ratio. Depending on actual operating conditions, ESP performance may improve with turndown.

Operating and maintenance costs for ESPs have proven to be very low compared to other particulate removal devices. The pressure drop across an ESP is typically less than 0.5 inch of water. Therefore, fan power requirements are very low, and even though the voltage used is high, the current used is very small, resulting in an overall low power consumption. Because no filter media are used, blinding is not a problem, and the expense of filter media replacement is avoided. Other maintenance expenses are also normally low, as the only moving parts are the rappers.

In terms of temperature limits, ESPs can normally handle temperatures up to 850°F. Higher temperatures may be achieved if special designs are used.

Disadvantages: The higher temperature range listed as an advantage above may actually be a disadvantage as well. In many ESP testing programs, it has been found that at these temperatures with the long residence times in the ESP, dioxins and furans are produced from other hydrocarbons remaining in the flue gas. The consequence is frequently a higher level of dioxins and furans in the ESP outlet gases when compared to the ESP inlet gases. This problem may be eliminated if the ESP is operated at a lower temperature.

Another disadvantage is the lack of a physical barrier between the inlet and the outlet. A physical barrier, such as a filter medium, prevents a release if there is a failure in the system. The ESP does not have this capability, therefore, releases may occur during abnormal operations or even while rapping if the collection electrodes have a heavy particulate coating.

Sticky fly ash may be a problem by contributing to a large buildup on the electrodes. An increase in flue gas particulate loadings without an increase in the cleaning frequency may also result in a heavy coating on the electrodes. Poor cleaning, resulting from too soft rapping, may also leave a heavy coating on the collection electrodes. When the electrodes become coated with a heavy layer of particulate, the buildup acts as an insulator, blocking the attractive force between electrode and charged particulate. This situation will have an obvious detrimental effect on ESP performance. ESP efficiency can also be affected by variations in flue gas temperature and moisture content and by particulate density and composition.

ESPs have high capital costs and can have excessive replacement costs when electrodes are damaged by rapping. Because of the high voltages involved, arcing and sparking can occur. These electrical discharges can damage the equipment, especially if the fly ash has a high carbon content, in which case a fire may result. ESPs also require a large space, which may add to building costs. Another problem, previously mentioned, is the formation of small quantities of ozone, which can, in some instances, be considered a pollutant.

High Efficiency Particulate Air Filters: HEPA filters are a standard in the nuclear industry as well as in other processes that require extremely high particulate removal. The filtering media in a HEPA filter is made from submicron glass fibers that are pressed into a high-density paper. Waterproof binders are used so that the media will be able to resist damage from small amounts of moisture. For nuclear applications, a radiation-resistant medium can be used, and the cell frame and the housing are typically made from stainless steel. The medium is shaped into deep pleats to maximize filtration surface area per unit of volume. Between the pleats are metal separators that hold the shape of the pleats as well as minimize the resistance to flow.

To prevent leakage, the filter medium is bonded to each wall of the cell frame. The frame, in turn, is mechanically locked in the housing and sealed with either a firm gel material or a gasket. Individual cells are available in a variety of sizes and are designed for a variety of face velocities and volumetric flow rates. Typical face velocities range from 125 to 500 ft/min.

HEPA filters consist of one or more HEPA filter cells in a common housing. When more than one cell is used, the cells are arranged in a parallel flow pattern. Because these filters intercept and trap the dust particles on the surface, the filters have a very limited capacity in terms of the amount of particulate that can be collected. Typically, only about 10 to 15 lb of particulate can be collected in a 24 x 24 x 12 inch HEPA filter, at which point the filter is spent and must be replaced.

Because HEPA filters are more expensive than other types of filters, they are normally used in series downstream of one or two lesser grade filters. In other words, a group of prefilters, arranged in a parallel flow, are encountered first and used for coarse particulate collection. These filters are followed by a group of intermediate filters that are also arranged in a parallel flow. These filters collect finer particulate. The HEPA filters, again in a parallel flow arrangement, are last in the series. By using prefilters and intermediate filters, the HEPA filters normally receive less than 10% of the particulate loading.

HEPA filters collect essentially all the particulate matter present in a gas stream. For this reason, they are frequently referred to as absolute filters. There are four collection efficiencies available for these type of filters. The standard grade of HEPA filters has a minimum collection efficiency of 99.97% for 0.3 μm and larger particles. The next grade has a 99.99% collection efficiency for 0.3 μm and larger particles.

There are two types of high-performance filters that are available. Ultra low penetration air (ULPA) filters have a minimum removal efficiency of 99.999% for 0.3 μm and larger particles. The highest grade filters available are referred to as laser tested, and have a minimum removal efficiency of 99.9995% for 0.12 μm and larger particles. High performance filters are normally only used in the manufacture of integrated circuits and other "clean room" applications because of their high cost.

Using a bank of prefilters, intermediate filters, and standard HEPA filters, the initial pressure drop across the HEPA filter bank is typically 1 to 1.5 inch of water. The HEPA filters will normally remain in service until the pressure drop exceeds 5 to 6 inches of water.

<u>Advantages:</u> There are no operational or maintenance costs other than filter replacement and fan energy requirements for the pressure drop across the HEPA filters. The manpower to replace the filters is minimal, but leak testing must be done after the installation of a new filter to verify that it is properly sealed and that there are no leaks.

Capital costs are moderate and depend on the options selected. The normal temperature limit is 250°F, but high-temperature options of 500° to 1000°F are available.

Housings with multiple, in situ test ports to individually test each filter in a bank are available. If desired, factory scan testing can be used to verify that there are no pinhole leaks in the media. For applications where gas adsorption is desired, the HEPA filter bank housing can be expanded, and activated carbon filters can be added. One of the options that some HEPA filter vendors can supply is a blowback feature. These HEPA filter housings have compressed air cleaning devices so that the particulate can be cleaned off and the filter may remain in service for a longer period. These housings have a hopper for dust collection and are available in modules that can be connected to handle the required gas volume.

Disadvantages: HEPA filters cannot be changed without taking the filter bank off-line. Replacement of prefilters can be necessary frequently, so it is common to use redundant HEPA filter banks in parallel, thereby doubling the cost of the equipment. With this arrangement, one filter bank is out of service while filters are being changed. Even though the bank may have been out of service for a long time before the filters are changed, the housing may be very hot if the flue gases are hot. Therefore, extreme caution must be used while changing filters. This may be difficult, as it may be necessary to be fully suited in anticontamination clothing and a respirator while changing the filters.

Metal Filters: With the exception that these filters are made out of metal, these filters are similar to other filtration systems, such as baghouses and ceramic candles. Metal filters are made from sintered metal powders, woven wire mesh with fibers sintered together at points of contact, or a combination of these. These metal filters are available in different grades ranging from coarse filtration to filtration equivalent to that of HEPA filters (i.e., 99.97% removal of 0.3 μm and larger particles). In one application, these filters were tested at a removal efficiency of 99.999997% for 0.1 μm particles.

The quality control used during the manufacturing of metal filters results in their high performance. During manufacturing, the size of particles used to make these filters is carefully controlled to ensure that the pore size is constant. Despite the high particulate removal efficiency, these filters have not been widely accepted in the nuclear industry or in the waste treatment industry. At this time, there are only approximately six metal filters installed on incineration systems and only a couple of installations of metal HEPA-type filters.

Metal filters can be pleated to form a high surface area per unit volume and can be made from 316L stainless steel or from a variety of alloys, such as Inconel. The metal filters are contained in housings with support grids that provide for thermal expansion but prevent the filters from rubbing or vibrating against the wall or other filters.

The metal filters remove particulate in a similar manner as other filters (i.e., a porous cake is built up on the outside wall of the filters). In operation, the initial pressure drop is typically 2 inches of water, and the filters are designed for a maximum pressure drop of 20 psi. The metal filters are normally operated with a face velocity of 5 to 7 ft/min but can go up to 20 ft/min. As with other filtration devices, metal filters are not affected by lower gas velocities and have essentially an infinite turndown. One of the advantages of metal filters over more conventional filters is their high temperature rating. Metal filters have a maximum operating temperature of 1650°F.

Metal filters are cleaned with compressed air in much the same manner as fabric and ceramic filters. Even though the metal filters may be rigid, their unique surface characteristics and thin walls provide for excellent cleaning with compressed air. If the particulate is light and re–entrainment is a concern, the dirty air inlet may be extended from the bottom side entrance up through the center of the housing to just below the tube sheet. The gas is then directed downward before passing through the filters. This configuration will not eliminate re–entrainment, but it will decrease the need for re–entrainment to some degree.

Advantages: Because of the similarity in concept, metal filters share the same types of advantages as baghouses. However, because of the metallic construction, these filters have other, unique advantages. Metal filters are mechanically durable and will withstand a variety of harsh environments. For example, the filters can be wetted and are not damaged by moisture. Metal filters can be cleaned with water or chemical solutions. This cleaning can be accomplished by removing the filters and cleaning. With a special design, the filters can be cleaned in situ. As previously mentioned, the filters can be operated at high temperatures and high pressure drops. Under these conditions, other filters would be distorted or their pore dimensions would be altered.

Because of the high structural integrity of these filters and the fact that the filters are cleanable, filter life is longer. Waste disposal volumes of spent media are also smaller. These characteristics allow the filters to handle high particulate loadings in excess of 0.1 lb of solids per cubic foot of gas. Metal filters also have a unique fail–safe feature that prevents the escape of particulate should one of the filters fail. A secondary filter element designed to be plugged off by particulate if the primary element is breached, is placed in the top of each filter. Once plugged, the gas flow through the damaged element is stopped, and little or no particulate is carried through the damaged element.

Disadvantages: There are two primary disadvantages to using metal filters. First, these filters have a large capital cost. However, the durability and resulting long life will make up for the higher cost in many cases. The second disadvantage to using metal filters is their weight. These filters and their housing are very heavy and may have special foundation requirements.

Cartridge Collectors: Cartridge collectors are another commonly used type of dust collector. Unlike baghouse collectors, in which the filtering media is woven or felt bags, this type of collector employs perforated metal cartridges that contain a pleated, nonwoven filtering media. Due to its pleated design, the total filtering surface area is greater than in a conventional bag of the same diameter, resulting in reduced air to media ratio, pressure drop, and overall collector size.

Cartridge collectors are available in single use or continuous duty designs. In single–use collectors, the dirty cartridges are changed while the collector is off. In the continuous duty design, the cartridges are cleaned by the conventional pulse–jet cleaning system.

Nested–Fiber Filters: Nested–fiber filter (NFF) technology is currently in the developmental phase, although commercialization is underway. Because of some unique features, NFFs have good potential in particulate control applications. NFFs are composed of a "nest" of thin, short metal fibers that form a randomly oriented interlocking bed. The interlocking of the fibers results in a dimensionally stable filter with 90 to 95% voids. The high amount of void space allows for low pressure drops with face velocities up to 200 ft/min; a face velocity for a baghouse is 2 to 5 ft/min. Consequently, NFFs are one–tenth the size of baghouses and ESPs.

Throughout the development of these filters, different shapes and configurations have been tested. The most promising of these shapes is a shallow bed approximately 7 to 10 inches deep or a horizontal cylinder with a hollow center core. In the shallow bed concept, a small vessel is filled with the fibers. The vessel has a dust collection hopper at the bottom with a screen above the hopper to support the fiber bed. Dirty flue gas then passes through the fibers exiting out the side of the housing just above the bed support screen.

The filter must be "regenerated" (i.e., cleaned) when 25 to 30% of the void space is filled or bleed–through and re–entrainment will become a problem. Unlike fabric filters, NFFs are not effectively cleaned by pulsing with compressed air. Rather, NFFs are best cleaned by vibrations. Although this can be accomplished using mechanical vibrators, better results are obtained using acoustic horns such as a natural gas–fired pulse combustor or a commercial sonic horn. The pulse combustor generates a series of pressure pulses that are formed when fuel gas is mixed with air and rapidly ignited in a unique, horn–shaped combustion device. Multiple ignitions and resulting pressure bursts, on the order of milliseconds in duration, are produced over a period of a few minutes until the filter is cleaned. Sonic horns also generate vibrations from acoustical energy, but, in this case, bursts of compressed air are used to drive the horn. However, for the horn to function properly, it must be acoustically tuned to the geometry of the NFF module.

In the horizontal cylinder concept, the filter is contained in a housing where the flue gas enters from the top and passes radially from the outer surface of the cylinder into the hollow core before exiting the side of the housing. When the filter requires cleaning, it is taken out of service by diverting the gas flow into a standby filter. The dirty cylinder will then be rotated about the horizontal axis, tumbling the fibers within the filter. This tumbling action will dislodge the particulate matter that has been collected, allowing the dust to fall into a collection hopper in the bottom of the housing.

The concept behind particulate collection with NFF technology is different than that used in other particulate collection devices. Although many of the larger particles (those greater than 10 μm) are initially collected by impaction and interception and diffusion, the particles (those less than 10 μm) are initially collected by interception and diffusion, the NFF does not rely upon a filter cake to form. Instead, after some particles begin to adhere to the base fiber, the following particles will preferentially agglomerate onto the particles that have already been collected. As a result, long branched chains of particulate matter will form as more and more particulate is collected.

For this type of particulate collection, face velocity is the most important process parameter with better performance resulting from slower velocities. Increasing bed thickness will also improve performance, but this parameter is not as important as face velocity. Fiber dimensions will impact performance and will affect the stability of the bed. An optimum fiber length has been found to be approximately 0.5 inch with a length–to–diameter ratio of 100.

<u>Advantages:</u> The NFF technology has demonstrated good filtration performance with the type of particulate encountered in thermal systems. Removal efficiency for particles of approximately 5 μm is 99.9%. Removal efficiencies greater than 99% have also been obtained for submicron particulate. Reasonable pressure drops (2 to 10 inches of water) are encountered during this service.

Furthermore, because alloys are used to make the needle fibers, the filters can be operated in a corrosive environment at high temperatures up to 1600°F. In addition, the housings for these filters can be made as a pressure vessel so that the filter can withstand pressure surges up to approximately 90 psi.

Another potential feature of this technology is the high degree of cleaning of the filter fibers that can be obtained. Because of the high void space, filter blinding and plugging is not likely. Furthermore, if the filters were to become so dirty that vibratory cleaning is ineffective, the durable nature of the metal fibers would allow the filters to be removed and cleaned in some other way, such as with high-pressure sprays, or chemical solutions. The final and perhaps most important feature will be the small size. The small size should significantly lower capital costs and space requirements.

Disadvantages: The principal concern with this technology has been the lack of experience. Although an NFF has been applied as a form of preliminary particulate removal in front of a HEPA filter in a hazardous waste incinerator off-gas train, there is not enough experience with this technology under the variety of conditions that can be encountered in several different processes. For example, with particulate larger than 44 μm, there was a slight decrease in performance. The cause of this performance decrease was investigated and the condition modified. As more operational experience is gained, other modifications may be necessary.

Gravity Settling Chambers: Gravity settling chambers are the simplest particulate removal device available. Settling chambers are empty chambers either vertically or horizontally inserted into the off-gas ducting. As the name implies, gravity settling chambers use the force of gravity to overcome the drag force from the gas flow that keeps particulate matter entrained in the off-gas. Settling chambers have large cross-sectional areas so that the gas velocity through the chamber is reduced and, consequently, the frictional force caused by gases flowing around the particulate is also reduced. In this manner, the particles fall downward and are collected in a hopper.

The cross-sectional area required is a function of the gas viscosity and density, the particle diameter and density, and the removal efficiency required. Normally, the cross-sectional area is sized to give a gas velocity of approximately 10 ft/sec. However, this size can be influenced by other considerations, such as particulate agglomeration, which results in larger diameters and, therefore, better collection.

Cooling the gas stream reduces the gas viscosity and volume which, in turn, lowers the gas velocity. These factors will allow a smaller cross-sectional area to be used. The length of the chamber depends on its orientation, horizontal or vertical. Horizontal chambers must provide a sufficient residence time to accommodate the rate of particle drop through the height of the chamber. Vertical chambers require a length such that fully developed flow is attained; however, the gas velocity at fully developed flow must be less than the settling velocity of the particulate.

Advantages: Gravity settling chambers have several advantages. They have low capital, installation, operating, and maintenance costs. The pressure drop through settling chambers is low, normally less than an inch of water. They can be operated at high temperatures, if needed, by lining the settling chambers with refractory materials. Also, because slower velocities produce less frictional drag on the particulate, the removal efficiency is increased with turndown of gas flow rate.

Disadvantages: Settling chambers offer high removal efficiencies only for large particles (typically greater than 50 μm). For this reason, settling chambers are not normally used in thermal treatment systems where the majority of the particulate is less than 50 μm.

Cyclone Separators: Cyclones separate solids from gas streams by centrifugal force. Cyclone separators are vertical, cylindrical vessels with a gas entrance designed to give a spiraling gas flow around and down the cyclone wall. This spiraling motion can be created by a tangential entrance, by vanes placed in the entrance, or by an involuted entrance that wraps around the outside circumference of the cylinder before opening up into the vessel.

Once the gas is in the cyclone, the downward spiraling flow of the gas stream imparts a centrifugal force on the particulate, which is thrown radially outward to the cyclone wall. When the particles hit the wall, much of their momentum is absorbed, and they fall to a cone–shaped section at the bottom of the cyclone. The particles are discharged out of the cone through a narrow neck.

The gas stream continues spinning along the wall inside the cylindrical vessel. When the gas flows into the converging section of the cone, the flow is forced back up through the center of the downward, spiraling gas flow. The gases exit through a tube that is mounted in the center of the top of the cyclone. This tube extends down into the center of the cyclone separator so that it is actually below the cyclone separator gas entrance, thus preventing gas flow from short circuiting the spiraling flow path.

The cyclone separator diameter has the greatest effect on process operations. Smaller diameters result in higher gas velocities and higher centrifugal force. These changes, in turn, result in a higher collection efficiency at a higher energy consumption.

Although cyclone separators are fairly simple in concept, there is a great variety in the models available. Some have continuously tapered diameters, and some have multiple sections with each section having a different diameter and a small tapered connection between each section. Multiple cyclone separators, arranged in parallel or series, are often used with a common housing and shared components, such as a gas inlet plenum and a dust collection hopper.

Advantages: Cyclone separators have some advantages associated with their simplicity. They have low capital, installation, operational, and maintenance costs. When refractory–lined, they can be operated at high temperatures. Cyclone separators can handle high dust loadings and, when compared to other particulate removal devices, cyclones have low to moderate pressure drops, ranging from approximately 1 to 5 inches of water.

Disadvantages: Cyclone separators have two major drawbacks that limit their use in thermal treatment systems. First, as previously mentioned, cyclones depend on gas velocity and are, therefore, sensitive to gas–flow turndown. Furthermore, even when operated at the proper gas velocity, most cyclone separators are not effective at removing particles smaller than 25 μm. In some cases, good removal efficiencies for particles as small as 5 μm can be obtained, but this particle diameter is still considered large for thermal treatment systems. For these reasons, cyclone separators are not commonly used in thermal systems. When they are used, it is normally in combination with other particulate removal devices.

Mechanical Centrifugal Separators: Mechanical centrifugal separators rely on centrifugal forces to cause the particulate to be thrown to the outer wall of the device, much like a cyclone separator. However, in the case of the mechanical centrifugal separator, the centrifugal force is developed via a rotating wheel composed of several blades. These units look and operate like a blower with the exception that dust is concentrated at the housing wall and then discharged through an opening in the bottom of the housing.

Advantages: With the added benefits that the units are compact and can provide motivational force to the gas stream, these separators have basically the same advantages and similar fractional collection efficiencies as cyclone separators.

Disadvantages: The mechanical centrifugal separators have approximately the same particulate removal efficiency as cyclone separators. Therefore, these devices are not normally used as a primary collection device for thermal treatment systems.

Impingement Separators: Impingement particulate removal devices consist of a structure, such as baffles or chevrons, inserted in the flue gas flow that causes the flue gases to change direction a number of times. In operation, the flue gas changes direction and continues around the constituents that make up the structure while the particulate, due to inertial forces, is impacted against the surface of the structure.

Because multiple changes in gas direction are used, there are multiple opportunities for the particles to impinge on the surface of the structure. The particles are then collected on the surface of the structure, or they drop to the bottom of the impingement separator where they are collected. Rappers are sometimes used on these devices to aid in removing particulate matter that is collected on the impingement surface.

Advantages: Impingement separators offer the advantages of high temperature operation and low pressure drop, normally between 0.1 and 1.5 inch of water. They are also low–cost devices and are easily adapted to existing systems.

Disadvantages: Impingement separators have low collection efficiencies on fine particulate and are normally used only when collecting particulate with diameters greater than 10 to 20 μm. For many thermal treatment systems, most of the particulate will remain in the off–gas if an impingement separator is used. Because these devices rely on inertial forces, gas velocity must be maintained to maintain removal efficiency. For these reasons, impingement separators are not used in thermal systems as primary collection devices.

Barrierless Ultrasonic Air Cleaners: The barrierless ultrasonic air cleaner is not an air pollution control technology that is used by itself, but this technology is rather an enhancement that can be applied to other particulate collection devices. The ultrasonic air cleaner will promote particulate agglomeration, which results in larger particles that are easier to collect. In principle, waves are created in the gas flow by applying ultrasonic energy to a flexural plate that will vibrate and form the waves. These waves will cause the lighter particles to collide with the heavier particles and consequently result in particle growth.

Two types of waves can be generated—a traveling wave and a standing wave. Both types of waves have been studied on a limited basis, and the enhancement of agglomeration has been proven. This research is at the development level, and a large effort will be required to commercialize this technology. The initial research into this technology concentrated on the traveling wave approach with later research conducted on the standing wave. Although research has not conclusively indicated whether the traveling or standing wave will be best for this service, the standing wave will likely require less energy and possibly less residence time, in addition to having fewer problems with attenuation effects.

The concept behind the standing wave system is to have the dust–laden flue gas enter in the side near the bottom of a vertical cylinder which serves as the coalescence chamber. The flexural plate and the electrical equipment required to vibrate the plate are located at the bottom of the cylinder below the gas inlet. The gas flow will travel up through the cylinder and exit out the side near the top. Within the cylinder, multiple standing waves are formed at nodule points where two waves intersect. As the gas flow passes through these standing waves, the differences in masses between various particles results in different particle velocities through the waves. This difference in velocities in turn leads to the collisions and agglomeration of particles.

The rate of coalescence depends directly on the ultrasonic power input, and the degree of coalescence is a direct function of the residence time in the coalescence chamber. The rate and degree of coalescence is dependent of both the particulate loading and the composition of the particulate matter.

Advantages: The benefit of this technology is the improvement of the collection of very fine particulate. As previously discussed, the fine particulate is the most difficult to collect and can cause the most difficulty in terms of blinding filtration media. By increasing the size of the small particulate, a standard particulate collection device can remove particulate matter.

In addition, with the minor exception of the inlet and outlet pressure drops, there is no pressure drop associated with the coalescence chamber. Therefore, flue gas flow is unaffected, and off–gas equipment, such as the ID fans, will not be required to deal with an increase in negative pressure.

Disadvantages: There are simply too many unknowns concerning this technology to adequately assess what problems will be encountered in its application. There are some areas of concern, however. This technology is limited in size and may require multiple units in parallel. Depending on what the flexural late construction materials are, this type of technology may have some temperature problems. The constant vibrational motion may also require extensive maintenance.

Baffle Chambers: Baffle chambers use a fixed baffle plate that causes the conveying gas stream to make a sudden change of direction. Large–diameter particles do not follow the gas stream but continue into a dead air space and settle. Baffle chambers are used as precleaners for more efficient collectors.

Unit Collectors: Unlike central collectors, unit collectors control contamination at its source. They are small and self–contained, consisting of a fan and some form of dust collector. They are suitable for isolated, portable, or frequently moved dust–producing operations, such as bins and silos or remove belt–conveyor transfer points. Advantages of unit collectors include small space requirements, the return of collected dust to main material flow, and low initial cost. However, their dust–holding and storage capacities, servicing facilities, and maintenance periods have been sacrificed.

A number of designs are available, with capacities ranging from 200 to 2,000 ft^3/min. There are two main types of unit collectors:

1. Fabric collectors, with manual shaking or pulse–jet cleaning—normally used for fine dust; and
2. Cyclone collectors, normally used for coarse dust.

Fabric collectors are frequently used in minerals processing operations because they provide high collection efficiency and uninterrupted exhaust airflow between cleaning cycles. Cyclone collectors are used when coarser dust is generated, as in woodworking, metal grinding, or machining.

The following points should be considered when selecting a unit collector:

1. Cleaning efficiency must comply with all applicable regulations.
2. The unit should maintain its rated capacity while accumulating large amounts of dust between cleanings.
3. The cleaning operation should be simple and should not increase the surrounding dust concentration.
4. The unit should be capable of operating unattended for extended periods of time, for example, 8 hours.
5. The unit should have an automatic discharge or sufficient dust storage space to hold at least one week's accumulation.
6. If renewable filters are used, they should not have to be replaced more than once a month.
7. The unit should be durable.
8. The unit should be quiet.

Use of unit collectors may not be appropriate if the dust-producing operations are located in an area where central exhaust systems would be practical. Dust-removal and servicing requirements are expensive for many unit collectors and are more likely to be neglected than those for a single, large collector.

6.10.2 Wet Particulate Removal

Wet particulate removal devices have a long history of use and have gone through several evolutions. As a result, there are more choices available in this category than in any other area of air pollution control. In all wet devices, the desired effect is to entrap the particulate in the liquid so that the particulate can be washed away. This entrapment is accomplished by designing the equipment to combine the use of a driving force, such as electrostatic or inertial forces, with intimate contact between liquid and particulate.

There are many advantages to using a wet particulate removal collection device. The primary liquid used in these devices is water, which will cool and saturate the gas flow by evaporation. By including a quenching section, hot gases exceeding 2000°F can be handled without using a heat exchanger or a boiler. Eliminating these components saves on capital costs and space, and results in less downtime for servicing.

The high degree of cooling accomplished by quenching, typically resulting in off-gas temperatures below 170°F, leads to condensation and collection of volatile pollutants, such as mercury and other toxic metals, as well as PICs. By adding a caustic compound to the scrubbing water, partial or even complete acid-gas removal is accomplished. The wetness of the device will also prevent fires, and by continually draining, solids are washed away from internal surfaces. Consequently, there is little or no solids buildup and the system internals are much cleaner.

When considering a wet particulate removal device, there are some negative attributes that must be considered. The potential spread of contamination from spills and leaks can be a serious problem. Corrosion problems are increased, so more expensive construction materials are often required. When the gases are saturated, these materials may be necessary throughout the entire system, as condensation occurs where there are cooler surfaces or where gas compression is encountered (as in the draft fan). In addition, when the saturated gases are exhausted from the stack, a steam plume will form. This steam plume can lead to problems as the plume is frequently mistaken for smoke and pollution by the public.

The liquid blowdown can also be a problem requiring the expense of additional treatment and potentially larger waste volumes. To avoid these drawbacks, scrubbing liquids are frequently recycled many times, resulting in a buildup of dissolved and suspended solids. These solids may plug nozzles, valves, and other components that have small orifices. The mists and droplets formed during operation may contain these solids. As such, exhaust gases may contain high levels of solids due to moisture carry–over. To minimize this problem, mist eliminators must be used.

There is a large variety of wet scrubbers; however, all have three basic operations:

1. Gas–Humidification: The gas–humidification process conditions fine particles to increase their size so they can be collected more easily.
2. Gas–Liquid Contact: This is one of the most important factors affecting collection efficiency. The particle and droplet come into contact by four primary mechanisms:
   (a) Inertial Impaction: When water droplets placed in the path of a dust–laden gas stream, the stream separates and flows around them. Due to inertia, the larger dust particles will continue on in a straight path, hit the droplets, and become encapsulated.
   (b) Interception: Finer particles moving within a gas stream do not hit droplets directly but brush against them and adhere to them.
   (c) Diffusion: When liquid droplets are scattered among dust particles, the particles are deposited on the droplet surfaces by Brownian movement, or diffusion. This is the principal mechanism in the collection of submicron dust particles.
   (d) Condensation Nucleation: If a gas passing through a scrubber is cooled below the dewpoint, condensation of moisture occurs on the dust particles. This increase in particle size makes collection easier.
3. Gas–Liquid Separation: Regardless of the contact mechanism used, as much liquid and dust as possible must be removed. Once contact is made, dust particulates and water droplets combine to form agglomerates. As the agglomerates grow larger, they settle into a collector.

The "cleaned" gases are normally passed through a mist eliminator (demister pads) to remove water droplets from the gas stream. The dirty water from the scrubber system is either cleaned and discharged or recycled to the scrubber. Dust is removed from the scrubber in a clarification unit or a drag chain tank. In both systems solid material settles on the bottom of the tank. A drag chain system removes the sludge and deposits it into a dumpster or stockpile.

Wet scrubbers may be categorized by pressure drop (in inches water gauge) as follows:

1. Low–energy scrubbers (0.5 to 2.5).
2. Low– to medium–energy scrubbers (2.5 to 6)
3. Medium– to high–energy scrubbers (6 to 15).
4. High–energy scrubbers (greater than 15).

Due to the large number of commercial scrubbers available, it is not possible to describe each individual type here. However, the following sections provide examples of typical scrubbers in each category.

Low–Energy Scrubbers: In the simple, gravity–spray–tower scrubber, liquid droplets formed by liquid atomized in spray nozzles fall through rising exhaust gases. Dirty water is drained at the bottom.

These scrubbers operate at pressure drops of 1 to 2 inches water gauge and are approximately 70% efficient on 10 μm particles. Their efficiency is poor—below 10 μm. However, they are capable of treating relatively high dust concentrations without becoming plugged.

Low– to Medium–Energy Scrubbers: Wet cyclones use centrifugal force to spin the dust particles (similar to a cyclone), and throw the particulates upon the collector's wetted walls. Water introduced from the top to wet the cyclone walls carries these particles away. The wetted walls also prevent dust re–entrainment.

Pressure drops for these collectors range from 2 to 8 inches water, and the collection efficiency is good for 5 μm particles and above.

Medium– to High–Energy Scrubbers: Packed–bed scrubbers consist of beds of packing elements, such as coke, broken rock, rings, saddles, or other manufactured elements. The packing breaks down the liquid flow into a high–surface–area film so that the dusty gas streams passing through the bed achieve maximum contact with the liquid film and become deposited on the surfaces of the packing elements. These scrubbers have good collection efficiency for respirable dust.

Three types of packed–bed scrubbers are:

1. Cross–flow scrubbers.
2. Co–current flow scrubbers.
3. Counter–current flow scrubbers.

Efficiency can be greatly increased by minimizing target size, i.e., using 0.003 inch diameter stainless steel wire and increasing gas velocity to more than 1,800 ft/min.

High–Energy Scrubbers: Venturi scrubbers consist of a venturi–shaped inlet and a separator. The dust–laden gases enter through the venturi and are accelerated to speeds between 12,000 and 36,000 ft/min. These high–gas velocities immediately atomize the coarse water spray, which is injected radially into the venturi throat, into fine droplets. High energy and extreme turbulence promote collision between water droplets and dust particulates in the throat. The agglomeration process between particle and droplet continues in the diverging section of the venturi. The large agglomerates formed in the venturi are then removed by an inertial separator.

Venturi scrubbers achieve very high collection efficiencies for respirable dust. Since efficiency of a venturi scrubber depends on pressure drop, some manufacturers supply a variable–throat venturi to maintain pressure drop with varying gas flows.

Devices: Many of the wet particulate removal devices commonly used in particulate removal have similar counterparts used in dry particulate removal. The primary difference in these systems is the addition of water for particulate entrapment and removal. These devices include spray chambers with gravity settling; wet impingement devices such as baffles, plates, or fibrous beds; wet cyclones; wet dynamic centrifugal devices; and wet electrostatic precipitators.

These devices rely on the same principles of operation as their dry counterparts but may have higher removal efficiencies and simpler operation. As with their dry counterparts, wet gravity settling chambers, wet impingement devices, wet cyclones, and wet dynamic centrifugal scrubbers do not have the efficiencies necessary for fine particulate removal, and are, therefore, rarely used as the primary removal device. However, these devices do have low pressure drops and high removal efficiencies for large particles. For these reasons, these devices are commonly used as mist eliminators designed to reduce moisture and solids carry–over from other wet devices which have larger particle sizes by condensation, entrapment, etc.

Wet electrostatic precipitators also perform similarly to their dry counterparts and are an excellent choice for particulate removal. However, because of the similarity between wet and dry electrostatic precipitators, no further discussion concerning the former will be presented. Likewise, wet scrubbers that are commonly used as mist eliminators will not be discussed because of their similarity to their dry counterparts, which have already been discussed. Wet particulate collection devices that do not have dry counterparts are described below.

Venturi Scrubbers: Venturi scrubbers typically consist of five sections: a converging approach, a narrow throat, a diverging section, a wetted elbow, and a mist eliminator. In operation, flue gas enters the converging approach section, which normally is a wetted wall design. In this design, liquid is continually injected into the venturi scrubber at the entrance of the converging section. In this manner, the entire section is washed to remove solids and the gas stream is quenched. If the off–gas has been previously cooled, a less expensive, nonwetted wall design can be used.

As the gases pass through the converging section, they are accelerated due to the continuously decreasing cross–sectional area. This acceleration will, in turn, result in a large drop in system pressure. When the gases enter the throat, they will have a velocity ranging from 100 to 600 ft/sec and will easily atomize the additional water that is injected into the venturi scrubber at this point. Depending on the gas velocity, the liquid droplets formed range in size from submicron up to approximately 2 μm. These small droplets will, due to the high velocity in the throat, impact with and entrap the particulate.

When the gases exit the throat, they pass through the diverging section where many of the droplets will combine and grow in size. In this section, momentum is reduced and converted back into system pressure. Not all of the pressure that was lost in the converging section is regained in the diverging section, and the difference is the scrubber pressure drop, which is a measure of the energy expended to remove the particulate and the resulting removal efficiency. Therefore, depending on the particle–size distribution and the removal efficiency required, venturi scrubber pressure drops range from approximately 20 to 70 inches of water. The magnitude of the pressure drop is affected by changes in volumetric gas flow. To account for this variation, variable area throats are frequently used so that the pressure drop can be automatically adjusted as the gas flow changes. With a variable throat, a 4:1 turndown can be obtained with a venturi scrubber.

After the particulate has been entrapped by the liquid droplets, the gases typically pass through an elbow. Although dry elbows are available, most elbows are wet (i.e., they contain a shallow pool) to protect against erosion that can occur due to the large velocities involved. This section also removes the large droplets and particles by impaction to the pool due to inertial forces from the gas flow turning in the elbow. The gases then enter a mist eliminator, such as a cyclone or a chevron, where the majority of the remaining droplets are removed by inertial forces.

Advantages: For small thermal systems, venturi scrubbers have historically been the chief competitor of baghouses and ESPs. This is due, in part, to the general advantages of wet systems—namely partial acid gas removal and low buildup of solids. However, venturi scrubbers have other significant advantages—they have lower capital costs, are simple to operate, and require only moderate maintenance.

Disadvantages: A disadvantage of venturi scrubbers is the large pressure drop required for high particulate removal. This pressure drop adds substantially to operating costs and may also impact the capital costs of other off–gas components in terms of vessel construction necessary to handle the large negative pressures required for the pressure drop. Even with the high pressure drop, venturi scrubbers have trouble removing particles less than 0.5 μm and, consequently, may have trouble achieving the more restrictive requirements for particulate emissions. The particulate that typically penetrates the venturi scrubber is the very fine particulate. Because of the high surface area–to–volume ratio, fine particulate can have a disproportionately large percentage of condensable heavy metals and PICs. Therefore, the ability of venturi scrubbers to remove these pollutants as effectively as other particulate removal devices is questionable.

Other High–Energy Scrubbers: The scrubbers in this category of particulate removal devices operate via the same basic principles as venturi scrubbers, but they are constructed somewhat differently. In general, these scrubbers remove particulate by accelerating the gas stream to very high velocities and then injecting a scrubbing liquid that is rapidly atomized and serves as an impaction target for the particulate. As with venturi scrubbers, this process creates a pressure drop that is a direct measure of the energy expended and the cleaning efficiency obtained. However, the pressure drop from gas acceleration is not the only energy input in these high–energy scrubbers, as additional removal techniques, such as other impingement surfaces or cyclonic sections, are normally combined into one unit.

There are several versions of high–energy scrubbers commercially available. Each version has unique features and characteristics.

One of the simplest variations of a high–energy scrubber is a bank of round tubes or rods. The tubes are typically mounted horizontally, and the gas flow is downward and perpendicular through the tubes. Scrubbing liquid is injected above the tubes at the gas inlet into the tube bank. As the gas flows through the bank of tubes, the gases are split into several streams that accelerate and decelerate as the gas passes through the narrow spacing between adjacent tubes and then flows into an open area between the next row of tubes.

In this manner, the gas flow experiences a series of smaller pressure drops, according to how many rows of tubes are used in the tube bank. In addition to the impaction of the particulate into the scrub liquor droplets, the numerous turns encountered as the gases flow around the tubes and the higher resultant velocities lead to some inertial separation and impingement of the droplets and particulate onto the tubes.

Another variation of a high–energy scrubber uses a combination of gas acceleration and cyclonic forces to generate the high pressure drop required for particulate removal. This scrubber is a vertical cylinder with a cone–shaped sump at the bottom. The scrubber contains one or more cylindrical cages made from vanes that form vertical slots. These narrow vertical slots in the vane cage accelerate the gas flow to achieve the velocity necessary to impact the particulate with the liquid droplets.

In this scrubber, dirty gases are introduced tangentially in the side to cause a cyclonic motion around the vane cage. Scrub liquor is injected into the gas stream through a port in the side of the scrubber wall that extends inward, almost touching the vane cage. As the gases pass through the vane cage, the high velocity and circular motion of the gas flow creates a revolving vortex cloud of very fine droplets inside the vane cage.

This cloud is held at this location by a balance of centripetal and centrifugal forces. As the droplets agglomerate with other droplets or particulate, their mass becomes larger and they are thrown out through the vane cage to the wall of the scrubber, where they are collected by the cyclonic action.

This design improves collection efficiency in many ways. The cyclonic flow provides particulate removal. The multitude of droplets thrown outward from the cage also perform a preliminary scrubbing. Because these droplets flow countercurrent to the gas stream through the vane cage slots, rather than cocurrent (as in a venturi scrubber), higher relative velocities between particulate and droplets are achieved. This higher relative velocity results in a more efficient impaction. Also, the cyclonic flow of the gas stream results in multiple passes around the cylinder and, therefore, more opportunities for impaction of particulate with droplets.

Advantages: In addition to the normal benefits of venturi scrubbers, the manufacturers of these modified high-energy venturi scrubbers report many additional benefits such as the capability of developing pressure drops up to 100 inches of water, less particulate and moisture carryover, and higher removal efficiencies for a given pressure drop.

Disadvantages: These scrubbers have disadvantages similar to those of venturi scrubbers. In addition, these scrubbers are more complicated and, therefore, can have higher capital costs, can be more difficult to service and maintain, and can require more space than normal venturi systems. These scrubbers are not as well known as the regular venturi scrubbers, so acceptance and confidence in their performance may pose problems.

Self-Induced Scrubbers: There are many designs of these self-induced scrubbers available in both low-energy (less than 10 inches of water pressure drop) and high-energy (greater than 20 inches of water pressure drop) versions. The concept behind these scrubbers is to eliminate the pump normally used to circulate and inject scrub liquor by letting the induced draft fan do the work. A description of one of these scrubbers follows. However, it should be noted that there are many other versions of these scrubbers.

This high-energy, self-induced scrubber is a vertical, cylindrical tower with a cone-shaped hopper. The hopper and bottom third of the cylinder are filled with scrub liquor. Located just above the surface of the scrub liquor is a vertical, cylindrical tube with an inverted cone placed partially in the bottom of the tube so that the cone does not touch the tube walls. In this way, an annulus is formed between the interior of the tube wall and the exterior of the cone.

Dirty gases enter the tower in the side at about mid-height. These gases flow down to the liquid surface and abruptly turn up through the annular opening. This action has three important effects. First, as the gases make the abrupt turn, inertial forces will throw some of the particulate into the scrub liquor, where it is collected. Second, as the gas flows at high velocity past the liquid surface, numerous droplets and particulate matter accelerate and impact against each other while flowing through the annulus. This action creates a scrubbing effect similar to that of a venturi scrubber.

The pressure drop and cleaning efficiency of this self-induced scrubber is controlled by raising and lowering the liquid level, which in turn controls the spacing between the vertical tube and the liquid surface that the gases must pass through. In addition, by moving the cone up or down, the size of the annulus is adjusted which, in turn, affects gas cleaning—as in a venturi scrubber.

Advantages: The elimination of the scrubber pump removes a component that can fail and need servicing. Since liquid circulation takes place in the interior of the scrubber, there are no leaky fittings, line pluggage, etc.

Disadvantages: The high-energy, self-induced scrubber is a technology that is not as widely applied as the venturi scrubber, so confidence and acceptance may be low. The liquid level control is a critical component and must be carefully maintained in order to maintain scrubber performance. Because of the similarity between this type of scrubber and venturi scrubbers, self-induced scrubbers share the same disadvantages as venturi filters.

Rotary Atomizing Wet Scrubbers: Rotary atomizing wet scrubbers are relative newcomers to the field of air pollution control. These scrubbers remove particulate by creating a dense curtain of small liquid droplets. The flue gas and particulate flow into the liquid curtain, impacting the particulate into the droplets. In operation, the flue gases are initially saturated and large particulate is removed in a quencher. The gases then flow into a small cylindrical chamber where the atomizer is mounted. Depending on the gas volume and the efficiency required, multiple atomizers can be used in parallel or series arrangements.

As with other wet particulate control devices, the amount of energy added to the system will dictate the removal efficiency obtained. In this case, however, the gas pressure drop is low at less than 4 inches of water, therefore, it is not the gas velocity that provides the energy for scrubbing; rather it is the energy supplied by the atomizer that controls cleaning efficiency. The atomizer typically rotates at speeds exceeding 10,000 rpm. This rotational speed imparts a great deal of energy into the liquid droplets that are formed. Because the rotational speed of the atomizer is constant, the amount of energy consumed in the atomizer is a function of how much liquid is atomized. As with a pump operating at a constant speed, the atomizer power input will increase with the amount of liquid handled.

Advantages: Rotary atomizers have some unique advantages when compared to other wet particulate scrubbers. For coarse particulate (above ~1 μm diameter), the energy input required to achieve a given removal efficiency is reported to be the same for a rotary atomizer and a venturi scrubber. However, for submicron particulate, the energy requirement is reported to be much less for a rotary atomizer. For example, with 0.5 μm particles, an atomizer can achieve the same removal efficiency at less than half the power. With 0.3 μm particles, an atomizer is reported to achieve the same removal efficiency at less than one-fourth the power.

In addition to the energy savings, the energy input for a rotary atomizer is not as limited as with a venturi scrubber. Rotary atomizers can supply energy inputs that are equivalent to pressure drops in excess of 100 inches of water in a venturi. This high input allows the atomizer to remove particles in the 0.1 μm range. Because the energy input is supplied by the atomizer, there is no need for maintaining the gas pressure drop. Therefore, performance is essentially unaffected by lowering gas flow.

Another advantage is the high acid–gas removal rates that are reported with this device. One manufacturer has indicated that 99% of the HCl and approximately 95% of the SO_2 is removed when using a caustic scrub liquor. These removal rates are as high as those for many acid–gas removal systems and may, therefore, eliminate the need for acid–gas scrubbing equipment.

Disadvantages: There are two primary disadvantages that must be considered before selecting a rotary atomizing scrubber. First, rotary atomizing scrubbers are mechanical devices with motors and moving components. As such, these devices can be subject to more frequent failures and maintenance.

A second concern is the lack of operating experience in thermal systems. Although particulate outlet loadings of less than 0.015 gr/dscf are reported, there are not enough systems in operation to warrant the confidence levels that other types of devices can provide. There is also a potential third disadvantage. These devices are intended to operate at a relatively low temperature, 200°F or less. During normal operations, this temperature limit should not be a problem since the flue gas is quenched to saturation before entering the rotary atomizer device. However, if a failure in the quencher should occur, or if a process upset should occur, then this temperature limit can be exceeded.

Free–Jet Scrubbers: Free–jet scrubbers have been in use for air pollution control since about 1970. The basic configuration is the same as that for other wet scrubbers (i.e., a quencher, the scrubber, and then a moisture separator). There are five models available, but there are basically combinations of two types of nozzles—a free–jet nozzle and an ejector nozzle. The model with the best performance and the one most likely to be used in radioactive waste treatment utilizes three nozzles in series, a steam– or air–driven ejector followed by two free–jet nozzles.

The ejector nozzle is located inside the flue gas duct, and the flue gases flow around the outside of the nozzle. Steam, air, or some other compressible fluid is discharged from the ejector nozzle at a speed of 3,200 ft/sec. At the nozzle exit, water is injected into the high velocity flow. The velocity through the ejector and the resulting shock waves create a very turbulent zone in the flue gas duct where the coarse water spray is broken up into a multitude of very small droplets ranging from submicronic to ~40 μm diameter. This turbulent flow of flue gases, water droplets, and steam (or other compressible fluid) then passes through the first free–jet nozzle.

The free–jet nozzle is mounted inside an expanded chamber followed by a narrow mixing tube. At the exit of the free–jet nozzle, additional water is injected into and broken up by the gas flow. Once the gases enter the expanded section of the free–jet nozzle, the gas velocity is lowered due to the larger cross–sectional area, and the pressure is lowered at the base of the nozzle due to the jet action of the emerging gas stream.

Under these conditions, another zone of turbulent motion with numerous swirling currents develops. This turbulence results in particulate matter impacting the water droplets. The gases and water droplets then enter the narrow mixing tube where the droplets agglomerate to form droplets that may contain hundreds of particles depending on how dirty the gas flow is. After passing through the mixing tube, the free water is drained from the system before the flow is passed through the second free–jet nozzle, where the process is repeated. After passing through the second mixing tube, the water droplets have grown to roughly 100 μm and are now easily removed in a moisture separator.

The ejector nozzle serves to create the optimum droplet size for maximum particulate collection. Without the steam jet, the liquid–to–gas ratio is 12 to 20 gal/min/1,000 acfm. With the steam jet, the ratio is 4 to 6 gal/min/1,000 acfm. This improved efficiency is reported to result in a 40 to 60% energy savings, in comparison to venturi scrubbers, and requires only 30 to 36 inches of gas pressure drop to collect 0.1 μm particles.

Advantages: In addition to the energy savings, there are several advantages to this type of scrubber. Most important is the particulate removal efficiency that is obtained. Test results have demonstrated that the fractional removal efficiency curve has a maximum removal of greater than 99.99% for 1 μm particle. For 0.1 μm particle, the removal efficiency is still greater than 99%, and for 0.01 μm particle the removal efficiency is greater than 80%.

Acid–gas removal efficiencies greater than 99% for HCl and greater than 70% for SO_2 can be achieved when using a caustic scrub liquid. The free–jet scrubber also has a unique advantage when the steam ejector is used. The steam ejector can provide the motivational force necessary to overcome the pressure drop produced in the scrubber. Therefore, if a heat recovery boiler is in use after the thermal treatment process, the steam from the boiler can be used to provide the required draft through the scrubber. As a result, energy consumption by the ID fan will be lower.

Disadvantages: In addition to the normal disadvantages associated with wet systems, free–jet scrubbers create the pressure drop necessary for gas cleaning using nozzles with fixed areas. Therefore, the free–jet scrubber is sensitive to reductions in flow. To mitigate this problem, a gas recirculation duct is supplied from the induced draft fan discharge to the inlet of the scrubber. Therefore, if the free–jet scrubber pressure drop should start to fall, the gas recirculation damper is adjusted to increase the flow of recirculation gases.

Ionizing Wet Scrubbers: Technologies discussed up to this point are normally used as individual components that are linked with other components to form a complete off–gas train. The ionizing wet scrubber (IWS) differs in that it utilizes multiple technologies that are designed to work together as an integrated package. In many cases, the IWS can function as the entire pollution control system, effectively removing acid gases, particulate, condensible organics, and toxic metals. The collection of particulate matter involves a combination of technologies using electrostatic particle charging, inertial impaction, and a phenomenon referred to as image force attraction.

In this process, dirty gases are first cooled and saturated in a quencher. The gas stream is then passed through a high voltage ionizing chamber to electrostatically charge the particulate, much like a conventional ESP. In an ESP, the charging electrodes can be either negative or positive and the electrodes with the opposite charge serve to complete the electrical circuit and as collection surfaces for the particles. An IWS differs from the ESP in that the charging electrodes are always negative and the positive electrodes are primarily used to complete the electrical circuit. Particulate collection on the positive electrodes, although beneficial for particulate removal, is not intended as the primary method of collection. As such, the positive electrodes are smaller, and they are sized to provide sufficient electrical current for proper ionization rather than being sized to provide a large collection area.

However, because some beneficial particulate collection will occur on the positive electrode surface, these electrodes are continuously flushed with water that overflows out of weirs at the top of the electrodes and drains down the electrode surfaces. This flushing is necessary to prevent the buildup of a resistive particulate layer that will impede collection performance. Because the positive electrodes are not the primary means for particulate collection, much of the particulate will pass through the ionizing section of the scrubber. This particulate matter enters a cross–flow, packed–bed scrubbing section where, along with acid–gas removal, the particulate removal is completed.

Particulate collection in the packed–bed section involves two different mechanisms occurring simultaneously. First, the packing used in the IWS is reported to have a shape that is very efficient for particle impaction. In addition, the shape of the packing generates a multitude of droplets that also serve as particle impaction sights. Consequently, particulate 5 μm and larger are removed by impaction as the gas stream passes through the chaotic flow paths in the packing bed. As with other wet systems, this particulate is then washed away in the continuous liquid flow.

The second method removes particulate less than approximately 5 μm by an electrical attraction referred to as image force attraction. Image force attraction occurs when a charged particle comes in close proximity to a neutral surface. The charged particle will induce an equal but opposite charge on the surface and will be drawn to the surface. The magnitude of this attractive force is equal to the force that would exist between the charged particle and an imaginary particle of equal charge but opposite polarity located at an equal distance behind the surface. In this manner, when the charged particles enter the packed tower, an induced image force attraction develops between the particles and the packing media and between the particles and liquid droplets.

Because of the large surface area of the packing media and the water droplets, an IWS has a large collection surface area. Furthermore, because of the intimate contact of the gas flow with the packing and the droplets, the particulate will need to migrate only a fraction of an inch as opposed to several inches, which may be the case in an ESP. As a result, the IWS maintains a high collection efficiency even for submicron particulate.

Advantages: In addition to the advantages associated with other wet systems, there are two principal advantages to this type of scrubber. First, the particulate removal efficiency is fairly constant for light or heavy loadings and for large or submicron particulate. If additional collection is needed, multiple units can be added in series. Then, each downstream unit will see the same particle–size distribution, rather than just fine material, and the removal efficiency will be approximately the same in each unit. In this manner, two scrubbers in series are reported to maintain a removal efficiency of 90 to 97% for 0.1 μm particle.

The second advantage of an IWS is the low energy requirement. The pressure drop per stage is approximately 2 inches of water. In comparison to a venturi scrubber, an IWS requires four to six times less energy for comparable removal.

Disadvantages: As with an ESP, an IWS has a large initial cost. There is also the danger of using high voltages with wet equipment. Extra precautions are necessary when operating and servicing this equipment.

Flux Force/Condensation/Collision Scrubbers: A unique device in particulate removal is the flux force/condensation/collision scrubber. This process combines technologies to form a complete pollution control system with enhancement of particulate removal.

In this process, the gases first enter a quencher and then pass through a packed–bed scrubber. This scrubber uses a structured packing that is reported to remove 99.9% of the

HCl and up to 99% of the SO_2.

An unusual feature of this scrubber is that the recycled scrub liquor is cooled to approximately 100°F before entering the scrubber to induce the flux force/condensation effect. The term "flux force/condensation" refers to the phenomenon that occurs when hot, saturated gases are brought into contact with cooler surfaces. At the cool surface, part of the water vapor in the gas stream condenses. If particulate is present in the gas stream near the cooler surface, the particulate serves as condensation nuclei for the moisture. In this manner, the particles grow in mass.

As the water vapor moves to the condensation nuclei, nearby particulate matter is swept by the condensing vapors and dragged into the cooler surfaces (e.g., droplets of scrubber liquid). This sweeping effect is referred to as flux force. The combination of flux force and condensation ultimately result in the growth of particulate 0.3 μm and smaller into larger particles/droplets with diameters of 0.6 μm and larger. As a result, approximately 15 to 40% of the particulate is removed by the scrubbing action of the packed tower. Furthermore, because of their increased size, the remaining particles are more efficiently removed in the collision scrubber which follows the packed-bed scrubber.

When the gas flow enters the collision scrubber, the flow is split into two separate streams. Each stream is then directed to a separate entrance of a unique venturi scrubber that has two inlets and one exit. The venturi is shaped like a T with an inlet at each side of the bar across the top of the T. There are two converging sections, one on each side of the top of the T, where the gas flows are accelerated before entering the venturi throat in the center of the top of the T.

At each entrance of the throat, water is injected and is immediately atomized into a flurry of small droplets. As the two gas streams rush down the throat towards each other, particulate is impacted into the water droplets as in a conventional venturi scrubber. When the gas streams collide into each other at the center of the throat, a turbulent zone is created where additional impaction and cleaning occurs.

Within this zone, many of the liquid droplets will collide, splitting into many smaller droplets that will enhance small particulate cleaning. Many droplets are also thrown into the opposing gas stream by their own momentum. When this happens, the initial velocity difference between the oncoming particulate and the droplets traveling in the opposite direction is very high. The collision scrubber makes use of the fact that the impaction scrubbing process becomes more efficient as the velocity difference between particles and droplets increases. The direct collision of gas streams results in a highly efficient cleaning process.

Eventually, the drag of the oncoming gas stream will slow and stop the droplets. Once the droplets are stopped and begin to reverse direction, flowing with the oncoming gas stream, the drag on the droplets tends to break the droplets into many smaller droplets, again enhancing collection. Eventually, the droplets will make the 90° turn with the gas stream and will flow down the leg of the T, which is a larger venturi scrubber throat, so that the combined gas stream will maintain the same velocity as the individual streams in the top of the T. At the end of the leg, the gases enter a diverging section as in a conventional venturi scrubber and then pass through a unique mist eliminator.

Advantages: An advantage to this system is the particulate removal efficiencies that are obtained. Because of the growth of the particles, outlet loadings down to 0.01 to 0.02

gr/dscf are achievable. If this low particulate outlet loading is not required, this system can obtain the same outlet loading as a conventional venturi scrubber but reportedly at approximately 75% of the pressure drop.

Besides improving particulate removal efficiency, the process of cooling the recycled scrub liquor is reported to have added benefits. The resulting flue gas temperature is 130°F or less. At this lower temperature, the flue-gas volume is decreased, and the induced fan and the fan power requirements are smaller.

The lower temperature results in better removal of condensible toxic metals and PICs. The lower gas temperature also results in lower gas saturation temperatures and lower amounts of moisture in the stack flue gas. As a result, the manufacturer of this system suggests that the heat of compression from the induced draft fan is sufficient to eliminate the stack steam plume in most weather conditions.

Disadvantages: There are two primary disadvantages associated with the flux force/condensation/collision scrubber technology. First, cooled scrubber liquor requires some type of cooling tower, chiller, or heat pump. Additionally, for radiological service, one or more heat exchangers would be required for cooling. This added equipment increases the capital and installation costs. Second, the added equipment increases the system complexity and, therefore, can result in additional operational problems and maintenance.

Froth Scrubbers: Froth scrubbers are a unique technology that can effectively and simultaneously quench flue gas temperature and provide subcooling if desired, remove particulate including submicron particles, and (when used with a caustic solution) remove acid gases. The following describes a froth and how it is created.

When mixing a liquid phase with a gas phase, four different regimes can be formed. With relatively high gas quantities and relatively low liquid quantities, the liquid is atomized and dispersed in the gas. The opposite condition occurs with relatively large quantities of liquid and relatively small to moderate quantities of gas. Under these circumstances, the gas will bubble through the liquid. When the quantities of gas and liquid are both relatively low, poor mixing can result and the phases will stratify (i.e., tend to remain separated). Between all of these regions at relatively moderate gas and liquid volumes, a froth zone can be formed by balancing the liquid and gas momentums. This froth zone results in a large interfacial area with intense mixing and turbulence and high mass and heat transfer.

This froth zone is utilized for gas scrubbing in a couple of devices. In the reverse jet scrubber, the dirty gas flow is directed through a cylindrical chamber. Pressurized scrub solution is injected in the opposite direction into the cylinder. As a result, a froth zone in the form of a standing wave is created at the point where the liquid begins to move radially outward. This standing wave can be induced to travel up or down through the cylinder by simply adjusting the gas or liquid relative power.

The second method used to generate a froth zone is a froth column. The froth column is similar to a standard tray scrubber, but the tray openings and liquid-to-gas ratio are designed to provide flooding in the top trays. In this way multiple froth zones are created in the column. The froth zones can be extended down the column by adjusting the liquid-to-gas ratio.

One to four reverse jets in series is the normal configuration for a froth scrubber. A froth column is normally used separately or in combination with one or more reverse jets when subcooling of the flue gas in the froth column is desirable to condense out volatile pollutants. In terms of liquid scrubbers, froth scrubbers use a large liquid–to–gas ratio ranging from 20 to 60 gallons per minute per 1,000 acfm. Reverse jet froth scrubbers have a relatively low gas pressure drop of 8 to 9 inches of water per stage. A froth column has a gas pressure drop of approximately 20 inches of water. For a typical application with multiple scrubbers, the overall gas pressure drop is approximately 35 inches of water.

In comparison to a venturi scrubber, this pressure drop is approximately one–half that required for removing submicron particulate. This low pressure drop is a consequence of 20 to 25% of the required scrubbing energy being supplied by the liquid flow. Also, because pump efficiencies are typically higher than fan efficiencies, froth scrubbers are reported to typically require 15 to 25% less total energy than a venturi scrubber.

Advantages: In addition to the lower energy requirements, froth scrubbers are simple to operate, require little maintenance, and are fairly reliable. These advantages stem from the fact that froth scrubbers have no moving parts, and the reverse jet scrubber uses injection nozzles with very large orifice diameters (typically 2 to 4 inches). With these large orifices, plugging is not a problem, and a high solids content in the scrubbing solution is not a problem.

As previously mentioned, the froth column can provide subcooling which can assist in particulate removal and condensation of volatile pollutants. Depending on the requirements of the scrubber, multiple staged units can achieve greater than 99% removal of particulate matter and acid gases. In addition, liquid/gas separation is not as difficult with froth scrubbers, in comparison to some other wet scrubbers, as the scrub solution is not atomized and fine mists are therefore minimized.

Disadvantages: As with all wet scrubbers, there is the problem of liquid effluent disposal. In addition, relatively little is known by the general engineering community about these scrubbers. These scrubbers were developed in the mid–1970's and held as proprietary technology until 1987. During this time, over 140 froth scrubbers were put into service for in–house use by the company that developed the technology. Unfortunately, very few companies knew about this technology or the operation of these units. Although these scrubbers are now commercially available, lack of information has kept them as relative mysteries. One limitation that is known about these scrubbers is a moderate turndown rate of approximately 2:1.

6.11 RETORTING

Retorting is a treatment technology applicable to wastes containing elemental mercury, as well as mercury present in the oxide, hydroxide, and sulfide forms, at levels above 100 parts per million, provided that the waste has a low total organic content (i.e., below 1%). For metals other than mercury, the typical retort operating temperatures (700° to 1000°F) are not high enough to decompose the metal compounds. High temperature metals recovery (HTMR) processes must be used to recover most other metals when they are not present in the pure metal form.

For most retorting processes, there is an additional requirement that the waste have a low water content, preferably below 20%. Dewatering reduces energy consumption by minimizing the amount of water evaporated and precludes problems involving the separation of recovered metals from large quantities of water.

Retorting is a process similar to that for high temperature metals recovery (HTMR) in that it provides for recovery of metals from wastes primarily by volatilization and subsequent collection and condensation of the volatilized components. Retorting yields a metal product for reuse and significantly reduces the concentration of metals in the waste residual, and, hence, the amount of treated waste that needs to be land disposed. This technology is different from HTMR in that HTMR includes a reduction reaction involving the use of carbon, while retorting does not use a reducing agent. Additionally, this process differs with regard to the form and, possibly, the leachability of the residue generated; HTMR generates a slag, while retorting generates a granular solid residue that may have lower leachability than a slag if mercury is the only constituent of concern present in the untreated waste.

The basic principle of operation of retorting is that sufficient heat must be transferred to the waste to cause elemental metals to vaporize. In the case of mercury present as an oxide, hydroxide, or sulfide compound, sufficient heat must be transferred to the waste to first decompose the compounds to the elemental form and then volatilize the mercury. In mercury wastes that are wastewater treatment sludges, mercury is most often present in the form of the sulfide (HgS) as a result of the use of sodium hydrosulfide treatment of mercury–bearing wastewaters. In a few instances, hydrazine is used to treat these same wastewaters; in such instances, a mercurous hydroxide sludge is generated. This latter compound can be more easily treated to yield elemental mercury because this reaction occurs more readily than the sulfide decomposition at the temperatures at which the process is normally operated. Preheated air is provided to the retort to supply the oxygen necessary for the sulfide decomposition and to enhance the heat transfer to the waste.

The retorting process generally consists of a retort (typically an oven, i.e., multiple hearth furnace or rotary kiln) in which the waste is heated to volatilize the metal constituents, a condenser, a metals collection system, and an air pollution control system.

Trays of wastes are placed in the retort, where they are heated, and decomposition of mercury compounds and volatilization of the metallic mercury and other volatile elemental metals occur. Although most commonly carried out in an oven, retorting can also be performed in a multiple hearth furnace.

The vapor stream from the retort is cooled in a condenser. If a scrubber is not used as an air pollution control device, an electrostatic precipitator is provided after the condenser to remove any residual metal in the exhaust vapor stream, as well as to control other potential emissions such as sulfur dioxide (SO_2), fly ash, and hydrogen chloride (HCl) vapors. Condensed metal is collected for reuse before the electrostatic precipitator.

Residual solids remaining in the retort, stripped of volatile metal contaminants, are collected and may be either directly land disposed or stabilized, to immobilize any remaining metal constituents, and then land disposed.

Because retorting is a recovery process, its product must meet certain purity requirements prior to reuse. If the waste contains other volatile constituents with boiling points equal to or below that of the metal(s) to be recovered, they will be volatilized and condensed along with the desired metal(s) present in the waste. These constituents may be difficult to separate from the recovered product and may affect the ability to reuse the product metal or refine it for subsequent reuse. Undesirable volatile constituents sometimes present in mercury–bearing wastes include (a) mercury chlorides, which distill unchanged in the retorting process, and other volatile metal halides; (b) arsenic oxide and arsenic trichloride; and (c) organomercury compounds, such as phenylmercuric acetate, which are not decomposed to elemental mercury by the retorting process.

For wastes containing significant levels (i.e., above 1%) of these contaminants, retorting may not be an appropriate technology. If the concentration of undesirable volatile constituents in the untested waste is significantly higher than that in the tested waste, the system may not achieve the same performance. Chemical pretreatment may be required to convert mercury chlorides and organomercury compounds to mercuric sulfide and/or elemental mercury and achieve the same treatment performance, or other, more applicable treatment technologies may need to be considered for treatment of the untested waste.

The ability to heat constituents within a waste matrix is a function of the heat transfer characteristics of the waste material. Mercury and other recoverable metals in the waste must be heated to near or above their boiling points in order to be volatilized and recovered. The rate at which heat will be transferred to the waste material is dependent on the material's thermal conductivity, which is the ratio of the conductive heat flow to the temperature gradient across the material. Thermal conductivity measurements, as part of a treatability comparison of two different wastes to be treated by a single retort system, are most meaningful when applied to wastes that are homogeneous (i.e., uniform throughout). As wastes exhibit greater degrees of nonhomogeneity, thermal conductivity becomes less accurate in predicting treatability because the measurement reflects heat flow through regions having the greatest conductivity (i.e., the path of least resistance) and not through all parts of the waste. Thermal conductivity may provide the best measure of performance of heat transfer. If the thermal conductivity of the untested waste is significantly lower than that of the tested waste, the system may not achieve the same performance and other, more applicable treatment technologies may need to be considered for treatment of the untested waste.

Temperature provides an indirect measure of the energy available (i.e., Btu/hr) to vaporize the metal of concern. The higher the temperature in the retort, the more likely it is that the metal will volatilize. The temperature must be equal to or greater than the boiling point of the metal. However, excessive temperatures could volatilize undesirable constituents into the product, possibly inhibiting its potential reuse, and also could cause sintering of the feed material. For mercury retorting, the retort temperature must be monitored to ensure that the system is operating at the appropriate design condition (a temperature at least equal to the boiling point of mercury 674°F but below 1000°F) and to diagnose operational problems.

The residence time impacts the amount of volatile metal volatilized and recovered. It is dependent on the retort temperature and the thermal conductivity of the waste. Typical residence times in retort systems for mercury range from 4 to 20 hours.

6.12 SOIL FLUSHING

Soil flushing is an in situ extraction of inorganic or organic compounds from soils and is accomplished by passing extractant solvents through the soils using an injection–recirculation process. This is to be distinguished from soil washing, which involves processing of excavated soil.

The use of soil flushing to remove soil contaminants involves the elutriation of organic and/or inorganic constituents from soil for recovery and treatment. The site is flooded with the appropriate washing solution, and the elutriate is collected in a series of shallow wellpoints or subsurface drains. The elutriate is then treated and/or recycled back into the site. During the elutriation process, contaminants are mobilized into the flushing solution by way of solubilization, formation of emulsions, or a chemical reaction with the flushing solution. Collection of elutriate is required to prevent uncontrolled contaminant migration through uncontaminated soil and into receiver systems, including ground and surface waters.

Flushing solutions may include water, acidic aqueous solutions (sulfuric, hydrochloric, nitric, phosphoric, and carbonic acid), basic solutions (e.g., sodium hydroxide), and surfactants (e.g., alkylbenzene sulfonate). Water can be used to extract water–soluble or water–mobile constituents. Acidic solutions are used for metals recovery and for basic organic constituents (including amines, ethers, and anilines); basic solutions for metals (including zinc, tin, and lead); and basic solutions for some phenols, complexing and chelating agents, and surfactants.

The addition of any flushing solution to the system requires careful management and knowledge of reactions that may adversely affect the soil system. For example, a sodium addition as sodium hydroxide to soil systems may adversely affect soil permeability by affecting the soil sodium absorption ratio. It is not only important to understand the chemical reaction(s) between the solvent and solute, but also between the solvent and the site/soil system.

At a site contaminated by organic constituents, recycling the elutriate back through the soil for treatment by biodegradation may be possible. Proper control of the application rate, based on hazardous waste land treatment principles, would provide for effective in situ treatment at soil concentrations that would allow controlled biodegradation of the waste constituents. This approach could eliminate the need for separate processes for treatment and disposal of the collected waste solution, or at least provide for a combination of pretreatment/land application that may be considerably more economical than the use of unit operations alone for treatment of elutriate.

For soils contaminated with inorganic and organic constituents, a combination of pretreatment that reduces or eliminates the metal constituent(s) in the elutriate by precipitation, followed by a land application of the elutriate, may be a feasible cost–effective method of treatment.

Soil flushing and elutriate recovery may also be appropriate in situations where chemical oxidizing or reducing agents are used to degrade waste constituents chemically and result in the production of large amounts of oxygenated, mobile, degradation products. The most conservative and safest approach may be to flush the soil after treatment to recover and possibly to reapply the elutriate in a controlled manner to the soil surface.

Both inorganics and organics are amenable to soil flushing treatment if they are sufficiently soluble in an inexpensive solvent that can be obtained in large volumes. Surfactant can be used for hydrophobic organics. Studies have been conducted to determine appropriate solvents for mobilizing various classes and types of waste constituents. Laboratory testing on site–specific soils should be undertaken to verify surfactant properties.

Success or failure also depends on the retention or inactivation of the solvent by the soil. The soil overlying the contaminated zone will need to be completely saturated before the solvent reaches the zone of contamination. The interaction of the solvent with the soil must be considered for successful application of the treatment.

The level of treatment that can be achieved will vary depending on the contact of the flushing solution with waste constituents, the appropriateness of the solutions for the wastes, the soil adsorption coefficients of the waste(s), and the hydraulic conductivity of the soil. This technology should produce the best treatment results in highly permeable soils with low organic content. Despite the varying level of treatment accomplished by soil flushing, however, once the waste components have been removed from the soil, results are not reversible and no retreatment is necessary.

The solutions used for the flushing may themselves be potential pollutants; they may have toxic and other environmental impacts to the soil system and water receiver systems. The soil system after treatment is altered from its original state. Its physical, chemical, and biological properties may be altered adversely (e.g., the pH may be lowered by the use of an acidic solvent) or the soil may be compacted as a result of being flooded. These soil properties may have to be restored to assure that other treatment processes can occur (e.g., biodegradation).

A potential exists for the flushing solvent to transport contaminants into an underlying aquifer. In almost all cases where extraction technologies are used, it is imperative that the groundwater flow rate and direction are fully understood before solvents are applied to the contaminated area. Also, withdrawal wells with sufficient capacity to draw all solvent/contaminant solutions to the surface for treatment are necessary.

Equipment used for soil flushing includes drains and a collection and distribution system. Reapplication of collected elutriate also may necessitate the construction of a holding tank for the elutriate. Solvents for flushing are required.

Recent research results from a field test have demonstrated that the surfactant flushing process is capable of rapid removal of dense nonaqueous phase liquids (DNAPLs) from a contaminated aquifer. The surfactant solution successfully removed perchloroethylene (PCE) at a rate far greater than it could have been removed by conventional pump–and–treat methods. Surfactants have the ability to greatly increase the solubility of organic compounds in water and thus to increase the efficiency of pump–and–treat operations.

Hot water or steam flushing can remove volatile and semi–volatile organic compounds such as TCE, TCA and dichlorobenzene (DCB) above and below the water table. Steam is forced into the aquifer by injection wells and the vaporized volatile components are removed by vacuum extraction. The technology uses readily–available components, such as extraction and monitoring wells, manifold piping and vacuum pumps. Hot water injection may be particularly useful at oil refineries, which often have oil–contaminated groundwater and waste heat that can be used in the recovery process.

A similar use of thermal technology can be used to recover oily wastes by adapting a technology presently used for secondary petroleum recovery and for primary production of heavy oil and tar–sand bitumen. Steam is injected below the oily wastes and condenses to cause rising hot water that dislodges the oils upward into more permeable soil regions. Hot water can then be injected adjacent to the now–floating oil bank to move the oil to extraction wells as it is contained and moved by barriers of hot water.

Advantages:

1. Removal of contaminants is permanent, no additional treatments are necessary if the soil flushing process is successful.
2. The technology is easily applied to permeable soils.
3. Costs are moderate, depending on the flushing solution chosen.

Disadvantages:

1. The technology introduces potential toxins (the flushing solution) into the soil system.
2. Physical/chemical properties of the soil system may be altered because of the introduction of the flushing solution.
3. A potential exists for solvents to transport contaminants away from the site into uncontaminated areas.
4. A potential exists for incomplete removal of contaminants due to heterogeneity of soil permeability.

6.13 SOIL VAPOR EXTRACTION

Soil vapor extraction (SVE) or vacuum extraction is an accepted cost–effective technique for removing volatile organic compounds (VOCs) and motor fuels from contaminated soil. This technology is known in the industry by various names, including soil vapor extraction, vacuum extraction, soil venting, aeration, in situ volatilization, and enhanced volatilization. In this book the term soil vapor extraction (SVE) is used. Vapor extraction systems have many advantages that make this technology applicable to a broad spectrum of sites:

1. SVE is an in situ technology that can be implemented with a minimum of site disturbance. In many cases, normal business operations may continue throughout the cleanup period.
2. SVE has potential for treating large volumes of soil at reasonable costs, in comparison to other available technologies.
3. SVE systems are relatively easy to install and use standard, readily–available equipment. This allows for rapid mobilization and implementation of remedial activities.
4. SVE effectively reduces the concentration of volatile organic contaminants in the vadose zone, which in turn reduces the potential for further transport of contaminants due to vapor migration and infiltrating precipitation.
5. SVE can serve as an integral component of a complete remedial program, which may include groundwater extraction and treatment.
6. Discharge vapor treatment options allow design flexibility required to satisfy site specific air discharge regulations.

The basic equipment for SVE systems consists of pumps or blowers to provide the motive force for the applied vacuum; the piping, valves, and instrumentation to transmit the air from the wells through the system and to measure the contaminant concentration and total air flow; vapor pretreatment to remove soil and water from the vapor stream; and an emission control unit to concentrate or destroy the vapor phase contaminants.

Contaminants removed from the subsurface through SVE are not destroyed but rather only transferred to a different location. Many states and localities realize that without some type of treatment of the off–gas, soil vapor extraction may be simply substituting an air quality problem for a soil and water quality problem. For this reason, regulations regarding the disposition of the vapor phase contaminants are becoming increasingly widespread. In general, treatment is now or soon may be required for all vapor streams except those that discharge minor amounts of contaminants into the atmosphere.

Some vendors now sell preassembled systems that can be hooked up to a well or manifold piping and incorporate all the equipment. Often these units are trailer– or skid–mounted and can be brought directly to the site. Such systems have air/water separation, emission control, and may be operated by computer. These systems can usually be rented, leased or purchased outright.

At the time of publication, soil vapor extraction is the most rapidly growing method of site remediation. It is particularly appropriate for compounds having high vapor pressure, such as chlorinated organics.

Cleaning Up the Nation's Waste Sites: Markets and Technology Trends reported the use of innovative technologies at 125 sites with Records of Decision (RODs) on the National Priorities List (NPL).

Soil vapor extraction was the method chosen to treat VOCs at 83 sites. Bioremediation was selected at 22 sites and thermal desorption was selected for use at 20 sites. In situ flushing was chosen for 11 sites, solvent extraction was chosen at four sites and soil washing was implemented at only one of the 125 NPL sites.

These methods often were used to treat not only VOCs, but also SVOCs and metals at a number of sites. Also, more than one treatment technology may have been used at a single site. For example, soil vapor extraction was used at two sites to enhance in situ bioremediation. Air sparging enhancement is discussed in Section 6.3.

A major advantage of soil vapor extraction technology is the relative simplicity of the design of these systems. In addition, the equipment that comprises the systems consists of commonly used and widely available devices such as PVC piping, valves, and pumps. These factors impart an advantage to soil vapor extraction over other techniques (e.g., biotreatment or soil flushing) that may require more complex design or single–purpose equipment. Simplicity of design, however, does not imply that a logical, reasoned, and informed design procedure has been followed for all site specific installations. Maximum system efficiency and contaminant removal will occur only through a thorough understanding of the site and the SVE process.

The objective of a well–thought out and reasoned design process is to construct a soil vapor extraction system that removes the greatest degree of contamination from the site in the most efficient, timely, and cost–effective manner. The attainment of that objective will occur through an understanding of the three main determinants of system effectiveness; the composition and characteristics of the contaminant; the vapor flow path and flow rate; and the location of the contamination with respect to the vapor flow paths. Design of an SVE system is basically a process to maximize the intersection of the vapor flow paths with the contaminated zone. Operation of the system should be done to maximize the efficiency of the contaminant removal and reduce costs.

Several options are available for layout of systems, including wells, trenches, and above ground soil piles. After selection of the appropriate system option, the number and placement of wells or trenches, the applied vacuum and pumping rate, the use of a surface seal or other types of air flow control, and the depth and size of the screened interval are all decisions that are made to maximize system effectiveness. Selection of the appropriate equipment type will also affect the system effectiveness.

6.13.1 Extraction System Options

Several options are available for extraction system layout: vertical wells, trenching or horizontal wells, and excavated soil piles.

Vertical Wells: Vertical wells are the most widely used SVE design method. This method is the only feasible option at sites where the contamination extends far below the land surface. Vertical wells are generally inappropriate for sites with a shallow water table due to the potential upwelling of the water table that may occur after the application of a high vacuum.

Extraction wells are similar in construction to monitoring wells and, in many cases, existing monitoring wells have been used as extraction wells. Construction of an extraction well is straightforward. The bore hole is augured or drilled, PVC casing and screening (usually 2 to 12 inches in diameter but depends on flow rate) are placed in the hole and the annular space is filled. Slots are usually sized as small as possible to reduce silt entrainment. A highly permeable sand or gravel packing is placed around the screen for optimal gas flow to the well. Above the pack, bentonite is used to seal the hole. A cement–bentonite grout is typically used to seal the annular space to the surface.

The extraction well is typically located to intercept the center of contamination. Where multiple wells are used, they are placed so that the flow zone intercepts the contaminated zone. The screened interval should also coincide with the depth of highest product concentration. Often, this is just above the water table for products lighter than water like petroleum. The screened interval should be extended into the water table to allow for the possibility of a fluctuating water table. Also, the application of a vacuum will result in upwelling of the water table; if not counteracted, the wells may remove less vapor and more water, depending on the magnitude of the upwelling.

Trenches/Horizontal Wells: Where the water table is near the surface, trenches or horizontal wells may be installed. Horizontal wells minimize the upwelling of the groundwater and allow coverage of a greater area than vertical wells. Installation of this type of well is accomplished quickly and easily where no surface or subsurface impediment exist. A PVC drain pipe, wrapped in filter fabric to prevent fine material from clogging the drain, is placed at the base of the trench and backfilled with gravel. The surface is typically sealed with bentonite, asphalt, or a manmade liner to prevent air short–circuiting and infiltrating rainwater. The result is simply a dry trench or french drain.

Horizontal wells can also be installed without excavation and backfill using special drilling techniques. Conventional drilling uses rigid drilling assemblies, whereas horizontal (lateral) drilling uses jointed, flexible drive pipe. While not widely practiced, there are companies such as Eastman Christensen (Houston, Texas) that specialize in horizontal drilling. Horizontal drilling possesses several advantages including those mentioned above, the ability to intercept specific zones, and the possibility of intercepting more vertical fractures.

Excavated Soil Pile: Soil vapor can also be extracted from above ground piles of excavated soil. This option is often used to remediate the soil removal from the leaking tank area at UST sites. The usual procedure is to excavate the contaminated soil and place it in a pile over one or more PVC pipes, which are packed in gravel and encased in filter fabric. An impermeable liner may be used to cover the excavated soil to prevent contaminants from volatilizing to the ambient air in an uncontrolled fashion and to prevent infiltration of precipitation. Prior to the system operation the cover is removed to allow air to be drawn into the pile.

6.13.2 Well Configuration

Two main issues must be addressed with regard to the configuration of the extraction well(s). First, the number of wells required and their proper spacing and placement must be determined. Second, the extraction vents need to be sized and placed for optimal removal.

Spacing and Placement: The number and locations of extraction wells required at a remediation site is highly site–specific and depends on many factors, including the extent of the zone of contamination, the physicochemical properties of the contaminants, the soil type and characteristics (especially the air permeability of the soil), the depth of contamination, and discontinuities in the subsurface. The radius of influence is the primary design variable and incorporates many of the above parameters. The radius of influence is the zone in which the effect of the vacuum is felt.

Extraction Vents: Extraction vents may be screened, slotted, or gravel packs. In addition to the vent construction, two other design decisions must be considered: the length of the vent and its location with respect to the unsaturated zone.

A goal of proper well design is to induce the air to flow through the zone of contamination to maximize cleanup efficiency. This is controlled by both well spacing and layout and by vent location. SVE operators use widely varying approaches to vent design, ranging from screening the entire depth from near the ground surface to the water table, to having a short interval at a particular depth corresponding to the zone of contamination. The location and length of screening will depend upon the stratification of the soil and the distribution of the contaminants in the soil. In many cases, the greatest concentration of petroleum vapors is immediately above the water table, especially at sites with a free product lens on the water table. Determination of the concentration gradient throughout the vadose zone can be accomplished with soil gas survey techniques.

In cases where the contaminant concentrations are greatest at the water table, the vents should be located close to or into the water table for optimal removal efficiency. In areas where the water table is expected to fluctuate throughout the period of venting, the screen length can be increased to ensure that venting can continue during periods of high water table.

The extraction wells are normally grouted or sealed with bentonite in the annular space to prevent ambient air from entering directly along the borehole/well interface. In some cases, the annular space is grouted down to the screened interval.

6.13.3 Air Flow Control

In addition to the placement of the vacuum extraction wells, several other methods are available to control the flow paths of the extracted vapor to result in more efficient contaminant removal. These methods include the use of air injection or passive inlet wells, impermeable surface seals, and groundwater depression pumping.

Air Injection Systems: To enhance air movement through the soil, inlet or injection wells can be placed in strategic locations. Inlet wells are open to the atmosphere and allow air to be drawn passively into the soil from the surface. Injection wells use forced air to control the movement of air through the soil.

Inlet or injection wells provide several advantages to SVE system design versus systems without air inlet wells. A major advantage is the ability to control the air pathways and thus, the zones of the soil to be affected by an SVE system. This allows the SVE system to be designed to give intensive treatment of a specific small area, rather than less intensive treatment of a larger area. The use of inlet or injection wells may also allow more rapid cleanup—by allowing greater flow rates—than would otherwise be possible. Injection of the extracted air may eliminate the need to obtain an air discharge permit. Also, if air is injected below the water table, volatiles may be "stripped" or volatilized from the dissolved phase into the soil gas. Disadvantages of their use include the added cost associated with construction of additional wells (although this would not be the case if inlet wells were just inactive extraction wells) and the added energy cost of the compressor for injection wells.

Typically, injection and inlet wells are similar in construction to extraction wells, although injection wells may have a longer screened interval to provide uniform air flow. In fact, a well designed SVE system allows wells to act as extraction, injection, and/or inlet wells depending on the system requirements. Injection wells should be placed so that contamination is not forced away from the extraction wells in a manner that will result in bypassing the vapor treatment system.

Surface Seals: An impermeable seal may be used where minimization of inflow from the surface is required. An impermeable surface seal prevents air from entering from near the extraction well (where the pressure gradient is the greatest) and forces air to be drawn from a greater distance and ultimately, to contact a greater volume of soil. Surface seals may also prevent infiltration of rainfall, reducing the amount of water removed by the extraction well, thereby minimizing the production of air–water separator sidestreams. Surface seals also reduce fugitive VOC emissions from the soil to the air.

Depending on the characteristics of the site, different materials can be used as an impermeable cap. A flexible membrane lining can be rolled out on the site and can easily be removed when the treatment is complete. These membranes are available in a variety of materials, with high density polyethylene (HDPE) being the most common. An alternative to a synthetic membrane is a clay or bentonite layer. A third alternative, the most common at commercial or industrial sites, is the use of concrete or asphalt as a cap.

6.13.4 Process Enhancement

Several technologies have been used to enhance the effectiveness of SVE systems, including physical methods such as ambient air, oxygen, or ozone injection, and biological methods such as nutrient or moisture addition. Another potentially useful method of enhancing SVE is elevation of soil temperature by heated air and steam injection.

Several methods have been suggested for elevating soil temperatures, including electrical means, radio frequency and conduction heating, injection of exhaust from combustion units, and steam injection.

Commercial processes have been developed for injection of steam and heated air underground, which evaporates the organics, causing them to rise to the surface.

In one process, steam is piped to the top of the drills and injected through the cutting blades. The steam heats the ground being remediated, which increases the vapor pressure of the volatile contaminants and the rate at which they can be stripped. Both the air and the steam serve as carriers to convey these contaminants to the surface. The shroud, a metal box designed to seal the process area above the rotating cutter blades from the outside environment, collects the volatile contaminants and ducts them to the process train. In the process train, the volatile contaminants and the water vapor are removed from the off-gas stream by condensation. The condensed water is separated from the organics by distillation, filtered through activated carbon beds, and subsequently used as makeup water in a cooling tower.

The Pacific Northwest Laboratory (DOE) has developed an extraction technique, called electrical remediation at contaminated environments (ERACE), enhances conventional soil vapor extraction by heating the soil with an in situ electrical field. The computer modeling developed for the ERACE technique helps predict the equipment and power requirements based on the characteristics of the site.

In the Dynamic Underground Stripping Process developed by the Department of Energy, injection wells are installed in permeable areas surrounding the concentrated plume, and one or more extraction wells are installed in the center. The extraction wells are pumped to depress the water table in the center of the pattern. Then, steam is injected at 50 to 60 psi. Injection pressure is controlled by depth, and would be lower in shallow applications.

As the steam is forced into the formation, the earth is heated to the boiling point of water. The advancing pressure front displaces ground water toward the extraction well. Near the steam–condensate front, organics are distilled into the vapor phase, transported to the front, and condensed there. The advancing steam zone displaces the condensed liquids toward the recovery well where they are pumped to the surface.

When the steam reaches the extraction well, vacuum extraction becomes the most important removal mechanism.

At this point in the process, electrode assemblies placed in the impermeable layers are turned on, passing 480 V current at several hundred amperes per electrode. This heats clay and fine–grained sediments, causing any water and contaminants trapped within to vaporize and be forced into the steam zones and toward the extraction well. This heating may be followed by one or more additional steam injection phases, for contaminant removal and to keep permeable zones hot as ground water returns.

Vacuum extraction can enhance biodegradation of volatile and semi–volatile chemicals in the soil by providing oxygen to the soil for use by microorganisms. Larger amounts of oxygen can be supplied per volume of air than per volume of water. This use of vacuum extraction to enhance biodegradation is also known as **bioventing.**

Also see: pneumatic fracturing (6.18.7) and air sparging (6.3).

6.14 SOIL WASHING

Soil washing is an ex situ process in which contaminated soil is excavated and fed through a water–based washing process. It operates on the principle that contaminants can be dissolved or suspended in an aqueous solution or removed by separating out clay and silt particles and the associated adhered contaminants from the bulk soil. Mineral processing techniques (hydrometallurgy) may be used. The aqueous solution containing contaminants may be treated by conventional wastewater treatment methods.

Most organic and inorganic contaminants bind chemically or physically to clay or silt soil particles, which in turn adhere to larger sand and gravel particles primarily by compaction and adhesion. Particle size separation by washing enables the contaminated clay and silt particles (and the bound contaminants) to be concentrated. Separating the sand and gravel from the small contaminated soil particles significantly reduces the volume of contaminated soil, making further treatment or disposal much easier. The larger particles may be returned to the site.

The wash water may be augmented with a basic leaching agent, surfactant, pH adjustment, or chelating agents such as ethylenediaminetetraacetic acid (EDTA) to help remove organics or heavy metals. Treated soil is cleaned of any residual additive compounds. The spent wash water is treated to remove the contaminants prior to recycling back to the treatment unit.

If soil washing lowers contaminant concentrations in the soil to acceptable levels, the only additional treatment to consider would be emissions controls. Soil washing also serves as a cost–effective pre–processing step for further treatment. It can potentially be effective for the remediation of soils with a small amount of clay and silt particles and with a wide variety of organic, inorganic and reactive contaminants. Large amounts of clay and silt particles mitigate the effectiveness of soil washing and make it inadequate as the only treatment method. Removal efficiencies range from 90 to 99% for volatile organic compounds (VOCs) and 40 to 90% for semi–volative compounds. Compounds with low water solubilities such as metals and pesticides sometimes require acids or chelating agents to assist in removal.

Particle size distribution is a key parameter in determining the feasibility of soil washing. The relative effectiveness of soil washing for various soil types are shown below.

Particle Size Distribution (mm)	Effectiveness
>2	Oversize pretreatment requirements
0.25–2	Effective soil washing
0.063–0.25	Limited soil washing
<0.063	Clay and silt fraction: difficult soil washing

Bench–scale and pilot–scale treatability tests are recommended before undertaking full–scale operation. Further concerns about feasibility include the fraction of hydrophobic contaminants that require surfactants or organic solvents for effective removal, the complexity and stability of the contamination that affect washing fluid formulation, and the effect of wash water additives on wastewater treatment.

Excavation and removal of debris and large objects precedes the soil washing process. Sometimes water is added to the soil to form a slurry that can be pumped. After the soil is prepared for soil washing, it is mixed with wash water and sometimes extraction agents.

At this point, three separation processes occur: (1) water–soluble contaminants are transferred to the wash water; (2) contaminants are suspended in the wash water; and (3) clay and silt particles to which contaminants are adhered separate from larger particles. After separation from the wash water, the soil is rinsed with clean water and may be returned to the site. The suspended soil particles are removed by gravity from the wash water as sludge. Sometimes flocculation is used to aid in sludge removal. This sludge is more highly contaminated than the original soil and undergoes further treatment or secure disposal. The spent wash water from which the sludge is removed is treated and recycled. Residual solids from the recycle process may require further treatment.

Soil washing generates four waste streams:

1. Contaminated solids separated from the wash water;
2. Wastewater;
3. Wastewater treatment sludges and residual solids; and
4. Air emissions.

Any of a number of treatments is feasible for the contaminated clay fines and solids. They may successfully undergo incineration, low temperature thermal desorption, solidification and stabilization, and biological or chemical treatment. It is recommended that as much blowdown water be recycled as possible. Blowdown water released to local wastewater treatment plants must meet local discharge standards. Sludges and solids from wastewater treatment require appropriate treatment and disposal. Collected air emissions from the waste site or soil washing unit can be treated as well.

Advantages of the soil washing process include:

1. Applicability to a wide variety of organic and inorganic compounds.
2. High removal efficiencies for certain soil types.
3. Minimal fire and explosion hazards.

Some disadvantages as compared to other remediation processes are that soil washing:

1. Is suitable for only certain soil types.
2. Does not destroy contaminants.
3. May require additives that improve removal but compromise treatment of the waste streams.

Combination Processes: A number of other processes can be used with soil washing. For example, chemical extraction and soil washing are physical transfer processes in which contaminants are washed from the soil, becoming dissolved or suspended in a liquid chemical. This liquid waste stream then undergoes subsequent treatment to remove the contaminants, and the solvent is recycled, if possible. Soil washing processes generally use water as the solvent to separate the clay particles, which contain the majority of the contaminants, from the coarser soil fractions. Chemical extraction processes generally use a solvent that separates the contaminants from the soil particles and dissolves the contaminant in the solvent. Numerous combinations of soil washing, chemical extraction, and other treatment technologies are available.

If the selection of the solvent is optimized with the addition of surfactants or chelating agents, chemical extraction and soil washing can successfully treat many organic and inorganic contaminants, particularly those that are more soluble in the solvent of choice.

The Biogenesis™ soil washing technology developed by BioGenesis Enterprises, Inc., utilizes biodegradation in the soil washing unit.

6.15 STRIPPING

Stripping, in general, refers to the removal of relatively volatile components from a wastewater by the passage of air, steam, or other gas through the liquid. The stripped volatiles are usually processed further by recovery or incineration.

Stripping processes differ according to the stripping medium chosen for the treatment system. Air and steam are the most common media, with inert gases also used.

Strippers are widely used in industry to remove a variety of materials, including hydrogen sulfide and volatile organics as well as ammonia, from aqueous streams. The basic techniques have been applied both in process and in wastewater treatment applications and are well understood. The use of steam strippers with and without pH adjustment is standard practice for the removal of hydrogen sulfide and ammonia in the petroleum refining industry and has been studied extensively in this context. Air stripping has treated municipal and industrial wastewater and is recognized as an effective technique of broad applicability. Both air and steam stripping have successfully treated ammonia-laden wastewater, both within the nonferrous metals manufacturing category or for similar wastes in closely related industries.

The major drawback of air stripping is the low efficiency in cold weather and the possibility of freezing within the tower. Because lime may cause scaling problems and the types of towers used in air stripping are not easily cleaned, caustic soda is generally employed to raise the feed pH. The two major limitations of steam strippers are the critical column design required for proper operation and the operational problems associated with fouling of the packing material.

6.15.1 Air Stripping

Air stripping (aeration) is essentially a gas transfer process in which a liquid containing dissolved gases is brought into contact with air and an exchange of gases takes place between the air and the solution. In general, the application of air stripping depends on the environmental impact of the resulting air emissions. If sufficiently low concentrations are involved, the gaseous compound can be emitted directly to the air. Otherwise, air pollution control devices may be necessary.

Factors important in removal of organics from wastewater via air stripping are temperature, pressure, air to water ratio and surface area available for mass transfer. Air to water volumetric ratios may range from 10:1 up to 300:1. The resulting residuals are the contaminated off-gas and the stripped effluent. Volatilized hazardous materials must be recaptured for subsequent treatment to preclude air pollution concerns.

This process is used to treat aqueous organic waste with relatively high volatility, low water solubility (e.g., chlorinated hydrocarbons such as tetrachloroethylene) and aromatics (such as toluene). Limitations include the fact that the process is temperature dependent so that stripping efficiency can be impacted by changes in ambient temperature and the presence of suspended solids may reduce efficiency. If the concentration of VOCs exceeds approximately 100 ppm, some other separation process (e.g., steam stripping) is usually preferred.

Types of air stripping devices include diffused aerators, mechanical surface aerators, coke tray aerators, sprays and spray towers, packed towers (countercurrent and cross-flow), as well as cascade and U-tube aeration. Countercurrent packed towers are best suited for leachate treatment applications because (1) they provide the greatest gas–liquid interfacial area for mass transfer, (2) they can be operated at higher air–to–water volume ratios than the other devices, and (3) emissions to the atmosphere are more easily controlled.

The applicability of air stripping for removal of a particular contaminant can be predicted by the use of vapor/liquid equilibria data. Because the vapor/liquid equilibrium behavior of a compound varies with temperature and the presence of other constituents, air–stripping efficiency should be determined experimentally in laboratory evaluations with actual leachate. Anticipated extremes of temperature and contaminant loading should be included in the treatability program.

High–temperature air stripping, in which the feed is preheated, has been applied to remove some chemicals (e.g., methyl ethyl ketone) that are not easily stripped at ambient temperatures.

Pretreatment requirements for air stripping include removal of suspended solids and separation of nonaqueous phases. In systems where chemical neutralization or precipitation/sedimentation precede the stripper, sodium–based reactants should be selected over calcium–based reactants, as the latter can lead to scale formation. Acidification of the stripper feed can also help prevent scaling.

Post–treatment also may be necessary to reduce corrosive properties that develop in the water due to the increased presence of dissolved oxygen during the aeration process. One solution is the use of a corrosion inhibitor.

Aeration provides a fixed percentage of contaminant removal regardless of influent concentration. To compensate for uncertainty, aeration systems can be designed to incorporate safety factors of two or three times the expected influent contaminant concentrations to ensure compliance with regulatory standards.

Aeration system performance is affected primarily by column size and airflow. Increases in airflow and column height improve removal efficiencies. Design considerations include:

1. Type of organic contaminant(s).
2. Concentration of contaminant(s).
3. Type of packing material.
4. Height of packing material.
5. Air–to–water ratio.
6. Water loading rate.
7. Water temperature.

Packed Tower Aeration: Packed tower aeration (PTA), or packed column aeration (PCA), is a waterfall aeration process that trickles raw water over a medium within a cylinder to mix water with air. The medium is designed to break the water into tiny droplets, a process enhanced by the introduction of air blown from underneath the medium in the tower or column.

The major process elements of PTA are the column (or tower), packing medium, blower, booster pump, and instrumentation. Columns can be constructed from fiberglass–reinforced plastic, aluminum, stainless steel, or concrete. Within the column are mist eliminators to prevent water from escaping in the vents, packing material, support grids for the packing material, and liquid distributors to separate the influent into many smaller streams. The four primary designs for liquid distributors are orifice plate, through–type distributor, orifice headers, and spray nozzles. Adding PTA to an existing plan will require (1) changes in the staging of the well pumps and (2) repumping treated water to the distribution system.

Housing the tower usually is not necessary because the water temperature remains fairly constant throughout the PTA–treatment process. Consequently, water rarely freezes during the process.

PTA systems vary: in some, water cascading over spillways imparts the necessary turbulence; in others, several layers of slats mix the water with air. Innovations in PTA are reflected in the newest additions to the latter type of aeration system. Emerging aeration techniques include the catenary grid and Higee systems which are discussed below.

Catenary Grid: Catenary grid systems are a variation of the PTA process. The catenary grid directs water through a series of parabolic wire screens mounted within the column, above which turbulence is created. The screens mix the air and water in the same way as the packing materials in PTA systems.

These systems can achieve VOC removal rates comparable to PTA systems. Catenary grid units require more airflow and, thus, have higher energy requirements than PTA systems. They also have shorter aeration columns with smaller diameters. Their more compact design lowers their capital cost relative to PTA.

Catenary grid systems, however, have limitations. Limited data are available concerning this system's removal effectiveness for a wide variety of organic compounds. Also, the procedure for scaling systems up from pilot plants to full–scale operations is not fully developed.

The principal design considerations for catenary grid systems are air–to–water ratio, number of screens, and hydraulic loading rate. Removal efficiency improves with increases in air–to–water ratios and increasing number of screens in the column.

Highee Aeration: Highee aeration systems are another variation of the PTA process. These systems pump water into the center of a spinning disc of packing material to achieve the necessary air and water mix. By design, the packing material has a large surface area per unit volume. Air is pumped countercurrently toward the center from the outside of the spinning disc. Simultaneously, water flows from the center of the disc and mixes with the air.

Highee units require less packing material than PTA units to attain equivalent removal efficiencies. They require smaller air volumes and can process high water flows in a compact space. The Higee unit's compact size permits its application within constrained spaces and heights.

Mechanical Aeration: Mechanical aeration systems use surface or subsurface mechanical stirring mechanisms to create turbulence to mix air with the water. These systems effectively remove VOCs but are generally used for wastewater treatment systems.

Mechanical aeration units consume large amounts of space because they demand long detention times for effective treatment. As a result, they often require open–air designs, which may freeze in very cold climates. These units also have high energy requirements. Mechanical aeration systems, however, are easy to operate and are less susceptible to clogging from biological growth than PTA.

Diffused Aeration: The diffused aeration system bubbles air through a contact chamber for aeration; the diffuser is usually located near the bottom of the chamber. The air introduced through the diffuser, usually under pressure, produces fine bubbles that impart water–air mixing turbulence as they rise through the chamber. Diffused aeration units are designed to serve either point–of–use or plant situations.

The main advantage of diffused aeration systems is that they can be created from existing structures, such as storage tanks. This type of aeration, however, is less effective than PCA and is generally used only in systems with adaptable existing structures.

Multiple Tray Aeration: Multiple tray aeration directs water through a series of trays made of slats, perforations, or wire mesh. Air is introduced from underneath the trays, either with or without added pressure.

Multiple tray aeration units have less surface area susceptible to clogging from iron and manganese precipitation than PTA. However, this type of aeration is not as effective as PTA and can experience clogging problems, in addition to biological growth and corrosion problems. Multiple tray aeration units are generally available as package systems.

The principal design considerations for multiple tray aeration are tray type, tray height, pressurized or unpressurized air flow, and air–to–water ratio. Trays are usually made from wood or plastic and range in stack height from 3.6 to 4.8 m (12 to 16 ft). Pressurized air flow is used to increase the air–to–water ratio, with the typical ratio being 30:1.

Spray Aerators: These aerators are used in drinking water plants, particularly in Florida, to remove organics from drinking water.

In Situ Volatilization: Air stripping related techniques can be utilized to remove volatile organic chemicals from soil, and is essentially a soil vapor extraction process, and discussed under that heading. It is also discussed in relation to horizontal wells.

In Situ Air Stripping: In situ air stripping using horizontal wells is designed to concurrently remediate unsaturated–zone soils and groundwater containing volatile organic compounds (VOCs). The in situ air stripping concept utilizes two parallel horizontal wells: one below the water table and one in the unsaturated (vadose) zone. The deeper well is used as a delivery system for the air injection. VOCs are stripped from the groundwater into the injected vapor phase and are removed from the subsurface by drawing a vacuum on the shallower well in the vadose zone. The technology is based on Henry's Law, and the affinity of VOCs for the vapor phase. The technology is probably most effective in soils with high permeability and likely works best in sandier units with no significant aquitards between the injection and extraction wells. Horizontal wells are used because they provide more surface area for injection of reactants and extraction of contaminants and they have great utility for subsurface access under existing facilities.

6.15.2 Steam Stripping

Steam stripping is essentially a fractional distillation of volatile components from a wastewater stream. The volatile component may be a gas or an organic compound that is soluble in the wastewater stream. More recently, this unit operation has been applied to the removal of water immiscible compounds (chlorinated hydrocarbons), which must be reduced to trace levels because of their toxicity.

Steam stripping is usually conducted as a continuous operation in a packed tower or conventional fractionating distillation column (bubble cap or sieve tray) with more than one stage of vapor/liquid contact. The preheated wastewater from the last exchanger enters near the top of the distillation column and then flows by gravity countercurrent to superheated steam and organic vapors (or gas) rising up from the bottom of the column. As the wastewater passes down through the column, it contacts the vapors rising from the bottom of the column. This contact progressively reduces the concentrations of volatile organic compounds or gases in the wastewater as it approaches the bottom of the column. At the bottom of the column, the wastewater is heated by the incoming steam, which also reduces the concentrations of volatile components to their final level. Much of the heat in the wastewater discharged from the bottom of the column can then be recovered by preheating the feed to the column.

Reflux (condensing a portion of the vapors from the top of the column and returning it to the bottom) may be practiced if it is desired to alter the composition of the vapor stream that is derived from the stripping column (e.g., increase the concentration of the stripped material for recovery purposes). There also may be advantages to introducing the feed to a tray below the top tray when reflux is used. Introducing the feed at a lower tray (while still using the same number of trays in the stripper) will have the effect of either reducing steam requirements, as a result of the need for less reflux, or yielding a vapor stream richer in the volatile components. The combination of using reflux and introducing the feed at a lower tray will increase the concentration of the volatile organic components in the overhead (vapor phase) beyond that obtainable by reflux alone and increase the potential for recovery.

Stripping of the organic (volatiles) constituents of the wastewater stream occurs because the organic volatiles tend to vaporize into the steam until its concentration in the vapor and liquid phases (within the stripper) are in equilibrium. The height of the column and the amount of packing material and/or the number of metal trays along with steam pressure in the column generally determine the amounts of volatiles that can be removed and the effluent pollutant levels that can be attained by the stripper. After the volatile pollutant is extracted from the wastewater into the superheated steam, the steam is condensed to form two layers of generally immiscible liquids—the aqueous and volatile layers. The aqueous layer is generally recycled back to the steam stripper influent feed stream because it may still contain low levels of the volatile. The volatile layer may be recycled to the process, incinerated on-site, or contract hauled (for incineration, reclaiming, or further treatment off-site) depending on the specific plant's requirements.

Steam stripping is an energy-intensive technology in which heat energy (boiler capacity) is required to both preheat the wastewater and to generate the superheated steam needed to extract the volatiles from wastewater. In addition, some waste streams may require pretreatment such as solids removal (e.g., filtration) prior to stripping because accumulation of solids within the column will prevent efficient contact between the steam and wastewater phases. Periodic cleaning of the column and its packing materials or trays is a necessary part of routine steam stripper maintenance to assure that low effluent levels are consistently achieved.

Steam strippers are designed to remove individual volatile pollutants based on a ratio (Henry's Law Constant) of their aqueous solubility (tendency to stay in solution) to vapor pressure (tendency to volatilize). The column height and diameter, amount of packing or number of trays, the operating steam pressure, and temperature of the heated feed (wastewater) are varied according to the strippability (using Henry's Law Constant) of the volatile pollutants to be stripped. Volatiles with lower Henry's Law Constants require greater column height, more trays or packing material, greater steam pressure and temperature, more frequent cleaning, and generally more careful operation than do volatiles with higher strippability.

Steam stripping is used to treat aqueous wastes contaminated with chlorinated hydrocarbons, aromatics such as xylenes, ketones such as acetone or MEK, alcohols such as methanol and high boiling point chlorinated aromatics such as pentachlorophenol. Steam stripping will treat less volatile and more soluble wastes than will air stripping and can handle a wide concentration range (e.g., from less than 100 ppm to about 10% organics). The steam stripping process requires some type of air pollution control (APC) mechanism to eliminate toxic emissions.

Although the process typically provides higher removal efficiencies than air stripping, it is considerably more expensive to operate.

In principle, a multistage steam stripper system can be designed to achieve almost any level of organic compound removal. In practice, the achievable VOC emission reductions and associated control costs are highly dependent on wastewater characteristics such as flow, organic concentration and composition, and the design of the collection and treatment systems.

Steam stripper systems may be operated in batch or continuous mode. Batch steam stripping is more prevalent when the wastewater feed is generated by batch processes. Batch strippers may also be used if the wastewater contains relatively high concentrations of solids, resins, or tars. Usually, batch steam strippers provide a single equilibrium stage of separation. Therefore, the removal efficiency is essentially determined by the equilibrium coefficients of the pollutants and the fraction of the initial charge distilled overhead.

Wastewater is charged to the receiver, or pot, and brought to the boiling temperature of the mixture. Heat is provided by direct injection of steam or by an external heat exchanger normally referred to as a reboiler. The overhead vapors are condensed and recovered. The solids, tars, resins, and other residue remaining in the pot are normally disposed. By varying the heat input and fraction of the initial charge boiled overhead, the same batch stripper can be used to treat wastewater mixtures with widely varying characteristics.

In contrast to batch strippers, continuous steam strippers are normally designed to treat wastewater streams with relatively consistent characteristics. Design of the continuous stripper system is normally based on the flow rate and composition of a specific wastewater feed stream or combination of streams. Multistage, continuous strippers normally operate at greater organic compound removal efficiencies than batch strippers. Continuous systems may also offer other advantages (over batch stripping) for applications involving wastewater streams with relatively high flows and consistent concentrations. These advantages include more consistent effluent quality, more automated operation, and lower annual operating costs.

The common uses for steam distillation can be summarized as follows:

1. To separate relatively small amounts of a volatile impurity from a large amount of material.

2. To separate appreciable quantities of low solubility, higher–boiling materials from nonvolatile wastes when the materials to be separated form low boiling azeotropes with water.
3. To recover material which is thermally unstable or reacts with other waste components at the boiling temperature.
4. To recover material which cannot be distilled by indirect heating even under low pressure, because of the high boiling temperature.
5. To recover material in instances where direct–fired heaters cannot be used because of ignition or explosion hazards.

Steam stripping is not well suited to treating wastes in which either the overhead or bottoms are difficult to separate from water. It is better utilized for separating organics which decant readily and have low solubilities in water (e.g., halogenated organics) and less applicable to treating water soluble wastes such as alcohols.

Steam stripping as an in situ enhancement technique is discussed under "Soil Vapor Extraction."

6.16 SUSPENDED SOLIDS TREATMENT / DEWATERING

There are numerous techniques available for removing wastewater solids. These solids are primarily inorganic, but can contain a high percentage of organics that originate from the source of the waste.

6.16.1 Centrifuges/Cylones/Hydrocyclones

Cyclones, hydrocyclones and centrifuges utilize centrifugal force to separate material of differing densities. The principle in the operation of these devices is that the heavier materials are thrown to the outside, and the lighter materials remain near the inside where they can be drawn off.

Usually, the solids are removed by the continuous displacement of the entering fluid or by a screw conveyor inside the unit. In filtration–type centrifugation the inner wall of the units is actually a filter. The solvent is forced through the filter and the solids are retained. Inside the unit, the solids are periodically removed by mechanical blades that scrape the filter surface as the unit rotates. Treatment of the residual solids depends on the composition of the wastestream.

Like filtration or gravity separation, centrifugation is used as a preliminary purification step before other recycling operations. Crystalline solids, for example, are often separated from spent dry cleaning solutions before distillation. For wastestreams that contain less than 1% sedimentable solids, tubular–bowl centrifuges can be used. A modification of the tubular–bowl model is the multichamber centrifuge. The multichamber centrifuge can achieve a better degree of separation. The disk centrifuge uses a stack of truncated cones to achieve centrifugal forces ranging from 4,000 to 14,000 times the force of gravity. This equipment is generally used for cream separation in the dairy industry. Improved efficiency can be obtained by using nozzle–discharge centrifuges, self–opening centrifuges, continuous decantor centrifuges, screen bowl decantors, and knife–discharge clarifiers. Available filter–type centrifuges include variable–speed or constant–speed units and a wide variety of basket designs.

A hydrocyclone consists of a cylindrical/conical shell with a tangential inlet for feed, an outlet for the overflow of slurry, and an outlet for the underflow of concentrated solids. Cyclones and hydrocyclones contain no moving parts. The slurry is fed to the unit with sufficient velocity to create a "vortex" action that forces the slurry into a spiral and, as the rapidly rotating liquid spins about the axis of the cone, it is forced to spiral inward and then out through a centrally located overflow outlet. Smaller-sized particles remain suspended in the liquid and are discharged through the overflow. Larger and heavier particles of solids are forced outward against the wall of the cone by centrifugal force within the vortex. The solids spiral around the wall of the cyclone and exit through the apex at the bottom of the cone.

In general, the larger the cyclone diameter and inlet, the coarser the separation and the greater the cyclone capacity. The smaller the diameter and inlet, the finer the separation and the lower the hydraulic capacity. In order to remove small particles from large volume slurries, it is necessary to use multiple, small-diameter cyclones connected in parallel. Banks of multiple cyclones, manufactured as a single unit with a single feed pipe, are commercially available.

Cyclones can also be connected in series or in various staging arrangements to accomplish different objectives. For example, a high degree of particle size separation can be achieved by employing a bank of cyclones in series with decreasing cyclone size and particle size removal in the direction of flow. It is also possible to achieve a higher underflow concentration and a more clarified overflow by staging the cyclones. The first stage of cyclones could be used to classify the solids according to the desired grain size. The second stage overflow cyclone could serve as a clarifier and the underflow cyclone could serve as the concentrator. However, the maximum underflow concentration achievable with cyclones is about 60%, since some liquid is necessary for solids discharge.

Cyclones are available for separation or classifying solids over a broad particle size range, from 2,000 microns down to 10 microns. However, in hazardous waste site applications they would be used primarily to remove smaller size particles from slurries and in situations where a sharp separation by particle size is needed. They are particularly applicable to situations where space is limited.

Cyclones are generally not effective for slurries with a solids concentration greater than 30%, for highly viscous slurries, or for separation of particle sizes with a specific gravity of less than about 2.5 to 3.2. Slurries with a high clay content exhibit high pseudoplasticity or high viscosity and cannot be effectively removed using cyclones or hydrocyclones.

Cyclone assemblies take up less space than most solids separation equipment and are well-suited for tight locations. Because of their compactness and simplicity of operation, cyclones are also well-suited for inclusion in mobile treatment systems.

There are three common types of centrifuges; disc, basket, and conveyer. All three operate by removing solids under the influence of centrifugal force. The fundamental difference among the three types is the method by which solids are collected in and discharged from the bowl.

In the disc centrifuge, the sludge feed is distributed between narrow channels that are present as spaces between stacked conical discs. Suspended particles are collected and discharged continuously through small orifices in the bowl wall. The clarified effluent is discharged through an overflow weir.

A second type of centrifuge which is useful in dewatering sludges is the basket centrifuge. In this type of centrifuge, sludge feed is introduced at the bottom of the basket, and solids collect at the bowl wall while clarified effluent overflows the lip ring at the top. Since the basket centrifuge does not have provision for continuous discharge of collected cake, operation requires interruption of the feed for cake discharge for a minute or two in a 10 to 30 minute overall cycle.

The third type of centrifuge commonly used in sludge dewatering is the conveyer type. Sludge is fed through a stationary feed pipe into a rotating bowl in which the solids are settled out against the bowl wall by centrifugal force. From the bowl wall, the solids are moved by a screw to the end of the machine, at which point they are discharged. The liquid effluent is discharged through ports after passing the length of the bowl under centrifugal force.

Virtually all industrial waste treatment systems producing sludge can use centrifugation to dewater it. Centrifugation is currently being used by a wide range of industrial concerns.

The performance of sludge dewatering by centrifugation depends on the feed rate, the rotational velocity of the drum, and the sludge composition and concentration. Assuming proper design and operation, the solids content of the sludge can be increased to 20 to 35%.

Sludge dewatering centrifuges have minimal space requirements and show a high degree of effluent clarification. The operation is simple, clean, and relatively inexpensive. The area required for a centrifuge system installation is less than that required for a filter system or sludge drying bed or equal capacity, and the initial cost is lower.

Centrifuges have a high power cost that partially offsets the low initial cost. Special consideration must also be given to providing sturdy foundations and soundproofing because of the vibration and noise that result from centrifuge operation. Adequate electrical power must also be provided since large motors are required. The major difficulty encountered in the operation of centrifuges has been the disposal of the concentrate which is relatively high in suspended, non-settling solids.

6.16.2 Clarification

Clarification is a physical process used to remove suspended solids from wastewater by gravity settling. Settling tanks, clarifiers, and sedimentation ponds or basins are designed to let wastewater flow slowly and quiescently, providing an adequate retention time to permit most solids more dense than water to settle to the bottom. The settling solids form a sludge at the bottom of the tank or basin. This sludge is usually pumped out continuously or intermittently from settling tanks or clarifiers, or scraped out periodically from sedimentation ponds or basins.

Settling is used alone or as part of a more complex treatment process. It is usually the first process applied to wastewaters containing high concentrations of settleable suspended solids. Settling is also often used in conjunction with other treatment processes such as removal of biomass after biological treatment or removal of metal precipitates after chemical precipitation. Clarifiers, in conjunction with chemical addition, are used to remove materials such as dissolved solids that are not removed by simple sedimentation.

Settling and sedimentation are discussed in later sections.

Coagulants can be added to clarifiers (chemically assisted clarifiers) to enhance liquid–solid separation, permitting solids denser than water to settle to the bottom and materials less dense than water (including oil and grease) to flow to the surface. Settled solids form a sludge at the bottom of the clarifier, which can be pumped out continuously or intermittently. Oil and grease and other floating materials may be skimmed off the surface.

Chemically assisted clarification may be used alone or as part of a more complex treatment process. It may also be used as:

1. The first process applied to wastewater containing high levels of settleable suspended solids.
2. The second stage of most biological treatment processes to remove the settleable materials, including microorganisms, from the wastewater; the microorganisms can then be either recycled to the biological reactor or discharged to the plant's sludge handling facilities.
3. The final stage of most chemical precipitation (coagulation/flocculation) processes to remove the inorganic flocs from the wastewater.

Clarification and sludge consolidation unit operations are typically applied as post–treatments to the majority of aqueous metals containing waste treatment systems.

Usually, wastewaters undergo chemical treatment and enter a clarifier where the flow is decreased to a point that allows solids with a specific gravity greater than that of the liquid settle to the bottom. For liquid/solid mixtures with a slight density difference, an organic polymer (flocculant) can be added to allow the solids to agglomerate and improve the settling characteristics. The supernatant in the overflow is drawn off and residual trace organics or solids are removed in a final polishing step such as carbon adsorption, ultrafiltration, or ion exchange. The solids in the underflow can then be discharged to a holding tank for subsequent dewatering.

Few, if any, sludges settle at a rate sufficient to utilize only clarifiers or thickeners to accumulate sludge for disposal on land. Therefore, the underflow from the clarifier is typically concentrated through the use of mechanical dewatering equipment such as centrifuges, rotary vacuum filters, belt filters, drying ovens, and recessed–plate filter presses. The obtainable degree of cake dryness can be determined by bench–scale tests by the equipment vendor to identify the suitability of a particular dewatering device. The low solids content of sodium hydroxide after sedimentation (3 to 10%) requires the use of a filter press. Conversely, suspended solids removal from lime neutralized sludges can be accomplished through use of a wider range of equipment including rotary vacuum or continuous belt filters.

6.16.3 Classification

Hydraulic: Hydraulic classifiers are commonly used to separate sand and gravel from slurries and classify them according to grain size. These units consist of elevated rectangular tanks with v–shaped bottoms to collect the material. Discharge valves which are located along the bottom of the tank are activated by motor–driven vanes that sense the level of solids as they accumulate. The principal of operation is simple. The slurry is introduced into the feed end of the tank. As the slurry flows to the opposite end, solids settle out according to particle size as a result of differences in settling velocity. Coarse materials settle out first near the feed end and materials are progressively finer along the length of the tank. Manually adjusted splitter gates below the discharge valves can be used to selectively direct materials of specific grain sizes to subsequent handling and treatment.

Because of the inability of hydraulic classifiers to handle large volumes of flow, a combination of solids separation methods may be advisable to reduce the number of hydraulic classifiers needed for a large solids handling operation. One possibility for reducing the number of classifiers needed would be to use these units to separate only those particles larger than 105 microns. Cyclones, hydrocyclones, or hydrosieves could then be used to remove the fine sand fraction.

Hydraulic classifiers offer an effective method for operating and classifying particles ranging in size from fine gravel to fine sands. Some fines are inadvertently removed with the sand and gravel, and the effectiveness of the separation can be improved by washing the collected solids in a spiral classifier to remove the fines.

Hydraulic classifier tanks are generally designed and sized to be truck mounted for mobile system applications. Start–up and shut–down can be accomplished quickly. Maintenance requirements are fairly simple. Use of hydraulic classifiers can be easily integrated with other solids separation methods and this is advisable where large flows are involved or where classification of fine–grained materials (clays, silts) is required.

Spiral: The spiral classifier consists of one or two long, rotating screws, mounted on an incline within a rectangularly shaped tub. It is used primarily to wash adhering clay and silt from sand and gravel fractions.

The screw conveys settled solids from a hydraulic classifier up an incline to be discharged through an opening at the top of the tub. Fines and materials of low specific gravity are separated from sand and gravel through agitation and the abrading and washing action of the screw, and are removed along with the wastewater overflow at the bottom of the tub. The tumbling and rolling action caused by the continuous screw grinds particles against each other and removes the deleterious material coating the sand particles. This tumbling action also aids in dewatering materials by breaking the moisture film on the sand particles. As the moisture is relieved of surface tension, it is free to drain from the material. The sands which are finally discharged are substantially dewatered.

In general, the greater the length of the tub the higher the degree of dewatering and the greater the screw diameter the larger the capacity of the spiral classifier.

Spiral classifiers are used primarily to wash, dewater, and classify sands and gravels up to 3/8 inch in diameter. They are not a singularly viable solids separation technology, but they are effective when used together with the hydraulic classifier. Spiral classifiers have a large capacity and are completely portable.

Spiral classifiers improve the efficiency of solids separation achieved with the hydraulic classifier by removing fine grained materials attached to coarser particles. Spiral classifiers are generally designed to be trailer mounted for use in mobile treatment systems. Start–up and shut–down can be accomplished quickly and maintenance requirements are simple.

6.16.4 Coagulation/Flocculation

Chemical coagulation and flocculation are terms often used interchangeably to describe the physiochemical process of suspended particle aggregation resulting from chemical additions to wastewater. Technically, coagulation involves the reduction of electrostatic surface charges and the formation of complex hydrous oxides. Coagulation is essentially instantaneous in that the only time required is that necessary for dispersing the chemicals in solution. Flocculation is the time–dependent physical process of the aggregation of wastewater solids into particles large enough to be separated by sedimentation.

The purpose of coagulation is to overcome electrostatic repulsive surface forces and cause small particles to agglomerate into larger particles, so that gravitational and inertial forces will predominate and affect the settling of the particles. The process can be grouped into two sequential mechanisms:

1. Chemically induced destabilization of the repulsive surface–related forces, thus allowing particles to stick together when contact between particles is made.
2. Chemical bridging and physical enmeshment between the non–repelling particles, thus allowing for the formation of large particles.

There are three different types of coagulants: inorganic electrolytes, natural organic polymers, and synthetic polyelectrolytes.

Inorganic electrolytes are salts or multivalent ions such as alum (aluminum sulfate), lime, ferric chloride, and ferrous sulfate. The inorganic coagulants act by neutralizing the charged double layer of colloidal particles and by precipitation reactions. Alum is typically added to the waste stream as a solution. At an alkaline pH and upon mixing, the alum hydrolyzes and forms fluffy gelatinous precipitates of aluminum hydroxide. These precipitates, partially as a result of their large surface area, act to enmesh small particles and thereby create large particles. Lime and ion salts, as well as alum, are used as flocculants primarily because of this tendency to form large fluffy precipitates of "floc" particles.

Natural organic polymers derived from starch, vegetable materials, or monogalactose act to agglomerate colloidal particles through hydrogen bonding and electrostatic forces. These are often used as coagulant aids to enhance the efficiency of inorganic coagulants.

Synthetic polyelectrolytes are polymers that incorporate ionic or other functional groups along the carbon chain in the molecule. The functional groups can be either anionic (attract positively charged species), cationic (attract negatively charged species), or neutral. Polyelectrolytes function by electrostatic bonding and the formation of physical bridges between particles, thereby causing them to agglomerate. These are also most often used as coagulant aids to improve floc formation.

The coagulation/flocculation and sedimentation process entails the following steps:

1. Addition of the coagulating agent to the liquid.
2. Rapid mixing to dispense the coagulating agent throughout the liquid.
3. Slow and gentle mixing to allow for contact between small particles and agglomeration into larger particles.

Coagulation and flocculation are used for the clarification of industrial wastes containing colloidal and suspended solids. Coagulants are most commonly added upstream of sedimentation ponds, clarifiers, or filter units to increase the efficiency of solids separation. This practice has also been shown to improve dissolved metal removal as a result of the formation of denser, rapidly settling flocs, which appear to be more effective in absorbing and adsorbing fine metal hydroxide precipitates. Coagulation may also be used to remove emulsified oil from industrial wastewaters. Emulsified oil and grease is aggregated by chemical addition through the processes of coagulation and/or acidification in conjunction with flocculation.

Coagulation/flocculation is a well-developed process widely used for many industrial wastewaters containing suspended and colloidal solids. The equipment used is relatively simple, readily available, and can often be skid mounted in a modular design. In many cases, coagulation/flocculation can be added to existing process trains with only minor modifications. For high volume applications, the cost of this technology drops dramatically improving economic viability. In addition, the process is often improved by high ionic strength and is applicable to high influent metal loadings.

Disadvantages and primary environmental considerations result from a metals laden high-water-content sludge which must be treated (i.e., solidification, encapsulation, etc.) and then disposed. In addition, the process is also not readily applied to small intermittent flows and many of the coagulants used form corrosive solutions. Finally, process efficiency is highly sensitive to initial contaminant concentration and the surface area of the primary floc formed in the rapid-mix chamber.

6.16.5 Evaporation/Drying

Evaporation is the physical separation of a liquid from a dissolved or suspended solid by the application of energy to volatilize the liquid. In hazardous waste treatment, evaporation may be used to isolate the hazardous material in one of the two phases, simplifying subsequent treatment. If the hazardous material is in the volatilized phase, the process is usually called "stripping."

Evaporation can be applied to any mixture of liquids and nonvolatile solids provided the liquid is volatile enough to evaporate under reasonable heating or vacuum conditions (both the liquid and the solid should be stable under those conditions). If the liquid is water, evaporation can be carried out in a large pond provided with solar energy. Evaporation of aqueous wastes can also be done in closed process vessels with energy provided by steam and the resulting water vapor can be condensed for possible reuse. Energy requirements are usually minimized by such techniques as vapor recompression or multiple effect evaporators. Evaporation is applied to solvent waste contaminated with nonvolatile impurities such as oil, grease, paint solids or polymeric resins. Mechanically agitated or wiped thin film evaporators are used. Solvent is evaporated and recovered for reuse. The residue is the bottom stream, typically containing 30 to 50% solids.

The basic types of evaporation include the following:

1. Spray Evaporator.
2. Rising (or Climbing) Film Evaporator.
3. Submerged Tube Evaporator.
4. Atmospheric Exhaust Evaporator.
5. Thermal Vapor Recompression (TVR).
6. Mechanical Vapor Recompression (MVR).
7. Thin Film Evaporator (see "Distillation").

The spray and rising film evaporators involve covering the heating surface with a thin film of waste liquid, whereas the submerged tube and the atmospheric exhaust evaporation systems transfer heat to a reservoir (tank) of waste liquid. The distillate is returned to the rinse tanks for each of these systems, with the exception of the atmospheric exhaust evaporator which vents the distillate to the air. The surface film evaporators (No. 1 and 2 above) are more efficient due to their higher heat transfer coefficients. Therefore, smaller corrosive-resistant surface areas are required for surface film evaporators, which in turn reduces capital costs.

Thermal vapor recompression evaporators use high-pressure steam to compress the vapor produced during evaporation, raising the vapor's pressure and temperature.

In the mechanical vapor recompression evaporator, the vapor leaving the separator is compressed mechanically to increase its pressure and condensing temperature so it may be used as the heating medium in the steam chest.

Advantages of the evaporation process are that it permits recovery of a wide variety of process chemicals, and it is often applicable to concentration or removal of compounds which cannot be accomplished by any other means. The major disadvantage is that the evaporation process consumes relatively large amounts of energy for the evaporation of water. However, the recovery of waste heat from many industrial processes (e.g., diesel generators, incinerators, boilers and furnaces) should be considered as a source of this heat for a totally integrated evaporation system. Also, in some cases solar heating could be inexpensively and effectively applied to evaporation units.

Capital costs for vapor compression evaporators are substantially higher than for other types of evaporation equipment. However, the energy costs associated with the operation of a vapor compression evaporator are significantly lower then costs of other evaporator types. For applications, pretreatment may be required to remove solids or bacteria which tend to cause fouling in the condenser or evaporator. The build-up of scale on the evaporator surfaces reduces the heat transfer efficiency and may present a maintenance problem or increase operating cost. However, it has been demonstrated that fouling of the heat transfer surfaces can be avoided or minimized for certain dissolved solids by maintaining a seed slurry which provides preferential sites for precipitate deposition. In addition, low temperature differences in the evaporator will eliminate nucleate boiling and supersaturation effects. Steam distillable impurities in the process stream are carried over with the product water and must be handled by pre- or post-treatment.

Spray Drying: Evaporation occurs when a solvent, usually water, vaporizes from a solution or slurry, and completion of the evaporation process results in drying. This technology can be used to vaporize off water, thereby concentrating the solute in the remaining solution, and is related to distillation, sublimation, and stripping, because they are all processes based on the common principles of vaporization.

In spray evaporation, or drying, a wet slurry is converted to a vapor, which is released, and a dry, free flowing powder, which may be recovered as product or disposed of as waste. A spray evaporation/drying treatment system normally consists of a drying chamber. The waste slurry is injected into the chamber through an atomizer which disperses the stream. A cyclone is created by injecting a high flow warm air stream countercurrent to the atomized slurry. In the spray drying chamber, the solids settle out of the air while the moisture is evaporated.

The solids which settle out of the primary and secondary chambers of the spray evaporation system may be either a product ready for formulation and packaging, or a solid waste stream requiring disposal or recycle. The water vapors are extracted from the primary chamber, filtered to further remove particulate in the secondary chamber, and then exhausted to the atmosphere, generating no wastewater. If the solvent is not water, it is necessary to condense or scrub the vapors to prevent hazardous air emissions.

This technology is appropriate for separation of non-volatile and insoluble materials from manufacturing wastewaters or from process solvents. It is not appropriate for wastewater streams containing volatile organic priority pollutants or cyanide, unless air pollution control devices are added to the exhaust prior to venting to the atmosphere.

Flash Drying: Flash drying is a system used for disposal of various sludge wastes. Sludge waste is dried and can then be burned or disposed of by another method or within the flash drying system. Its principle elements are a hot gas heater, sludge mixer, cage mill, cyclone collector, vapor fan, and dry product conveyor.

Thermal Drying: A wide range of batch and continuous dryers is available. One commonly used continuous type is the screw–flight dryer. The screw–flight dryer consists of a screw surrounding a hollow shaft enclosed in a trough. Heat transfer fluid is heated to temperatures as high as 750°F and circulated, usually countercurrent to the flow of waste, through the hollow shaft. Heat transfers from the shaft to the screw blades, and then into the feed material, causing water and organics to be driven off in a vapor form. The dried cake is discharged from the dryer.

The dryer is designed to create good contact between the screw and feed material. The screw is usually equipped with breaker bars to ensure proper shearing of the input materials and to prevent the screw surfaces from fouling.

Vapors emerging from this system are managed in one of two ways, depending on their composition. If the vapors contain only water, they are usually directly vented to the atmosphere. However, if the vapors contain volatile organics, they are generally passed through a water–cooled condenser system. The recovered organic liquids from the condenser unit are then forwarded to another process for treatment or recovery.

Sludge Bed Drying: As a waste treatment procedure, sludge bed drying is employed to reduce the water content of a variety of sludges to the point where they are amenable to mechanical collection and removal to landfill. These beds usually consist of 15 to 45 cm (6 to 18 inches) of sand over a 30 cm (12 inch) deep gravel drain system made up of 3 to 6 mm (1/8 to 1/4 inch) graded gravel overlying drain tiles.

Drying beds are usually divided into sectional areas approximately 7.5 meters (25 ft) wide x 30 to 60 meters (100 to 200 ft) long. The partitions may be earth embankments, but more often are made of planks and supporting grooved posts.

To apply liquid sludge to the sand bed, a closed conduit or a pressure pipeline with valved outlets at each sand bed section is often employed. Another method of application is by means of an open channel with appropriately placed side openings which are controlled by slide gates. With either type of delivery system, a concrete splash slab should be provided to receive the falling sludge and prevent erosion of the sand surface.

Where it is necessary to dewater sludge continuously throughout the year regardless of the weather, sludge beds may be covered with a fiberglass reinforced plastic or other roof. Covered drying beds permit a greater volume of sludge drying per year in most climates because of the protection afforded from rain or snow and because of more efficient control of temperature. Depending on the climate, a combination of open and enclosed beds will provide maximum utilization of the sludge bed drying facilities.

Sludge drying beds are a means of dewatering sludge from clarifiers and thickeners. They are widely used both in municipal and industrial treatment facilities.

The main advantage of sludge drying beds over other types of sludge dewatering is the relatively low cost of construction, operation, and maintenance. Its disadvantages are the large area of land required and long drying times that depend, to a great extent, on climate and weather.

6.16.6 Filtration

Filtration is an established unit operation for achieving the removal of suspended solids from wastewaters. The removal is accomplished by the passage of water through a physically restrictive medium (e.g., sand, coal, garnet, diatomaceous earth, fabric or screen) with resulting entrapment of suspended particulate matter by a complex process involving one or more removal mechanisms, such as straining, sedimentation, interception, impaction, and adsorption. Continued filtration reduces the porosity of the bed as particulate matter removed from the wastewater accumulates on the surface of the grains of the media and in the pore spaces between grains. This reduces the filtration rate and increases the head loss across the filter bed. The solids must be removed by "backwashing" when the head loss increases to a limiting value. Backwashing involves forcing wash water through the filter bed in the reverse direction of the original fluid flow so that the solids are dislodged from the granular particles and are discharged in the spent wash water. When backwashing is completed, the filter is returned to service.

The filter medium may be precoated with a filtration aid such as ground cellulose or diatomaceous earth. Fluid flow through the filter medium may be accomplished by gravity, by inducing a partial vacuum on one side of the medium, or by exerting a mechanical pressure on a dewaterable sludge enclosed by filter media.

Filtration is used for the dewatering of sludges and slurries as a pretreatment for other processes. Granular media filtration is typically used after gravity separation processes for additional removal of suspended solids and oils prior to the other treatment processes. It is also used a polishing step for treated waste to reduce suspended solids and associated contaminants to low levels. Pretreatment by filtration is appropriate for membrane separation processes, ion exchange, and carbon adsorption in order to prevent plugging or overloading of these processes. Filtration of settled wastes is often required to remove undissolved heavy metals which are present as suspended solids. Filtration does not reduce the toxicity of the waste although sometimes powdered activated carbon may be used as a combination adsorbent and filter aid. Filtration should not be used with sticky or gelatinous sludges, due to the likelihood of filter media plugging.

As a simplified form of filtration, strainers are often employed to remove large or coarse particles from liquid. Equipment includes stationary or moving screens, perforated plates, metal mesh baskets, belts, or chains.

Filters can be classified by the following factors:

1. The driving force (i.e., the manner by which the filtrate is induced to flow, either by gravity or pressure.
2. The function (i.e., whether the filtrate or the filtered material is the product of greater value).
3. The operating cycle (i.e, whether the filter process occurs continuously or batchwise).

4. The nature of the solids (i.e., the size of the particles being filtered out); and
5. The filtration mechanism (i.e., whether the filtered solids are stopped at the surface of the medium and pile up to form a filter cake or are trapped within the pores or body of the filter medium).

Some of the process-related factors important to the proper selection of filtration equipment are particle size distribution, viscosity, production throughput, process condition, performance requirements, and permissible materials of construction. Equipment-related factors include type of operation (batch or continuous), driving force, production rate, separation effectiveness, washing capability, dependability, material of construction, and cost.

Filtration is an established wastewater treatment technology currently in full-scale use for industrial waste treatment. Filtration has several applications: (1) pretreatment to remove suspended solids prior to processes such as activated carbon adsorption, steam stripping, ion exchange, and chemical oxidation; (2) removal of residual biological floc from settled treatment process effluents; (3) removal of residual chemically coagulated floc from physical/chemical treatment process effluents; and (4) removal of oil from oil separation and dissolved air flotation effluents.

Adsorptive filtration is discussed in the section entitled "Adsorption."

This section does not discuss filtration as applied to drinking water facilities.

Belt Filter Presses: Belt filter presses (BFP) have been used in Europe since the 1960s and in the United States since the early 1970s. They were initially designed to dewater paper pulp and were then modified to dewater sewage sludge. The European models were the first used in the United States. However, the difference between U.S. sludge and European sludge led to performance problems (low cake solids and poor solids capture). American manufacturers began building and selling presses in the United States.

Belt filter presses are designed on the basis of a very simple concept. Sludge sandwiched between two tensioned porous belts is passed over and under various diameter rollers. For a given belt tension, as roller diameter decreases, an increased pressure is exerted on the sludge, thus squeezing out water. Although many different designs of belt filter presses are available, they all incorporate the following basic features.

1. Polymer conditioning zone.
2. Gravity drainage zone.
3. Low pressure zone.
4. High pressure zones.

New developments in belt filters combine gravity, pressure, or vacuum pressing steps to improve throughput.

Other Filtration Equipment:

1. Drum filters: The most widely used continuous filter, and can be operated as either a pressure filter, or a vacuum filter.
2. Disk filters: A vacuum filter with vertical disks attached to a continuously rotating horizontal hollow central shaft.
3. Plate and frame press: An alternate assembly of plates covered on both sides with a filter medium, usually a cloth, and hollow frames that provide space for cake accumulation during filtration.

4. Recessed–plate filter press: Similar to the plate and frame press, but consists of only plates.
5. Tubular filters: The tubes can be constructed of wire cloth, porous ceramic, plastic or metal and a filter cloth can be used on the outside for finer particle separations.
6. Pressure leaf filters: Sometimes called tank filters, these consist of flat filtering elements (leaves) supported in a pressure shell. The leaves are circular, arc–sided or rectangular and they have filtering surfaces on both faces.

Polishing Filtration: Polishing filtration is a treatment technology applicable to wastewaters containing relatively low concentrations of solids (usually less than 1,000 mg/ℓ). This type of filtration is typically used as a polishing step for the supernatant liquid following chemical precipitation and settling/clarification of wastewaters containing metal and other inorganic precipitates. Polishing filtration removes particles that are difficult to settle because of their size and/or density, as well as precipitated particles from an undersigned settling system.

During polishing filtration, wastewater may flow by gravity or under pressure to the filter. The two most common polishing filtration processes are cartridge and granular bed filtration. Both processes remove particles that are much smaller than the pore size of the filter medium by straining, adsorption, and coagulation/flocculation mechanisms; these processes are also capable of producing an effluent with a low level of solids (typically less than 10 mg/ℓ).

Sludge Filtration: Sludge filtration, also known as sludge dewatering or cake–formation filtration, is a technology used on wastes that contain high concentrations of suspended solids, usually higher than 1% (10,000 mg/ℓ). Sludge filtration is commonly applied to waste sludges, such as clarifier solids, for dewatering. Typically, these sludges can be dewatered to 19 to 50% solids concentration using this technology.

For sludge filtration, waste is pumped through a cloth–type filter medium (also known as pressure filtration, such as that performed with a plate–and–frame filter); drawn by vacuum through the cloth or metal mesh medium (also known as vacuum filtration, such as that performed with a vacuum drum filter); or gravity–drained and mechanically pressed through two continuous fabric belts (also known as belt filtration, such as that performed with a belt filter press). In all cases, the solids "cake" builds up on the filter medium and acts as a filter for subsequent solids removal.

For a plate–and–frame filter, removal of the solids is accomplished by taking the unit off–line, opening the filter, and using mechanical or manual methods to scrape off the solids (a batch process). For the vacuum filter, cake is removed continuously by using an adjustable knife mechanism to scrape the sludge from the vacuum drum as the drum rotates. For the belt filter, the cake is continuously removed by a discharge roller and blade, which dislodge the cake from the belt. For a specific sludge, the plate–and–frame filter will usually produce the driest cake (highest percentage of solids). The belt filter produces a drier cake than does a vacuum filter, but usually not as dry as that produced by a plate–and–frame filter. Dewatered solids are further treated in processes such as sludge drying, incineration, solvent extraction (if treatable levels of organics are present), stabilization (if treatable levels of leachable metals are present), and/or disposal. The liquid filtrate that penetrates the filter medium is further treated in processes such as polishing filtration, carbon adsorption, and biological treatment, and/or it is disposed of.

Cartridge Filtration: Cartridge filters can be used for relatively low waste feed flows. In this process, a hollow, cylindrically shaped cartridge with a matted cloth–type filter medium is placed within a sealed vessel. Wastewater is pumped through the cartridge wall until the flow drops excessively or until the pumping pressure becomes too high because of plugging of the filter medium. The sealed vessel is then opened, and the plugged cartridge is removed and replaced with a new cartridge. The plugged cartridge is then cleaned and/or disposed of. Cartridge filters can be assembled in a parallel arrangement to increase the overall system flow.

Vacuum Filtration: In wastewater treatment plants, sludge dewatering by vacuum filtration generally uses cylindrical drum filters. These drums have a filter medium which may be cloth made of natural or synthetic fibers or a wire–mesh fabric. The drum is suspended above and dips into a vat of sludge. As the drum rotates slowly, part of its circumference is subject to an internal vacuum that draws sludge to the filter medium. Water is drawn through the porous filter cake to a discharge port, and the dewatered sludge, loosened by compressed air, is scraped from the filter mesh. Because the dewatering of sludge on vacuum filters is relatively expensive per kilogram of water removed, the liquid sludge is frequently thickened prior to processing.

Vacuum filters are frequently used both in municipal treatment plants and in a wide variety of industries. They are most commonly used in larger facilities, which may have a thickener to double the solids content of clarifier sludge before vacuum filtering.

The function of vacuum filtration is to reduce the water content of sludge, so that the solids content increases from about 5 to about 30%.

Although the initial cost and area requirement of the vacuum filtration system are higher than those of a centrifuge, the operating cost is lower, and no special provisions for sound and vibration protection need be made. The dewatered sludge from this process is in the form of a moist cake and can be conveniently handled.

Granular Bed Filtration: Filtration occurs in nature as surface and groundwaters are cleansed by sand. Silica sand, anthracite coal, and garnet are common filter media used in water treatment plants. These are usually supported by gravel. The media may be used singly or in combination. The multi–media filters may be arranged to maintain relatively distinct layers by virtue of balancing the forces of gravity, flow, and buoyancy on the individual particles. This is accomplished by selecting appropriate filter flow rates (gpm/ft^2), media grain size, and density.

Granular bed filters may be classified in terms of filtration rate, filter media, flow pattern, or method of pressurization. Traditional rate classifications are slow sand, rapid sand, and high rate mixed media. In the slow sand filter, flux or hydraulic loading is relatively low, and removal of collected solids to clean the filter is therefore relatively infrequent. The filter is often cleaned by scraping off the inlet face (top) of the sand bed. In the higher rate filters, cleaning is frequent and is accomplished by a periodic backwash, opposite to the direction of normal flow.

A filter may use a single medium such as sand or diatomaceous earth, but dual and mixed (multiple) media filters allow higher flow rates and efficiencies. The dual media filter usually consists of a fine bed of sand under a coarser bed of anthracite coal. The coarse coal removes most of the influent solids, while the fine sand performs a polishing function. At the end of the backwash, the fine sand settles to the bottom because it is denser than the coal, and the filter is ready for normal operation. The mixed media filter operates on the same principle, with the finer, denser media at the bottom and the coarser, less dense media at the top. The usual arrangement is garnet at the bottom (outlet end) of the bed, sand in the middle, and anthracite coal at the top. Some mixing of these layers occurs and is, in fact, desirable.

The flow pattern is usually top–to–bottom, but other patterns are sometimes used. Upflow filters are sometimes used, and in a horizontal filter the flow is horizontal. In a biflow filter, the influent enters both the top and the bottom and exits laterally. The advantage of an upflow filter is that with an upflow backwash, the particles of a single filter medium are distributed and maintained in the desired coarse–to–fine (bottom–to–top) arrangement. The disadvantage is that the bed tends to become fluidized, which ruins filtration efficiency. The biflow design is an attempt to overcome this problem.

The classic granular bed filter operates by gravity flow; however, pressure filters are fairly widely used. They permit higher solids loadings before cleaning and are advantageous when the filter effluent must be pressurized for further downstream treatment. In addition, pressure filter systems are often less costly for low to moderate flow rates.

Wastewater treatment plants often use granular bed filters for polishing after clarification, sedimentation, or other similar operations. Granular bed filtration thus has potential application to nearly all industrial plants. Chemical additives which enhance the upstream treatment equipment may or may not be compatible with or enhance the filtration process.

The principal advantages of granular bed filtration are its comparatively (to other filters) low initial and operating costs, reduced land requirements over other methods to achieve the same level of solids removal, and elimination of chemical additions to the discharge stream. However, the filter may require pretreatment if the solids level is high (over 100 mg/ℓ). Operator training must be somewhat extensive due to the controls and periodic backwashing involved, and backwash must be stored and dewatered for economical disposal.

Pressure Filtration: Pressure filtration works by pumping the liquid through a filter material which is impenetrable to the solid phase. The positive pressure exerted by the feed pumps or other mechanical means provides the pressure differential which is the principal driving force.

A typical pressure filtration unit consists of a number of plates or trays which are held rigidly in a frame to ensure alignment and which are pressed together between a fixed end and a traveling end. On the surface of each plate, a filter made of cloth or synthetic fiber is mounted. The feed stream is pumped into the unit and passes through holes in the trays along the length of the press until the cavities or chambers between the trays are completely filled. The solids are then entrapped, and trays are completely filled. The solids are then entrapped, and a cake begins to form on the surface of the filter material. The water passes through the fibers, and the solids are retained.

At the bottom of the trays are drainage ports. The filtrate is collected and discharged to a common drain. As the filter medium becomes coated with sludge, the flow of filtrate through the filter drops sharply, indicating that the capacity of the filter has been exhausted. The unit must then be cleaned of the sludge. After the cleaning or replacement of the filter media, the unit is again ready for operation.

The pressures which may be applied to a sludge for removal of water by filter presses that are currently available range from 5 to 13 atmospheres. As a result, pressure filtration may reduce the amount of chemical pretreatment required for sludge dewatering. Sludge retained in the form of the filter cake has a higher percentage of solids than that from centrifuge or vacuum filter. Thus, it can be easily accommodated by materials handling systems.

As a primary solids removal technique, pressure filtration requires less space than clarification and is well suited to streams with high solids loadings. The sludge produced may be disposed without further dewatering, but the amount of sludge is increased by the use of filter precoat materials (usually diatomaceous earth). Also, cloth pressure filters often do not achieve as high a degree of effluent clarification as clarifiers or granular media filters.

Two disadvantages associated with pressure filtration in the past have been the short life of the filter cloths and lack of automation. New synthetic fibers have largely offset the first of these problems. Also, units with automatic feeding and pressing cycles are now available.

For larger operations, the relatively high space requirements, as compared to those of a centrifuge, could be prohibitive in some situations.

Colloidal Filtrations: A colloid filtration process removes inorganic heavy metals and non–tritium radionuclides from industrial wastewater and groundwater. The filter unit has an inorganic, insoluble filter bed material (Filter Flow–1000) contained in a dynamic, flow–through configuration resembling a filter plate. The pollutants are removed from the water via sorption, chemical complexing, and physical filtration. By employing site–specific optimization of the water chemistry prior to filtration, the methodology removes the pollutants as ions, colloids, and colloidal aggregates. A three–step process is used to achieve heavy metal and radionuclide removal. First, water is treated chemically to optimize formation of colloids and colloidal aggregates. Second, a prefilter removes the larger particles and solids. Third, the filter bed removes the contaminants to the compliance standard desired. By controlling the water chemistry, water flux rate, and bed volume, the methodology can be used to remove heavy metals and radionuclides in low to high volume waste streams.

6.16.7 Flotation

Flotation is a process by which suspended solids, free and emulsified oils, and grease are separated from wastewater by releasing gas bubbles into the wastewater. The gas bubbles attach to the solids, increasing their buoyancy and causing them to float. A surface layer of sludge forms, and is usually continuously skimmed for disposal. Flotation may be performed in several ways, including foam (froth), dispersed air, vacuum flotation, and flotation with chemical addition. The principal difference between these variations is the method of gas bubbles generation.

Flotation is used primarily in the treatment of wastewater streams that carry heavy loads of finely divided suspended solids or oil. Solids having a specific gravity only slightly greater than water, which would require abnormally long sedimentation times, may be removed in much less time by flotation. Thus, it is often an integral part of standard clarification.

Flotation is used primarily in the treatment of wastewater streams that carry heavy loads of finely divided suspended solids or oil. Solids having a specific gravity only slightly greater than 1.0, which would require abnormally long sedimentation times, may be removed in much less time by flotation. Dissolved air flotation is of greatest interest in removing oil from water and is less effective in removing heavier precipitates.

The primary variables for flotation design are pressure, feed solids concentration, and retention period. The suspended solids in the effluent decrease, and the concentration of solids in the float increases with increasing retention period. When the flotation process is used primarily for clarification, a retention period of 20 to 30 minutes usually is adequate for separation and concentration.

Some advantages of the flotation process are the high levels of solids separation achieved in many applications, the relatively low energy requirements, and the adaptability to meet the treatment requirements of different waste types. Limitations of flotation are that it often requires addition of chemicals to enhance process performance and that it generates large quantities of solid waste.

Froth Flotation: Froth flotation is based on differences in the physiochemical properties in various particles. Wettability and surface properties affect the particles' ability to attach themselves to gas bubbles in an aqueous medium. In froth flotation, air is blown through the solution containing flotation reagents. The particles with water repellent surfaces stick to air bubbles as they rise and are brought to the surface. A mineralized froth layer, with mineral particles attached to air bubbles, is formed. Particles of other minerals which are readily wetted by water do not stick to air bubbles and remain in suspension.

Dispersed Air Flotation: In dispersed air flotation, gas bubbles are generated by introducing the air by means of mechanical agitation with impellers or by forcing air through porous media. Dispersed air flotation is used mainly in the metallurgical industry.

Dissolved Air Flotation: In dissolved air flotation, bubbles are produced by releasing air from a supersaturated solution under relatively high pressure. There are two types of contact between the gas bubbles and particles. The first type is predominant in the flotation of flocculated materials and involves the entrapment of rising gas bubbles in the flocculated particles as they increase in size. The bond between the bubble and particle is one of physical capture only. The second type of contact is one of adhesion. Adhesion results from the intermolecular attraction exerted at the interface between the solid particle and gaseous bubble.

Vacuum Flotation: This process consists of saturating the wastewater with air either directly in an aeration tank, or by permitting air to enter on the suction of a wastewater pump. A partial vacuum is applied, which causes the dissolved air to come out of solution as minute bubbles. The bubbles attach to solid particles and rise to the surface to form a scum blanket. which is normally removed by a skimming mechanism. Grit and other heavy solids that settle to the bottom are generally raked to a central sludge pump for removal. A typical vacuum flotation unit consists of a covered cylindrical tank in which a partial vacuum is maintained. The tank is equipped with scum and sludge removal mechanisms. The floating material is continuously swept to the tank periphery, automatically discharged into a scum trough, and removed from the unit by a pump also under partial vacuum. Auxiliary equipment includes an aeration tank for saturating the wastewater with air, a tank with a short retention time for removal of large bubbles, vacuum pumps, and sludge pumps.

6.16.8 Gravity Sludge Thickening

In the gravity thickening process, dilute sludge is fed from a primary settling tank or clarifier to a thickening tank where rakes stir the sludge gently to densify it and to push it to a central collection well. The supernatant is returned to the primary settling tank. The thickened sludge that collects on the bottom of the tank is pumped to dewatering equipment or hauled away.

Thickeners are generally used in facilities where the sludge is to be further dewatered by a compact mechanical device such as a vacuum filter or centrifuge. Doubling the solids content in the thickener substantially reduces capital and operating cost of the subsequent dewatering device and also reduces cost for hauling. The process is potentially applicable to almost any industrial plant.

Organic sludges from sedimentation units of 1 to 2% solids concentration can usually be gravity thickened to 6 to 10%; chemical sludges can be thickened to 4 to 6%.

The principal advantage of a gravity sludge thickening process is that it facilitates further sludge dewatering. Other advantages are high reliability and minimum maintenance requirements.

Limitations of the sludge thickening process are its sensitivity to the flow rate through the thickener and the sludge removal rate. These rates must be low enough not to disturb the thickened sludge.

6.16.9 Grit Chambers

Grit removal is achieved in specially designed chambers. Grit consists of sand, gravel, cinders, or other heavy solid materials that have subsiding velocities or specific gravities substantially greater than those of the organic putrescible solids in wastewater. Grit chambers are used to protect moving mechanical equipment from abrasion; to reduce formation of heavy deposits in pipelines, channels, and conduits; and to reduce the frequency of digester cleaning that may be required as a result of excessive accumulations of grit in such units.

Normally, grit chambers are designed to remove all grit particles with a 0.21 mm diameter, although many chambers have been designed to remove grit particles with a 0.15 mm diameter. There are two types of grit chambers:

1. Gravity grit chambers; and
2. Aerated grit chambers.

The major advantage of the aerated grit chamber is that it occupies much less space.

6.16.10 Heavy Media Separation

Heavy media separation is a process for separating two solid materials that have significantly different densities. The mixed solids to be separated are placed into a fluid with a specific gravity chosen (or adjusted) to allow the lighter solid to float while the heavier sinks. Usually, the separating fluid (the heavy media) is a suspension of magnetite in water. The specific gravity of the fluid is, thus, adjustable by varying the amount of magnetite powder used. Magnetite can be easily recovered magnetically from rinsewaters and spills and then reused.

It is readily used for separating two insoluble solids having different densities. Limitations include the possibility of solids dissolving and ruining the heavy media, the presence of other solids of similar density, and the inability to separate magnetic materials cost–effectively (because of the need to recovery magnetite). The process can probably be used to separate and recover used blasting grit.

Commonly used in the mining industry to separate ores from tailings.

6.16.11 Jigging

Jigging is a separation process in which material of similar size but different densities can be separated by immersing them in a fluid that is less dense than either material. Pulsating the material up and down in the fluid is done in such a manner that different buoyancy factors are imparted to particles of differing densities. The lighter materials separate to the top; the heavier materials settle to the bottom. This process can also be used to separate material of the same density but different particle size. The jigging process is used extensively in the mining industry, particularly coal mining and uranium mining.

The Campbell Centrifugal Jig (CCJ) separates very fine heavy metal particles from waste material. The CCJ combines two widely used methods of heavy particle separation: jiggling and centrifuging. Standard jigs have been used to separate solids of different specific gravities in a fluid medium through gravity–induced differential settling. Standard jigs have the advantage of high capacity and continuous material flow. However, under gravity forces alone, separation is relatively ineffective for particles smaller than about 150 microns. Because the motion of these particles is governed more by hydrodynamics than gravity, they tend to remain fluidized and pass through the jig bed and screen. Centrifuges are very effective in separating solids from liquids but do not effectively separate solids of different specific gravities in a slurry.

6.16.12 Lagoons/Air Drying

Dewatering lagoons use a gravity or vacuum assisted underdrainage system to remove water. The base of the lagoon is lined with clay plus a synthetic liner or other appropriate liner material to prevent migration of contaminants into the underlying soils and groundwater. At a minimum, the liner consists of a low permeability clay layer which is several feet thick. When the lagoon is no longer in use, the clay liner is excavated and properly disposed of. In some instances this design may not be adequate to protect groundwater supplies. A combination clay/synthetic liner and a secondary leachate collection system are required in some instances.

The underdrainage system can be designed and operated using one of the following approaches:

1. Gravity underdrainage.
2. Vacuum pumping.
3. Vacuum assisted drying beds.
4. Electroosmosis

A distinction must be made between sludge drying lagoons and sludge lagoons primarily intended for storage. Some drying occurs in storage lagoons but the primary intent is to provide temporary or semi-permanent storage.

Dewatering lagoons are best suited to large-scale dewatering operations where the volume of sludge or sediment would require an inordinately large number of mechanical dewatering units (e.g., filters or centrifuges). Lagoons are one of the more effective dewatering methods. A gravity dewatering system is capable of achieving 99% solids removal and a dewatered cake of 35 to 40% solids after 10 to 15 days (based on municipal sludges). Vacuum assisted systems may be capable of achieving a dry cake in a shorter retention time.

The major limitations on the use of dewatering lagoons is that they require large land areas and long set-up times. Because of their large surface area they may not be well suited to areas with heavy rainfall or to areas where long periods of freezing would prevent dewatering.

Gravity drainage systems have the lowest operating costs. However, dewatering is achieved at a relatively slow rate and this may result in the need for more land area than required with the other methods. Gravity drainage systems are also more prone to clogging, particularly if the system is not carefully designed.

Vacuum pumping or vacuum assisted dewatering beds are capable of dewatering at a much more rapid rate than gravity systems. Vacuum assisted dewatering beds reportedly increase the rate of dewatering by about 50% (with a negative pressure of 8 psi or less) However, they require a higher degree of maintenance and are considerably more costly to operate than gravity systems.

Electroosmosis is a very costly technique which would be limited to dewatering of very fine grained (2 to 10 microns), very hazardous and difficult to dewater solids.

Dewatering lagoons provide an effective means of dewatering solids. They are also versatile in that they can provide storage capacity for solids prior to disposal. Of all the dewatering technologies they require the largest time to implement and have the greatest potential for secondary impacts due to localized air pollution and groundwater contamination. Operating costs are higher than other dewatering technologies because of the need to remove the solids with mechanical dredging equipment.

Air drying refers to those dewatering techniques by which the moisture is removed by natural evaporation and gravity or induced drainage. There may be some mechanical assistance, such as turning and mixing the sludge on paved beds, or some vacuum assistance but the movement of water is controlled by natural forces.

In addition to sludge lagoons these techniques include:

1. Sand beds.
2. Freeze assisted sand bed dewatering.
3. Vacuum assisted beds.
4. Wedgewire beds.
5. Paved beds.
6. Reed beds.

6.16.13 Screening

The screen is a simple device used for the grading or separating of particles by size. Vibrating and oscillating screens are used for the same purpose, but the passing of the material is enhanced by a vibrator or oscillator, respectively. The centrifugal screen enhances the passing of the material by making use of centrifugal force and the density of the material.

Screening is the process of removing coarse and/or gross solids from wastewater before subsequent downstream treatment, and is usually accomplished by passing wastewater through drum– or disk–type screens. Typically, coarse screens are stainless steel or nonferrous wire mesh with openings from 6 to 20 mm. Fine screens have openings that are less than 6 mm. Solids are raised above the liquid level by rotation of the screen and are backflushed into receiving troughs by high–pressure jets.

Screening has proven to be a very reliable process when properly designed and maintained.

Screening equipment can be divided into the following five general categories:

1. Grizzly screens—Sets of parallel bars set at predetermined spaces (can be stationary or vibrating).
2. Revolving screens (trommel screens)—A revolving cylindrical frame surrounded by wire cloth, open at both ends.
3. Shaking screens—A rectangular frame lined with wire cloth (often used in conjunction with conveying system).
4. Vibrating screens—Used for high capacity and efficiency (may be mechanically or electrically powered).
5. Oscillating screens—Characterized by low–speed oscillation (often used with silk cloth).

Grizzlies are used primarily for scalping, i.e., removing a small amount of oversized material from material that is primarily fines. Moving screens (i.e., vibrating, shaking, revolving, and oscillating) are used to separate particles by grain size, typically in the size range of 0.125 to 6 inches.

6.16.14 Sedimentation

Sedimentation is a gravity settling process which allows heavier solids to collect at the bottom of a containment vessel resulting in its separation from the suspending fluid. This type of operation can be accomplished using a batch process or a continuous removal process. There exist several physical arrangements in which the sedimentation process can be applied. The first is a settling pond wherein aqueous waste flows through while the suspended solids are permitted to gravitate and settle out. Occasionally the settled particles (sludges) are removed so this system would be considered a semi–batch process.

The second is a circular clarifier equipped with a solids removal device to facilitate clarification on a continuous process basis resulting in a lower solids content outlet fluid.

The third type is a sedimentation basin. It uses a belt–type solids collector mechanism to force the solids to the bottom of the sloped edge of the basin where they are removed. The efficiency of sedimentation treatment is dependent upon the depth and surface area of the basin, settling time (based on the holding time), solid particle size and the flow rate of the fluid.

Sedimentation is considered to be a separation process only, and typically, some type of treatment process for the aqueous liquid and the sludges will follow. Its use is restricted to solids that are more dense than water and it is not suitable for wastes consisting of emulsified oils.

Sedimentation is widely used for the removal of settleable solids and immiscible liquids, including oil and grease and some organics. Although hazardous waste leachate typically contains only small loadings of suspended solids, sedimentation may be included as a pretreatment step because of the sensitivity of many downstream processes to fouling and interference from suspended solids.

Frequently, sedimentation is included in leachate treatment process trains for separation of solids generated by chemical and biological processes. It is an integral part of every precipitation/flocculation process, activated sludge process, and powdered activated carbon treatment (PACT) process.

The wet sludge underflow produced by sedimentation may be hazardous and requires further treatment and disposal. Sedimentation may also generate a nonaqueous organic liquid phase, which must be recovered and disposed of.

6.16.15 Settling

Settling is used alone or as part of a more complex treatment process. It is usually the first process applied to wastewaters containing high concentrations of settleable suspended solids. Settling is also often used in conjunction with other treatment processes such as removal of biomass after biological treatment or removal of metal precipitates after chemical precipitation. Clarifiers, in conjunction with chemical addition, are used to remove materials such as dissolved solids that are not removed by simple sedimentation.

Settling is a process which removes solid particles from a liquid matrix by gravitational force. This is done by reducing the velocity of the feed stream in a large volume tank or lagoon so that gravitational settling can occur.

Settling is often preceded by chemical precipitation which converts dissolved pollutants to solid form and by coagulation which enhances settling by coagulating suspended precipitates into larger, faster settling particles.

If no chemical pretreatment is used, the wastewater is fed into a tank or lagoon where it loses velocity and the suspended solids are allowed to settle out. Long retention times are generally required. Accumulated sludge can be collected either periodically or continuously and either manually or mechanically. Simple settling, however, may require excessively large catchments, and long retention times (days as compared with hours) to achieve high removal efficiencies. Because of this, addition of settling aids such as alum or polymeric flocculants is often economically attractive.

In practice, chemical precipitation often precedes settling, and inorganic coagulants or polyelectrolytic flocculants are usually added as well. Common coagulants include sodium sulfate, sodium aluminate, ferrous or ferric sulfate, and ferric chloride. Organic polyelectrolytes very in structure, but all usually form larger floc particles than coagulants used alone.

Following this pretreatment, the wastewater can be fed into a holding tank or lagoon for settling, but is more often piped into a clarifier for the same purpose. A clarifier reduces space requirements, reduces retention time, and increases solids removal efficiency. Conventional clarifiers generally consist of a circular or rectangular tank with a mechanical sludge collecting device or with a sloping funnel-shaped bottom designed for sludge collection. In advanced settling devices, inclined plates, slanted tubes, or a lamellar network may be included within the clarifier tank in order to increase the effective settling area, increasing capacity. A fraction of the sludge stream is often recirculated to the inlet, promoting formation of a denser sludge.

Settling is based on the ability of gravity (Newton's Law) to cause small particles to fall or settle (Stokes' Law) through the fluid they are suspended in. Presuming that the factors affecting chemical precipitation are controlled to achieve a readily settleable precipitate, the principal factors controlling settling are the particle characteristics and the upflow rate of the suspending fluid. When the effective settling area is great enough to allow settling, any increase in the effective settling area will produce no increase in solids removal.

Therefore, if a plant has installed equipment that provides the appropriate overflow rate, the precipitated solids (including toxic metals) in the effluent can be effectively removed. The number of settling devices operated in series or in parallel by a facility is not important with regard to suspended solids removal. Rather, it is important that the settling devices provide sufficient effective settling area.

Another important facet of sedimentation theory is that diminishing removal of suspended solids is achieved for a unit increase in the effective settling area. Generally, it has been found that suspended solids removal performance varies with the effective up–flow rate. Qualitatively the performance increases asymptotically to a maximum level beyond which a decrease in upflow rate provides incrementally insignificant increases in removal. This maximum level is dictated by particle size distribution, density characteristic of the particles and the water matrix, chemicals used for precipitation and pH at which precipitation occurs.

Settling and clarification are used in the nonferrous metals manufacturing category to remove precipitated metals. Settling can be used to remove most suspended solids in a particular waste stream; thus it is used extensively by many different industrial waste treatment facilities. Because most metal ion pollutants are readily converted to solid metal hydroxide precipitates, settling is of particular use in those industries associated with metal production, metal finishing, metal working, and any other industry with high concentrations of metal ions in their wastewaters. In addition to toxic metals, suitably precipitated materials effectively removed by settling include aluminum, iron, manganese, cobalt, antimony, beryllium, molybdenum, fluoride, phosphate, and many others.

A properly operating settling system can efficiently remove suspended solids, precipitated metal hydroxides, and other impurities from wastewater. The performance of the process depends on a variety of factors, including the density and particle size of the solids, the effective charge on the suspended particles, and the types of chemicals used in pretreatment. The site of flocculant or coagulant addition also may significantly influence the effectiveness of clarification. If the flocculant is subjected to too much mixing before entering the clarifier, the complexes may be sheared and the settling effectiveness diminished. At the same time, the flocculant must have sufficient mixing and reaction time in order for effective set–up and settling to occur. Plant personnel have observed that the line or trough leading into the clarifier is often the most efficient site for flocculant addition. The performance of simple settling is a function of the movement rate, particle size and density, and the surface area of the basin.

The major advantage of simple settling is its simplicity as demonstrated by the gravitational settling of solid particulate waste in a holding tank or lagoon. The major problem with simple settling is the long retention time necessary to achieve complete settling, especially if the specific gravity of the suspended matter is close to that of water.

Some materials cannot be practically removed by simple settling alone.

Settling performed in a clarifier is effective in removing slow–settling suspended matter in a shorter time and in less space than a simple settling system. Also, effluent quality is often better from a clarifier. The cost of installing and maintaining a clarifier, however, is substantially greater than the costs associated with simple settling.

Inclined plate, slant tube, and lamella settlers have even higher removal efficiencies than conventional clarifiers, and greater capacities per unit area are possible. Installed costs for these advanced clarification systems are claimed to be one half the cost of conventional systems of similar capacity.

6.16.16 Tabling

Tabling is a type of separation that utilizes the difference in densities of materials. The process can be performed either wet or dry, with dry tabling having the broader particle size range where it can be used. The device uses slotted incline planes, and the surface oscillates to move the solids. In the case of wet tabling, washing fluid removes the less dense solid fractions, and the heavier solids are collected in the grooves and moved to the collection point by the table oscillation. For example, dry tabling is used extensively in the coal industry. The undesired mineral content of mined coal is much more dense than the actual bituminous anthracite, or lignite fractions of the coal. The ash content of the mineral–containing fraction is much higher than that of the coal fraction making separation desirable.

Another type of the tabling process is called agglomerate tabling. Here, pretreatment is involved to precipitate a particular constituent of the material so that it can be separated by the tabling process. This type of tabling is particularly useful for sludges that require pretreatment.

6.17 THERMAL DESORPTION

In the thermal desorption process, volatile and semi–volatile contaminants are removed from soils, sediments, slurries, and filter cakes. This process typically operates at temperatures of 350° to 700°F but may operate in the 200° to 1200°F temperature range. It is often referred to as low temperature thermal desorption to differentiate it from incineration. At these lower temperatures, thermal desorption promotes physical separation of the components rather than combustion. Contaminated soil is removed from the ground and transferred to treatment units, making this an ex situ process. Direct or indirect heat exchange vaporizes the organic compounds producing an off–gas that is typically treated before being vented to the atmosphere.

After it is excavated, the waste material is screened to remove objects greater than 1.5 to 3.0 inches in diameter. In general, four desorber designs are used: rotary dryer, asphalt plant aggregate dryer, thermal screw, and conveyor furnace. The treatment systems include both mobile and stationary process units designed specifically for treating soil and asphalt aggregate dryers which can be adapted to treat soils.

At five NPL sites, thermal desorption has been selected to separate organics from metals prior to solidification/stabilization of the metal–containing residuals.

Because thermal desorbers may operate near or above 1000°F, some pyrolysis and oxidation may occur in addition to the vaporization of water and organic compounds.

Collection and control equipment such as afterburners, fabric filters, activated carbon, or condensers prevent the release of the contaminants to the atmosphere. Various types of thermal desorption systems can produce up to nine process residual streams: treated soil, oversized media rejects, condensed contaminants, water, particulate control dust, clean off-gas, phase separator sludge, aqueous phase spent carbon, and vapor phase spent carbon.

The final temperature is a function of residence time and heat transfer and is the principle variable in controlling effectiveness. A study by the Hazardous Waste Research and Information Center and the Gas Research Institute showed that temperatures and residence times effective in bench-scale systems also proved effective in pilot-scale systems. Such data support the use of a bench-scale test to determine the best residence time and temperature variables as well as whether the thermal desorption process will suitably treat the waste. The typical treatment temperature range for petroleum fuels from leaking underground storage tanks (LUST) sites is 400° to 900°F. For the treatment of soils containing pesticides, dioxins, and PCB's, temperatures should exceed 850°F.

Thermal desorbers effectively treat soils, sludges and filter cakes and remove volatile and semi-volatile organic compounds. Some higher boiling point substances such as polychlorinated biphenyls (PCB's) and dioxins may also be removed (if present). Inorganic compounds are not easily removed with this process, although some relatively volatile metals such as mercury may be volatilized. Temperatures reached in thermal desorbers generally do not oxidize metals.

The soil is most effectively treated if it contains a moisture level within a specified range due to the cost of treating waste with a high water content. Typical acceptable moisture ranges for rotary dryers and asphalt kilns are 10 to 30%, while thermal screw systems can accommodate a higher water loading of 30 to 80%. For VOC removal, soils ideally should contain 10 to 15% moisture since water vapor carries out some of the VOCs.

High molecular weight organic compounds may foul or plug baghouses or condenser systems, therefore, the types of petroleum products that may be treated by specific technologies may be limited. Rotary dryers typically can treat soils that have an organic content of less than 2%. Thermal screw units may treat soils that contain up to 50% organics.

Thermal desorption has several advantages over other treatment processes. It treats a wide range of organic contaminants, is mobile, commercially available, and enjoys more public acceptance than other thermal treatment methods. Also, because thermal desorbers operate at lower temperatures than incinerators, significant fuel savings may result. They also produce smaller volumes of off-gases to treat than incinerators. Thermal desorption also differs from incineration in the regulatory and permitting requirements and in the partitioning of metals within the process residual streams.

Potential limitations of the treatment process exist as well. Foremost, thermal desorption does not destroy contaminants; it merely strips them from the solid or liquid phase and transfers them to the gas-phase. Therefore, devices to control VOC emissions are necessary. Since metals (e.g., Pb) tend to remain in the soil after treatment, further treatment of the soil, such as stabilization, may be required. The efficiency of the thermal desorption process will vary with the chemical and physical properties of the specific contaminants.

Commercial Processes: Four processes are mentioned below.

The Low Temperature Thermal Aeration (LTTA) process was developed by Canonie Environmental Services, Inc. (Canonie), as a treatment system that desorbs organic contaminants from soils by heating the soils up to 800°F. The main components of the process include the following: (1) a materials dryer, (2) a pug mill, (3) two cyclonic separators, (4) a baghouse, (5) a wet Venturi scrubber, (6) a liquid–phase granular activated carbon (GAC) column, and (7) two vapor–phase GAC beds.

A front–end loader transports contaminated soils to feed hoppers, which release the soil onto a conveyor belt. The conveyor belt transports the contaminated soils into the materials dryer. Contaminated soils in the materials dryer are heated by a parallel–flow hot air stream heated by a propane/fuel oil burner. The materials dryer is a rotating drum 8 ft in diameter and 40 ft long equipped with longitudinal flights for soil mixing.

Processed soil is discharged to an enclosed pug mill, where water is added to cool it and to control fugitive dust emissions. Treated soil is released onto a discharge conveyor and stockpiled. The stockpiled soil is tested onsite to confirm that the treated soil meets clean–up goals and then disposed of onsite or retreated, as required.

The exhaust air stream from the materials dryer, containing vaporized organic contaminants and airborne soil particulates, is treated with a series of standard air pollution control devices before being vented to the atmosphere.

X*TRAX™ Model 200 Thermal Desorption System developed by Chemical Waste Management, Inc. (CWM), is a low–temperature process designed to separate organic contaminants from soils, sludges, and other solid media. The X*TRAX™ Model 200 is fully transportable and consists of three semitrailers, one control room trailer, eight equipment skids, and various pieces of movable equipment. The equipment requires an area of about 125 ft by 145 ft.

Contaminated solids are fed into an externally heated rotary dryer where temperatures range from 750° to 950°F. Evaporated contaminants are removed by a recirculating nitrogen carrier gas that is maintained at less than 4% oxygen to prevent combustion. Solids leaving the dryer are sprayed with treated cooling water to help reduce dusting when the treated solids are returned to their original location and compacted in place. The nitrogen carrier gas is treated to remove and recover dust particles, organic vapors, and water vapors.

In the **Remediation Technologies, Inc. (Retec)** process, waste from the feed hopper is fed to the thermal processor, which consists of a jacketed trough that houses two intermeshing, counter–rotational screw conveyors. The rotation of the screws moves material through the processor. A molten salt eutectic, consisting primarily of potassium nitrate, serves as the heat transfer media. This salt melt has heat transfer characteristics similar to those of oils and allows maximum processing temperatures of up to 850°F. The salt melt is noncombustible, it poses no risk of explosion, and its potential vapors are nontoxic. The heated transfer media continuously circulates through the hollow flights and shafts of each screw and also circulates through the jacketed trough. An electric or fuel oil/gas–fired heater is used to maintain the temperature of the transfer media. Treated product is cooled to less than 150°F for safe handling.

The **HT–6 process** developed by Seaview Thermal Systems, subjects the contaminated soil to high temperatures (up to 2000°F), using electrically generated heat. The targeted organics are vaporized in a nitrogen–rich environment, and then condensed in a water–cooled condenser. The resulting condensate is separated into water and oil, both of which are recovered.

6.18 UNDERGROUND DELIVERY/RECOVERY SYSTEMS

Delivery/recovery systems refer to technologies that either deliver remediating materials into environmental compartments or recover contaminating materials from these compartments. Delivery technologies generally involve the transport of remediating materials into soils or groundwater. The delivered materials are usually liquids; however, newer technologies involve the delivery of solids, gases, and vapors as well. Recovery technologies generally remove contaminants from these same environmental compartments. This latter category includes technologies that expedite removals by altering the physical or chemical attributes of contaminants or pathways. Recovery technologies typically involve fluid flows driven by hydraulic gradients, thermal methods, or chemical reactions.

Delivery/recovery technologies have been used in hazardous waste site remediation for several years. Most of these technologies involve pumping groundwater from recovery wells, treating it and then reintroducing it at injection wells. Such "pump and treat" technologies vary in effectiveness with variations in site and contaminant properties. In addition, dense soil formations (with hydraulic conductivities of less than 10^{-4} cm/sec) severely limit the application of recovery techniques to contaminated hazardous waste sites.

Other problems associated with the implementation of delivery and recovery systems at waste sites include the presence of contaminants with low solubilities; adsorption of contaminants onto clayey soils; the existence of fractured soils or rocks, which create pathways of high conductivity separated by a matrix block of low conductivity; and the absence of an underlying impermeable layer to preclude the possibility of delivered materials migrating into the groundwater.

Some of the delivery/recovery techniques are found in other sections of this book. These are (1) colloidal gas aphrons, (2) soil vapor extraction, (3) steam stripping, (4) radiofrequency heating, (5) electrokinetics, (6) ground freezing, and (7) ultrasonics.

6.18.1 Carbon Dioxide Injection

Implementation of this technology in the petroleum industry involves injecting carbon dioxide (CO_2) into oil–bearing rock formations to maintain pressure and to displace the oil. The two principal mechanisms for mobilizing the oil by carbon dioxide injection are the reduction of the oil viscosity upon solution of the gas into the oil and an increase in the volume of the reservoir.

Use of this technology for the recovery of groundwater contaminants probably would be limited to applications where CO_2 is either dissolved in water or contained in aphrons. In either case, the injection of carbon dioxide could decrease the viscosity and increase the recovery of hydrocarbons. It has been reported that injecting CO_2 at high pressures (approximately 1,200 psi) increased its solubility in oil and dramatically reduced the viscosity of the oil. The feasibility of applying this technique to a site where contamination is near the surface is questionable. The pressure exerted by the injection stream may be sufficient to displace (with great force) the soil overburden.

Reduction of the viscosity of the oil results in two phenomena that should increase recovery. A decrease in viscosity results in an increase in mobility and thereby decreases the head gradient required to sustain flow. A decrease in the difference in the viscosities of the oil and the fluid used to displace it tends to minimize the formation of viscous instabilities or fingers.

The most effective method for introducing CO_2 involves following a CO_2 slug with alternating water and CO_2 injections. Other methods involve the injection of water saturated with CO_2 and the application of high-pressure injections of the gas itself.

6.18.2 Cyclic Pumping

Cyclic pumping (both a delivery and recovery technique) varies the rates of either injection or extraction in an effort to minimize pumping costs. Optimizing pumping activities could conceivably reduce remediation costs at contaminated waste sites.

In typical pump and treat operations, the pumping rates are usually held constant, as opposed to a cyclic pumping operation, in which the rates are variable and the pumps are turned on during active cycles and turned off during rest cycles. Pumping in this manner (active and rest cycles) can be accomplished for either the injection of treating solutions or the extraction of contaminated groundwater.

The purpose of a rest cycle is to permit sufficient time to elapse for diffusion between high-permeability channels (fractures of large pores) and the comparatively low-permeability blocks between them. Pumped treatment solutions diffuse from the pathways into the low-permeability areas, and contaminants diffuse from these areas to the high-permeability channels during a rest cycle. The active cycle of cyclic pumping is designed to deliver the necessary volume or reactants or to recover the necessary volume of reaction products.

It appears that cyclic pumping would be most effective in soils composed of preferred high-permeability channels and low-permeability blocks. The petroleum industry uses this technique to enhance oil recovery.

6.18.3 Funnel and Gate System

The Waterloo Centre for Groundwater Research has developed Funnel-and-Gate systems that isolate contaminant plumes in groundwater and funnel the plumes through in situ bioreactors. The Funnel-and-Gate consists of low hydraulic conductivity cutoff walls with gaps that contain in situ reactors (such as reactive porous media), which remove contaminants by abiotic or biological processes. The cutoff walls (the funnel) modify flow pattern so that groundwater flows primarily through high conductivity gaps (the gates).

Groundwater plumes are thus directed through the in situ reactors in the gates where physical, chemical or biological processes remove contaminants from groundwater. Remediated groundwater exits the downgradient side of the reactor. Funnel-and-Gate systems can be installed at the front of plumes, to prevent further plume growth, or immediately downgradient of contaminant source zones to prevent contaminants from developing into plumes.

6.18.4 Hot Brine Injection

It is possible to stimulate contaminant removal by use of hot brine. Hot–brine injection could be used at uncontrolled hazardous waste sites to recover certain contaminants whose dissociation temperatures decrease with an increasing salinity of pore fluid. The applicability of this method will require a salinity dependence on the dissociation temperature of the contaminant species to be recovered.

Natural gas deposits typically occur as liquids or solids at existing pressures in many reservoirs. Because extraction of these liquid or solid deposits is more difficult than extraction of the gas in vapor phase, petroleum engineers have developed methods for converting the gas from the solid or liquid phase to the vapor phase.

The hot brine injection method for recovering natural gas hydrates appears to be limited to depths greater than 150 m because the hydrate equilibrium curves indicate lower dissociation temperatures are needed than the ambient earth temperatures that occur at shallow depths. The dissociation temperature could be reduced through the artificial increase of salinity of the pore solution, however, to facilitate the recovery of contaminating chemicals with a dissociation temperature exceeding the temperature at shallow depths. For this method to be used successfully, the dissociation temperature of the contaminant to be recovered must have a salinity dependence.

At depths of approximately 150 to 1,000 meters, the temperature of hydrate dissociation from a solution with no salinity exceeds the ambient temperature of the surrounding earth by almost 10°C. Increasing the salinity of the pore solution to 15 wt % reduces the temperature of hydrate dissociation by about 11°C. The relationship between salinity and dissociation temperature reduction is approximately linear within the range of 0 to 15 wt % salinity.

The amount of thermal energy required to dissociate gas hydrates into a pure gas vapor phase and liquid water decreases as dissociation temperatures decline. Increasing the salinity of the pore solution from 0 to 15 wt % reduces the energy of hydrate dissociation by nearly 8%.

6.18.5 Hydraulic Fracturing

Hydraulic fracturing is a technology that could be used to increase the rates of either delivery or recovery. It is used extensively within the petroleum industry both to stimulate the recovery of hydrocarbons from low permeability reservoirs, and to enhance the delivery of fluids used to displace petroleum in sweeping operations.

The technology involves the injection of a fluid (typically water) at pressures exceeding the confining pressures at the bottom of a borehole. This process generates a single fracture (either horizontal or vertical) that propagates away from the borehole. Sand is introduced into the formed fracture to hold it open and to create a highly permeable channel suitable for either the delivery of remediating materials or the recovery of contaminants.

Preliminary investigations consisting of theoretical calculations and comparative investigations on applications developed by the energy industry suggest that this technology can be used with soil and rock types commonly found at contaminated waste sites. Possible applications include increasing the efficiency of pump and treat systems, stimulating the extraction of vapor phases from dense soils, or forming a horizontal drain to capture leachate.

Hydraulic fracturing is a permeability–enhancement technique used in conjunction with other soil remediation methods, such as soil vapor extraction, or bioremediation. Therefore, overconsolidated silty clays that have low in situ permeabilities are best suited for the use of hydraulic fracturing. In overconsolidated clays, the horizontal stress is greater than the vertical stress, and this stress condition permits fractures to propagate in a horizontal direction. Fractures that remain horizontal can grow to significant lengths, thereby enhancing flow in the subsurface.

Hydraulic fracturing is ineffective in normally consolidated clays where the horizontal stress is less than the vertical stress. Demonstrations of hydraulic fracturing in such clays created fractures that were steeply dipping and vented to the surface. Also, the presence of water decreases the efficiency of soil extraction (SVE), hence, the use of hydraulic fracturing to enhance SVE should be limited to unsaturated clays.

Horizontal fractures increase the permeability of soil in their immediate vicinity. However, the permeability of soil between the fractures is not affected unless vertical or inclined fractures are created between the horizontal fractures. Further work at UC is being planned to study this. In SVE applications, changes in soil vacuum (suction head) applied to horizontal fractures may induce communication between the fractures.

Bioremediation was determined to be a viable method of degrading the hydrocarbon contaminants at a fuel distribution and storage facility in Dayton, Ohio. Laboratory tests done by the on site contractor indicated that percolating water containing oxygen and nutrients through the soil would result in biodegradation of the contaminants. The site is underlain by silty clay till of relatively low hydraulic conductivity, so conventional methods of delivery were expected to result in either slow rates of percolation, and thus slow rates of remediation, or excessive drilling costs. Therefore, the site was selected as a candidate for hydraulic fracturing, a technique of creating high permeability channel ways in tight soils.

Hydraulic fracturing was shown to be a feasible procedure at this contaminated site. The size and shape of the fractures were as planned. Enough information was gathered at the site in the initial assessment to evaluate the post treatment effects of the hydrofracturing as compared to a conventional injection well arrangement. The site conditions make it conducive to bioremediation (e.g., pH about neutral, significant hydrocarbon degrading microbial population).

The technology developed by the Risk Reduction Engineering Laboratory (RREL) and the University of Cincinnati, creates sand–filled horizontal fractures up to 1 inch in thickness and 20 ft in radius. These fractures are placed at multiple depths ranging from 5 to 30 ft below ground surface (bgs) to enhance the efficiency of treatment technologies such as soil vapor extraction, in situ bioremediation, and pump–and–treat systems.

The fracturing process begins by using a hydraulic jet to cut a disk–shaped notch extending 0.5 ft from the borehole wall. Water is injected into the notch until a critical pressure is reached and a fracture is formed. A proppant composed of a granular material (sand) and a viscous fluid (guar gum and water mixture) is then pumped into the fracture at a rate of 16 to 24 gal/min. After pumping, the sand holds the fracture open while an enzyme additive breaks down the viscous fluid. The process is repeated at greater depths to create a stack of multiple, sand–filled hydraulic fractures.

The technology is designed for use in low permeability silty clays contaminated with organic compounds. This technology enhances other in situ remediation techniques such as vapor extraction and bioremediation.

Hydraulic fracturing has applications for pump and treat, vapor extraction, soil flushing, and bioremediation. It already has been used for some pump–and–treat operations in crystalline rock in New York. It also is used in some cases to increase the area over which a monitoring well can sample, and it can be used to recover leachate. The flat–lying pancake–shaped geometry created by fracturing is ideal for any of these applications.

The advantages of hydraulic fracturing include the ability to increase well yield from tight materials, increase the radius of influence, form a flat–lying planar sink thus avoiding some geometry problems, and inject solids. The technique is easy to perform, and is similar to injection grouting.

The disadvantages are that this technology is currently limited to rock or overconsolidated soil, there are limited performance data on soils, the technology causes surface deformation of about 1 inch, and availability is limited for soil.

6.18.6 Jet–Induced Slurry

The jet–induced slurry method is a technique used to excavate an ore body by fragmenting subsurface material with a high–velocity hydraulic jet and then pumping the slurry through a borehole to the surface. This technology offers the unique capability of recovering solid material from the subsurface at depth without removing the overburden.

The use of high–pressure water jets to slurry ore–bearing rock is not a new technique; it has been used for more than 100 years to mine placer gold and gilsonite at the ground surface. The application of this technology to subsurface formations, however, has only occurred within the past few years.

The jet–induced slurry method (called "in situ borehole slurry mining" by the mining industry) is facilitated by drilling an 18 inch borehole to roughly 6 ft below the bottom horizon of an ore–bearing formation and installing a sump pump at the bottom of the borehole, below the hydraulic jet. The jet slurries the rock in the ore zone and the slurry flows to the sump, where it is pumped to the surface for processing. The tailings from the processing are used to backfill the cavity left by the excavation.

This could be a potential technique for assisting in the remediation of hazardous waste sites.

6.18.7 Kerfing

Kerfing (or borehole notching) is a technique to cut a slot either normal or parallel to the axis of a borehole. Although recent interest in kerfing has been as a technique to help stop the migration of pollutants from uncontrolled waste sites, it may also have application as a recovery technique.

Kerfing was developed in the United States and Europe; however, it was first applied in Japan. Kerfing is used primarily as a method of placing barriers of low permeability beneath hazardous waste sites to intercept leachate and to prevent further groundwater contamination.

This technique uses a high–pressure water jet and an abrasive material (e.g., sand) to cut a slit in the wall of a borehole. The high–pressure water jet is placed in an existing borehole, where it is rotated to cut a disk–shaped cavity, moved along the axis of the hole to create an axial slit, or kept in place to cut a cylindrical hole. The water jet advances the kerf at a rate of several centimeters per second, and the final slit or hole is 1 to 3 m long. The slot created is subsequently filled with a permeable material to create a drain, an impermeable material to create a barrier, or a remediating material to facilitate cleanup.

Kerfing may also be suitable as a technique for improving recovery processes. A potential method of improving recovery would be to fill the kerf with a highly permeable material (much as a hydraulic fracture is filled with sand), which would improve the yield of a recovery well.

As a recovery technique, kerfing appears to be most applicable to materials of low permeability, such as clay, silt, or rock, where most contaminants would be immobile. Using kerfing to increase the rate of recovery from recovery wells is undocumented. The rate could be expected to increase by a factor of 2 to 10 compared with the rate before fracturing.

Kerfing is widely used in the petroleum industry (where it is referred to as borehole notching) to initiate hydraulic fractures. Some oil field service companies provide borehole notching services.

6.18.8 Pneumatic Fracturing

Pneumatic fracturing involves injecting high pressure air at controlled flow rates and pressures into fine grained soils such as clay and silt. This opens the clays up to either withdrawing contamination by vacuum extraction or, in the case of bioremediation, improving microbial access to nutrients or air.

After drilling a small diameter hole in the ground, the NJIT (New Jersey Institute of Technology) group lowers a packer–like device down the hole. Upon inflation, this device seals off a two foot section. At this point, air or gas is injected into that two foot zone. The injected air forms a plane of fracture that, for the most part, is horizontal because of the in situ geologic stresses that exist. As a result, at the fracture plane at a level of 10 to 12 ft below grade, the next step would be to move the device down to 14 to 16 ft and repeat the injection. So the procedure is to move the device up and down the bore hole injecting air at whatever depth it is desired to create a fracture network.

To date, most of the work on this technology has been in the unsaturated zone but the NJIT devlopers believe that the removal of contaminated groundwater can also be enhanced by pneumatic fracturing.

The Pneumatic Fracturing Extraction™ (PFE)™ process developed by Accutech Remedial Systems, Inc. makes it possible to use vapor extraction to remove volatile organics at increased rates from a broader range of vadose zones. In the PFE process, fracture wells are drilled in the contaminated vadose zone and left open bore (uncased) for most of their depth. A packer system is used to isolate small (2 ft) intervals so that short bursts (~20 sec) of compressed air (less than 500 psig) can be injected into the interval to fracture the formation. The process is repeated for each interval. The fracturing extends and enlarges existing fissures and/or introduces new fractures, primarily in the horizontal direction.

6.18.9 Polymer Injection

Water–soluble polymers are added to waterflooding solutions in the petroleum industry to (1) increase the viscosity of flooding fluid, thereby decreasing the viscosity contrast between the oil and flooding fluids, or (2) form highly viscous gels in situ (to homogenize reservoir permeability).

The homogenization of permeability is one possible application of polymers for the recovery of contaminants. Temporarily filling high-permeability channelways with polymer gels could be used to induce flow and transport contaminants out of local zones of low permeability.

6.18.10 Pump and Treat

Pump and treat technology may, in many cases, be used to aid in the removal of light non-aqueous phase liquid (LNAPL) and/or DNAPL (dense NAPL) which may be present. Recovery of LNAPL residing as free product on the surface of the water table, for example, can be facilitated by using pumping wells to create cones of depression. DNAPL residing as large pools in topographical lows at the bottom of aquifers can be recovered by pumping from wells screened over the thickness of the pools. In cases where recovery is not feasible (e.g., DNAPL resides in fractures or is present as spatially discontinuous free product within an aquifer), alternative measures such as physical containment (e.g., cement-bentonite walls) should be considered.

Pump and treat remediation technology is applicable to the saturated zone and refers to the extraction of contaminated groundwater from the subsurface and subsequent treatment of the extracted groundwater at the surface. Extraction of contaminated groundwater is accomplished through the use of extraction (pumping) wells which are completed at specified locations and depths to optimize contaminant recovery. Determination of the locations and depths of extraction wells requires prior delineation of the contaminant plume and knowledge of the aquifer properties. Injection wells may be installed to enhance contaminant recovery by flushing contaminants toward extraction wells.

Pumping technology may also be used as a means of containing or controlling contaminant plumes. This is accomplished through control of hydraulic gradients by selectively locating pumping wells in the area of the plume. Control of hydraulic gradients should be considered in conjunction with physical containment options.

The surface treatment of extracted groundwater will vary depending on the contaminants present. Typical actions include air stripping, activated carbon adsorption and biological treatment. In some cases, treated groundwater may be amended with nutrients and oxygen and reinjected into the subsurface to aid in stimulating biodegradative processes.

Pulsed pumping is a modification of standard pump and treat technology which involves regular or periodic cessation of pumping activities to optimize groundwater cleanup. Pulsed pumping may be necessary or more cost-effective in cases where extraction wells can not sustain yields (e.g., in bedrock and unconsolidated deposits of low permeability), where desorption and/or dissolution of contaminants in the subsurface is relatively slow, or where hydraulic conductivity heterogeneity is high.

A potential concern associated with implementation of pulsed pumping is the uncontrolled migration of the contaminant plume during non-pumping phases. Nearby water supply wells or irrigation systems may significantly impact the behavior of the contaminant plume during non–pumping phases and thereby create a potentially more serious contamination scenario.

Reinjection, which often is used in combination with pump and treat or pulsed pumping, generally refers to injection of treated groundwater back into the subsurface. Reinjection may be accomplished through the use of injection wells or other means such as infiltration galleries. Reinjected groundwater can be used to help remove contaminants residing in the unsaturated zone by forcing these contaminants towards extraction wells. Reinjection also may be used in the stimulation of biodegradative processes in the saturated zone, thereby enhancing cleanup of the saturated zone. In such cases, the injectate is amended with nutrients and an oxygen source. In special cases, the injectate may be amended with surfactants or other compounds (i.e., chemical extraction) to facilitate removal of adsorbed and residual organics in the unsaturated and/or saturated zones.

The process can be enhanced by injecting air into the groundwater via wells to vaporize organic compounds. The increased oxygen also escalates biodegradation, cutting down treatment time without expanding the contamination plume.

The most difficult problem associated with reinjection is plugging and can be caused by:

(a) Suspended solids.
(b) Entrained air through physical blocking or the Jamin effect.
(c) Microbial fouling.
(d) Chemical incompatibility between injection and formation water.
(e) Compaction of the gravel/sand pack around the well.

Limitations: Reducing groundwater concentrations to standards required by the Safe Drinking Water Act or Land Disposal Restriction is difficult using available technology for many contaminants. There are several inherent limitations that hinder effective pump–and–treat site remediation. These include the potentially long time necessary to achieve the remediation goal; system designs failing to contain the contaminants as predicted, allowing the plume to migrate; and failure of surface equipment.

Research has substantiated other limitations with the use of pump–and–treat technology. These limitations include contaminant residual saturation, chemical sorption of the contaminant, and low hydraulic conductivity causing tailing effects.

Almost all remediation of groundwater at heavily contaminated sites is based on groundwater extraction by wells or drains, usually accompanied by treatment of the extracted water prior to disposal. This often causes an initial decrease in contaminant concentrations in the extracted water, followed by a leveling of concentration, and sometimes a gradual decline that is generally expected to continue over decades. In such cases, the goal of reaching stringent health–base cleanup standards is very remote and the ultimate cost of cleanup very high. The process is often expensive, inexact, and extremely lengthy. Also, in some cases, after pumping has ceased, there has been a rise in contaminant levels in the groundwater.

Scientists have so far identified the following potential reasons for the failures of the pump and treat methodology.

1. **Binding to Soil:** Particularly where contaminants have been present for a long time, they tend to adhere to particles of soil and to resist being drawn up by the groundwater flow created by the pumping process.

Because the contaminant cannot be removed from the groundwater any faster than it is released from the soil, the attainment of cleanup levels is rate limited by the process of diffusion. Under these circumstances the cleanup process can take thousands of years.

2. **Channelization:** The pumping process causes the groundwater to flow through those soils that have the highest permeability (i.e., those which impede the flow the least) and contaminants in soils with lower permeability are not picked up by the groundwater extraction.
3. **Non-Aqueous Phase Liquids:** Certain contaminants, called non-aqueous phase liquids (NAPLs), do not mix with water. If they are denser than water (DNAPLs), they will sink downward through groundwater and will be more responsive to gravity than to the flow established by pumping. If they are lighter than water (LNAPLs), they will tend to float on top of the groundwater and pumping may simply remove the groundwater from beneath them while they remain in place.
4. **Gradual "Recontamination":** Whether bound to the soil, floating, or sunk to the bottom of the groundwater, NAPLs do dissolve very gradually, over very long periods of time. Thus, after pumping stops, groundwater that has reached a certain level of cleanliness can become recontaminated, perhaps even to its original levels.

Although some success has been achieved, the Oak Ridge National Laboratory has concluded flatly: "When large pools of DNAPLs are present at the bottom of an aquifer, meeting drinking-water standards is unachievable at any cost."

EPA has had to agree. Regarding DNAPLs, the EPA acknowledged in January 1992 that "to date there have been no field demonstrations where sufficient DNAPL has been successfully recovered from the sub-surface to return the aquifer to drinking water quality."

Besides not achieving what it is supposed to achieve, pumping and treating can sometimes have negative effects on the area's ecology, and where mistakes are made, can actually exacerbate the problem. Thus, dewatering from pumping can cause serious land subsidence and other ecological damage. For example, pumping at the IBM site in San Jose, California, is reported to have resulted in the dewatering of an aquifer. Additionally, drilling too deep or lowering a water table too much can cause a NAPL pool of contaminants to migrate to a deeper aquifer. For example, at the Department of Energy's Savannah River site in South Carolina, where pumping has gone on for five years, evidence of movement of contamination to deeper aquifers is reported.

Optimizing groundwater pump-and-treat system performance is typically achieved by altering the groundwater extraction well schedule, or modifying the groundwater extraction system by installing additional extraction wells or implementing an alternative technology.

Methods to improve system performance, and suggested system modifications have been presented by Simon and Thiell.

6.18.11 Subsurface Drains

Subsurface drains include any type of buried conduit used to convey and collect aqueous discharges by gravity flow. Subsurface drains essentially function like an infinite line of extraction wells. They create a continuous zone of influence in which groundwater within this zone flows towards the drain.

There are two categories of drains: interceptor and relief drains. Interceptor drains are installed perpendicular to groundwater flow and are used to intercept groundwater from an upgradient source. Relief drains are installed parallel to the direction of flow or around the perimeter of a site where the water table is relatively flat. Relief drains are used primarily to lower the water table beneath a site.

The major components of a subsurface drainage system are:

1. Drain pipe or gravel bed—for conveying flow to a storage tank or wet well. Pipe drains are used most frequently at hazardous waste sites. Gravel bed or french drains and tile drains are used to a more limited extent.
2. Envelope—for conveying flow from the aquifer to the drain pipe or bed.
3. Filter—for preventing fine particles from clogging the system, if necessary.
4. Backfill—to bring the drain to grade and prevent ponding.
5. Manholes or wet wells—to collect flow and pump the discharge to a treatment plant.

Since drains essentially function like an infinite line of extraction wells, they can perform many of the same functions as wells. They can be used to contain or remove a plume, or to lower the groundwater table to prevent contact of water with the waste material. The decision to use drains or pumping is generally based on a cost–effectiveness analysis.

The most widespread use of subsurface drains at hazardous waste sites is to intercept a plume hydraulically downgradient from its source. Frequently, these interceptor drains, as they are commonly called, are used together with a barrier wall. There are two primary reasons for the interceptor drain/barrier wall combination. In the case where a subsurface drain is to be placed just upgradient of a stream, the drainage system would reverse the flow direction of the stream and cause a prohibitively large volume of clear water to be collected. The addition of a barrier wall would prevent infiltration of clean water from the stream thereby reducing treatment costs.

In another application, where the primary remedial action involves installation of a downgradient barrier wall to contain wastes, an interceptor drain can be installed just upgradient of the wall to prevent overtopping and to minimize contact with wastes which may degrade the wall.

Subsurface drains can also be placed around the circumference of a waste site in order to lower the groundwater table or to contain a plume. A circumferential subsurface drain may be part of a total containment system which consists of a barrier wall and a cap in addition to the subsurface drain.

In addition to depth, other limitations to the use of subsurface drains include the presence of viscous or reactive chemicals which could clog drains and envelope material. Conditions which favor the formation of iron manganese or calcium carbonate deposits may also limit the use of drains.

For hazardous waste site applications, pipe drains are most frequently used. French or gravel drains can be used where the amount of water to be drained is small and flow velocities are low. If used to handle high volumes or rapid flows, these drains are likely to fail due to excessive siltation, particularly in fine grained soils. Tile drains have not been widely used in hazardous waste site applications.

Evaluation of the suitability of subsurface drains as a remedial technology is generally made by comparing the cost–effectiveness of this alternative with pumping. Relative to pumping, subsurface drains can be difficult and costly to install particularly where extensive hard rock excavation and dewatering is required. They are also time consuming to install and may not be an appropriate alternative where immediate remediation is required. Safety of field workers is also more of a concern with subsurface drains because of the need for extensive trench excavation.

However, there are several advantages of drains relative to pumping. They are generally more cost–effective than pumping in areas with low hydraulic conductivity particularly where pumping would be required for an extended period of time. They are easier to operate since water is collected by gravity flow. They are also more reliable from the standpoint that there are no electrical components which can fail. However, when drains fail due to clogging, breaks in the pipes, or sinkhole formation, they can be costly and time consuming to rehabilitate.

6.18.12 Wells and Trenches

Wells are dug at remediation sites for a number of purposes: (1) pump and treat injection and extraction wells, (2) soil vapor extraction technology, (3) delivery of remediating and enhancement materials, etc.

Groundwater Pumping: Groundwater pumping techniques involve the active manipulation, removal, containment, and management of groundwater, in order to contain or remove a plume or to adjust groundwater levels in order to prevent formation of a plume. Types of wells used in management of contaminated groundwater include wellpoints, suction wells, ejector wells, and deep wells. The selection of the appropriate well type depends upon the depth of contamination and the hydrologic and geologic characteristics of the aquifer.

Pumping is most effective at sites where underlying aquifers have high intergranular hydraulic conductivity. It has been used with some effectiveness at sites with moderate hydraulic conductivities and where pollutant movement is occurring along fractured or jointed bedrock. In fractured bedrock, the fracture patterns must be traced in detail to ensure proper well placement.

Where plume containment or removal is the objective, either extraction wells or a combination of extraction and injection wells can be used. The use of a line of extraction wells serves to halt the advance of the leading edge of a contaminant plume and thereby prevent contamination of a drinking water supply. Use of extraction wells alone is best suited to situations where contaminants are miscible and move readily with water; where the hydraulic gradient is steep and hydraulic conductivity high; and where quick removal is not necessary. Extraction wells are frequently used in combination with slurry walls to prevent groundwater from overtopping the wall and to minimize contact of the leachate with the wall in order to prevent wall degradation. Slurry walls also reduce the amount of contaminated water that requires removal, so that costs and pumping time are reduced.

A combination of extraction and injection wells is frequently used in containment or removal where the hydraulic gradient is relatively flat and hydraulic conductivities are only moderate. The function of the injection well is to direct contaminants to the extraction wells. This method has been used with some success for plumes which are not miscible with water. One problem with such an arrangement of wells is that dead spots (i.e., areas where water movement is very slow or nonexistent) can occur when these configurations are used. The size of the dead spot is directly related to the amount of overlap between adjacent radii of influence; the greater the overlaps the smaller the dead spots will be. Another problem is that injection wells can suffer from many operational problems, including air locks and the need for frequent maintenance and well rehabilitation.

Determination of the proper spacing of wells to completely capture a groundwater plume is probably the most important item in system design. Field practitioners have long had a standing "rule of thumb" for estimating well spacing: adjacent cones of depression should overlap (i.e., radii of influence should overlap). This method is reasonably accurate for aquifers that have low natural flow velocities but will not be valid for aquifers with high natural flow velocities.

For a confined aquifer, capacity and therefore pumping rate is directly proportional to the drawdown as long as the aquifer is not unwatered. Increasing the pumping rate will not affect the radius of influence but will affect the amount of time pumping is necessary. Therefore, pumping rates can be selected to suit the situation. In situations where the contaminated plume floats, drawdowns and pumping rates will probably be small. Large drawdowns and high pumping rates are desirable where contaminants are dispersed throughout the aquifer, quick removal is desired, and natural groundwater flow rates are large.

For an unconfined aquifer it has been found that maximum efficiency for well operation occurs at about 67% of the maximum drawdown; pumping rates should be adjusted accordingly. This is because part of the formation within the cone of depression is actually unwatered during pumping and the specific yield decreases with increased drawdown. Optimum operating conditions are achieved when the product of specific yield and capacity are greatest and this occurs at about 67% of maximum drawdown.

The major components of a **deep well** include: casing, screen, filter pack and seal, and pump.

Wellpoint systems consist of a group of closely spaced wells connected to a header pipe and pumped by a suction pump. Wellpoints are best suited for groundwater extraction in stratified soils where total life or drawdown will not exceed 22 ft. The advantages to using wellpoints are that the system design is flexible and the wellpoints are relatively inexpensive even when closely spaced.

A suction (vacuum) pump is typically used in wellpoint systems to lift water. Suction pumps (either oil sealed or water sealed) accomplish lift by developing a negative pressure head at the pump intake rather than by applying force to the water source as in ejector pump systems. The maximum lift attainable by suction pump is about 22 ft.

Ejector wells have certain advantages over wellpoints and deep wells because they are not typically depth limited as wellpoints are and they are less expensive than deep wells when close spacing is required. The biggest drawback to the use of ejector wells is that they are very inefficient (typically less than 15% efficiency). Ejector wells can be used independently of each other or arranged so that they utilize a common pumping system.

Soil Vapor Extraction: Several options are available for extraction system layout. The most common methods are vertical wells, trenching or horizontal wells, and excavated soil piles.

Vertical wells are the most widely used SVE design method. This method is the only feasible option at sites where the contamination extends far below the land surface. Horizontal wells or trenches may be more practical than vertical wells where the depth to groundwater is less than 12 ft. Vertical wells are generally inappropriate for sites with a shallow water table due to the potential upwelling of the water table that may occur after the application of a high vacuum.

Soil vapor extraction is discussed in Section 6.13.

Horizontal Wells: Horizontal wells drilled radially outward from a central borehole can be used to enhance access to a contaminated formation or aquifer. They can be used for either delivery or recovery. Many radial wells can be placed at the same level or on multiple levels in the same borehole. The potential applications of radial wells include groundwater control and in situ treatment. Horizontal wells have the ability to intersect fracture zones that are perpendicular to the ground surface or to intercept groundwater moving horizontally, allowing contaminant plumes to be entered laterally rather than vertically. Applications of this technology include groundwater extraction and delivery of in situ remediating materials.

Initial applications of this technology involved drilling a central downward core sufficiently wide to allow a worker to descend into the borehole and to drill radially outward. Recent developments made by the petroleum industry have dramatically improved the method of drilling horizontal wells. This new system includes a jet nozzle fixed to 1.25 inch diameter tubing (used to create the horizontal well) and a whipstock, which is placed at the bottom of the vertical borehole to redirect the jet nozzle and tubing horizontally into the formation. Radial wells are created by pumping high–velocity water out of the jet nozzle. The water for this process is supplied from the surface through steel tubing. This equipment cuts a borehole with a diameter several times larger than the tubing. In addition, static pressure on the inside of the nozzle results in a force that pulls the nozzle and attached tubing down the drill stem, through the whipstock, and into the radial well. This force is responsible for keeping the radial progressing in a straight line.

Radial wells applied to hazardous waste sites can be positioned in both saturated and unsaturated media and can facilitate the remediation of contaminated sites by increasing the available delivery/recovery routes for delivering remediating materials or recovering contaminated groundwater.

The Department of Energy (DOE) has developed a remediation technology, known as in situ air stripping, to remove chlorinated solvents from soil and groundwater. The process combines vapor extraction and air injection using horizontal wells. The combination of vapor extraction and air injection will allow for concurrent remediation of both the groundwater and the overlying soil.

The geometry of horizontal wells may improve the performance of in situ remediation technologies. Horizontal wells can provide more contact area with the contaminant plume. Because many water–bearing formations are deposited as relatively thin but extensive zones, the use of horizontal wells may improve the efficiency of delivery of reactants to or recovery of contaminants from these formations. In addition, horizontal wells can be used along linear sources of contamination and under existing facilities.

A study was conducted at the Savannah River site by the Department of Energy in 1988, and found the following factors that should be considered:

1. Trips in and out of the borehole should be minimized to maintain hole stability. No reaming to enlarge the hole should be attempted.
2. Drilling fluid performance should be maximized by utilizing a low solids, low weight, high lubricity fluid. Interruption of drilling fluid circulation should be minimized.
3. Well materials should possess adequate flexibility to negotiate the curve. A flexible guide should be attached to the front of the well screen to guide the screen downhole.
4. Sands containing a minor amount of clay are recommended for completion targets, as better drilling control in the laterals was obtained in these sections.

Directional drilling is another useful technique. If the remedial manager desires the linear geometry of a trench, but is constrained by some of the disadvantages, one option is directional drilling. Directional drilling, which installs horizontal wells, will improve access and will have some geometric advantages such as greater working surface area. It also can address problems related to heterogeneities, such as preferred flow paths induced by vertical fractures. This method of drilling is gaining in popularity.

Directional drilling, as defined here, requires a rig that is surface launched and has a drilling head that can be steered in any direction. It also must include an electronic device to monitor location. Two systems meet these requirements: one from the oil industry and one from the utility industry.

Trenches: The general applications for trenches are (1) to intercept groundwater, (2) to lower the water table, (3) to stabilize contaminant migration, and (4) to use with a barrier. An example of the fourth application is using a trench on one side of the contaminants and a barrier on the other to prevent uncontaminated water from entering the trench. Uncontaminated water would dilute the contamination, increasing the volume of contaminant to be treated.

Trenches have a better capture effectiveness than vertical wells and are relatively simple to construct, especially the shallow ones. Trenches also have a greater yield in tight soils than wells and, as a result, can have lower operating and maintenance costs. The disadvantages are that excavated contaminated material from trenches could require disposal as hazardous waste, and access can be difficult, especially with a large backhoe. In addition, the depth limit for conventional excavation techniques is about 15 ft, although more specialized trenches can be built to a depth of 40 to 65 ft. Another disadvantage of trenches is the difficulty of abandoning them.

6.19 UNDERGROUND INJECTION AND DISPOSAL

There are two methods utilized for underground disposal of hazardous material: (1) deep-well injection, and (2) solution-mined salt caverns, or abandoned potash mines.

6.19.1 Deep-Well Injection

U.S. EPA regulations (53 *Federal Register* 28118–28157, July 26, 1988) stipulate that deep–well injection of hazardous wastes is allowed only if either of two no–migration standards is met:

1. Fluid movement conditions are such that the injected fluids will not migrate within 10,000 years: vertically upward out of the injection zone; or laterally within the injection zone to a point of discharge or interface with an Underground Source of Drinking Water (USDW) as defined in 40 CFR Part 146, Subpart G.
2. Before the injected fluids migrate out of the injection zone or to a point of discharge or interface with USDW, the fluid will no longer be hazardous because of attenuation, transformation, or immobilization of hazardous constituents within the injection zone by hydrolysis, chemical interactions, or other means.

There are two important considerations relating to the avoidance of groundwater contamination. Firstly, the well should be dug through a thick layer of impermeable materials such as shale to avoid upward migration of the hazardous material. Secondly, there should be double casing through the saturated zone and some distance into the impermeable zone to avoid leaks into the groundwater.

The use of wells for disposal of industrial wastes dates back to the 1930s, but this method was not used extensively until the 1960s, when it was implemented primarily in response to more stringent water pollution control regulations.

In 1986 there were a total of 263 hazardous waste injection wells in the United States. The injection wells are concentrated in Texas (112 wells) and Louisiana (70 wells). In 1986 there were also 165 non–hazardous injection wells.

Gulf Coast injection wells are typically between 4,000 and 7,000 ft deep with temperatures up to 80°C. A typical injection zone is an arenaceous horizon containing up to 70 wt % detrital quartz, together with 15 wt % of detrital plagioclase and potash feldspars, with the remainder clay minerals with secondary calcite. Confining shale horizons typically consist of about 70 wt % clays with smaller amounts of other detrital minerals and secondary pyrite. A significant although minor amount of organic detritus is present in both shales and sandstones.

For a more complete description of deep–well injection, please see the author's previous book: *Handbook of Pollution Control Processes (1991).*

6.19.2 Underground Disposal

The construction of solution–mined salt caverns is a technique that has been an accepted method of storage for the past two decades. More than 1,000 salt caverns are currently in use worldwide, storing primarily crude oil, oil products, and natural gas.

Salt makes an excellent material as a geologic repository because it exhibits plastic flow. This property allows the salt to conform to changes in pressure or movement, inhibiting cracking of the dome. Also, it is assumed that since the salt domes have existed for millions of years, their environment is not threatening in any way, and they will continue to be extremely stable for millions of years to come.

Salt caverns are constructed in salt formations by solution mining. Access wells are installed with standardized deep drilling techniques, down to the projected final cavern depth. The last cemented casing typically extends 100 to 200 m into the salt formation.

In Germany, abandoned potash mines are being utilized for waste disposal.

REFERENCES

1. Albertson, O., et al, *Dewatering Municipal Wastewater Sludges,* Noyes Data, 1991.
2. Berger, B., et al, *Control of Organic Substances in Water and Wastewater,* Noyes Data, 1987.
3. Berkowitz, J., et al, *Unit Operations for Treatment of Hazardous Industrial Wastes,* Noyes Data, 1978.
4. Bravo, J., et al, *Fluid Mixture Separation Technologies for Cost Reduction and Process Improvement,* Noyes Data, 1986.
5. Breton, M., *Treatment Technologies for Solvent Containing Wastes,* Noyes Data, 1988.
6. Burton, D., *Treatment of Hazardous Petrochemical and Petroleum Wastes,* Noyes Data, 1989.
7. Chambers, C., et al, *In-Situ Treatment of Hazardous Waste-Contaminated Soils,* Noyes Data, 1991.
8. Cheremisinoff, P., *Oil/Water Separation,* Nat. Env. Jnl., 5-6/93.
9. Chironna, R., *Dry/Wet Scrubbers for Clean Air Compliance,* Poll. Engr., 11/1/92.
10. Dalton, J., et al, *An Assessment of Off-Gas Treatment Technologies,* Dept. of Energy DOE/MWIP-1, 9/92.
11. DOE, *Environmental Restoration and Waste Management Five Year Plan, 1994-1998,* DOE, 1/93.
12. Dosani, M., et al, *Materials-Handling Technologies Used at Hazardous Waste Sites,* Noyes Data, 1992.
13. Doty and Travis, *The Effectiveness of Groundwater Pumping as a Restoration Technology,* Oak Ridge National Laboratory (ORNL/TM-11866) 5/91.
14. EPA, *Accutech Pneumatic Fracturing, Extraction, and Hot Gas Injection,* EPA/540/SR-93/509, 8/93; and EPA/540/AR-93/509, 7/93.
15. EPA, *Advances in Air Sparging Design,* Haz. Waste Conc., 1-2/93.
16. EPA, *Air Emissions from the Treatment of Soils Contaminated with Petroleum Fuels and Other Substances,* EPA 68-D1-0117, 5/92.
17. EPA, *Alternative Control Technology Document—Organic Waste Process Vents,* EPA-450/3-91-007, 12/90.
18. EPA, *Biogenesis™ Soil Washing Technology,* EPA/540/R-93/510, 9/93.
19. EPA, *Development Document for Best Available Technology, Pretreatment Technology, and New Source Performance Technology for the Pesticide Chemical Industry,* EPA 821 R-92-005, 4/92.
20. EPA, *Construction Quality Management for Remedial Action and Remedial Design Waste Containment Systems (2nd),* EPA/540/R-92/073, 10/92.
21. EPA, *Corrective Action: Technologies and Applications,* EPA/625/4-89/020, 9/89.
22. EPA, *Development Document for Effluent Limitations Guidelines and Standards for the Inorganic Chemicals Manufacturing Point Source Category,* EPA 440/1-82/007, 6/82.
23. EPA, *Development Document of Effluent Limitations Guidelines and Standards for the Nonferrous Metals Manufacturing Point Source Category,* EPA 440/1-89-019.1.
24. EPA, *Development Document for Effluent Limitations Guidelines and Standards for the Organic Chemicals, Plastics and Synthetic Fibers, Vols. 1 & 2,* EPA 440/1-87/009, 10/87.
25. EPA, *EPA Surveys Alternatives to Conventional Pump-and-Treat Measures,* Haz. Waste Cons., 3-4/93.
26. EPA, *Estimating Potential for Occurrence of DNAPL at Superfund Sites,* Publication 9355.4-07FS, Office of Emergency and Remedial Response, 1/92.
27. EPA, *Ground Water Currents,* (OS-110W), EPA/542/N-93/006, 6/93.
28. EPA, *A Guide to Pump and Treat Groundwater Remediation Technology,* OS-110, 9/90.
29. EPA, *Hydraulic Fracturing Technology,* EPA/540/R-93/505, 9/93.
30. EPA, *Industrial Wastewater Volatile Organic Compound Emissions,* EPA-450/3-90-004, 1/90.
31. EPA, *Innovative Hazardous Waste Treatment Technologies: Domestic and International,* EPA/540/2-89/056, 9/89.
32. EPA, *Innovative Hazardous Waste Treatment Technologies: Domestic and International,* EPA/540/2-90/009, 9/90.
33. EPA, *Innovative Hazardous Waste Treatment Technologies, Domestic and International, Fourth Forum,* EPA/540/R-92/081, 12/92.
34. EPA, *Innovative Treatment Technologies: Overview and Guide to Information Sources,* EPA/540/9-91/002, 10/91.

35. EPA, *In–Situ Treatment of Contaminated Groundwater,* OSW, 9/92.
36. EPA, *Metals Treatment at Superfund Sites by Adsorptive Filtration,* EPA/540/F–92/008, 9/92.
37. EPA, *Pneumatic Fracturing Extraction™ and Hot Gas Injection,* EPA/540/MR–93/509, 4/93.
38. EPA, *Proceedings of the Symposium on Soil Venting,* EPA/600/R–92/174, 9/92.
39. EPA, *RCRA Corrective Action Stabilization Technologies,* EPA/625/R–92/014, 10/92.
40. EPA, *Reactor Processes in Synthetic Organic Chemical Manufacturing Industry—Background Information for Proposed Standards,* EPA–450/3–90–016a, 6/90.
41. EPA, *Remedial Action, Treatment, and Disposal of Hazardous Waste, RREL, (16th),* EPA/650/9–90/037, 8/90.
42. EPA, *Resources Conservation Company B.E.S.T.™ Solvent Extraction Technology,* EPA/540/AR–92/079.
43. EPA, *Risk Reduction Engineering Laboratory Research Symposium (18th),* EPA/600/R–92/028, 4/92.
44. EPA, *RREL's Mobile Volume Reduction Unit,* EPA/540/R–93/508, 9/93.
45. EPA, et al, *Solvent Waste Reduction,* Noyes Data, 1990.
46. EPA, *Summary of Treatment Technology Effectiveness for Contaminated Soil,* 9355.4–06, 6/90.
47. EPA, *The Superfund Innovative Technology Evaluation Program, (5th),* EPA/540/R–92/077, 11/92.
48. EPA, *Synopses of Federal Demonstrations of Innovative Site Remediation Technologies, 2nd ed.,* EPA/542/B–92/003, 9/92.
49. EPA, *A Technology Assessment of Soil Vapor Extraction and Air Sparging,* EPA/600/R–92/173, 9/92.
50. EPA, *Treatment Technology Background Document,* OSW, 1/91.
51. Heller, E., *Remediation Survey,* Fla. Spec., 4/93.
52. HWCP, *The Limits of Technology in Dealing with Hazardous Waste Site Cleanups,* Hazardous Waste Cleanup Project, 6/93.
53. Illman, D., *Waste Acid Recovery System Cited for Technology Transfer,* C&EN, 4/5/93.
54. Krishnan, E., et al, *Recovery of Metals from Sludges and Wastewaters,* Noyes Data, 1993.
55. Lutzow, T., *Selecting Sorbents for Cost Savings,* Env. Prot., 3/93.
56. Macek, S., *Putting a Lid on Evaporation Costs,* Chem. Eng., 12/92.
57. Martin, R., et al, *Selecting the Most Appropriate HAP Emission Control Technology,* Air Poll. Conc., 3–4/93.
58. McArdle, J., et al, *Treatment of Hazardous Waste Leachate, Unit Operations and Costs,* Noyes Data, 1988.
59. Melvold, R., et al, *Sorbents for Liquid Hazardous Substance Cleanup and Control,* Noyes Data, 1988.
60. Mody, V., et al, *Dust Control Handbook,* Noyes Data, 1988.
61. Moore, J., *In–Situ Air Sparging,* Fla. Spec., 4/93.
62. Noyes, R., *Handbook of Pollution Control Processes,* Noyes Data, 1991.
63. Palmer, S., et al, *Metal/Cyanide Containing Wastes, Treatment Technologies,* Noyes Data, 1988.
64. Pedersen, T., et al, *Soil Vapor Extraction Technology,* Noyes Data, 1991.
65. Pirocanae, D., *Soil Vapor Extraction is a Popular Technology Choice,* Poll. Eng., 8/93.
66. Roberts, R., et al, *Hazardous Waste Minimization Decision Report,* Naval Civil Eng. Lab., 6/88.
67. Simon, J., et al, *Recent Developments in Cleanup Technologies,* Remediation, Winter 93/94.
68. Sircor, S., *Novel Applications in Adsorption Technology,* AICHE, Front. in C60 Eng., 10/92.
69. Sittler, S., et al, *Thermal–Enhanced Soil Vapor Extraction Accelerates Cleanup of Diesel–Affected Soils,* Nat. Env. Jnl., 1–2/93.
70. Smith, J., et al, *Upgrading Existing or Designing New Drinking Water Treatment Facilities,* Noyes Data, 1991.
71. Surprenant, N., et al, *Halogenated–Organic Containing Wastes, Treatment Technologies,* Noyes Data, 1988.
72. Unterberg, W., *How to Respond to Hazardous Chemical Spills,* Noyes Data, 1988.
73. Vance, D., *Groundwater Injection and Problem Prevention,* Nat. Env. Jnl., 9–10/93.
74. Wagner, K., et al, *Remedial Action Technology for Waste Disposal Sites,* Noyes Data, 1986.
75. Weismantel, G., *Sludge Dewatering Technology,* Poll. Engr., 4/1/93.
76. West, C., et al, *Surfactants and Subsurface Remediation,* Env. Sci. Tech., 5.26, N. 12, 1992.
77. Wilk, L., et al, *Corrosive–Containing Wastes, Treatment Technologies,* Noyes Data, 1988.

7

Radiation and Electrical Technology

There are a number of processes utilized for detoxification that utilize radiation and electrical energy, and are included in this chapter. However, there are some exceptions: infrared energy, for thermal destruction is discussed in Chapter 8; electrical energy used for immobilization is discussed in Chapter 4; electrodialysis is discussed in Chapter 5; and electrostatic precipitation is discussed in Chapter 6.

7.1 ACOUSTIC/ULTRASONIC PROCESSES

7.1.1 Soil Processes

The current applications of ultrasonic vibration suggest at least three ways the technique could be used to increase the recovery of contaminants from soils. One is the dispersion or disaggregation of clay particles, which could be used to clean the pores and increase permeability adjacent to boreholes by reducing the so-called "skin effect." Another is an increased ability to remove chemical species adhered to the solid particles, which could be used to improve recovery efficiency. A third is the sterilization of wells and elimination of microorganisms that clog neighboring pore spaces. This application could potentially eliminate the need to use antibacteriological chemicals to clear wells used in bioreclamation.

Electroacoustic decontamination is used to remediate soils by applying electrical and acoustical fields. The electrical field is used to transport liquids through soils. The acoustic field can enhance the dewatering or leaching of waste such as sludges. Electroacoustic decontamination is effective on soils contaminated by inorganic, organic, and/or heavy metal liquids. Because this technology depends on surface charge to be effective, fine-grained clay soils are an ideal medium for application.

Ultrasonic methods (ultrasonic vibrations) can be used to increase the efficiency of recovery wells. Extrapolations from current applications of ultrasonic methods can be used to describe the mode of action at sites requiring remediation. At least three potential applications to hazardous waste sites have been identified: (1) the dispersion or disaggregation of clay particles during cleaning of pores or well screens by enhancing the removal of chemicals adhered to solid particles (which could improve recovery efficiency); (2) the sterilization of wells; and (3) the elimination of microbes that clog pore spaces. This technology could eliminate the need for antibacterial agents in the cleaning of wells used in bioreclamation.

Soil scientists have used this technology extensively to disperse clay and silt particles. Ultrasonic methods have also proved to be effective for removing mineral films and clay aggregates from sand grains. Laboratory tests involving the use of ultrasonic methods have shown that they often reduce the time required to extract humic acids, to leach various metallic ions from soils, and to clean MoS_2 from steel surfaces. Ultrasonic methods may be used to increase recovery volumes from wells clogged with clay particles or microorganisms and to separate contaminants from clay particles near well sites.

The Battelle In Situ Electroacoustic Soil Decontamination Process is used for in situ decontamination of soils containing hazardous organics by applying electrical (direct current) and acoustic fields. The direct current facilitates the transport of liquids through soils. The process consists of electrodes (an anode and a cathode) and an acoustic source.

The double–layer boundary theory is important when an electric potential is applied to soils. For soil particles, the double layer consists of a fixed layer of negative ions that are firmly held to the solid phase and a diffuse layer of cations and anions that are more loosely held. Applying an electric potential to the double layer displaces the loosely held ions to their respective electrodes. The cations drag water along with them as they move toward the cathode.

Besides the transport of water through wet soils, the direct current produces other effects, such as ion transfer, development of pH gradients, electrolysis, oxidation and reduction, and heat generation. The heavy metals present in contaminated soils can be leached or precipitated out of solution by electrolysis, oxidation and reduction reactions, or ionic migration. The contaminants in the soil may be (1) cations, such as cadmium, chromium, and lead; and (2) anions, such as cyanide, chromate, and dichromate. The existence of these ions in their respective oxidation states depends on the pH and concentration gradients in the soil. The electric field is expected to increase the leaching rate and precipitate the heavy metals out of solution by establishing appropriate pH and osmotic gradients.

When properly applied in conjunction with an electric field and water flow, an acoustic field can enhance the dewatering or leaching of wastes such as sludges. This phenomenon is not fully understood. Another possible application involves unclogging of recovery wells. Since contaminated particles are driven to the recovery well, the pores and interstitial spaces in the soil can become plugged. This technology could be used to clear these clogged spaces.

Trinity Environmental Technologies, Inc. is developing an ultrasonically assisted PCB destruction chemical process for PCB–contaminated soils. The technology uses an aprotic solvent and other reagents in the presence of ultrasonic irradiation to dehalogenate the PCBs to inert biphenyl and chloride. Gas chromatograph/mass spectrograph (GCMS analysis of processed PCB materials shows that the process has no toxic or hazardous by–products. The commercial process is expected to be less costly than incineration but more costly than land disposal. There are no stack emissions associated with this process.

The process begins by sizing the solid material to allow better contact between the solvents and the PCBs. The PCBs are extracted from the contaminated material in a multiple–stage, counter–current extraction process. The extraction solvent of choice is not aqueous. Therefore, the process avoids separation problems typically associated with water and clay soils. The solvent can be completely removed from the soil by using various unit operations, but solvent residue should not pose an environmental problem. The solvent has a high boiling point and is recyclable.

After the solvent is separated from the treated solids, it is pumped to an ultrasonic flow reactor. Reagent is added to the PCB–laden solvent prior to ultrasonic irradiation. During ultrasonic treatment, the PCBs are dehalogenated, and chlorine is completely stripped from the biphenyl structure. After treatment, the reaction products can be removed from the solvent, and the solvent is recycled to the extractor.

The Excalibur Enterprises, Inc. technology is designed to treat soils with organic and inorganic contaminants. The technology is a two–stage process: the first stage extracts the contaminants from the soil, and the second stage oxidizes contaminants present in the extract. The extraction is carried out using ultrapure water and ultrasound. Oxidation involves the use of ozone and ultraviolet light. The treatment products of this technology are decontaminated soil and inert salts.

7.1.2 Other Processes

A dewatering process is being developed by the Department of Energy to determine if it may be possible to apply the necessary amount of energy in the form of cavitation to the liquid–mass on a screen for purpose of disengagement of the liquid from the material and to release and recover the liquid that has passed through a screen. The technique utilizes sonically induced cavitation to enhance the dewatering process. Results show potential for extraction of liquid from waste products containing 0.5 to 2% (5,000 to 20,000 mg/ℓ) suspended solids.

According to recent studies, sulfur dioxide (SO_2) and nitrogen oxides (NO_x) can be removed from flue gas by reacting them with hydroxyl radicals (OH). Researchers at the U.S. Department of Energy are investigating the use of ultrasound technology to generate OH radicals in water or other aqueous solutions through which flue gases are bubbled. In this manner, SO_2 is oxidized to sulfate (SO_4) compounds, and NO_x is oxidized to nitrite (NO_2) and nitrate (NO_3) compounds, by the hydrogen peroxide (H_2O_2) formed via the recombination of OH radicals. The ultrasound process is unique because it may integrate NO_x removal into a conventional wet–scrubbing SO_2 process. It also may provide a relatively cold environment for this high–energy chemical process.

Sandia is developing a process to destroy VOCs in a two–stage (thermal and catalytic) destruction process with a pulse combustor. This could lead to compact designs for the transportation industry.

Pulse combustion can be applied to incineration. A rotary kiln incinerator simulator was retrofitted with a frequency–tunable pulse combustor to enhance the efficiency of combustion. The pulse combustor excites pulsations in the kiln and increases the completeness of combustion by promoting better mixing within the system. Tests were performed using toluene sorbed onto a ground corn cob sorbent and placed in cardboard containers. The burner was operated in a non–pulse mode as a baseline condition, and then in a pulse mode in which the frequency of the pulse combustor was adjusted to the natural frequency of the combustion chamber, creating resonant pulsations of large magnitude. The test was also performed using polyethylene tube bundles to simulate a solid waste and to investigate a surrogate which produces different puff characteristics.

The addition of turbulence in the rotary kiln due to high amplitude acoustic pulsations has a strong tendency to reduce the amount of soot and/or semivolatile and non-volatile hydrocarbons. Mass emissions of soot were consistently reduced in all tests. Carbon monoxide increased during acoustic pulsations in the toluene tests. Sonotech has developed a turnable pulse combustor used for the EPA test.

Pulsed combustion is also being investigated for use in black liquor gasification in the paper industry, in a DOE project. It would also be beneficial to energy-intensive industrial processes.

Both ultrasound and certain metal ions can catalyze the ozonation reaction. The catalytic metal ions are nickel, titanium, chromium and iron. By using sonocatalytic ozonation, removal of 85% of COD (chemical-oxygen-demand) can be obtained in two hours, whereas only 70% of COD removal can be achieved by catalytic ozonation, indicating that ultrasound has a substantial influence on the ozonation reaction rate. Ozonics Technology has patented a process using ultrasound in ozonation.

Ultrasonic cavitation is a new technology being developed by Argonne National Lab for treating groundwater contaminants. High-frequency sound waves create tiny bubbles which, when they collapse, generate high temperatures and pressures that reduce contaminants as much as 99%. Tests on slurried soil samples reduced CCl_4 levels to less than 1 ppm as the halogen was converted to CO_2 and HCl.

7.2 ALTERNATING CURRENT ELECTROCOAGULATION

Alternating current electrocoagulation (ACE Technology) was originally developed as a treatment technology in the early 1980s to break stable aqueous suspensions of clays and coal fines in the mining industry. The technology was developed as a replacement for primary chemical coagulant addition to simplify effluent treatment, realize cost savings, and facilitate recovery of fine-grained products that would otherwise have been lost. The traditional approach for treatment of such effluents entails addition of organic polymers or inorganic salts to promote flocculation of fine particulates and colloidal-sized oil droplets in aqueous suspensions. These flocculated materials are then separated by sedimentation or filtration. Unfortunately, chemical coagulant addition generates voluminous, hydrous sludges which are difficult to dewater and slow to filter. As an alternative to chemical conditioning and flocculation, ACE Technology agglomerates the particulates without adding any soluble species to produce a sludge with a lower contained water content and which will filter more rapidly. Another disadvantage of chemical coagulation is the high susceptibility to filter shear of the particulates and emulsion droplets entrained in the sweep flocs. Through separation of the hazardous components from an aqueous waste, the volume of potentially toxic pollutants requiring special handling and disposal can be minimized. Waste reduction goals may be accomplished by integrating this technology into a variety of operations which generate contaminated water. Laboratory-scale testing has also indicated the ACE Technology is capable of effecting removal of soluble and insoluble metals and anions from aqueous streams.

The process is being developed by Electro-Pure Systems, Inc.

ACE technology offers an alternative to metal salts or polymer and polyelectrolyte addition for breaking stable emulsions and suspensions. The technology is also effective at removing certain metals and other soluble pollutants in the polishing step of effluent treatment. For example, the approximate maximum efficiency removal rates in percent are: lead–56%, copper–96%, zinc–91%, phosphate–97%, and fluoride–56%. Electrocoagulation introduces highly–charged polyhydroxide aluminum species into aqueous media that prompt the flocculation of colloidal particles and destabilization of oil–in–water emulsions. Liquid/liquid, and solid/liquid phase separations are achieved with production of sludges that can be filtered and dewatered more readily than those formed through chemical flocculant addition. The technology can be used to break stable aqueous suspensions containing submicron–sized particles of up to 10% total solids and stable aqueous emulsions containing up to 5% oil.

Electrocoagulation is performed in either batch or continuous (one–pass) mode in an ACE Separator™ apparatus of one of two designs: (1) cylindrical chambers containing fluidized beds of aluminum alloy pellets entrained between a series of noble metal electrodes, or (2) an upright box containing aluminum plate electrodes spaced at nominal distances of ½ to 2 inches. Both ACE Separator™ units are small; the working volume of the parallel plate unit is 70 liters and that of the fluidized bed cell, excluding the external plumbing, is 1.5 liters. They have no moving parts, and can be easily integrated into a process treatment train either for effluent, pretreatment, or polishing.

Coagulation and flocculation occur simultaneously within the ACE Separator™ and continue within the product separation step. Charge neutralization and the onset of coagulation occur within the ACE Separator™ as a result of exposure of the effluent to the electric field and dissolution of aluminum from the electrodes. This activity occurs rapidly (often within 30 seconds) for most aqueous suspensions. After charge neutralization and the onset of coagulation, treatment is complete and the suspension and emulsion may be transferred by gravity flow to the product separation step.

Product separation is accomplished in conventional gravity–separation, decant vessels or by means of pressure or vacuum filtration. Coagulation and flocculation continue until complete phase separation is achieved. Waste is removed by using surface skimming, bottom scraping, and decanting.

After the product separation step, each phase (oil, water, and solid) is removed for reuse, recycling, further treatment, or disposal. The technology can be employed in conjunction with conventional water treatment systems, including those relying on metal precipitation, membrane separation technologies, mobile dewatering and incineration units, and soil extraction systems. A typical decontamination application, for example, would produce a water phase that could be discharged directly to a stream or local wastewater treatment plant for further treatment. The solid phase, after dewatering, would be shipped off–site for disposal, and the dewatering filtrate would be recycled. Any floatable material would be reclaimed, refined, otherwise recycled or disposed of.

7.3 COMBINED FIELD PROCESSES

These processes are particularly applicable to dispersion of aqueous phases. The combined field approach will work well in dilute systems because the energy can be focused on the solute and not on the solution as a whole. Several types of fields may be used including electrical, magnetic and gravitational. The energy costs of these processes would be reasonable.

7.4 CORONA DESTRUCTION

Since 1988, EPA's Air and Energy Engineering Research Laboratory has conducted research in the area of corona destruction of VOCs and air toxics. EPA's interest in corona destruction of molecular species started with modeling of a point–plane reactor for destroying toxic organic compounds. The emerging concern for excessive concentrations of ambient ozone, for which many VOCs are precursors, is the need to develop technology to control low concentration streams, and the economic advantages of ambient temperature operation provided impetus for the work on high intensity corona reactor devices.

Two corona destruction processes have been evaluated for their potential in destroying VOCs and air toxics. One of the corona destruction processes uses high dielectric barium titanate pellets in a packed bed reactor across which a high voltage alternating current (AC) is applied. The micro–electric fields developed in the interstitial spaces between the pellets form a multiplicity of corona sites which generate electrons. These electrons initiate the reactions that lead to destruction of the challenge gas species. The second process consists of a wire–in–tube reactor which is energized by high voltage nanosecond pulses. These techniques have the potential of generating very energetic electrons without wasting power by accelerating ions.

The corona processes operate at ambient temperature. The corona is generated in the packed bed of barium titanate pellets or along the wire in the pulsed reactor. The necessity of heating the contaminated air streams to the temperature required for a catalyst or for thermal incineration to work is avoided.

The corona destruction processes were also evaluated as a means to control very low concentrations of contaminants in air streams. Experiments with contaminant streams using 10 ppmv single component VOCs in air demonstrated the ability to destroy the contaminant beyond the detection limit of the analytical equipment (<10 ppbv). Corona destruction may be an alternative control technology for low concentration streams where conventional control devices such as catalytic incineration and carbon adsorption have disadvantages, either economical or technical. Preliminary power estimates for the corona process indicate that the power requirement for VOC destruction is approximately 3 W/ft^3 (106 $J/sec/m^3$). The annual operating costs of corona destruction could fall below the costs of catalytic incineration and thermal incineration. For these reasons, corona destruction may be well suited to destroy VOCs and air toxics in the outlet stream from various operations.

Corona destruction is an effective alternative control method for a variety of VOCs and air toxics. Its destruction efficiency for VOCs such as benzene, cyclohexane, ethanol, hexane, hexene, methane, methylene chloride, methyl ethyl ketone, styrene, and toluene may be predicted from the ionization potential and types of bonds in the molecules.

During the destruction of VOCs and air toxics by the corona process, no products of incomplete reaction are formed. This means that, when compounds are treated with corona, molecules either are completely destroyed or pass through the reactor unaffected. The percentage of molecules destroyed can be predicted and controlled by appropriate reactor design and specification of operational parameters.

The performance of the corona destruction system can be enhanced by either increasing the power input or adjusting the residence time in the reactor to achieve a high degree of destruction for VOCs.

Advantages of corona destruction include:

1. Performs effectively and economically at very low concentrations.
2. Operates at ambient temperature.
3. Eliminates disposal or treatment problems associated with carbon adsorption.
4. Eliminates sensitivity to poisoning by sulfur or halogen containing compounds.
5. Requires no auxiliary fuel.
6. Requires low maintenance.

In a corona/catalytic reduction process more than 98% of the nitric oxide in flue gas streams can be converted to nitrogen and oxygen in a technique being developed at the University of New Hampshire. The route uses two electrodes to set up an intense electric field in the presence of ordinary glass wool as a catalyst. The electric field breaks down the NO_x molecules, and the glass wool helps to distribute the field uniformly.

Another method under investigation in Germany for coal-fired power plants is to remove sulfur dioxide and nitric oxide in the flue gas by electrical discharges. In the corona discharge process, electrons are energized, which results in formation of radicals, which in turn oxidizes SO_2 and NO_x to the corresponding acids (H_2SO_4 and HNO_3). Ammonia is injected, producing ammonium nitrate and ammonium sulfate. These salts are then removed by filtration.

Pacific Northwest Laboratory is developing its ERACE which is a high energy corona reactor for destruction of VOCs, and the High Energy Corona System.

The corona process is also being investigated for use in contaminated soils. Use of a high energy corona is an innovative thermal treatment process that does not require high temperatures or additives. Electrodes/vents are placed in the contaminated soil. Peripheral electrodes/vents are used as air inlets, while a center electrode/vent is used as an off-gas vent. A form of corona developes a higher voltages to generate energetic electrons and robust oxidants from soil gases. The high energy corona technology is used to treat organic contaminated soils, sludges, slurries and sediments.

A related technology, silent discharge plasma, is discussed in Chapter 8.

7.5 ELECTROKINETICS/ELECTRO-OSMOSIS

Electro-osmosis as a civil engineering technique was originally developed as a method for dewatering low permeability active clay soils. The principle is that if an electric potential is applied to a saturated porous material, electrolyte will flow from the anode to the cathode. The movement of cations towards the cathode is accelerated by the movement of water in this direction. If moisture content is maintained in the vicinity of the anode as an electrolyte, anions will move in the other direction.

Electrokinetic soil processing is an in situ separation and removal technique for extracting heavy metals from soils. The technology uses electricity to affect chemical concentrations and groundwater flow. In electro-osmosis (EO), the fluid between the soil particles moves, because a constant, low direct current is applied through electrodes inserted into a soil mass.

Studies of the electrochemistry associated with the process indicate that an acid front is generated at the anode. This acid front eventually migrates from the anode to the cathode. Movement of the acid front by migration and advection results in desorption of contaminants from the soil. The concurrent mobility of the ions and pore fluid under the electrical gradients decontaminates the soil mass. These phenomena provide an added advantage over conventional pumping techniques for in situ treatment of contaminated fine–grained soils.

Recent bench–scale data indicate that the process may be applied to both saturated and partially saturated soils. The process will lead to temporary acidification of the treated soil. However, equilibrium conditions will be rapidly reestablished by diffusion when the electrical potential is removed.

Studies have indicated that metallic electrodes may dissolve as a result of electrolysis, and introduce corrosion products into the soil mass. However, if the electrodes are made of carbon or graphite, no residue will be introduced into the treated soil mass as a result of the process.

Electrokinetics, Inc. has developed an in situ separation technique for extracting heavy metals, radionuclides, other inorganic contaminants, and some organics. Bench–scale laboratory data demonstrate the feasibility of removing arsenic, benzene, cadmium, chromium, copper, ethylbenzene, lead, nickel, phenol, trichloroethylene, toluene, xylene, and zinc from soils. Recent bench–scale tests demonstrated the feasibility of removing uranium and thorium from kaolinite. Limited field tests demonstrated that the method removed zinc and arsenic from both clays and saturated and unsaturated sandy clay deposits. Lead and copper were also removed from dredged sediments. The treatment efficiency was dependent upon the specific chemicals and concentrations.

The technique proved 85% efficient in the removal of 500 ppm phenol from saturated kaolinite. In addition, the removal efficiency of lead, chromium, cadmium, and uranium, at levels up to 1,000 ppm ranged between 75 and 95%.

The application of an electric field appears to have a greater potential for the removal of solubilized metals than for the removal of organics. This method shows promise because the metal ions are positively charged and the electroosmotic flow is enhanced by the electromigration of these cations to the cathode, at least when the soil is negatively charged.

Organics can be removed from contaminated soils by dc electric fields, provided they have some degree of solubility.

Advantages:

1. The remediation of contaminated sites is permanent.
2. The contaminated soil solution is easily extracted from the point of collection.
3. Could be particularly effective for fine grained soils.

Disadvantages:

1. The technology is primarily used for sites contaminated with metals.
2. Electrical power requirements could be excessive.
3. The effects on the soil matrix itself are unknown.
4. Further treatments may be required for sites contaminated with organics or other waste types.
5. Precipitation of salts and secondary minerals could decrease the effectiveness of this technology.

7.6 ELECTROLYTIC PROCESSES

Electrolytic processes are used extensively for recovery of metals from industrial wastewaters. The electrolytic cell is the basic device used in electrolytic recovery operations. The cell consists of an anode and a cathode immersed in an electrolyte. When current is applied, dissolved metals in the electrolyte are reduced and deposited on the cathode. Because the metal(s) removed from solution can be reused, the technology, termed *electrowinning,* is considered a recovery process.

The electrolytic cell is the basic device used in electroplating operations. The cell consists of an anode and a cathode immersed in an electrolyte. When current is applied, dissolved metals in the electrolyte are reduced and deposited on the cathode. This process is attractive for pollution control because of its ability to remove specific contaminants from the waste stream without the addition of chemicals which produce large quantities of sludge. In addition, it is often possible to reuse the metal which is removed from solution, thereby making the technology a recovery process as opposed to an end–of–pipe treatment process.

A commonly used configuration for electrolytic recovery is to connect the electrolytic unit to the dragout tank that follows metal plating or etching baths and precedes the running rinse. The solution is the dragout tank, which contains diluted plating chemicals, is circulated through the electrolytic reactor and back into the dragout tank. In this way, the concentration of metals in the dragout tank is maintained at a low level. Instead of being carried into the running rinse and eventually into the wastewater treatment system, the metals are recovered by the electrolytic reactor.

Electrolytic treatment is not effective in removing all contaminants. It is most effective in removing the noble metals such as gold and silver. These metals have high electrode potentials and are easily reduced and deposited on the cathode. Other metals, such as aluminum and magnesium, cannot be removed by this type of process because their electrode potentials favor oxidation rather than reduction. Compounds such as cadmium, tin, lead and copper can be removed, but a greater amount of current is required, particularly when the metal concentration is low, e.g., less than 1,000 ppm.

In addition to the type of metal, the type of solution also has an effect on the practicality of electrolytic recovery. Extremely corrosive solutions (e.g., certain etchants) may pose problems for electrolytic recovery because the metal that is plated on the cathode is etched off as quickly as it is plated. In addition, solutions with chelated metals, such as electroless copper as quickly as it is plated. In addition, solutions with chelated metals, such as electroless copper plating solutions, may be more difficult for electrolytic recovery than solutions containing free metal ions such as acid copper electroplating solutions.

For dilute metal–containing solutions, electrolytic recovery can be extremely difficult, particularly when using standard flat plate electrodes. One of the primary limitations of this type of electrode is that high mass–transfer rates are difficult to achieve. When plating metals form a solution, the layer of solution next to the cathode becomes depleted in metal ions. Since there are fewer ions present in dilute solutions, diffusion into and across the depleted layer is much slower and the layer becomes thicker and more depleted. Mass transfer rates can be enhanced both by agitation and by increasing the effective surface areas of the electrodes, particularly the cathode. Both of these actions will increase the rate of movement of metal ions to the cathode, which is equivalent to an increase in the current passed between the electrodes.

Electrolytic recovery can be more economical than other treatment methods in certain situations. However, the unit cost of reducing hexavalent chromium electrolytically is greatly affected by the inlet concentration. If the concentration is low, it may be more economical to use chemical reduction. Another major factor to consider is the cost of disposing of the metal–bearing sludge generated by most chemical treatment methods. The land disposal ban is likely to cause further increases in secure landfilling costs. Thus, situations for which recovery is not yet economically feasible may be more cost–effectively managed through recovery in the future.

Electrolytic recovery is applicable for certain metal/cyanide waste streams. It is a particularly attractive process for metal–bearing waste streams because it allows for metal recovery, thereby precluding the generation of metal bearing sludge. A number of different types of electrolytic reactors are currently manufactured. Simple, parallel–plate reactors can be used to recover noble metals such as gold and silver. More complex units with porous or granular electrodes may be required to remove metals such as copper, tin, and lead, particularly when these metals are present in low concentrations, e.g., <100 mg/ℓ.

Electrolytic oxidation is a treatment technology with demonstrated applicability to the treatment of wastes containing high concentrations of cyanide in solution. Because of excessive retention time requirements, the process is often applied as preliminary treatment for highly concentrated cyanide wastes, prior to more conventional chemical cyanide oxidation.

Since most rinsewaters requiring treatment contain metals at concentrations of less than 1,000 ppm, a number of electrolytic reactors have been designed with electrodes that either enhance mixing or have large surface areas. Some of the electrode designs are:

1. Concentric cylinder;
2. Parallel, porous plates;
3. Rotating cylinder;
4. Packed bed;
5. Fluidized–bed; and
6. Carbon fiber.

Although the electrodes used in these reactors are more effective in the removal of metals from solution, their design makes it difficult to remove the metal once it has been plated onto the cathode. For example, the use of a reactor with parallel stainless steel cathodes generally allows for the production of a compact layer of metal that can be mechanically removed and sold as scrap. Conversely, the use of a reactor with a high electrode area results in the deposition of metal within pores of a cathode, which generally makes mechanical removal of the metal impossible. In this case, recovery of the metal must be accomplished by leaching the deposited metal out of the cell by anodic dissolution.

Electrowinning is a form of electroplating. Electrowinning uses an electrochemical cell to plate an electrode with metals as the result of an applied current. The process uses electricity to pass a current through the metal–bearing waste stream between a cathode plate and an insoluble anode. Positively charged metal ions migrate to the negatively charged cathode, which creates a metal deposit that is recovered by scraping.

Electrowinning is most effective for recovery of noble metals such as gold and silver. These metals have high electrode potentials and are easily reduced and deposited on the cathode. Other metals, such as aluminum and magnesium, cannot be removed from solution by this type of process because their electrode potentials favor oxidation rather than reduction. Metals such as cadmium, copper, chromium, lead, tin and zinc can be removed, but a greater amount of current is required.

The metal coatings industry widely uses electrowinning to reclaim metals from waste generated by electroplating, anodizing, electrolysis plating, and printed-circuit board manufacturing. Because the metal can etch off as quickly as it is plated, extremely corrosive solutions will pose problems for electrowinning. Electrowinning is very effective for plating solutions used in printed circuit boards, which contain chelated metals that are difficult to remove by other means.

Complete oxidation of organic matter can be achieved, however, the energy requirements are quite high, and the process can be used only in special situations. AEA Technology uses indirect electrolysis at elevated temperatures, using Ag(II) in concentrated nitric acid. Chevron uses Fe(III) as a redox reagent. There are at least two processes for the dehalogenation of chlorinated organic compounds.

Electrolysis can also be used for nuclear decontamination.

The Electrochemical In Situ Chromate Reduction and Heavy Metal Immobilization process (Andco Environmental Processes, Inc.) uses electrochemical reactions to generate ions for removal of hexavalent chromium and other metals from groundwater. As contaminated water is pumped from an aquifer through the treatment cell, electrical current passes from electrode to electrode through the process water. The electrical exchange induces the release of ferrous and hydroxyl ions from opposite sides of each electrode. A small gap size coupled with the electrode potentials of hexavalent chromium and ferrous ion causes the reduction of hexavalent chromium to occur almost instantaneously. Depending on the pH, various solids may form. They include chromium hydroxide, hydrous ferric oxide, and a chromium-substituted hydrous iron complex.

7.7 ELECTRON BEAM IRRADIATION

E–Beam technology has potential applications in treating aqueous matrices such as solutions and sludges for ultimate disposal of hazardous organic contaminants. Research to this effect is being conducted at Virginia Key, Miami, FL, under EPA sponsorship, with Florida International University, and the University of Miami. Also, the Department of Energy is developing an advanced E–Beam flue gas scrubber for SO_2 and NO_x removal. Germany and Japan have been working on a similar process. An E–Beam process is also being investigated by Nutek Corp. for disinfecting medical waste, and reducing organic chemicals.

High energy electron beams provide an effective, non-selective method of destroying hazardous organic compounds in aqueous solution. The process of E–Beam irradiation is best understood in aqueous solutions in which sizable quantities of free radicals (e^-_{ag}, $H^\cdot$, and $OH^\cdot$) and the more stable oxidant hydrogen peroxide are produced. These highly reactive species decompose the organic contaminants to produce carbon dioxide, water and salts, which are no longer hazardous.

Electron–beam destruction has advantages over existing industrial waste treatment technologies because it is not sensitive to the water quality. This technology is particularly attractive for halocarbons and waste streams containing up to 10% solids. The E–Beam technology also shows high potential for application to mixed wastes, i.e., the destruction of radionuclide–contaminated organic ion exchange resins.

At the Florida facility, at one end of the system, an aqueous solution containing the hazardous organic chemical(s) is directed over a weir where it falls in a thin sheet (approximately 4 millimeters thick). At the other end of the system, a 1.5–million volt insulated core transformer (ICT) electron accelerator generates electrons and accelerates them to about 97% the speed of light. These accelerated electrons are propelled in a concentrated beam down a high–vacuum tube toward a scanner that scans the beam to a rectangular shape and directs it toward the aqueous solution that is flowing over the weir. It is at this point, when the electrons penetrate the waste stream, that treatment occurs. Studies were conducted at 120 gal/min and can be easily scaled up for larger applications. The process is essentially pH independent in the range 3–11.

The electron beam and waste flow are adjusted to deliver the necessary dose of electrons. Although this is a form of ionizing radiation, there is no residual radioactivity; the system is "cold" within seconds after leaving the beam. Both oxidation and reduction can be accomplished by this process.

This process can treat complex mixtures of organic hazardous chemicals in drinking water, groundwater, wastewater, and water containing up to 5% suspended solids.

For coal fired power plants, the Department of Energy is developing an E–Beam process. A mixture of water and ammonia or lime is injected into the combustion gases, then the gases are sent through a beam of electrons. When the electrons strike molecules of SO_2 and NO_x, they cause chemical reactions that convert the gaseous pollutants to solid, dry particles that can be captured.

In this process, incoming flue gas is cooled and humidified in a water quench tower, resulting in a gas moisture content of about 10%. Ammonia is injected into the cooled gas and the gas is passed through an E–beam reactor. In the reactor, oxygen and water are ionized to form the radicals [HO], [O], and [HO_2] by the application of electrons at a dose of 1 to 3 Mrads (1 Mrad is equivalent to 10 joules/g of flue gas). These radicals react with SO_2 and NO_x to form sulfuric acid and nitric acid. The acids are neutralized by ammonia and water in the flue gas to form solid ammonium sulfate and ammonium sulfate nitrate. The reaction time for formation of the sulfate and nitrate salts is less than one second. Product solids are collected in a hopper below the E–beam reactor or in a downstream particulate collector.

In the other version of the E–beam process, the water quench tower is replaced with a lime–based spray dryer. Reactions in the E–beam reactor occur in the same manner as above except that the products formed are calcium salts instead of ammonium salts.

For over 10 years, work has been proceeding in Japan to develop the electron beam NO_x abatement process. This process has also been studied in Germany for approximately 5 years. In this process, particulate matter is first removed from the flue gas which is then cooled, by heat exchange or water injection, down to 160° to 250°F. Ammonia is injected into and mixed with the flue gas following cooling. At this point in the process, the flue gas is passed through a reactor where the flue gas is irradiated by 1 to 3 megarad of electrons.

In the reactor, the SO_2 and the NO_x molecules go through a series of intermediate steps before finally reacting with the ammonia to produce ammonium salts. The salts that are generated in the process are collected in a particulate removal device such as an ESP.

Research has demonstrated that up to 70% of all nitric oxide (NO_x) and 40% of all SO_2 are removed in this process. This research has also shown that NO_x removal is slightly enhanced when SO_2 is present in the gas stream. The removal efficiency obtained by this process has been shown to be dependent on the dose of radiation that is applied. Based upon the research that has been done and some projections of improvements that can be made, it is estimated that when this process is used on a coal–fired power plant, 1 to 3% of the energy produced by the power plant will be needed for NO_x abatement. By comparison, selective catalytic reduction and acid–gas scrubbing need approximately 2% of the energy produced.

The utilization of an electron accelerator to produce x–rays is discussed in Section 7.18.

7.8 ELECTROPHORESIS

Electrophoresis is the transport of electrically charged particles under the influence of a d.c. electric field. The particles may be complex macromolecules and colloids or particulate matter, either living cells (such as bacteria or erythrocytes) or inert material such as oil emulsion droplets or clay). Almost any particulate can be made to electrophorese. This section considers electrophoresis for removing electrically charged particles from liquids.

The general principle is based on the migration of charged particles, as a result of the applied d.c. electric field, to a collecting plate for subsequent removal. Because the particles are not allowed to reach the electrode, reactions at the electrode (with the material) do not take place. Particles without charge, isoelectric particles, are not collected. The net result of electrophoretic treatment is thus a reduction of the volume of liquid in which a hazardous material is contained. The membranes which stop the migration of particles are essentially dialyzing membranes (e.g., cellulose) which allow only water and small ions to pass through.

In order for migration to occur, suspended material must carry an electrical charge, and the liquid phase must carry an opposite charge to preserve electrical neutrality. The source of this apparent charge on the suspended matter is not always evident; it may arise from a partial ionization of one substituent group on a molecule or it may be due to the sorption of ions or ionizable substances on to the suspended material.

Electrodecantation (electrogravitation) is based on the difference in density created by transport depletion of any impurity and the difference in resistivity caused by the transport depletion of ionic species. The first effect causes a film of less dense liquid to rise along the membrane surface on the depleted side and a film of more dense material to fall on the concentrated side. This density difference is amplified by heating effects associated with differences in conductivity of the two films. (The less dense film has a lower conductivity and thus increases in temperature which further decreases its density.) Thus a filtration, which is essentially continuous, is effected by collecting the less dense (purified) liquid at the top of the membranes and the more dense (concentrated liquid at the bottom of the membranes. (Electrodecantation is sometimes classified as a variation on conventional electrodialysis, rather than a form of electrophoresis.) Electrodecantation was used commercially for a time by Dunlop Rubber Company, for creaming of rubber latex; and DuPont has used the process for the concentration of teflon latex.

Forced–flow electrophoresis is essentially electrophoresis with a superimposed hydraulic pressure. Another essential change with respect to electrodecantation is the addition of a frictional boundary between the downward flowing colloid–enriched layer and the upward flowing colloid–depleted layer. This filter allows the flow of liquids within the two compartments to be controlled at will by means of pumps thereby avoiding dependence on the uncontrollable element of decantation.

Potentially treatable waste forms are:

1. Aqueous Wastes
2. Non–aqueous Wastes
3. Liquids
4. Slurries (if not too thick)

The most promising application would be in water purification, including reclamation of sewage and industrial wastes, separation of emulsions, color removal, water sterilization (removal of viruses), silt removal, and algae removal.

Development of practical applications is not expected to be straightforward because there are several critical aspects of equipment design including:

1. Electrode reactions, which may have to be suppressed or circumvented;
2. pH, adjustments may be necessary;
3. Conductivity, which may have to be below 300 micromhos/cm in aqueous systems;
4. Flow, to avoid turbulent mixing in the cells;
5. Temperature; and
6. Factors affecting membrane plugging, including backwash techniques, the most serious problem.

Even with proper control of these variables there is no assurance that high removal efficiencies can be obtained at acceptable flow rates without careful piloting of the proposed system.

7.9 GAMMA RADIATION

A technology for treating infectious waste involves the use of ionizing radiation. Experience being gained from irradiation of medical supplies, medical components, food, and other consumer products is providing a basis for the development of practical applications for treatment of infectious waste.

The advantages of ionizing radiation sterilization for treatment of infectious waste relative to other available treatment methods include:

1. Nominal electricity requirements
2. No steam requirements
3. No residual heat in treated waste
4. Performance of the system

The principal disadvantages of a radiation sterilization facility are:

1. High capital cost
2. Requirement for highly trained operating and support personnel
3. Large space requirement

4. Problem of ultimate disposal of the decayed radiation source

Gamma ray radiation can be used to disinfect municipal wastewater sludge. Radiation destroys certain organisms by altering the colloidal nature of the cell contents (protoplasm).

Commercial sources of gamma radiation generally are unstable isotopes of cesium and cobalt. Gamma rays from cesium and cobalt-60 sources have energies in the range of 0.40 to 1.33 MeV. In this range, the primary mechanism of gamma ray interaction with matter is the Compton effect.

One gamma radiolysis program is the Idaho National Engineering Laboratory (INEL) component of a joint collaborative effort with Lawrence Livermore National Laboratory (LLNL). The purpose of this effort is to demonstrate a viable process for breaking down hazardous halogenated organic wastes to simpler, non-hazardous waste using high energy ionizing radiation. The INEL effort focuses on the use of spent reactor fuel gamma radiation sources to decompose complex wastes such as PCBs.

Atomic Energy of Canada Ltd., is also investigating dechlorination of organic compounds with gamma radiolysis. Advantages of the process related to PCBs includes:

1. The process is carried out in the absence of air, eliminating the formation of benzodifurans and dioxins.
2. On-line monitoring of the dechlorination process eliminates the possibility of incomplete detoxification.
3. Applicable to bulk PCBs, as well as PCB-contaminated material, such as transformers.

7.10 MAGNETIC SEPARATION

Magnetic separation is usually thought of as the art of separating one solid from another utilizing a magnetic field.

Magnetic separation is also applicable to the removal of magnetic particles from liquid streams. A waste stream is fed into a magnetic field where magnetic particles are collected on filters, generally woven steel fabric or compressed steel wool. The magnetic field is then shut off, and the collected waste material is washed from the filter bed.

Magnetic separators designed for field use are available, most notably the Dynactor. Soluble, non-magnetic contaminants can often be removed from a waste stream by first using powdered activated carbon to adsorb the contaminants. The carbon suspension is then thickened by the formation of a magnetic floc when the proper amounts of magnetic material are mixed with the carbon in the presence of polyelectrolyte flocculating agent such as aluminum sulfate. Magnetite is a commonly used magnetic material.

High-gradient magnetic separation (HGMS) is a process for removing magnetic and paramagnetic particles from liquid and slurry streams. It is also possible to use HGMS for removing non-magnetic particles, but this is a complicated process requiring chemical seeding of the waste stream, and the same results can nearly always be achieved by conventional filtration and separation techniques. Therefore, in its present state of development, HGMS should be considered only when the substance to be removed is magnetic or is of sufficient value to justify investment for its recovery.

HGMS can operate at high flow rates, compared to other filtration and separation processes, and thus has potential for applications where large volume can offset costs. For example, work has been done in using HGMS to remove non–magnetic particles and dissolved metal ions from water.

It can separate weakly paramagnetic materials and other non–magnetic suspended solids (down to colloidal particle size) from gas or liquid streams, on a large scale and at flow rates over one hundred times faster than the flow rates possible in ordinary filtration, and at lower costs and energy requirements. Magnetic separators capable of removing highly magnetic (i.e., ferromagnetic) particulates have existed for sometime; HGMS extends the capabilities of magnetic separation to include particles that are only weakly magnetic.

HGMS may be operated as either a cyclic or a continuous operation. Both methods are in commercial use today. Cyclic units are preferable when (1) the material being removed is a small percentage of the total volume passing through the system and/or (2) high operating pressures are needed. Continuous operation may be preferable when (1) high operating pressures are not needed and (2) the material being removed is a large percentage of the total volume passing through the system.

Under optimum conditions, HGMS can achieve very high removal efficiencies, sometimes above 95%, yielding hazardous waste concentrations in the purified stream of less than 10 ppm. In less than optimum conditions, the removal efficiencies can be quiet low. Removal efficiencies will increase with (1) increasing magnetic field strength (especially for paramagnetic material; (2) increasing magnetic susceptibility of the material; (3) increasing particle size; (4) decreasing feed velocity; (5) decreasing filter matrix loading; and (6) sometimes, increasing hazardous material concentration.

For the future, superconducting magnets are seriously being considered for energy saving with very large (yet to be built) units.

Commercial applications of HGMS include removal of iron oxides from waste streams, and iron ore beneficiation.

Four applications are being investigated:

1. Beneficiation of iron ores and other ores
2. Coal desulfurization
3. Removal of flue dusts in air streams from blast furnaces
4. Water purification

Generally, HGMS can be used if: (1) the wastes can be made to flow through a relatively open filamentary matrix (95% void space); (2) the wastes are finely divided; and (3) the feed stream is not too corrosive to iron or steel. Potentially treatable waste forms are: (1) liquids (aqueous and non–aqueous); (2) slurries; and (3) dry powders. Not treatable are sludges, tars and gases.

Removable compounds fall into three groups: (1) paramagnetic (or ferromagnetic) elements and compounds; (2) non–magnetic suspended solids; and (3) dissolved material in water.

HGMS is probably best suited for the removal of magnetic hazardous wastes that are present in low concentrations in a liquid waste stream. Removal of non–magnetic material may be effected by seeding with a magnetic material plus a flocculant, but this is not likely to be cost competitive with processes such as froth flotation and centrifugation.

Open–Gradient Magnetic Separation (OGMS) using superconducting quadrupole magnets offers a novel beneficiation technology for removing pyritic sulfur from pulverized dry coal. It is estimated to have a power demand 75% lower than techniques using conventional electromagnets, while achieving higher separation forces. Additionally, the system operates in a continuous mode and uses no chemicals. Because OGMS is specifically applicable to finely ground coal (120 to 325 mesh), its development could encourage the commercialization of other unconventional coal technologies, such as coal–water slurries, fluidized–bed combustion, and synfuels. The process is being developed at Argonne National Laboratory.

In the Actimag process, the use of magnets to agitate a bed of iron granules has been applied to extracting valuable or toxic metals from solution. The technique is being used by Thomson CSF to extract highly toxic byproducts of chromium plating, and may also be used to recover copper and other metals from chemical solutions. Actimag's process is designed to enable safe iron to be substituted for chromium in the solution.

In contrast to conventional contact precipitation techniques, the solution is passed through moving iron granules, not static ones.

7.11 NON–THERMAL PLASMAS

The essence of non–thermal plasma techniques for pollution control is the efficient use of electrical energy through selective dissociation of the toxic molecules. "Non–thermal" plasmas as the name implies, are plasmas in which the electron temperatures are considerably higher than those of the components of the ambient gas. Non–thermal plasma techniques are particularly efficient when the toxic materials are present in very small concentrations, as is the case for flue gas emissions. This contrasts with the use of plasma furnaces or torches and several chemical techniques in which the whole gas is heated in order to break up the undesired molecules. This contrasts also with the use of thermal plasma arcs for liquid and solid waste disposal.

Penetrante and his associates at Lawrence Livermore National Laboratory have explored three particular applications of non–thermal plasmas: (1) decomposition of hydrogen sulfide (H_2S), (2) removal of trichloroethylene (TCE), and (3) removal of NO_x. They have explored three basic device configurations: (1) the metal–electrode pulsed corona, (2) the dielectric barrier, and (3) the barrier flashover.

It is generally agreed that the useful energy for catalyzing the chemical process which leads to the removal of pollutant molecules is only a fraction of the energy needed for producing the discharge. Several experiments show a strong correlation between the amount of energy dissipated in the discharge, the number of streamers formed, and the rise time of the voltage pulse. They believe the use of fast rising voltage pulses provides a more efficient method of energizing a non–thermal discharge device. This is true even in cases where a dielectric barrier is present to self–terminate each streamer to prevent spark breakdown. With a fast–rising voltage pulse, much more streamers are produced initially, and thus the radical production per injected energy is larger. It is important to maximize the number of streamers because most of the electrical energy is dissipated after the streamer channels have already been formed.

7.12 MICROWAVE TREATMENT

Microwaves are being used, or investigated in a number of applications including:

1. Treatment of medical waste
2. Destruction of VOCs

3. Detoxification of contaminated soils
4. Separation of organic materials from mixed hazardous waste
5. Decomposition of inorganic gases

Scouting experiments have shown that microwave energy can induce chemical reactions to detoxify organics at lower temperatures and with greater efficiency than conventional heating. An experimental program is underway to evaluate this technology with respect to various mixed waste matrices to determine reaction rates and mechanisms, and other parameters that could be used for an economic assessment with respect to alternative technologies.

The major advantages of microwave heating over conventional thermal heating are:

1. Chemical reactions are enhanced (higher rates at lower bulk temperature) due to direct microwave activation at the solid–vapor interface.
2. Waste materials can be heated without heating containers.
3. Waveguide windows can effectively isolate the microwave generator from the reactor to minimize contamination and human exposure.
4. More efficient and faster processing is possible.

A number of catalyzed reactions have been investigated experimentally. Preliminary analysis of the results indicate that rates of chemical reactions taking place at solid surfaces can be increased several decades by microwave absorption. Also such reactions proceed to completion at temperatures which are 100° to 200°C lower than that required by conventional heating.

Soils contaminated with toluene and p–xylene are readily decontaminated at low temperature without combustion when heated with microwave energy under vacuum conditions. Findings indicated that the solvent removal rate was increased several times if the soil samples contain moisture in the form of 3.00 wt % water. The combination of moisture and vacuum yielded the best results. This observation can be attributed to the enhancement of microwave absorption by the water molecule and by partial pressure effects of the water vapor which is generated upon heating.

The feasibility of utilizing the differential heating characteristics of microwave energy (MW) to aid in the chemical extraction and separation process of hazardous organic compounds from mixed hazardous waste, was studied at the INEL. The objective of this work was to identify a practical method of separating or enhancing the separation process of organic hazardous waste components from mixed waste using microwave (MW) frequency radiation. Methods using MW energy for calcination, solidification, and drying of radioactive waste from nuclear facilities is becoming more attractive. In order to study the effectiveness of MW heating, samples of several organic chemicals simulating those which may be found at the Radioactive Waste Management Complex at the INEL were exposed to MW energy. Vapor collection and analysis was performed as a function of time, signal frequency, and MW power throughout the process. Signal frequencies ranging from 900 to 8,000 MHz were used. Although the signal frequency bandwidth of the selectivity was quite broad, for the material tested an indication of the frequency dependence in the selectivity of MW heating was given. Greater efficiency in terms of energy used and time required was observed. The relatively large electromagnetic field intensities generated at the resonant frequencies which were supported by the cavity sample holder demonstrated the use of cavity resonance to aid in the process of differential heating.

Argonne National Laboratory has been investigating a microwave discharge plasma process to destroy low-concentration VOC emissions. Plasma is essentially a high-temperature ionized gas state that is extremely reactive. The application of plasma technology to chemical destruction has focused on two methods of plasma generation: arc discharge and high-frequency discharge. In an arc discharge unit, an electrical field is used to produce a discharge, or arc, between two electrodes. The intense energy from the arc is used to generate and sustain plasma. High-frequency discharge units, on the other hand, use microwave or other radio frequency radiation (i.e., typically greater than 1 MHz) to create and sustain the plasma state. One advantage to this type of unit is that no electrodes, which are possible sources of contamination, are used.

At this point, only laboratory-scale experiments have been conducted; however, some commercial interest in the process has been expressed. Previous work was conducted in an argon gas atmosphere, but current research is focused on reactions in air. Preheating the incoming gas and saturating it with water vapor provides wet oxidation conditions that appear to work well. Furthermore, the addition of UV radiation to assist the reaction is being investigated.

The EG&G Rocky Flats plant is testing the use of microwaves for reducing and solidifying radioactive waste. The process reduced waste volume and weight by 87% in several earlier experiments. The Rocky Flats method uses microwaves to melt sludge-type waste at temperatures of up to 2800°F. The result is vitrification of the waste into a glass matrix that is denser and more leach-resistant than the usual sludge byproduct.

Research aimed at adapting a Russian-developed microwave technology for U.S. oil and gas company operations is under way. The technology would be used to split the hydrogen sulfide in natural gas into sulfur and hydrogen. Thus, hydrogen could be recovered along with sulfur, which conventional technologies now recover alone, and used as fuel or in oil refining. The U.S. could save as much as the energy equivalent of 70 billion cubic feet of natural gas annually if such hydrogen were recovered, according to John Harkness, a research scientist at Argonne National Laboratory, one of the groups involved in the international project. The technology's compatibility with operating practices in the U.S. oil and gas industries and with OSHA regulations will be evaluated by Argonne; the Russian Scientific Center Kurchatov Institute, Moscow, Russia's leading research center for plasma-chemical technology and nuclear and thermonuclear power engineering; and Wavemat Inc., Plymouth, Michigan, which designs and makes proprietary microwave equipment for processing advanced materials. A new company, Acid Rain Control Inc., Detroit, has been started to commercialize the technology.

A process that not only destroys sulfur and nitrogen oxides, but also generates useable products has been developed by Cha Corp. of Laramie, Wyoming. It was described at an American Chemical Society meeting in Washington, DC by Soung Kim, a project manager at the DOE Pittsburgh Energy Technology Center, which funded the research. First, sulfur dioxide is absorbed on coal char. This is then heated by microwaves, the char catalyzing the reduction of SO_2 to sulfur and oxygen. The O_2 reacts with the char, producing carbon dioxide and activated charcoal. The microwaves decompose the NO_x directly to nitrogen and oxygen.

Microwaves are being used to treat medical wastes. Using this technique, wastes are first ground and shredded to improve the effectiveness of the treatment system. Next, the wastes are sprayed with water. An auger moves the wastes past a series of microwave power packs which subject the waste to microwaves. The microwaves heat the waste to 200°F, and volatile materials and water are driven off during the process.

The grinding reduces the volume and recognizability of waste to be treated by as much as 80%. However, the weight is essentially unaffected. Portable microwave treatment facilities are commercially available. The main disadvantage of microwave treatment systems is that they are not capable of treating pathological wastes such as body parts or animal carcasses. Also, the potential for the release of volatile material may exist.

7.13 PHOTOLYSIS/PYROLYSIS

The Morse Photolysis Processor, designed by Chips Morse, basically combines light (photolytic) and heat (pyrolytic) treatment to convert solid wastes into useable energy. It is a linear continuous energy conversion process consisting of multiple chambers of modular design and utilizing catalysts and accelerators to convert carbonaceous materials into oil, gases and char.

Solid waste is rough sorted so that salvageable and non–carbon materials are segregated. The remainder of the waste, consisting of combustibles and noncombustibles, is shredded to a size of approximately one inch. That portion of the waste going to the Morse Energy Processor is treated as wet feed material, though moisture does not cause a problem for system operation. Before material is delivered to the first reactor, it is heated to approximately 200°F using waste heat and evacuated. The purpose is to pull off highly volatile compounds and use them in the combustion gas. At least two feedbins are provided so that the processes of filling, evacuating, and feeding can be undertaken continuously. The material delivery system and the reactor chambers operate under a vacuum of approximately 4" of mercury.

The feed material is delivered to the top of the first reactor chamber, which operates at a minimum temperature of 250°F. At this point, the photon generator (with proprietary characteristics) sends its beam along an internal rotor in the chamber, striking the feed material and breaking molecular bonds, causing gases to form. The feed material is moved along in the reactor by the rotor and heated. The turning of the rotor continuously exposes new feed material to the photons. The gases are drawn out of the reactor by the vacuum pump into the process gas heater. The remaining feed material drops into the next reactor where the temperature has been increased to 400°F. The process is repeated for as many chambers as desired. It has been determined that by the time material reaches 850°F, calling for a five chamber system, all the feed material would be decomposed into gases, leaving only sterile char to be disposed off. The char is conveyed to an elevated holding tank where it can be dumped into trucks for disposal.

7.14 RADIO FREQUENCY/ELECTRICAL SOIL HEATING

Radio frequency (RF) heating is a technique for rapid and uniform in situ heating of large volumes of soil. This technique heats the soil to the point where volatile and semivolatile contaminants are vaporized into the soil matrix. Vented electrodes are then used to recover the gases formed in the soil matrix during the heating process. The concentrated extracted gas stream that is recovered can be incinerated or subjected to other treatment methods.

Radio frequency heating is accomplished by the use of electromagnetic energy in the radio frequency band. The energy is introduced into the soil matrix by electrodes inserted into drill holes. The mechanism of heat generation is similar to that of a microwave oven. The heating process does not rely on the thermal conductivity of the soil. A modified radio transmitter serves as the power source, and the industrial, scientific, and medical (ISM) band provides the frequency at which the modified transmitter operates. The exact operational frequency is obtained from an evaluation of the areal extent of contamination and the dielectric properties of the soil matrix.

The frequencies used during RF heating remedial work may be as low as 45 Hz or as high or higher than 10 GHz. The frequency range for most RF heating applications at hazardous waste sites is between 6.78 MHz to 2.45 GHz.

Radio frequency heating is a new technology for cleanup of hazardous waste sites. It is currently in the pilot- and field-scale demonstration stage and has been tested at the Volk Field ANGB, Wisconsin in cooperation with the U.S. Department of Defense, U.S. EPA, and the Illinois Institute of Technology Research Institute. Final soil sampling showed excellent removal rates of fuel contaminants. Over 99% of the aromatic compounds, such as benzene, were removed from the soils. Less volatile aliphatics were reduced by over 95%.

The high soil temperatures associated with this technology would inhibit or destroy existing colonies of microbes in the soil matrix. The high temperatures could also have an adverse effect on humic matter within the soil matrix (Devetal, 1988)

Advantages of Radio Frequency Heating:

1. It offers a potentially high level of treatment.
2. No other treatments are necessary if the contamination consists solely of volatile and semivolatile organics.
3. It is economically feasible if soil consists primarily of sand and loams.
4. It provides permanent remediation.

Disadvantages of Radio Frequency Heating:

1. Has limitations related to various soil types and contaminants, such as high moisture content, or large buried metal objects.
2. Could be difficult to apply.
3. Does not remediate nonvolatile organics, metals, or other inorganic contaminants.
4. Must be supplemented with other treatment methods if nonvolatile contaminants are present.
5. Very deep contamination would require more costly solutions.

This process is also referred to as electromagnetic (EM) heating. The only major difference between RF and EM is in the choice of frequency of the applied power. The EM technology is suitable for heating soils only to the boiling point of water.

Electrical soil heating is being developed in at least three locations. Battelle Pacific Northwest Laboratories (PNL) is using electrodes to apply an electrical current to the soil. Moisture within the soil boils and forms steam, stripping volatile and semi-volatile contaminants from soil particles. The technique: Electrical Remediation at Contaminated Environs (ERACE), will remove 99.99% of soil contaminants such as chlorinated solvents, polychlorinated biphenyls (PCBs), pesticides and industrial fuel oils and lubricants.

Lawrence Livermore National Laboratory completed a small scale test of electrical heating combined with vapor extraction. Removal of trichloroethylene more than doubled by electrically heating the soil. A large scale test will add steam injection to clean up a gasoline spill.

At the IIT Research Institute (IITRI), Chicago, scientists are working on a microwave-like radio frequency (RF) process that uses electromagnetic energy to heat volatile and semi-volatile chemicals in the vadose zone. Perforated metal pipes act as in situ electrodes, and also collect the vaporized organics with a vacuum. A plastic or silicone sheet above the work site acts as a vapor barrier and captures fugitive emissions.

7.15 SOLAR ENERGY

This section discusses the use of natural sunlight. Artificial ultraviolet radiation is discussed in Section 7.17.

The sun is a low intensity source of energy for soil heating; under ideal conditions the power density available from the sun is about one kilowatt per square meter of collector. Parabolic collectors coupled to optical fibers are another method that is being tested for using solar energy to heat soil in situ. Collectors have been used in the past for direct heating of air and water. There are significant heat losses during transmission that would make this process undesirable for in situ heating of soil. An approach to overcoming these heat losses during transmission is being tested in a research project by Brown and Murdoch that was started in July 1990.

Compound parabolic concentrators, which have the advantage of collecting scattered sky light as well as direct sunlight, are being modified for coupling to a cable of optical fibers. The optical fiber has the potential to transfer the solar radiation with high efficiency over long distances. By conducting the solar energy as light, rather than as a hot fluid, thermal losses along the transmission line are eliminated. The collector surfaces also remain relatively cool and radiate away little energy because conversion to heat occurs at the radiator or the end of the optic cable, rather than at the collector.

The "Solar" sludge drying beds used in the arid southwestern United States differ little in concept from drying lagoons or paved beds. In all cases the key to success is the mixing and turning of the sludge and break up of the surface crust. The evaporative losses will be very rapid in arid climates. There are two other processes that are unique; one incorporates additives of various types and the other utilizes an undrained sand bed and growing reeds or bulrushes to dewater the sludge.

Dellinger, Graham, Berman and others at the University of Dayton Research Institute have recently demonstrated that the rate of many gas-phase photochemical reactions can be increased by initiating these reactions at elevated temperatures (e.g., >400°C). The development of very high-temperature photochemistry has raised exciting possibilities for applications such as the destruction of toxic organic wastes. Since concentrated sunlight contains a considerable quantity of near-UV photons (λ>300 nm) that can be used to initiate photochemical reactions, as well as infra-red (IR) photons that can serve as a source of considerable thermal energy, the solar induced thermal/photolytic destruction of hazardous organic wastes appears to be technically feasible.

Their initial laboratory studies using simulated, broad-band, solar radiation (filtered xenon arc emission) in conjunction with a thermoelectrically heated flow reactor have clearly shown that the destruction rates of target compounds can be significantly increased and the production of stable reaction intermediates reduced as compared to identical thermal exposures.

The Department of Energy's National Renewable Energy Laboratory (NREL) in Golden, CO is investigating industrial uses of solar energy, including destruction of hazardous wastes. It was field tested at Lawrence Livermore National Laboratory. NREL utilizes a parabolic trough collector. This type of collector could be used to decontaminate groundwater by photocatalysis. An example is a project under way that removes trichloroethylene (TCE) from groundwater.

The basic idea behind this NREL project is to immerse a semiconductor catalyst, such as titanium dioxide, in the contaminated water and then expose it to concentrated sunlight. The ultraviolet light in the solar radiation activates the catalyst to produce hydroxyl radicals, which decompose the TCE to carbon dioxide and water, with some residual hydrogen chloride.

Similar work is being undertaken at Sandia National Laboratories.

IT Corporation has developed a Photolytic and Biological Soil Detoxification Process that can use either natural sunlight or artificial UV light. This technology is a two–stage, in situ photolytic and biological detoxification process for shallow soil contamination. The first step in the process is to degrade the organic contaminants by using ultraviolet (UV) radiation. Degradation is enhanced by adding detergent–like chemicals (surfactants) to mobilize the contaminants. Photolysis of the original contaminants is expected to convert them to less–resistant compounds. Biological degradation, the second step, is then used to further destroy the organic contamination and detoxify the soil. The rate of photolytic degradation is several times faster with artificial UV light than with natural sunlight.

When using sunlight for soil with shallow contamination, the soil is tilled with a power tiller and sprayed with surfactant. Tilling and spraying are repeated frequently to expose new surfaces. Water may also be added to maintain soil moisture. UV lights with parabolic reflectors are suspended over the soil to irradiate it. After photolysis is complete, biodegradation is enhanced by adding microorganisms and nutrients and by further tilling of the soil.

The use of solar energy utilizing a titanium dioxide photocatalyst is being investigated at the State University of New York, University of New Mexico, and Sandia National Laboratories.

The Hydrolytic Terestial Dissipation (HTD) process, developed by engineers at Dames & Moore, HTD was designed to clean up very high concentrations of toxaphene, which is commonly used to control insects on cotton, corn, grains, vegetables and fruits.

In HTD, heat from sunlight raises the temperature of the soil to the point where the toxaphene begins to volatilize. As the chemical volatilizes, the sun's ultraviolet rays complete dehalogenation. Byproducts include small quantities of chlorine gas released to the air and camphene, which is ultimately degraded to carbon dioxide and water.

At the same time, hydrolysis breaks the chlorine bonds in the soil, a process catalyzed by naturally occurring iron in the soil or by adding metals such as copper and aluminum.

There appears to be several advantages of solar destruction over thermal destruction which include:

1. Increased destruction efficiency of the parent and by–products;
2. Control of vaporization of toxic metals through lower operating temperatures;

3. Control of NO_x formation through lower operation temperatures;
4. Recovery of excess thermal energy that can be used for thermal desorption of solids and sludges;
5. Control of CO, CO_2, and toxic organic emissions through substitution of solar energy for conventional fuels;
6. Cost savings due to lower fuel costs, increased materials lifetime, and reduced size and complexity of air pollution control devices;
7. Increased public acceptance through use of a renewable, non–polluting energy source for a non–incineration waste disposal technology.

Apparent major disadvantages include:

1. The unreliable availability of solar radiation;
2. Cost of collection and concentration of solar radiation;
3. Lack of an off–the–shelf technology to construct a working pilot– or full–scale system.

Their task has been to develop an approach that utilizes the advantages and to minimize the disadvantages.

One approach that has been previously proposed is to develop a hybrid two–stage system targeted for detoxification of contaminated soils and other solids. With this concept, a hybrid primary unit (possibly an indirectly–fired rotary drum design) may be used to thermally desorb organics from solids, while a secondary solar reactor would be used to thermal/photolytically destroy the desorbed organics. An auxiliary heat source is necessary to operate the process continuously during intermittent cloud cover and maintain night–time operation. The desorbed organic matter during dark operation may be stored by cryogenic trapping or sorption on carbon for destruction during light periods. Since the total volume of material desorbed is small, the photolytic reactor should readily handle the stored off–gases during light operation. This approach maintains the previously listed advantages for solar based waste destruction while minimizing two of the three disadvantages. The hybrid primary unit allows continuous operation, thus eliminating the concern over the unreliability of sunlight.

7.16 TRANSMUTATION

Interesting work is being undertaken at three National Laboratories to transmute minor actinides and long–lived fission wastes into fuel grade material, more stable isotopes, or isotopes with shorter half–lives.

At Hanford the ability of a small fast reactor to destroy hazardous long–lived minor actinide and fission product wastes was evaluated. It was determined that by using a novel technique wherein high energy neutrons leaking from the active core of the reactor are moderated by yttrium hydride located in the target assemblies, substantial amounts of long–lived fission products can be destroyed and useful quantities of the beneficial isotope Pu_{238} can be produced by transmutation of the Np_{237} and Am_{241} minor actinide waste components. In addition, it was shown that minor actinides recovered from spent Light Water Reactor fuel can be used to fuel such a reactor, increasing the amount of hazardous minor actinide and fission product wastes that can be destroyed.

At Los Alamos there is a new concept for an accelerator–driven transmutation system. The central feature of the concept is generation of intense fluxes of thermal neutrons. In the system all long-lived radionuclides comprising high-level nuclear waste can be transmuted efficiently. Transmutation takes place in a unique, low material inventory environment. Presently two principal areas are being investigated for application of the concept. The first is associated with cleanup of defense high-level waste at DOE sites such as Hanford. The second, longer term area involves production of electric power using a coupled accelerator-multiplying blanket system. This system would utilize natural thorium or uranium and would transmute long-lived components of high-level waste concurrently during operation.

At Brookhaven a means of transmuting key long-lived nuclear wastes, primarily the minor actinides (Np, Am, Cm) and iodine, using a hybrid proton accelerator and sub-critical lattice, is proposed. By partitioning light water reactor (LWR) spent fuel and by transmuting key elements, such as the plutonium, the minor actinides, and a few of the long-lived fission products, some of the most significant challenges in building a waste respository can be substantially reduced. The proposed machine would transmute the minor actinides and the iodine produced by 75 LWRs, and would generate usable electricity (beyond that required to run the large accelerator) of 850 MW_e. The high flux particle bed reactor is also being investigated for this purpose.

7.17 ULTRAVIOLET RADIATION

Ultraviolet radiation (UV) from artificial sources is used both for disinfection purposes, and for destruction of toxic organic compounds.

7.17.1 Disinfection

The inactivation of microorganisms by ultraviolet radiation is a physical process, relying on the photochemical changes brought about when far-UV radiation is absorbed by the genetic material of the cell (deoxyribonucleic acid, or DNA). The wavelengths for optimum effectiveness correspond, as expected, to the maximum absorption spectrum for nucleic acids, between 250 and 265 nanometers (nm).

Ultraviolet (UV) disinfection systems are being widely considered for application to treated wastewaters, for both new plants and retrofitting existing plants in lieu of conventional chlorination facilities. The technology is relatively new, with most systems installed over the past three to four years. It has generally been successful, although there had been many problems with the systems installed in the early to mid-eighties. Subsequent "second generation" designs have resolved many of the earlier issues, resulting in a higher degree of reliability and a more rapid acceptance of the technology. These use modular, open-channel configurations in place of the fixed, closed shell arrangements typical of the earlier designs.

Ultraviolet disinfection is now being widely applied to wastewaters, with greater than 500 operating facilities, as compared to an estimated 50 facilities in 1984. Whereas closed shell and pipe systems were typical in the early to mid-eighties, the modular, gravity flow, open-channel systems now comprise essentially all new installations. The configuration is found in greater than two-thirds of active plants, as compared to less than 5% in 1985. It is comprised of horizontally or vertically placed lamp modules, placed in open, relatively narrow channels, with the lamps fully submerged in the wastewater. Horizontal systems represent approximately 85% of the open channel facilities.

The UV source used with essentially all systems is the low pressure mercury arc lamp. Alternate lamps are being actively investigated and are in use at several operating plants. These include medium pressure lamps and modifications of the conventional low pressure lamps. A recent advance has been the introduction of an efficient electronic ballast, which is lighter and is incorporated into the modules themselves.

UV is effective and has been demonstrated to be capable of meeting existing disinfection criteria. This includes secondary fecal coliform limits (200 fecal coliforms/100 mL) and shellfish limits (14 fecal coliforms/100 mL). An exception may be the California total coliform limit of 2.2/100 mL for discharge to shellfish waters. Filtration is generally required if UV is to meet the lower shellfish standards.

UV light leaves no residual disinfectant. Therefore, regrowth by photoreactivation may occur when water is exposed to light. Since the water in a distribution system beyond the point of disinfection may be subject to bacterial contamination, a residual disinfectant is recommended in drinking water supplies. If UV radiation is used in such cases, it must be followed by halogen treatment to provide the necessary residual.

UV radiation (254 nm) disinfection of bacteria and viruses has several important advantages: (1) it is readily available; (2) it produces no toxic residuals; (3) required contact times are relatively short; and (4) the equipment is easy to operate and maintain, although maintenance must be performed on a regular basis to prevent fouling of certain components.

Since UV radiation is ineffective against *Giardia* cysts, but effective against viruses and bacteria, it is a good candidate for disinfecting groundwater not directly influenced by surface water. If the amount of radiation received by a target organism is not a lethal dose, however, reconstitution of the organism and reinfection of the water can occur.

7.17.2 Photolysis

Photolysis (or photodegradation) is a process that breaks down a chemical by light energy, usually in a specific wavelength range. Ultraviolet (UV) radiation is defined as electromagnetic radiation with a wavelength shorter than visible light but longer than x-ray radiation. The energy content of light increases as the wavelength decreases. The energy of the wavelengths in the UV region is sufficient to break down chemical bonds and cause rearrangement or dislocation of molecular structures.

Ultraviolet photolysis is appropriate for difficult-to-treat chemicals (e.g., pesticides, dioxins, chlorinated organics), nitrated wastes, and those chemicals in media which permits photolyzing the waste. The waste matrix can often shield chemicals from the light (e.g., ultraviolet light absorbers, suspended solids, solid wastes). The photolysis process typically requires pretreatment to remove suspended materials.

A number of processes have been developed utilizing ultraviolet radiation to degrade toxic organics, some processes utilizing additional oxidants such as ozone or hydrogen peroxide. In these cases the UV energy reacts with the oxidant to create a hydroxyl radical, which in turn causes reactions that destroy the organic contaminants.

A newer process for photo-oxidation of volatile organic compounds (VOCs) in air uses an advanced ultraviolet (UV) source, and a pulsed xenon flashlamp. The flashlamps have greater output at 200 to 250 nm than medium-pressure mercury lamps at the same power and, therefore, cause much more rapid direct photolysis of VOCs.

Ozonation Enhancement: One of the modifications of the ozonation process is the simultaneous application of ultraviolet light and ozone for the treatment of wastewater, including treatment of halogenated organics. The combined action of these two forms produces reactions by photolysis, photosensitization, hydroxylation, oxygenation, and oxidation. The process is unique because several reactions and reaction species are active simultaneously.

Ozonation is facilitated by ultraviolet absorption because both the ozone and the reactant molecules are raised to a higher energy state so that they react more rapidly. In addition, free radicals for use in the reaction are readily hydrolyzed by the water present. The energy and reaction intermediates created by the introduction of both ultraviolet and ozone greatly reduce the amount of ozone required compared with a system using ozone alone. A system to treat mixed cyanides requires pretreatment that involves chemical coagulation, sedimentation, clarification, equalization, and pH adjustment.

The ozone–UV radiation process was developed primarily for cyanide treatment in the electroplating and color photo–processing areas. It has been successfully applied to mixed cyanides and organics from organic chemicals manufacturing processes. The process is particularly useful for treatment of complexed cyanides such as ferricyanide, copper cyanide, and nickel cyanide, which are resistant to ozone alone.

The influent to the system is mixed with ozone and then enters a reaction chamber where it flows past numerous ultraviolet lamps as it travels through the chamber. Flow patterns and configurations are designed to maximize exposure of the total volume of ozone–bearing wastewater to the high energy UV radiation. Although the nature of the effect appears to be influenced by the characteristics of the waste, the UV radiation enhances oxidation by direct dissociation of the contaminant molecule or through excitation of the various species within the waste stream. In industrial systems, the system is generally equipped with recycle capacity. Gases from the reactor are passed through a catalyst unit, destroying any volatiles, replenished with ozone, and then recycled back into the reactor. The system has no gas emissions.

Hydrogen Peroxide Enhancement: In the presence of UV radiation, the rate of oxidant recomposition (such as H_2O_2) is accelerated, with a corresponding increase in the rate of hydroxyl radical formation. Organic molecules that have adsorbed UV energy are in an excited state and are more susceptible to attack. Therefore, the rate at which organic compounds are oxidized is significantly higher than that attained by using UV radiation or chemical oxidants alone.

The UV dosage is the critical parameter for UV/H_2O_2 systems. The UV dosage is adjusted by installing more or fewer lamps in the reactor. Power consumption can be optimized in the field by turning lamps on and off.

Organic carbon, soluble iron and manganese, and general turbidity can reduce the efficiency of the UV/H_2O_2 process by reducing the amount of ˆUV energy available for adsorption by the organic contaminants and the chemical oxidant. Organic carbon will compete with the constituent of concern by adsorbing UV energy and consuming oxidant. Soluble iron and manganese will oxidize to their soluble form, thereby directly competing with the contaminant for UV energy and oxidant. Highly turbid H_2O will reduce UV intensity in a similar manner.

7.17.3 Commercial Processes

Energy and Environmental Engineering Inc.: This is a laser–induced photochemical oxidation process. This technology is designed to photochemically oxidize organic compounds in wastewater by adding a chemical oxident and applying ultraviolet (UV) radiation with an Excimer laser. The photochemical reactor can destroy low concentrations of organics in water. The energy is sufficient to fragment the bonds of organic compounds, and the radiation is not absorbed to any significant extent by the water molecules in the solution. The process is envisioned as a final treatment step to reduce organic contamination in groundwater and industrial wastewaters to acceptable discharge limits. The overall reaction uses hydrogen peroxide as the oxidant in the reaction.

IT Corporation: This is a two–stage, in situ photolytic/biological detoxification process for shallow soil contamination. The first step in the process is to degrade the organic contaminants using ultraviolet (UV) radiation. Degradation is enhanced by adding detergent–like chemicals (surfactants) to mobilize the contaminants. Photolysis of the original contaminants is expected to convert them to less resistant compounds. Biological degradation, the second step, is then used to further destroy the organic contamination and detoxify the soil. The rate of photolytic degradation is several times faster with artificial UV light than with natural sunlight.

Magnum Water Technology Corporation: The CAV–OX process utilizes hydraulic cavitation, a process by which waste liquid undergoes a dynamic pressure reduction while under constant temperature. As a result, gas bubbles form and collapse, generating heat that decomposes the solution into highly reactive hydrogen atoms and hydroxyl radicals. When these components recombine, they form hydrogen peroxide and molecular hydrogen. The flow is then passed through a UV reactor, which also generates hydroxyl and hydroperoxyl radicals. The end products are water, carbon dioxide, halides and organic acids in some cases.

Nutech Environmental: Nulite™ Technology uses illuminated titanium dioxide to destroy organic pollutants and detoxify inorganic pollutants in water.

The photoreactor comprises a stainless steel jacket, a lamp and a photocatalytic sleeve. The lamp emits ultraviolet (UV) light in the 300 to 400 nm range and is coaxially mounted within the jacket. Around the lamp is a Fiberglass mesh sleeve coated with titanium dioxide (anatase). Contaminated water flows through the photocatalytic sleeve. Because of its open pore configuration and large surface area, the mesh creates turbulent mixing. The photoreactors can be linked in series or in parallel or both. The process can be operated both in continuous flow and batch operational modes. After passing through the photoreactors, water returns to the reservoir.

Another titanium dioxide catalytic process has been developed at the University of Florida, and has been licensed to American Energy Technologies, Inc.

Peroxidation Systems, Inc.: The perox–pure™ technology is designed to destroy dissolved organic contaminants in groundwater or wastewater through an advanced chemical oxidation process using ultraviolet (UV) radiation and hydrogen peroxide. Hydrogen peroxide is added to the contaminated water, and the mixture is then fed into the treatment system. The treatment system contains four or more compartments in the oxidation chamber. Each compartment contains one high intensity UV lamp mounted in a quartz sleeve. The contaminated water flows in the space between the chamber wall and the quartz tube in which each UV lamp is mounted.

Purus, Inc.: This process is based on photolytic oxidation for the destruction of volatile organic compounds (VOC) in soil and groundwater. The treatment system design embodies the use of a xenon pulsed-plasma flashlamp that emits short wavelength ultraviolet (UV) light at very high intensities. The process strips the contaminants into the vapor phase, where the UV treatment converts the VOCs into less hazardous compounds.

An innovative feature of the Purus, Inc. technology is the ability to shift the UV spectral output to better optimize the photolysis. Hydrogen peroxide enhancement is available.

Ultrox International: This radiation/oxidation process uses UV radiation, ozone (O_3) and hydrogen peroxide (H_2O_2) to destroy toxic organic compounds, particularly chlorinated hydrocarbons, in water. The process oxidizes compounds that are toxic or refractory (resistant to biological oxidation) in concentrations of parts per million or parts per billion.

The Ultrox system consists of a reactor module, an air compressor/ozone generator module, and a hydrogen peroxide feed system. It is skid-mounted and portable, and permits on-site treatment of a wide variety of liquid wastes, such as industrial wastewaters, groundwaters, and leachate.

Influent to the reactor is simultaneously exposed to UV radiation, ozone, and hydrogen peroxide to oxidize the organic compounds. Off-gas from the reactor passes through an ozone destruction (Decompozon) unit, which reduces ozone levels before air venting. The Decompozon unit also destroys gaseous volatile organic compounds (VOC) stripped off in the reactor. Effluent from the reactor are tested and analyzed before disposal.

Ultraviolet Energy Generators, Inc.: The UVERG Portable UV Detoxification System, developed by Ultraviolet Energy Generators, Inc., Oakland, California, is a compact, rugged unit that houses a single high-powered lamp and a combination of UV, hydrogen peroxide and titanium oxide to destroy organic contaminants. The single lamp and mobile design allow the unit to be transported by truck to remediation sites.

Waste Management, Inc.: Based on short wave-length ultraviolet radiation, this technology combines the effects of ozone generation, free radical formation (OH^- and OH^-_e) and photolysis of the contaminants to effectively control the volatile organic compound emissions. These are two basic equipment configurations. The gas phase reactor is generally used for the production of oxidants or the treatment of contaminated air, such as that found in off-gases from an air stripping tower. The liquid phase reactor is used when a simultaneous production of oxidants and photolysis of contaminants in the liquid phase is desirable.

7.18 X-RAY TREATMENT

Pulse Sciences, Inc. has developed an x-ray treatment process. X-ray treatment of organically contaminated soil and water products is based on the in-depth deposition of ionizing radiation. The collision of energetic photons with matter generates a shower of energetic secondary electrons within the contaminated waste material. The electrons break up the complex molecules and form radicals that react with contaminant materials to form compounds such as water, carbon dioxide, and oxygen. Direct electron beam processing has been established as highly effective for the destruction of organic compounds in aqueous solutions, with residual organic contaminant levels in the μg/ℓ range. However, the electrons do not penetrate deeply and, thus, material handling can be a problem. X–ray treatment does not have this problem.

The physical mechanism by which volatile organic compounds (VOC) and semivolatile organic compound (SVOC) contaminants are removed is primarily dependent on the substrate. For example, in oxygenated water the primary reactant is the hydroxyl (OH) radical. This kinetic mechanism is also expected to play an important role in nonaqueous matrices, because of the presence of moisture in contaminated soil, sludge, and sediments. It is expected that the complete mineralization of contaminants at sufficiently high dose levels can be achieved with the elimination of undesirable air emissions and waste residuals.

A linear induction accelerator (LIA) is used to generate the x–rays used in the treatment process. The LIA can accelerate electron beams to energies of 1 to 5 million electron volts (MeV). A pulse of electrons, 55 nanoseconds in duration, is directed onto a converter to generate x–rays. The x–rays penetrate the waste material. The penetration depth of the x–rays is tens of centimeters long. Large volumes can, therefore, be easily treated, and standard container walls will not absorb a significant fraction of the ionizing radiation. Either flowing waste or waste contained in disposal barrels can be treated. No additives are required for the process; therefore, sealed containers can also be accommodated. The cost of x–ray processing is estimated to be competitive with alternative processes. Moreover, electron accelerators offer a high level of safety; the x–ray output of the LIA is easily turned off by disconnecting the electrical power.

Matthews and his co–workers at Lawrence Livermore National Laboratory have shown that radiolytically induced destruction of halogenated and nonhalogenated VOCs dissolved in groundwater is possible using accelerator–produced bremsstrahlung (x–rays).

7.19 SILENT ELECTRIC DISCHARGE

Los Alamos National Laboratory is developing silent electric discharge reactors for hazardous organics destruction. Silent electrical discharges are used to product highly reactive free radicals that destroy hazardous compounds entrained in gaseous effluents at ambient gas temperatures and pressures.

REFERENCES

1. Arienti, M., et al, *Dioxin–Containing Wastes, Treatment Technologies,* Noyes Data, 1988.
2. Berkowitz, J., et al, *Unit Operations for Treatment of Hazardous Industrial Wastes,* Noyes Data, 1978.
3. Bramlette, T., et al, *Pulse Combustion–Based VOC Destruction Techniques,* Sandia National Laboratory, SAND–92–8429C, 1992.
4. Burton, D., et al, *Treatment of Hazardous Petrochemical and Petroleum Wastes,* Noyes Data, 1989.
5. Chambers, C., et al, *In Situ Treatment of Hazardous Waste–Contaminated Soils, 2nd ed.,* Noyes Data, 1991.
6. Dalton, J., et al, *An Assessment of Off–Gas Treatment Technologies for Application to Thermal Treatment of Department of Energy Wastes,* DOE/MWIP–1, 9/92.
7. Dellinger, B., et al, *Destruction of Organic Wastes Using Concentrated Solar Radiation,* University of Dayton Research Institute.
8. EPA, *Application of Pulse Combustion to Solid and Hazardous Waste Incineration,* EPA/600/D–91/158, 1991.

9. EPA, *Control Technologies for Hazardous Air Pollutants,* EPA/625/6-91/014, 1991.
10. EPA, Emerging Technology Bulletin, *Electron Beam Treatment,* EPA/540/F-92/009, 1992.
11. EPA, *Innovative Hazardous Waste Treatment Technologies, Domestic and International,* EPA/540/2-89/056, 9/89.
12. EPA, *Innovative Hazardous Waste Treatment Technologies, Domestic and International (Second),* EPA/540/2-90/009, 9/90.
13. EPA, *Innovative Hazardous WAste Treatment Technologies, Domestic and International (Third),* EPA/540/2-91/015, 1991.
14. EPA, et al, *Medical Waste Management and Disposal,* Noyes Data, 1991.
15. EPA, *Remedial Action, Treatment, and Disposal of Hazardous Waste, (15th),* EPA/600/9-90/006, 1990.
16. EPA, *Remedial Action, Treatment, and Disposal of Hazardous Waste, (16th),* EPA/600/9-90/037, 1990.
17. EPA, *Remedial Action, Treatment, and Disposal of Hazardous Waste, (17th),* EPA/600/9-91/002, 1991.
18. EPA, *Risk Reduction Engineering Laboratory Research Symposium, (18th),* EPA/600/R-92/028.
19. EPA, *Superfund Innovative Technology Evaluation Program (Fourth),* EPA/540/5-91/008, 1991.
20. EPA, *Superfund Innovative Technology Evaluation Program, Fifth),* EPA/540/R-92/076, 11/92.
21. EPA, *Treatment Technology Background Document,* 1/91.
22. EPA, *Ultraviolet Disinfection Technology Assessment,* EPA 832-R-92-004, 1992.
23. Haggin, J., *Current Directions of Research on Solar Energy Look Promising,* C & EN, 11/2/92.
24. Humprey, J., et al, Advances in Industrial Separations Technologies, CONF-9003250, DE 91 010/01.
25. Humphrey, J., et al, *Separation Technologies—Advances and Priorities,* Dept. of Energy, DOE/ID/12920-1, (1991).
26. Jamshidi, M., et al, *Environmentally Conscious Manufacturing, Recent Advances,* ECM Press, 1992.
27. Krause, T., et al, *Microwave Discharge Plasma Reactor Destroys Chlorinated Organics,* Air Pollution Consultant, 11-12/92.
28. Krishnan, E., et al, *Recovery of Metals from Sludges and Wastewaters,* Noyes Data, 1993.
29. Krukowski, J., *New Hazardous Waste Solutions,* Poll. Engr., 5/15/93.
30. Matthews, S., et al, *Radiolytic Decomposition of Environmental Contaminants and Site Remediation Using an Electron Accelerator,* Remediation, Autumn 1993.
31. Nichols, A., *A Market In the Fields,* Env. Prot., 5/93.
32. Noyes, R., *Handbook of Pollution Control Processes,* Noyes Data, 1991.
33. Noyes, R., *Pollution Prevention Technology Handbook,* Noyes Data, 1993.
34. Nunez, C., et al, *Corona Destruction: An Innovative Control Technology for VOCs and Air Toxics,* EPA/600/A-92/162, 1992.
35. Nunno, T., et al, *International Technologies for Hazardous Waste Site Cleanup,* Noyes Data, 1990.
36. Palmer, S., et al, *Metal/Cyanide Containing Wastes, Treatment Technologies,* Noyes Data, 1988.
37. Penetrante, B., et al, *Application of Non-Thermal Plasmas to Pollution Control,* Second International Plasma Symposium, 2/93, DOE, UCRL-JC-112689, 12/93.
38. Pletcher, D., *The Green Potential of Electrochemistry,* Chem. Engr. 9 & 11/92.
39. Pollution Engineering, 10/15/92.
40. Probstein, R., et al, *Removal of Contaminants from Soils by Electric Fields,* Science, Vol. 260, 4/23/93.
41. *Science/Technology Concentrates,* p. 17, C & EN, 11/23/92.
42. Shajaie, R., et al, *A Novel Approach for SO_2/NO_x Control Using Ultrasound,* DOE, Pittsburgh, 1992.
43. Smith, J., et al, *Upgrading Existing or Designing New Drinking Water Treatment Facilities,* Noyes Data, 1991.
44. Suprenant, N., et al, *Halogenated-Organic Containing Wastes, Treatment Technologies,* Noyes Data, 1988.
45. *Technology Update,* National Environmental Journal, p. 24, 11-12/92.
46. Unterberg, W., et al, *How to Prevent Spills of Hazardous Substances,* Noyes Data, 1988.
47. Virden, J., et al, *High-Energy Corona for Destruction of Volatile Organic Contaminants in Process Off-Gases,* PNL, 10/92.
48. Walker, S., et al, *An Overview of In-Situ Waste Treatment Technologies,* Idaho National Engineering Laboratory, EGG-M-92-342, 1992.

8

THERMAL DESTRUCTION TECHNOLOGY

This chapter discusses those processes that destroy organic materials by the application of heat. These include the incineration processes, as well as those flameless technologies such as wet air oxidation, and supercritical water oxidation.

Thermal processes used for transferring organic materials from one phase to another, without destroying them, are essentially physical processes and discussed in Chapter 7. Thermal processes utilized primarily for immobilizing metals and inorganics in the slag, even though they destroy organics, are discussed in Chapter 4.

All types of wastes can be incinerated including solids, sludges, liquids and gases. There are many types of incineration equipment utilized for municipal waste, medical waste, and hazardous wastes.

Municipal wastes are normally burned in mass burn facilities, by starved air (modular) combustion, and in refuse-derived fuel (RDF) combustors. The three most common types of medical incinerators are starved air (modular) incinerators, rotary kilns and, retort or batch incinerators. The five most common types of hazardous waste incinerators are hearth, rotary kiln, liquid injection, infrared, and fluidized beds. In addition to these incinerator types there are a number of others currently in use, and others under development.

An incinerator is a device in which wastes are burned at a high temperature (typically greater than 1800°F) with a proper amount of air, and with adequate time to ensure destruction of the wastes. A state-of-the-art incinerator is equipped with operating controls and monitoring systems which assure good combustion to destroy the waste, and an effective method of cleaning the air and the water which are by-products of the process.

Incineration processes are often considered an attractive management alternative for hazardous wastes because they possess many advantages over other technologies, including the following:

1. Thermal destruction by incineration provides the ultimate disposal of hazardous wastes, minimizing future liability from land disposal.
2. Toxic components of hazardous wastes can be converted to harmless or less harmful compounds.
3. The volume of waste material may be reduced significantly by

incineration; and

4. Resource recovery, i.e., heat value recovery, is possible through combustion.

Of all of the 'permanent' treatment technologies, properly designed incineration systems are capable of the highest overall degree of destruction and control for the broadest range of hazardous waste streams. Substantial design and operation experience exists in this area and a wide variety of commercial systems are available.

In considering whether an incinerator can combust a specific hazardous waste stream, one must take into account the waste feed mechanism of the incinerator, the size and configuration of the furnace itself, the nature of the furnace's refractory material and the design of its ash handling mechanism. Many operators and hazardous waste incineration experts consider that the feed mechanism is the most critical aspect of an incinerator. This is because experience has shown that the feed mechanism is one of the major sources of problems in actual operation of incinerators. It is obvious that, if a waste is to be burned, there must be a viable mechanism for introducing the waste into a combustion system. The limitations of the feed mechanism as well as limitations of the ash removal mechanism set the requirements for any preprocessing of the waste. In fact, because Superfund sites often contain such a varied mix of waste types and matrices, preprocessing requirements may be so extreme that the preprocessing could turn out to be the most extensive (and possibly the most expensive) operation in the entire remediation process.

Incineration is the most frequently selected of any technology for treating soil, sludge, and sediment in Superfund and was the first technology available for treating organic contaminants in these matrices. The major advantage of incineration is it is able to achieve stringent cleanup standards for highly–concentrated mixtures of organic contaminants.

Even where public opposition is not mounted against on–site incineration, there are other site–specific factors that are often overlooked in the remedy selection process but that can impede the efficiency of the process and/or the achievement of the desired cleanup standards. Some of these factors include:

1. Foreign objects, e.g., rocks, drums, auto bodies, in what was expected to be soil, or liquids, of uniform consistency able to be fed to the incinerator without interruption or special treatment.
2. Styrene tars in a lagoon, seemingly pumpable directly to the incinerator while covered with liquid, but which turn into a stringy, stretchy, non–pumpable "mess" when exposed to air.
3. Metals, e.g., arsenic and lead, in contaminated media that is being incinerated to destroy organic contaminants, with the risk that these metals will volatilize during the incineration process and escape through the incinerator stack to the atmosphere. Even if they do not volatilize, they will remain in the incinerator ash, requiring further treatment.

Any discussion of incineration as an option for treatment of wastes, usually includes mention of the three T's of incineration: time, temperature, and turbulence. Specifically, these are the temperature at which the furnace is operated, the time during which the combustible material is subject to that temperature, and the turbulence required to ensure that all the combustible material is exposed to oxygen to ensure complete combustion. Different combustion systems utilize different mechanisms for addressing the three T's.

Incineration can be applied to most organic–bearing wastes under various conditions. Wastes may be burned which contain relatively high water content, are largely inorganic in nature, i.e., they possess high ash content, or are in the solid or semi–solid state. The primary question is: When should incineration be chosen for application as opposed to other technologies? To decide whether incineration is the best technology for a specific waste, consideration must be given to the following issues:

1. Limitations which arise from the quantity or nature of the waste;
2. The environmental impact of incineration, including stack and fugitive emissions;
3. The requirements for disposal of residues, i.e., ash and air pollution control (APC) residues;
4. Permitting issues.

Incineration is an oxidative process which is used for:

1. Detoxification and sterilization;
2. Volume reduction;
3. Energy recovery;
4. By–product chemical recovery.

The incineration process may be viewed as consisting of four parts: (1) preparation of the feed materials for placement in the incinerator (pretreatment); (2) incineration or combustion of the material in a combustion chamber; (3) cleaning of the resultant air stream by air pollution control devices (APCDs) which are suitable for the application at hand; and (4) disposal of the residues from the application of the process (including ash, and air pollution control system residues).

Waste reduction efforts, as well as the trend to wastes with a higher solids content (less water) would have a tendency to reduce incinerator volume. However, additional materials will be regulated, and land banning could move more volume into incinerators. Oxygen enhancement will allow increased throughput on existing equipment, and at the same time reduce incinerator emissions. The biggest problem is public opposition to the building of new incinerators.

8.1 OPERATING INFORMATION

8.1.1 Data Needs

Important Thermal Treatment Data Needs*

Data Need	Purpose
Heat Content (HHV and LHV)	Combustibility
Volatile Matter Content	Furnace Design
Ash Content	Furnace Design, Ash Handling
Ash Characteristics	Furnace Design
Halogen Content	Refractory Design, Flue Gas Ductwork Specification, APC Requirements

(continued)

Data Need	Purpose
Moisture Content	Auxiliary Fuel Requirements
Heavy Metal Content	Air Pollution Control

*Generally, the data needs for evaluating thermal processes includes the data needed for physical treatment for the purpose of feed mechanism design.

8.1.2 Combustion Zone Temperature

Liquid injection incinerators generally operate at high combustion zone temperatures than the other types of incinerators because of the low excess air requirements. Liquid wastes are relatively easy to atomize and do not require the large amount of combustion air used to ensure complete oxidation of solid wastes. Hazardous wastes containing organically bound chlorine are generally incinerated at temperatures higher than 2200°F. Aqueous wastes containing salts are seldom incinerated at temperatures higher than 1900°F in order to prevent refractory damage.

Rotary kilns generally operate at lower combustion temperatures than liquid injection units and at approximately the same temperatures as hearth units. Wastes are volatilized in the kiln, and some combustion does occur in the kiln. The general practice in this country has been to operate the kiln temperatures lower than the slagging temperature of steel when incinerating wastes in steel drums. European units are often operated above the steel slagging temperature and several recent kilns have been designed by American manufacturers to operate above the steel slagging temperature, which is 2400° to 2800°F, depending on eutectic materials present in the waste. Afterburner temperatures are correspondingly higher for steel slagging units. Liquid and gaseous wastes are often incinerated in the afterburner to provide the necessary thermal input that might otherwise be provided by auxiliary fuel.

Hearth and fluidized bed incinerators have not been designed to operate under steel slagging conditions. Combustion zone temperatures are relatively low because high overall excess air is used to ensure turbulence. Turbulence is attained in fluidized beds through contact between the waste and the bed particles. The high turbulence permits oxidation at relatively low combustion zone temperatures in fluidized beds.

8.1.3 Residence Time

Residence time may be defined as the time a parcel of combustion gas remains at the combustion zone temperature. Residence times cannot be measured directly and are most often estimated from the combustion gas volume flow rate at combustion zone conditions and the combustion chamber volume. In fact, the design residence time is attained by adjusting the size of the secondary combustion chamber of kilns and hearths and the primary chamber of liquid injection and fluidized bed units. Gas residence times in rotary kilns and the primary chamber of hearths are usually not included in residence time computation. Solids retention times in these chambers are designed to obtain sufficient burn out of the ash.

Incinerator manufacturers have used several methods to increase residence times and turbulence in the combustion chamber. The most common method is to induce cyclonic flow by the design and location of waste and fuel burners. Baffles and mixing chambers are also used to promote turbulence. The effectiveness of the various methods must be evaluated from performance test results of the entire incineration system.

8.1.4 Air Usage

Excess air usage varies widely among incinerator manufacturers. The amount of excess air is usually minimized if energy recovery equipment is installed in order to ensure the maximum temperature difference between the heat exchanger inlet and outlet and, therefore, the highest energy recovery efficiencies. Liquid injection incinerators always operate with an excess of air, generally 20 to 60% in excess of stoichiometric air requirements. The only exception is the incineration of organic wastes containing nitrogen, when two chamber incinerators are used. The primary chamber is operated under starved air conditions (less than stoichiometric air) to minimize the formation of nitrogen oxides. Organics are oxidized in the presence of excess air in the secondary chamber. Liquid injection incinerators may operate under slightly positive pressure if a forced draft blower is used, or slightly negative combustion chamber pressures if an induced draft blower is used. In either case, pressure differentials in the combustion chamber are generally less than two inches water column from atmospheric pressure.

The primary chamber of rotary kiln incinerators may be operated either in a starved air or excess air mode. Generally, smaller rotary kilns (5 to 20 million Btu/hr) may be operated in a starved air mode and the larger units are always operated using excess air. Because rotary kiln incinerators are always equipped with an oxidizing secondary combustion chamber, there is always overall excess air usage. Rotary kilns operate at a slight vacuum (less than two inches water column) in the combustion chambers, maintained by an induced draft blower. As a result, air may enter through charging doors, seals, and ports as well as combustion air forced draft blowers. Allowances for air entering through leaks are incorporated in kiln designs.

Similarly, hearth incinerators commonly operate under starved air conditions in the primary chamber and with excess air in the secondary chamber. Overall excess air usage ranges from 30 to 100%. Hearth incinerators require higher excess air usage than rotary kilns to ensure turbulence. The rotation of the kiln agitates the waste to improve turbulence. Methods used to agitate solid wastes in hearths and promote turbulence include the use of reciprocating grates, rakes over the grates, and hydraulic rams to push the waste along the hearth. Combustion air may enter the incinerator both above and below the hearth and combustion chamber pressures are usually slightly negative.

Fluidized bed incinerators are able to operate with very little excess air because of the excellent mixing of air and waste in the bed. The majority of the air entering the incinerator is used to fluidize. Some air is introduced in the headspace above the bed in order to ensure complete oxidation. Negative combustion chamber pressures are usually maintained at less than one inch water column.

8.2 OXYGEN ENRICHMENT

Successful hazardous waste incineration requires an intensive, complete destruction/oxidation of waste molecules with oxygen. Some current incinerators require about 150% excess air to provide enough oxygen for oxidation, and require that the so–

called 3–T factors (temperature, turbulence, and residence time) be adequate to insure efficient destruction. Because 79% of air is nitrogen, the majority of any excess air used will not contribute to the effectiveness of incineration and will only result in extra energy required to raise the nitrogen to combustion temperature, and additional product gas handling and cleaning requirements. As a matter of fact, two of the 3–T's (turbulence and residence time) are essentially the physical parameters used to promote the contact of hazardous waste particles with oxygen. Therefore, it seems logical that increased oxygen concentration should improve incineration or destruction efficiency.

Advantages include:

1. Throughput increase
2. Fuel savings of over 60%
3. Kiln puff reduction
4. Less costly pollution control of the flue gas
5. High oxygen content could reduce NO_x formation
6. Reduced flue gas value
7. Lower particulate carryover

Disadvantages include:

1. Cost of oxygen supply
2. Higher flame temperatures can result in higher NO_x formation, and local overheating.

Oxygen enrichment systems have been developed by Union Carbide (Linde OCS System), and American Combustion, Inc. (Pyretron™).

8.3 WASTE CHARACTERISTICS AFFECTING PERFORMANCE

In determining whether incineration will achieve the same level of performance on an untested waste that it achieved on a previously tested waste and whether performance levels can be transferred, EPA examines the following waste characteristics: (a) the thermal conductivity of the waste, (b) the component boiling points, (c) the component bond dissociation energies, (d) the heating value of the waste, (e) the concentration of explosive constituents, and (f) the concentration of noncombustible constituents.

The contaminant content coupled with the physical and chemical waste properties of the matrix must be considered to determine which thermal destruction process is most appropriate for treatment. These considerations include the actual compounds present in the waste, their concentrations, and the combustion temperature of the contaminant.

The combustion temperature of a contaminant defines the minimum operating temperature of a thermal destruction unit. The specific compounds and concentration ranges present in the waste can also affect treatment performance. Alkali metal salts, particularly sodium and potassium sulfate, and elevated levels of organic phosphorous compounds cause refractory attack and slagging at high temperatures. High halogen concentrations in the presence of oxygen and moisture in the gas stream, form acids which are extremely corrosive and attack refractory metals and metallic pollution control devices. Elevated chlorine concentrations decrease the heating value of the waste and increase the emission of HCl. Volatile metals produce emissions that are difficult to remove with conventional air pollution control equipment. Thermal destruction is not useful for wastes containing non–volatile metals because these inorganics are not destroyed and remain in the ash residue. At high temperatures (>1374°C), these metals may slag, and the generation of oxides of nitrogen can increase significantly.

Heterogeneity of the waste matrix and debris content are two other factors that greatly affect the treatment performance. A thermal destruction process is selected and optimized based on an expected contaminant concentration. However, a non–homogeneous waste such as a Superfund soil, often contains "hot spots" or high contaminant concentrations localized in the matrix. A thermal destruction unit may not be capable of handling the surges created by this phenomenon. As a result, there may be heavy particulate carryover into the afterburner or the particulate removal systems. Stack emissions may rise above acceptable limits. Debris that enters the thermal destruction unit may cause these same results by "occluding" (shielding) the waste from treatment. Debris such as drums, polyurethane bags, and other materials may also interfere with the process by lowering the operating temperature or by slagging and fouling the equipment.

Preprocessing can compensate for the effects of heterogeneity and debris content. Preprocessing may include screening and mixing as well as crushing to provide a consistent particulate size and homogeneity more suitable for treatment. Although extensive preprocessing will appear to increase capital and O&M costs, the tests performed have demonstrated the economic advantage of these preliminary operations compared to the costs of operating the primary process over a broader range of conditions. The extensive equipment repair and replacement costs and the ripple effects caused by equipment downtime, experienced at some hazardous waste sites to date, strongly support the use of extensive preprocessing of the soil wastes.

Other waste properties that affect treatment performance include moisture content, heating value, and special properties such as explosive content. The moisture content affects treatment performance by decreasing the heating value of a waste. Therefore, more energy has to be added to the process. The heating value is defined by the amount of energy released when a waste is oxidized. Some of this energy is used to fire subsequent waste as it enters the combustion chamber. Thus, once combustion is started in the chamber, enough energy must be added to the unit to make up the difference between the energy released during combustion (heat of combustion) and the energy needed to maintain the operating temperature. Explosives also present a problem because a waste containing high explosive concentrations may produce excessive heat or even explosions during incineration.

8.4 DESIGN AND OPERATING PARAMETERS

In assessing the effectiveness of the design and operation of an incineration system, EPA examines the following parameters: (a) the incineration temperature, (b) the concentration of excess oxygen in the combustion gas, (c) the concentration of carbon monoxide in the combustion gas, (d) the waste feed rate, and (e) the degree of waste/air mixing. In addition, incineration of hazardous waste must be performed in accordance with the incineration design and emissions regulations in 40 CFR 264, Subpart 0. For many hazardous organic constituents, analytical methods are not available or the constituent cannot be analyzed in the waste matrix. Therefore, it would normally be impossible to measure the effectiveness of the incineration treatment system. In these cases EPA tries to identify measurable parameters or constituents that would act as surrogates to verify treatment.

For organic constituents, each constituent contains a measurable amount of total organic carbon (TOC). Removal of TOC in the incineration treatment system indicates removal of organic constituents. Hence, TOC analysis is likely to be an adequate surrogate analysis where the specific organic constituent cannot be measured.

However, TOC analysis may not be able to adequately detect treatment of specific organics in matrices that are heavily organic–laden (i.e., the TOC analysis may not be sensitive enough to detect changes at the milligrams per liter (mg/ℓ) level. In these cases other surrogate parameters should be sought. For example, if a specific analyzable constituent is expected to be treated as well as the unanalyzable constituent, the analyzable constituent concentration should be monitored as a surrogate.

Temperature: The optimum operating temperature must be high enough to maintain combustion. Thus, it must remain above the contaminant ignition temperature. The temperature must also be high enough for complete combustion of the waste components to occur. Because some intermediate products are more stable than the initial products, the temperature must also be high enough to combust these products. This is often accomplished in a two–phase system consisting of primary and secondary combustion chambers. The maximum temperature must also be controlled. Combustion chamber temperatures should not reach the point at which ash turns into a molten agglomerate.

Residence Time: The residence time also affects the degree of combustion. For cost–effective operation, the residence time must be minimized, but maintained long enough to ensure complete combustion. This time is a significant factor affecting system capacity, throughput, and cost.

Turbulence (Degree of Mixing): The most difficult variable to quantify is the degree of mixing. Sufficient mixing (or turbulence), with temperature and residence time, is necessary to effectively ensure that the entire matrix is efficiently treated by the process.

Quantity of Excess Oxygen: The quantity of oxygen that is theoretically required to complete combustion is the stoichiometric requirement. Because no process is 100% efficient, excess oxygen, usually as air, must be provided beyond the stoichiometric amount to ensure complete combustion. Otherwise, undesirable products of incomplete combustion are formed, such as carbon monoxide. The amount of air introduced into the combustion chambers must also be closely monitored to ensure that the presence of too much air does not lower the temperature or "choke" the combustion process.

Air Handling Design: Air handling for thermal destruction includes the particulates catch and scrubber design. Each component must be designed not only to remove particulates and gases, but also to handle surges in the process. Careful attention should be given to air handling design to ensure adequate emissions treatment prior to release of the final product gases to the atmosphere. The further treatment and disposal requirements of the particulates catch and effluent scrubber water must also be considered.

8.5 ASH GENERATION AND DISPOSAL

The incineration of soils generates large amounts of ash and residue. Ash characteristics will depend on the type of thermal destruction process. Very little information is available in the literature on the type of ash generated by different incineration technologies.

When treating soils by incineration, the characteristics of the treated soil will provide important information on the ultimate disposal of the ash. Bench scale tests should be performed to determine the characteristics of the ash that may be generated during full scale treatment tests. A muffle furnace could be used to generate ash that would be similar in heavy metals content to the ash produced by some full scale systems. Attempts should be made to simulate full scale incineration temperatures in the laboratory. However, a muffle furnace may not be accurate to simulate high temperatures and short residence times in some afterburners. Sample quantities should be large enough to allow for subsequent total waste constituent, EP Toxicity, and TCLP testing. Information gained from testing ash and treated soils will be very useful during feasibility study preparation, since test results will reveal whether delisting of the waste is possible. If the ash still meets hazardous waste criteria and requires further treatment, the costs of the incineration alternative can increase substantially.

8.6 METAL PARTITIONING

In recent years, metal emissions have become one of the main concerns surrounding waste incineration. As a result, recent regulations on boilers and industrial furnaces (BIFs) burning hazardous wastes are designed to control emissions of ten metals. Furthermore, it is anticipated that future regulations on both municipal waste and hazardous waste incinerators will address metal emissions.

One aspect of incineration that is not fully understood involves the distribution, or "partitioning," of metals between bottom ash, flyash, scrubber water, and stack gas.

Metal emissions can be affected by a number of factors, including the amount of metal in the feed, kiln temperature, flue gas temperature, pollution control devices used, and the amount of chlorine in the feed. Typically, metals will become part of stack emissions if they volatilize within the kiln or boiler and either (1) solidify into particulate, or (2) bond to other particulate matter that is carried by the flue gas. Thus, the volatilization temperature, rather than the melt temperature, gives a good indication of the amount of feed metal that will be entrained in gases leaving the kiln.

A number of conclusions were made from the results of the first test series at the EPA Incineration Research Facility. Cadmium, lead and bismuth are relatively volatile, based on normalized discharge distribution data; less than 32% of their discharge was accounted for by kiln ash. Barium, copper, strontium, chromium and magnesium are relatively nonvolatile; more than 75% of their discharge was in the kiln ash. Average apparent scrubber efficiency for the individual metals ranged from 32 to 88%; scrubber efficiency for the three volatile metals was lower than that for five of the six nonvolatile metals. Both kiln ash partitioning and scrubber efficiency appear to be impacted negatively by increases in feed chlorine content and, to a lesser extent, increases in kiln temperature. With the exception of arsenic, discharge distributions of the metals correlate strongly with volatility temperatures.

Unfortunately, because metals are not destroyed by incineration and have no heat recovery value there is little incentive to treat most metal bearing wastes by incineration. An exception might be an organometallic compound containing waste such as tetraethyl lead or a cyanide complex which is highly toxic and not readily treated by more conventional methods.

Several problems must be faced when incinerating metal wastes. A primary consideration is the extent to which air emissions of toxic heavy metal particles or vapors will be generated. Certain metals and their oxides such as mercury, lead, selenium, and arsenic are volatile, particularly at the elevated temperatures of incineration. A significant percentage of the input of these relatively volatile metals will be emitted as a vapor or as fine particulates which are difficult to control.

A second problem involved in the incineration of such wastes is the generation of an incinerator ash or sludge containing metals or metal oxides, which will require safe disposal. Third, wastes containing high concentrations of noncombustible materials require greater energy input via auxiliary fuel combustion, thus increasing processing costs significantly. Finally, such wastes may be difficult to handle in certain incineration systems. Liquid injection incinerators may not be used, for example, should the solids content of the waste be such that the injectors will become clogged.

Although numerous studies of incinerator performance have been conducted in which organic wastes containing metals were burned, the available data are limited in content relative to the effect of metals on combustion. Based on the available data, it does not appear as though the presence of metals in small concentrations will hinder the destruction of organics. The data does show, however, that certain metal species may present more of a concern relative to potential air emissions than do others.

Changes in waste disposal patterns prompted by newly enacted legalization has resulted in a significant change in the composition of hazardous wastes presented for incineration. Metal–containing wastes that were historically landfilled are now being incinerated with increasing frequency.

Overall, it may be concluded that incineration appears to be a limited and potentially costly alternative for the treatment of hazardous wastes containing heavy metals. The wastes which may be handled in this manner are limited to organic wastes (including organometallic compounds such as cyanides and tetraethyl lead) which contain metals in fairly low concentrations. Most commercial incineration facilities will handle such wastes, but will charge a premium based on metals concentration.

Most of the pyrometallurgical processes identified for metal waste treatment are classified as "calcination" or "smelting" operations. Calcination processes are generally those which form metal oxides, while smelting produces pure metal.

Drying and calcination are usually carried out in various types of kilns such as rotary kilns, shaft furnaces, and rotary hearths. Smelting operations are conducted in blast or reverberatory furnaces as described in reports and texts dealing with metal processing. Many nonferrous metals can be extracted by reduction smelting: copper, tin, nickel, cobalt, silver, antimony, bismuth, and others. Blast furnaces are sometimes used for the smelting of copper or tin, but reverberatory furnaces are mote common for most metals.

Overall, the key element in evaluating the economic attractiveness of pyrometallurgical systems is the value which may be derived from recovery of metals. However, systems which can not produce reusable materials may be attractive in terms of providing good volumetric reduction of wastes, but may not be viable economically. Immobilization of metals is discussed in Chapter 4.

8.7 CHLORINE CONTENT

Thermal incineration can effectively treat both halogenated and nonhalogenated hydrocarbons in gas streams. However, due to the flame–inhibiting properties of halogenated hydrocarbons, effective incineration requires systems using more auxiliary fuel, higher operating temperatures, and longer residence times than similar systems designed for nonhalogenated compounds. Furthermore, additional auxiliary fuel is necessary to sustain a flame if the gas stream contains low concentrations of the target hydrocarbons.

Incinerator emissions are of particular concern with chlorinated hydrocarbons. For example, a chlorinated feed (defined as containing at least 0.5% chlorine) produces corrosive hydrochloric effluent gas. But probably more importantly, if combustion is incomplete, organic constituents are only partially oxidized and products of incomplete combustion (PICs) are yielded.

In order to avoid the problems in systems technology and apparatus involved in the possible occurrence of free halogens in the flue gas, it has been found advantageous to incinerate halogen–containing waste together with sulfur–containing waste. Apparently the SO_2 produced during this process will reduce the halogens present to the corresponding halogenides to a degree where not even traces of halogen can be identified by analysis. Therefore, the prerequisite for stopping the halogens in the flue–gas flow is to maintain a constant waste mix and an adequately great SO_2 concentration. As a result of this mode of operation it has been possible so far to avoid any corrosion damage by halogens.

Certain polychlorinated compounds may appear problematic from the point of view of incineration in that they are characterized by their increased thermal stability. Consequently, their complete incineration may require higher temperatures and longer residence times in the combustion zone.

It is hypothesized that chlorine may act to inhibit NO_x formation through its interaction with free radical species.

8.8 SLAG FORMATION

Slag formation in hazardous waste incinerators causes major operating problems, including increased downtime, maintenance costs, and energy costs. Slag formation can often be controlled by lowering process temperatures, increasing waste screening practices, and increasing the melting point of the feed via more effective waste blending. However, operators of hazardous waste incinerators may not have the freedom to make such changes due to limitations under their RCRA and/or TSCA permits.

Alternatively, slag formation can be reduced by using chemical additives that increase the melting point of the wastes, as proposed by Schofield.

Slag forms when low–melting point inorganic materials are heated above their fusion or melting temperature. When the melted materials cool, they attach to the inside surfaces of process equipment as slag deposits. Elements that tend to form slag due to their low melting or fusion temperatures include sodium, potassium, lithium, boron, vanadium, phosphorous, and antimony.

Slag formation can occur in several situations, such as when ashen material is melted and then cooled. For example, ash is commonly heated to above its melting temperature in rotating kilns. As the ash cools, it physically attaches to the refractory. Since the kiln is rotating, it forms a ring or circular dam around the inside of the kiln. If the slag dam is not removed, it can eventually blind the entire kiln cross section or overload the weight capacity of the kiln and its supports. Refractory damage may also result due to chemical attack from the molten slag and/or physical damage when the slag must be removed via jackhammers or other severe methods.

Another example of slag formation is when inorganic particles melt in the burner or other high–temperature zones of boilers, furnaces, and incinerators, e.g., secondary combustion chambers or boiler tubes. If these particles impact the inside surfaces before they cool, they will stick to metal, refractory, or previously formed slag. Inorganic particles typically present a slag formation problem in situations where (1) the unit burns liquid waste containing suspended or dissolved solids, or (2) inorganic particles are entrained in the combustion gases.

Slag formation can have a serious impact on operating costs. The major costs incurred due to slag formation are: (1) downtime, (2) maintenance, and (3) energy costs.

There are two principal ways of minimizing slag formation: (1) reducing the ash temperature, or (2) increasing the fusion temperature by changing the composition of the inorganic material(s) in the feed. Ash temperature can be reduced by:

1. Reducing process temperatures (without significantly affecting waste treatment, boiler, or furnace performance);
2. Reducing ash residence time in the hot zone;
3. Substantially increasing the inorganic throughput rates; and/or
4. Changing burner flame size, temperature, pattern, or luminosity.

There are also several ways to minimize slag formation by increasing the fusion temperature. These include:

1. Identifying waste streams containing more than trace amounts of low–melt materials, e.g., sodium, potassium, lithium, boron, vanadium, phosphorus, and antimony, and not accepting them for incineration;
2. Improving waste blending to obtain a higher minimum fusion temperature; and/or
3. Using chemical additives to increase the fusion temperature.

Other less common options for minimizing slag formation include using nonwetting ceramic coatings on the inside surfaces of the hot zone, and changing the equipment geometry or gas flow patterns to reduce impaction of molten particles on the interior surfaces.

8.9 CENTRAL WASTE INCINERATORS

Central waste incinerators are those which accept waste from several external sources for destruction in a central facility. They are usually large [in excess of 50 t/d (55 T/d)], continuously operated installations equipped with heat recovery equipment. Waste is burned in these incinerators without pre–processing.

The features of large central mass burning incinerators are distinguished by the design of the grate system. The grate must transport refuse through the furnace and promote combustion by providing adequate agitation without contributing to excessive particulate emissions.

European systems are typically equipped with inclined grates that move the waste through the furnace by reciprocating or rotating drum action. Stationary circular grates with rotating rabble arms, travelling grates, and rocking grates are also used.

As the waste moves progressively through the furnace, it is dried, burned, and combusted to ash. Approximately 40 to 60% of the total air entering the furnace is provided as underfire air to cool the grates and prevent ash slagging. The balance is supplied as overfire air to completely combust the flue gas and particulate rising from the grates.

Many central waste incineration systems are built with waterwall construction in addition to boiler tubes within the flue gas steam (convection sections) to maximize energy recovery from the incinerator. Refractory-lined combustion chambers with a separate downstream boiler section may also be used in lieu of waterwall construction. If refractory walls are used, higher excess air is required to control the operating temperature.

Central waste incineration systems are used primarily for refuse incineration. The practice has been more prevalent and more long-standing in Europe than in North America.

Mass burning of refuse in central incineration facilities is a commercially demonstrated technology. Due to the relatively large size of such units and unique features of each system with respect to waste handling, energy recovery, etc., these facilities are normally designed and built to meet each customer's specific needs.

Operating temperatures in mass-burning central waste incinerators are normally maintained in the order of 1000°C (1832°F) and refuse residence time on the grate ranges from 20 to 45 minutes. Refractory wall systems normally require 100 to 150% excess air to maintain operating temperatures, whereas waterwall systems require only about 80% excess air. This offers the advantages of a smaller furnace volume and reduced NO_x formation with the latter system, due to lower airflow. Waterwalls extract heat from the burning waste. Without waterwalls, where the furnace chamber is lined with refractory, the furnace temperature must be controlled by the injection of cool air. In refractory or waterwall furnaces the maximum temperatures should be below 1100°C (2010°F), the temperature at which slagging problems will begin to occur.

Underfire air is provided beneath the grates to prevent overheating of the grate system and to supply part of the waste combustion air requirement. Air is also provided above the grates (overfire air) to burn off the products of combustion of the waste and to properly direct flue gas flow within the furnace. Underfire air will comprise from 40 to 60% of the air flow to the furnace, with the over-fire air flow inserting the balance of the air required for incineration.

White goods (stoves, refrigerators, etc.) must be removed from the waste feed. With some systems there is limited ability to handle waste tires, and they must be removed or distributed throughout the daily feed.

The mass burn waterwall furnace is designed to generate steam (or hot water) as well as to incinerate waste. Its design must include provisions to minimize boiler tube corrosion.

8.10 MOBILE INCINERATION

Mobile incineration grows with future bans on the landfill disposal of certain wastes. Mobile systems include rotary kiln, circulating fluidized bed, infrared, and liquid injection incineration, equipped with secondary combustion chambers and environmental controls. These mobile incinerators are capable of handling a variety of wastes including PCBs, carbon tetrachloride, other hazardous wastes and soils. The primary advantage of the mobile incinerator is its ability to treat on–site and thus eliminate the need for off–site transport of waste. Mobile incinerators must meet all applicable state requirements which typically include air emission permits.

The EPA mobile unit is mounted on four heavy–duty semi–trailers which can be transported to a treatment site and connected in series. The system includes a rotary kiln, a secondary combustion chamber, and a scrubber, along with the following support equipment: bulk fuel storage, waste blending and feed equipment for both liquids and solids, scrubber solution feed equipment, and receiving drums, stack monitoring equipment, and an auxiliary diesel power generator. ENSCO has developed a modified version of the EPA mobile rotary kiln incinerator.

The EPA and ENSCO mobile incinerators are able to handle both solid and liquid hazardous waste streams. Bulky solids need to be fed through a specially designed solids feed system prior to being fed to the kiln. Wastes with a low heating value, i.e., less than 8,000 Btu/lb, may require blending with kerosene prior to being fed to the combustor.

Mobile/transportable incineration has been shown to be effective in treating soils, sediments, sludges, and liquids containing primarily organic contaminants such as halogenated and nonhalogenated volatiles and semivolatiles, polychlorinated biphenyls (PCBs), pesticides, dioxins/furans, organic cyanides, and organic corrosives. The process is applicable for the thermal treatment of a wide range of specific Resource Conservation and Recovery Act (RCRA) wastes and other hazardous waste matrices that include pesticides and herbicides, spent halogenated and nonhalogenated solvents, chlorinated phenol and chlorinated benzene manufacturing wastes, wood preservation and wastewater sludge, organic chemicals production residues, pesticides production residues, explosives manufacturing wastes, petroleum refining wastes, coke industry wastes, and organic chemicals residues.

As a rule of thumb, transportable incineration is cost effective for sites containing more than 1,000 tons of soil. For smaller sites, the mobilization and demobilization costs are significant when compared to the actual operational costs.

Depending on the requirements of the incinerator type for soils and solids, various equipment is used to obtain the necessary feed size. Blending is sometimes required to achieve a uniform feed size and moisture content or to dilute troublesome components.

The waste feed mechanism which varies with the type of the incinerator, introduces the waste into the combustion system. The feed mechanism sets the requirements for waste preparation and is a potential source of problems in the actual operation of incinerators if not carefully designed.

Different incinerator designs use different mechanisms to obtain the temperature at which the furnace is operated, the time during which the combustible material is subject to that temperature, and the turbulence required to ensure that all the combustible material is exposed to oxygen to ensure complete combustion.

At-sea incineration usually utilizes a liquid injection unit mounted on a ship, to destroy hazardous waste far away from populated areas and shipping lanes. No acid gas pollutant removal system is applied. The wastes treated include toxic organochloride compounds, herbicides, and Agent Orange. The basic advantage of at-sea incineration is the distance from populated areas and the high efficiency of combustion. The disadvantages are problems with monitoring an at-sea process, the danger of spills, and the need to operate on-shore auxiliary facilities. It is estimated that 30,000 tons of special waste per year were incinerated by Germany on board ships in the North Sea, half of it highly-chlorinated material.

8.11 WASTE TO ENERGY SYSTEM

The two major processing and conversion technologies currently being used by commercial-scale facilities to recover energy from municipal waste are mass burning and refuse derived fuel (RDF) technologies.

The combustion of MSW occurs in four stages: (1) drying, in which heat is used to raise the temperature of the moisture in MSW and evaporate it; (2) devolatilization, in which the combustible volatiles in MSW are released, between 350° and 980°F; (3) ignition, in which combustion begins as the volatiles reach ignition temperature in the presence of oxygen; and (4) combustion of fixed carbon, in which combustion of the volatile matter is completed (with the fixed carbon being oxidized to carbon dioxide).

The key to the success of an energy recovery system is controlling the combustion process so that the heat produced can be transferred from the hot combustion gases to some other medium—almost always water in some kind of boiler. In order to transfer heat in a boiler burning prepared or unprepared refuse fuel without damaging the boiler, certain conditions must be met. These conditions include the following:

1. The temperature of the combustion gases entering the boiler's main heat transfer section should not exceed 1600°F to avoid high temperature corrosion;
2. The temperature of the combustion gases leaving the boiler must be maintained above 300°F to prevent the corrosion that results from the condensation of acids present in the gas stream; and
3. The volatile gases released during combustion must be well mixed with air and completely burned before the gas stream enters the boiler section, because corrosion can occur if the boiler environment alternates between oxidizing and reducing conditions.

RDF technology attempts to overcome the limitations associated with the combustion of MSW for energy by improving the quality of the fuel. By removing noncombustible material and increasing the homogeneity of the remaining combustible fraction, greater control of the combustion process can be achieved. RDF systems require less excess air than systems that require complete combustion, which increases furnace efficiency. Other advantages of RDF systems include:

1. RDF boilers can be smaller than those for mass burning since a considerable amount of noncombustible material is removed from raw MSW;

2. The handling and storage characteristics of RDF are significantly better than MSW;
3. RDF can be burned in existing fossil fuel boilers, which can greatly reduce capital costs;
4. RDF burned in a dedicated boiler can displace more expensive fossil fuels, i.e., natural gas and oil;
5. RDF can be produced at a remove site and transported to the conversion facility—an important advantage if land is scarce or expensive, if truck traffic is undesirable near the intended energy user, or if the energy user is far from the source of MSW; and
6. The recovery and sale of reusable materials from MSW can reduce landfill requirements and produce revenues for the project.

8.12 AIR POLLUTION CONTROL

Air pollutants from the incineration of hazardous wastes may arise both as a result of incomplete combustion and from the products of combustion of constituents present in the wastes and combustion air. The products of incomplete combustion include carbon monoxide, carbon, hydrocarbons, aldehydes, amines, organic acids, polycyclic organic matter (POM), and any other waste constituents or their partially degraded products that escape thermal destruction in the incinerator. In well designed and operated incinerators, these incomplete combustion products are emitted in insignificant amounts. The primary overall end products of combustion are in most cases carbon dioxide (CO_2) and water vapor (H_2O), but there are also a multitude of other products formed, depending on the composition of the waste material incinerated and combustion conditions. Hydrogen chloride (HCl) and small amounts of chlorine (Cl_2), for example, are formed from the incineration of chlorinated hydrocarbons. Hydrogen fluoride (HF) is formed from the incineration of organic fluorides, and both hydrogen bromide (HBr) and bromine (Br_2) are formed from the incineration of organic bromides. Sulfur oxides, mostly as sulfur dioxide (SO_2), but also including 1 to 5% sulfur trioxide (SO_3), are formed from the sulfur present in the waste material and auxiliary fuel. Phosphorus pentoxide (P_2O_5) is formed from the incineration of organophosphorus compounds. In addition, nitric oxide (NO) is formed by thermal fixation of nitrogen from the combustion air and from nitrogen compounds present in the waste material. Particulate emissions include particles of mineral oxides and salts from the mineral constituents in the waste material, as well as fragments of incompletely burned combustible matter.

Organic pollutants emitted as a result of incomplete combustion of waste material are often present in effluents from the primary combustion chamber at low concentration levels well under the lower flammability limit. The control of the emission of these organic pollutants can be handled by continued combustion at high temperatures using afterburners (also termed secondary combustion chambers).

Several factors affect the installation of air pollution control equipment on hazardous waste incinerators, including:

1. Federal, state, and local regulations regarding emissions
2. Properties of the waste being incinerated
3. Type of incinerator used

4. Customer preference
5. Equipment cost

Generally, both gaseous and particulate emissions are controlled. Air pollution control equipment is located downstream of the combustion chamber and energy recovery equipment and consists of the following components:

1. Quench chamber
2. Particulate collection device
   (a) Venturi scrubber
   (b) Baghouse
   (c) Electrostatic precipitator
   (d) Cyclone
   (e) Ionizing wet scrubber
3. Gas absorbing device
   (a) Packed tower scrubber
   (b) Plate or tray scrubber
   (c) Spray tower scrubber
   (d) Ionizing wet scrubber
4. Mist eliminator
5. Flue gas handling equipment

8.13 SOLIDS/LIQUIDS INCINERATION PROCESSES

8.13.1 Catalytic Extraction Processing (CEP)

This technique uses molten metal baths to recover marketable components from waste streams. The technology was developed by researchers at US Steel (now USX) who were looking for a way to derive additional value from waste products. They discovered that by dissolving wastes in a modified metal bath process, not only could toxics be completely destroyed but valuable products could be retrieved as well.

Key to the process is the step in which other reactants, such as limestone or oxygen, are co-fed with the wastes to create a chemical transformation. The process is being commercialized by Molten Metal Technology, Inc.

The 3000°F metal bath causes waste to dissociate into its constituent elements in just seconds. The technique can handle both organics and inorganics, in all forms—liquid, gaseous and solid, including sludge, says its developers.

During operation, the dissociated elements segregate within the reactor. Heavy metals collect in the metal bath; inorganics form a slag in the center; and organics are broken down to simple gases, such as carbon monoxide and hydrogen, which rise to the top. The inorganic slag can be recovered, vitrified and used as an aggregate or an abrasive, while the gases are recovered for industrial use, or are burned onsite for their Btu value.

Once separated, the gases and inorganic slag are purged continuously. Metals are either removed continuously, or are allowed to collect until there is a change in the system's catalytic properties. At that time, the bath is tapped, and returned to a metal processor for purification or use as an alloy.

8.13.2 Circulating Bed Combustion

Circulating bed combustion (CBC) systems constitute an innovation in fluidized bed incineration technology. These systems utilize high air velocities and recirculating granular bed materials to maintain and achieve combustion of waste under fluidized bed conditions. The circulating bed material, serves to not only transfer heat energy and increase turbulence, but can be chosen for its chemical characteristics to bring about reaction and neutralization of certain products of combustion such as sulfur dioxides and hydrochloric acid. CBC systems are applicable to solids, liquids, slurries, and sludges, over a wide range of heat values and ash contents. Numerous performance tests have been conducted which indicate that circulating bed combustion can achieve very high destruction and removal efficiencies, while limiting other pollutant emissions to acceptable levels. CBC systems can offer both technological and economic advantages over established fluidized–bed incineration systems primarily due to the increased turbulence of the system. CBC systems operate at higher air velocities, and are not limited, as are fixed bed units, to the narrow range of design velocities needed to maintain fluidization while, at the same time, limiting entrainment and carry over of bed material.

Solids, slurries, or liquids can be introduced into the chamber loop where they contact hot bed material recirculating through the cyclone. When introduced into the primary combustion zone, the waste heats rapidly and continues to be exposed to high temperatures (up to 1800°F) throughout its residence time. High velocity air entrains the circulating soil, which travels upward through the combustor and into the cyclone. The cyclone separates the combustion gases from the hot solids. The solids then are returned to the combustion chamber via a proprietary non–mechanical seal. Temperatures around the entire combustion loop are uniform to within ±50°F. The hot flue gases and fly ash pass through a combustive flue gas cooler into a baghouse filter which traps the ash. Filtered flue gas then exits to the atmosphere. Heavier particles of purified soil remaining in the combustor lower bed are removed slowly by a water–cooled bed ash conveyor system.

Acid gases and sulfur oxides formed during combustion are captured by limestone added directly into the combustor. Emissions of CO and NO_x are controlled to low levels by the turbulent mixing, low temperatures (1425° to 1800°F), and staged combustion achieved by injecting secondary air at sequenced locations in the combustor.

The CBC process may be applied to liquids, slurries, solids, and sludges contaminated with corrosives, cyanides, dioxins/furans, inorganics, metals, organics, oxidizers, pesticides, polychlorinated biphenyls (PCB), phenols, and volatiles.

Industrial wastes from refineries, chemical plants, manufacturing site cleanups, and contaminated military sites are amenable to treatment by the CBC process. The CBC is permitted by EPA, under the Toxic Substance Control Act (TSCA), to burn PCBs in all ten EPA regions, having demonstrated a 99.9999% destruction removal efficiency (DRE).

Waste feed for the CBC must be sized to less than 1 inch. Metals in the waste do not inhibit performance and become less leachable after incineration. Treated residual ash can be replaced on–site or stabilized for landfill disposal if metals exceed regulatory limits.

The EPA SITE program concluded that a recent test successfully achieved the desired goals, as follows:

1. Obtained DRE values of 99.99% or greater for principal organic

hazardous constituents (POHC) and minimized the formation of products of incomplete combustion (PIC).

2. Met the OES Research Facility permit conditions and the California South Coast Basin emission standards.
3. Controlled sulfur dioxide emissions by adding limestone, and determined that the residual materials (fly ash and bed ash) were nonhazardous. No significant levels of hazardous organic compounds left the system in the stack gas or remained in the bed and fly ash material. The CBC was able to minimize emissions of sulfur oxide, nitrogen oxide, and particulates. Other regulated pollutants were controlled to well below permit levels.

The CBC, according to the developer, is an advanced fluidized bed incineration system. When compared to other conventional incinerators, the CBC system operates at much higher turbulence and combustion particle burn up. Combustion air moves 15 to 20 times faster than the conventional incinerators. This difference leads to two main advantages: lower temperatures than normal are needed to destroy contaminants, and lower operating costs are incurred. The lower temperature makes it possible to control acid gases by adding limestone in the combustor. Lower temperatures also result in less nitrogen oxides being formed. This technology can treat all types of halogenated hydrocarbons (containing chlorine, bromine and fluorine), including PCBs and other aromatics. It can treat solids, liquids, sludges and slurries directly in the combustion loop. Atomizing and multiple feed ports are not required. Recovery of energy is efficiently accomplished in both the combustor zone and the flue gas cooler.

Advantages:

1. Because of its intimate contact with heated bed particles, wastes can be combusted at lower temperatures than that of conventional incinerators.
2. Temperatures in the vessel are reportedly high enough to destroy wastes, but low enough to prevent formation of significant amounts of NO_x.
3, The bed material acts as a scrubber to capture acid gas from the process, reportedly creating a non–toxic solid residue.

Disadvantages:

1. Removal and disposal of the inert residual bed materials could be a problem.
2. Relatively large amounts of fine particulate matter entrained in the exhaust gases may require elaborate pollution control devices.
3. Waste feed particle size must be controlled to maintain a uniform feed rate.
4. Accurate control is needed to ensure that retention time in the bed is sufficient for complete combustion, and that radical increases in the waste's heat value will not drastically boost bed temperatures and adversely affect bed operation.

8.13.3 Detonation

The Industrial Research Institute of Kanagawa Perfecture (IRI; Yokohama, Japan), is the developer of a continuous pilot system to convert CFCs into harmless gases. Depending on CFC type, 5 to 10 kg of the material is destroyed with 99.99% conversion efficiency in eight hours.

CFCs, propane, and oxygen in a ratio of 3:1:6 are pumped to a cylindrical stainless–steel detonation chamber measuring 35 mm (inside diameter) x 1 m (length). An ignition coil and spark plug initiate the explosion that generates an instantaneous temperature of 3000°C and a shockwave with peak pressure of 43 kgf/cm^2 on the walls of the detonation chamber.

8.13.4 Fluidized Bed Incineration

Fluidized bed incinerators utilize a very turbulent bed of inert granular material (usually sand) to improve the transfer of heat to the waste streams to be incinerated. Air is blown through the granular bed materials until they are "suspended" and able to move and mix in a manner similar to a fluid, i.e., they are "fluidized." In this manner, the heated bed particles come in intimate contact with the wastes being burned. The process requires that the waste be fed into multiple injection ports for successful treatment. Advantages of this technology include excellent heat transfer to the material being incinerated and a long residence time. An off–shoot of this technology is a circulating bed combustor (discussed earlier).

A representative fluidized bed reactor will have the following basic system components: fluidized bed reactor, fluidizing air blower, waste feed system, auxiliary fuel feed system, and air pollution control device system.

A typical reactor has an inside diameter of 26 ft (8 m) and an elevation of 33 ft (10 m). Silica beds are commonly used and have a depth of 3 ft (1 m) at rest and extending up to 6.5 ft (2 m) in height when fluidizing air is passed through the bed. Waste and auxiliary fuel are injected radially into the bed and reacted at temperatures from 840° to 1500°F (450° to 710°C). Further reaction occurs in the volume above the bed at temperatures up to 1800°F (980°C). A typical residence time for liquid hazardous waste is 12 to 14 seconds.

Reactor heat release rates of up to 15 million kcal/hr and waste, input feed rates of up to 48 ft^3/hr (1,360 ℓ/hr) for liquids over 10,000 Btu/lb (5,560 kcal/kg) in heat content, and up to 270 ft^3/hr (7,570 ℓ/hr) for liquids with a heat content of 3,000 Btu/lb (1,670 kcal/kg), are reported.

Liquid wastes can be pumped directly from a tank truck into the reactor by a recirculating pump system. Wastes are injected radially into the reactor bed through a nozzle. Flow rates are determined by recording waste liquid level changes in the calibrated tanker as a function of time. Auxiliary fuel is often fed radially into the bed through a number of bed nozzles manifolded around the reactor circumference.

Atmospheric emissions from the combustion of liquid hazardous wastes have been controlled by a venturi scrubber. Recirculating water is injected into the venturi to scrub particulate matter from the combustion gas stream and quench the gas temperature from 1500° to 175°F (about 820° to about 80°C) prior to emission into the atmosphere through the stack. Spent scrubber liquid is sent to a wastewater treatment plant for processing.

Sludge, liquids, and prepared solid waste can be fed directly to the bed of the fluid bed reactor. Sludges with a moisture content in excess of 80% and aqueous liquids with high water contents can be fed in the top of the furnace at the furnace ceiling rather than into the bed. Top injection tends to reduce the bed area otherwise required for sludge moisture evaporation.

The temperature within the bed should be maintained at approximately 760°C (1400°F) for organic sludges. The use of sludges or other wastes containing inorganic materials should be subject to a test burn before injecting into a fluid bed furnace.

The freeboard should be maintained in the range of 760° to 870°C (1400° to 1600°F). This temperature should never be allowed to exceed 870°C (1600°F) when a heat exchanger is used, unless the heat exchanger is specifically designed for operation at a higher gas inlet temperature. Water sprays should be provided at the ceiling of the furnace for cooling the gas stream if its temperature approaches 870°C (1600°F).

Gaseous waste or fuel should not be injected into the bed. It can be fired in the windbox.

The oxygen content of the flue gas exiting the furnace should be continuously measured at the breeching leaving the furnace upstream of the recuperator (if one is provided). A sample should be continuously extracted from the gas stream, passed through a water bath to clean the sample and to reduce its moisture content, and then measured for oxygen content. The oxygen content should be in the range of 4 to 8% by volume, on a dry basis.

In general, industrial waste liquids and gases should not be introduced into a fluid bed incinerator if they require a temperature in excess of 870°C (1600°F) for destruction. The air heater normally provided with these systems cannot withstand greater than 870°C (1600°F) inlet gas temperature. An afterburner is normally not provided for a fluid bed incinerator system.

A major issue associated with a fluid bed furnace is its ability to handle a wide variety of waste streams. It is sensitive to waste composition. Certain wastes, particularly those containing clays, inorganics (salts), or high quantities of lime, will tend to seize the bed, preventing fluidization. Test burns on materials that have not previously been fired in a fluid bed furnace must be performed to determine the bed reaction to those particular materials. If bed seizure does occur, bed additives may be available to help eliminate this problem by changing the physical properties of the waste feed. If not, a fluid bed unit may not be the one to use. A related issue is agglomeration. Waste materials may build up on individual particles within the bed. With changes in operation (such as maintaining a higher bed residence time) this could be controlled. If not addressed, agglomeration could result in bed seizure.

Advantages:

1. General applicability for the disposal of combustible solids, liquids, and gaseous wastes.
2. Simple design concept, requiring no moving parts in the combustible zone.
3. Compact design due to high heating rate per unit volume (100,000 to 200,000 Btu/hr–ft^3) (900,000 to 1,800,000 kg–cal/hr–m^3) which results in relatively low capital costs.
4. Relatively low gas temperatures and excess air requirements which tend to minimize nitrogen oxide formation and contribute to smaller, lower cost emission control systems.
5. Long incinerator life and low maintenance costs.
6. Large active surface area resulting from fluidizing action enhances the

combustion efficiency.

7. Fluctuation in the feed rate and composition are easily tolerated due to the large quantities of heat stored in the bed.
8. Provides for rapid drying of high-moisture-content material, and combustion can take place in the bed.
9. Proper bed material selection suppresses acid gas formation; hence, reduced emission control requirements.
10. Provides considerable flexibility for shockload of waste; i.e., large quantities of waste being dumped in the bed at a single time.

Disadvantages:

1. Difficult to remove residual materials from the bed.
2. Requires fluid bed preparation and maintenance.
3. Feed selection must avoid bed degradation caused by corrosion or reactions.
4. May require special operating procedures to avoid bed damage.
5. Operating costs are relatively high, particularly power costs.
6. Possible operating difficulties with materials high in moisture content.
7. Formation of eutectics is a serious problem.
8. Not well suited for irregular, bulky wastes, tarry solids, refractory wastes, or wastes with a fusible ash content.

A hybrid design that simultaneously operates in both the bubbling bed and circulating bed modes is termed the multi-solid fluid bed combustor (MSFBC). This design, developed by Battelle Columbus Laboratories, utilizes an entrained bed of fine ash and limestone particles superimposed in a dense bed of large particles.

Another hybrid design developed by Energy & Environmental Research Corporation is the Hybrid Fluidized Bed (HFB) system which treats contaminated solids and sludges by (1) incinerating all organic compounds, and (2) extracting and detoxifying volatile metals. The system consists of three stages: a spouted bed, a fluidized afterburner, and a high temperature particulate soil extraction system.

The Institute of Gas Technology has developed the Fluidized-Bed Cyclonic Agglomerating Incinerator, which is discussed in Chapter 4.

Lurgi AG (Frankfurt) has developed a circulating fluidized-bed (CFB) furnace that recovers lead and zinc from blast-furnace dust and sludge.

Waste Tech Services, Inc. has developed a Low-Temperature Fluidized Bed that functions similarly to the conventional fluidized bed except that a higher air volume is forced through the bed material. Also, the bed is composed of a mixture of a granular combustion catalyst and limestone. Limestone is continuously added to the bed and the bed material is periodically drained from the vessel. A multicyclone system employing a baghouse to clean the flue gas is used for air-pollution control. The Waste-Tech fluidized bed is able to operate at lower temperatures than conventional fluidized beds and also has reduced supplemental fuel requirements.

Another variation is the rotary reactor which consists of a hollow, three-compartment cylinder that rotates from 10 to 30 revolutions per minute. The rotary reactor acts as a fluidized bed with a hot inert medium, e.g., sand, with solid or semi-solid wastes mechanically lifted on internal radial fins and cascaded through the combustion gases in the combustion zone. The solid cascading action provides effective mass transfer and high rates of heat transfer. This results in destruction efficiencies of greater than 99.99% at 870°C (1598°F).

Pedco states the following advantages of the fast rotary reactor:

1. Reliability and dependability, being firmly grounded in well-established rotary kiln and dryer technology.
2. High thermal efficiency resulting from effective heat recovery.
3. Ease of adjustment and control of incineration conditions.
4. Handles wide variations in feed rate; e.g., high turndown ratio.
5. Easy to cofeed limestone or other adsorbent for gas phase emission control.

Another variant, the Circulating Bed Combustor is mentioned earlier in this chapter.

The Energy & Environmental Research Corporation has developed the Spouted Bed Reactor (SBR) which utilizes the unique attributes of the "spouting" fluidization regime, which can provide heat transfer rates comparable to traditional fluid beds, while providing robust circulation of highly heterogeneous solids, concurrent with very aggressive comminution particle size reduction through abrasion). The primary spouted bed provides a zone for volatilization, pyrolysis, and gasification reactions. The gaseous products can then be applied to highly efficient oxidation/incineration in conventional combustion equipment, used for power production in prime movers or, alternatively, chemical products can be recovered. Thus, gasification provides greater opportunity for product recovery through Advanced Recycling. The SBR Advanced Recycling technology is primarily applicable to waste with significant heat content that are contaminated with toxic organic compounds and heavy metals.

8.13.5 Industrial Boilers and Furnaces

Waste materials disposed of in industrial boilers include (1) waste lubricating, and other oils, (2) waste chemical pulping liquors, (3) petroleum refining on-site waste burning, (4) industry waste solvents, (5) plywood hydrocarbon residues, 6) wood processing sludge, (7) textile liquid wastes, and (8) petrochemical plant liquids.

Those industries that use high operating temperatures are candidates for hazardous waste disposal. These industries include: brick, carbon black, cement, primary copper, primary lead, primary zinc, iron and steel, lime, lightweight aggregate, glass, and clay.

A boiler or process heater can be used for organic vapor destruction. The organic vapor stream is either (1) premixed with a gaseous fuel and fired using the existing burner configuration, or (2) fired separately through a special burner or burners that are retrofitted to the combustion unit. Industrial boilers and process heaters currently are being used to burn vent gases from chemical manufacturing, petroleum refining, and pulp and paper manufacturing process units.

Industrial boilers and process heaters are located at a plant site to provide steam or heat for a manufacturing process. Because plant operation requires these combustion units to be on-line, boilers and process heaters are suitable for controlling only organic vapor streams that do not impair the combustion device performance (e.g., reduce steam output) or reliability (e.g., cause premature boiler tube failure).

While in most cases hazardous wastes will be burned as a fuel at the generator site, the disposal of hazardous wastes by combustion in a permitted commercial kiln or boiler site exists as an alternative to destruction in a hazardous waste incinerator. Because these facilities derive an economic value from burning the wastes, the prices they charge for hazardous waste disposal are well below those of hazardous waste incinerators. The types of wastes which may be burned, however, is somewhat limited.

Both boilers and process heaters are essential to the operation of a plant. As a result, only streams that are certain not to reduce the device's performance or reliability warrant use of a boiler or process heater as a combustion control device. Variations in vent stream flowrate and/or heating value could affect the heat output or flame stability of a boiler or process heater and should be considered when using these combustion devices. Performance or reliability may be affected by the presence of corrosive products in the vent stream. Since these compounds could corrode boiler or process heater materials, vent streams with a relatively high concentration of halogenated or sulfur containing compounds are usually not combusted in boilers or process heaters. When corrosive VOC compounds are combusted, the flue gas temperature must be maintained above the acid dewpoint to prevent acid deposition and subsequent corrosion from occurring.

The introduction of a reactor process vent stream into the furnace of a boiler or heater could alter the heat transfer characteristics of the furnace. Heat transfer characteristics are dependent on the flowrate, heating value, and elemental composition of the process vent stream, and the size and type of heat generating unit being used. Often, there is no significant alteration of the heat transfer, and the organic content of the process vent stream can in some cases lead to a reduction in the amount of fuel required to achieve the desired heat production. In other cases, the change in heat transfer characteristics after introduction of a process vent stream may affect the performance of the heat generating unit, and increase fuel requirements. For some process vent streams there may be potential safety problems associated with ducting reactor process vents to a boiler or process heater. Variation in the flowrate and organic content of the vent stream could, in some cases, lead to explosive mixtures within a boiler furnace. Flame fluttering within the furnace could also result from variations in the process vent stream characteristics. Precautionary measures should be considered in these situations.

When a boiler or process heater is applicable and available, they are excellent control devices since they can provide at least 98% destruction of VOC. In addition, near complete recovery of the vent stream heat content is possible. However, both devices must operate continuously and concurrently with the pollution source unless an alternate control strategy is available in the event that the heat generating capacity of either unit is not required and is shut down.

Industrial Boilers: Some industrial boilers can use limited amounts and types of wastes as supplemental fuels so that the wastes are destroyed while recovering the available heat from the waste. Hazardous waste is used as supplementary fuel to coal, oil or natural gas in fire–tube and water–tube industrial boilers. Hazardous waste fuel (generally limited to liquid waste) can be fed into a boiler with the primary fuel or it can be fed separately into the furnace. If a facility is burning its own wastes as fuel, it can control "fuel quality" to a great extent. If wastes are imported for use as fuel, then it is common to blend incoming wastes to an "optimum" supplemental fuel for that facility's boilers.

Chlorine and sulfur must be limited in Hazardous Waste Fuel (HWF) to minimize corrosion of boiler materials of construction and to avoid increases in HCl and sulfur oxide air emissions. Solid hazardous wastes such as contaminated soil are not applicable for use as HWF in boilers. Industrial boilers are particularly useful for the disposal of hazardous waste generated on site.

Theoretically boilers appear to have the potential for safe waste destruction, and the following conclusions can tentatively be made:

1. Cofiring of hazardous wastes at a small percentage of the base fuel (about 5 to 10%) appears to be a viable method of disposing of most hazardous organic material.
2. Cofiring many wastes, or firing some high Btu content wastes entirely may produce lower levels of criteria pollutant and trace element emissions than either traditional coal or oil combustion.
3. The conditions found in many types of watertube boilers appear to be sufficient to achieve at least 99.99% destruction of most hazardous organic compounds.
4. The conditions found in firetube boilers do not appear to be sufficient to destroy all hazardous organic materials. There is too great a likelihood that cold tube–wall quenching of the waste degradation reactions is possible before complete destruction can occur.
5. Watertube boilers whose furnace exit temperatures are greater than 1100K (1500°F) and whose furnace mean residence times are greater than 1 second appear to be best candidates for the destruction of simple hazardous organic waste streams.
6. Complex organic waste streams are likely to require approximately 200K (360°F) higher temperatures.
7. Insufficient data are available to predict whether hazardous products of combustion could be released from a boiler. Laboratory data should, however, be able to provide some conservative estimates of the likelihood of the release of such materials. The formation of hazardous products of combustion must be addressed by further research.
8. In the absence of other data, the autoignition temperature appears to be a possible predictor of the relative case of a compound's thermal destruction. This needs further confirmation.

Industrial Kilns (Cement, Lime, Aggregate, Clay): Industrial kilns are used to incinerate liquid wastes while recovering heat value. The system consists of rotary kilns constructed of steel casings lined with refractory brick. These kilns are much longer than rotary kiln incinerators and have much longer retention times. Blended feed material (a waste/air mixture) is fed into the hot end of the kiln as a supplement to the primary fuel (coal, gas, or oil). Kiln temperatures are about 3000°F for cement and lime kilns and less than 2000°F for aggregate and clay drying kilns. Organics are destroyed while the ash is assimilated into the kiln product. Waste blending is necessary to obtain desired fuel characteristics to control product quality. The kiln should contain a precipitator or baghouse in order to remove suspended particulates in the flue gases.

Kilns have generally been limited to liquid waste. Heavy metals, ash, chlorine and sulfur content of waste fuel must be controlled to prevent kiln operating and product quality problems. Contaminated soils are not good candidates for treatment in industrial kilns because of concern for product quality. The system should be equipped with air pollution control devices.

Blast Furnaces (Iron and Steel): Blast furnace temperatures may reach up to 3400°F and are generally above 3000°F. High heat content hazardous wastes can be used to supplement the fuel requirements for blast furnaces. A blast furnace produces molten iron from iron ore and other iron bearing feed materials. Iron ore, carbon (coke) and limestone are fed to the top of the furnace, and iron product and slag are removed in different layers from the bottom. Hazardous wastes used as fuels can be injected above the slag layer.

The composition (trace elements) of the waste must be controlled to avoid product quality problems. Waste oils were fired in a blast furnace in HWERL test programs. Some concerns have been expressed that the reducing atmosphere in a blast furnace could result in reduced DREs.

Guidelines: The general guidelines established for identifying suitable high temperature industrial processes (HTIPs) are that they should be capable of achieving levels of performance which are consistent with the requirement established for hazardous waste incinerators; these requirements are:

1. At least 99.99% destruction and removal efficiency (DRE) for each principal organic hazardous constituent (POHC) in the waste feed;
2. At least 99% removal of hydrogen chloride from the exhaust gas if hydrogen chloride stack emissions are greater than 4 lb/h; and
3. Particulate emissions not exceeding 0.08 grains per dry standard cubic foot (dsef), corrected to 7% oxygen in the stack gas.

Solvent hazardous wastes exhibit certain characteristics which limit their application to specific high temperature industrial process technologies. The high temperature industrial processes in which hazardous wastes may be burned as a fuel are, in general, more limited in the types of waste streams they can handle effectively than are hazardous waste incinerators. Generally, they are not equipped with extensive air pollution controls or ash recovery and handling systems. Other technical limitations to burning hazardous wastes in HTIPs include:

1. Possibly more frequent shutdown for boiler cleaning due to fouling of the boiler tubes;
2. High flue gas exit temperatures needed to prevent condensation of acidic components; and
3. Safety problems associated with low boiling point/ignitable solvents.

Problems based on the characteristics of the wastes are:

Physical Form: The physical form of a waste dictates the manner in which it may be input to the system, and the relative ease with which it will burn.

Btu (Heat) Content: Wastes must exhibit high heats of combustion to be considered as a fuel. A common standard used to determine whether a waste may sustain combustion adequately for this purpose is 8,500 Btu/lb, thus a waste with a Btu value below 8,500 probably cannot be used as a fuel without blending.

Chlorine Content: Chlorine presents a limitation to the process both due to the general low combustibility of highly chlorinated substances and due to the composition of by-products of combustion. Most HTIPs are not equipped with air pollution controls which can adequately handle acid gases produced when chlorinated wastes burn, nor can they withstand the corrosive attack of hydrochloric acid on linings and internal surfaces. A chlorine content of 3% is considered a maximum.

Metals/Ash/Organic Salt Content: Wastes which contain high levels of solid, thermally inert materials are not generally good candidates for usage as a fuel, due to their negative environmental impact (particulate emissions), and possible impact on the quality of the products. Solids in the fuel feed also tend to have a deleterious effect upon HTIP equipment, for example, fouling of boiler tubes.

Water Content: Water is a hindrance to effective combustion, and may also affect product quality. High moisture content wastes, therefore, are generally not good candidates for combustion as a fuel.

Flash Point: Safety considerations require that highly ignitable components be removed prior to their storage and introduction into combustion systems.

8.13.6 Infrared Incineration

Infrared Incineration is of particular interest for mobile units, in that electric power may be more easily attainable at some hazardous waste sites than liquid fuels. It also allows a more compact unit.

In a typical system, a woven metal conveyor belt transports the waste under the infrared heating elements. The elements are equally spaced and can be located along the entire length of the unit. At the end of the unit, ash falls into a collection hopper for disposal. The off-gases are sent to a burner for complete combustion. Advantages of this system include controlled residence time by varying belt speed, controlled temperature by varying electrical input to the heating units, and high thermal efficiency. Other advantages include a wide turndown range, versatility, rapid start-up and shut-down, and continuous or intermittent operation. Portable or fixed systems, designed to handle loads of 10 lb/hr up to 100 tons/day, are available from Shirco Infrared Systems in Dallas, Texas.

The Shirco Infrared portable unit was in operation at many hazardous research facilities of the Missouri State Department of Natural Resources and the U.S. EPA. Though the infrared technology successfully demonstrated destruction and decontamination capabilities, Shirco, Inc. went out of business and the company was dissolved in early January 1988. The technology was subsequently acquired by ECOVA, Inc. of Dallas, Texas.

The Shirco Infrared System features 20 major components. Conditioned waste material is fed to the furnace by means of the feed system. Waste passes through the rotary air-lock and onto a metering conveyor. There the material is spread and leveled in the metering section before entering the incinerator feed module. The incinerator conveyor moves the waste material through fiber blanket insulated heating modules where it is brought to high combustion temperature (1500° to 1800°F) by infrared heating elements and gently turned by rotary rakes. Ash (or processed material) passes from the discharge module into the ash discharge system to a receptacle. A blower forces air through a combustion air preheater to extract energy from the exhaust gases as it enters the discharge module. Exhaust gases exit the furnace through the exhaust duct. At this point, the gases may go to the secondary process chamber to incinerate any combustibles remaining at 2300°F, and on to a heat recovery device such as a combustion air preheater or waste heat boiler. Gases are then cooled and cleaned in the scrubber and exhausted by a blower through the exhaust stack.

Waste feed materials to the infrared system include a wide range of hazardous wastes such as dioxins, PCBs, petroleum sludges, contaminated soils, and coke oven and blast furnace sludges. The temperature of the incinerator in the primary chamber ranges between 1500° and 1800°F, while the secondary chamber operates at a higher temperature range of 2200° to 2400°F. Average materials residence time is about 20 minutes, and the gas phase time is approximately 2 seconds.

Commercial scale infrared systems of 100 to 250 tons per day were designed to process hazardous wastes. The infrared system consistently provided destruction removal efficiencies greater than 99.9999% especially in the case of PCBs and dioxins. Average HCl gas emission is less than 180 mg/hr. Carbon monoxide and carbon dioxide emissions are below the EPA limit. However, heavy metals like nickel, chromium, copper, lead and zinc remain in ash, and the scrubber effluent contains mercury and cadmium.

This technology is suitable for soils or sediments with organic contaminants. Liquid organic wastes can be treated after mixing with sand or soil. Optimal waste characteristics are as follows:

1. Particle size, 5 microns to 2 inches
2. Moisture content, up to 50% by weight
3. Density, 30 to 130 lb/ft^3
4. Heating value, up to 10,000 Btu/lb
5. Chlorine content, up to 5% by weight
6. Sulfur content, up to 5% by weight
7. Phosphorus, 0 to 300 parts per million (ppm)
8. pH, 5 to 9
9. Alkali metals, up to 1% by weight

Advantages: Major advantages of the Shirco Infrared Incineration System, according to the developer, are: (1) relatively very high combustion efficiency of the burners (99%); (2) DRE, especially in the case of dioxin contaminants and PCBs are extremely high (>99.9999%); and (3) transportability of the system to any waste location, and easy system assembly and dismantling in less than a week.

Limitations: Limitations of the system include: (a) liquid waste cannot be processed (unless mixed with sand or soil) in the Shirco Infrared System; only sludges and solid wastes are treatable; (b) heavy metals remain in the ash, scrubber effluent and stack; this involves additional disposal costs; and (c) capital and operating costs are high and may not be economical for small job sizes.

8.13.7 Hearth Incineration

Two of the most common modular technologies are (1) Vertical-fixed-hearth incinerators, and (2) Horizontal-fixed-hearth incinerators.

Vertical Fixed-Hearth Incinerator: Size—handles from several hundred pounds per day to 2.5 tons of waste per day; Applications—pathological, industrial and general refuse; Design—two-chamber, two-stage combustion process to ensure high temperature; Benefit—promises smokeless and odorless operation, flexible, easy-to-maintain.

Horizontal Fixed-Hearth Incinerator: Size—small to mid-size applications, between 3,000 and 100,000 pounds of waste per day; Applications—general, bio/hazardous and industrial plant waste; Design—flat or stepped with internal hydraulically operated ash rams; starved air process in the primary chamber, with or without energy recovery; Benefit—Most common incinerator for industrial users. In most cases, the system will provide complete payback in less than three years.

Single fixed hearth incineration systems usually consist of a single steel shell lined with a refractory material. The overall design of the unit is simple and the units are seldom custom designed. Incineration of the waste takes place in stages in the primary and secondary combustion chambers. The first chamber operates in a starved air mode and at temperatures ranging from 600° to 1600°F. Vortex–type burners are used to inject liquid wastes into the primary chamber. Solid wastes are fed onto grates located above the chamber. Gaseous combustion products travel upward into the secondary chamber where more air is added to ensure complete combustion. Solid combustion products (ash) fall through the grate and are discharged from below the unit. To ensure complete incineration, the secondary combustion chamber operates at a temperature range of 1200° to 1800°F.

The **multiple hearth incinerator** (Herreshoff furnace) can be used for the disposal of all forms of combustible industrial waste materials, including sludges, tars, solids, liquids and gases. The incinerator is best suited for hazardous sludge destruction. Solid waste often requires pretreatment such as shredding and sorting. It can treat the same wastes as the rotary kiln provided that pretreatment of solid waste is applied. The principal advantages of multiple hearth incineration include high residence time for sludge and low volatile materials; ability to handle a variety of sludges; ability to evaporate large amounts of water; high fuel efficiency and the utilization of a variety of fuels. The greatest disadvantages of the technology include susceptibility to thermal shock; inability to handle wastes containing ash, which fuses into large rock–like structures, and wastes requiring very high temperatures. Also control of the firing of supplemental fuels is difficult. The multiple hearth incinerator has high maintenance and operating costs. The operating cost may be reduced by utilizing liquid or gaseous combustible wastes as secondary fuel.

Furnaces range from 6 to 25 ft (1.8 to 7.6 m) in diameter and from 12 to 75 ft (3.6 to 23 m) in height. The diameter and number of hearths are dependent on the waste feed, the required processing time, and the type of thermal processing employed. Generally, the uppermost hearth is used as an afterburner. Normal incineration usually requires a minimum of six hearths, while pyrolysis applications require a greater number.

Normally, waste material enters the furnace by dropping through a feed port located in the furnace top. Rabble arms and teeth, attached to a vertically positioned center shaft, rotate counterclockwise to spiral the waste across the face of the hearth to the drop holes. The waste drops from hearth to hearth through alternating drop holes located either along the periphery of the hearth or adjacent to the central shaft. Ultimately, the residual ash falls to the furnace floor. Air and combustion products flow countercurrently to the feed from the bottom to the top of the combustion chamber.

The rabble arms and teeth located on the central shaft all rotate in the same direction; additional agitation of the waste (back rabbling) is accomplished by reversing the angles of the rabble teeth. Waste retention time is controlled by the design of the rabble tooth pattern and for rotational speed of the central shaft.

Liquid and/or gaseous combustible wastes may be injected into the unit through auxiliary burner nozzles. This utilization of liquid and gaseous waste represents an economic advantage because it reduces secondary fuel requirements, thus lowering operating costs.

The furnace is particularly well suited to the combustion of sludge, which is dried on the top hearths and starts to burn toward the center of the furnace. It burns out to ash at the bottom of the furnace where it is discharged.

Multiple hearth incinerators are designed to burn wastes with low heating value such as sewage sludge and other high–moisture–content wastes. Its design includes drying as well as burning sections. Materials with less moisture content, such as coal or solid waste, will start to burn too high in the furnace. There would be insufficient residual time in these cases for effective burnout.

Where sludge contains grease or other volatile components, an afterburner may be required for effective burnout, i.e., elimination of smoke and odor. The maximum off–gas temperature from the incinerator is below 700°C (1291°F). If higher temperatures are required, a separate afterburner must be provided. The afterburner can be an expanded top hearth or it can be provided as a separate piece of equipment.

Excess air of 100 to 125% must be provided to ensure complete burnout of the sludge to ash. Since approximately 20% of the ash can be entrained in the flue gas, extensive gas–cleaning equipment must be provided for its capture. This incinerator will handle sludges in the range of 50 to 85% moisture. It is generally not applicable to the incineration of solid materials.

The multiple hearth furnace is a flexible piece of equipment. There is a limitation in the sludge consistency that it can process (generally from 15 to 50% solids content) but the nature of the sludge is not necessarily limiting as with, for instance, the fluid bed furnace. Another feature of the multiple hearth furnace is that it has a relatively constant fuel use curve, i.e., the use of fuel is directly proportional to the amount of sludge burned, its moisture and combustible content, and its heating value. If the sludge feed is doubled, the fuel required to incinerate the sludge is doubled. This is not necessarily true with other types of incinerator systems, such as the fluid bed incinerator.

There are a number of disadvantages in the use of this equipment. Generally, the multiple hearth furnace cannot accommodate a temperature in excess of 1000°C (1830°F) without damage. If higher temperatures are required, an afterburner must be provided, which represents higher capital costs and higher operating (fuel) costs.

It is virtually impossible to maintain heat in a multiple hearth furnace without firing supplemental fuel. This furnace, as noted above, has many areas of leakage and, therefore, heat cannot be effectively maintained within the units as in, for example, a fluid bed furnace.

Fuel or gas can be used as supplemental fuel. Generally solid fuels, such as coal or wood chips, should not be placed on a hearth and used as supplemental fuel. They are relatively dry and will start burning on the top hearth, encouraging premature release of volatiles from the waste stream and inadequate burnout can result.

Sludge or other wastes deposited on the hearth of a multiple hearth furnace should have a solids content of 15 to 40% for proper movement and rabbling through the furnace. The temperature above at least two hearths should be maintained at approximately 870°C (1600°F) at all times when burning sludge cake. The off–gas temperature can range from 425° to 760°C (800° to 1400°F).

Generally, grease (scum) should not be added to sludge feed. If grease is to be incinerated in the multiple hearth furnace it should be added at a lower hearth (a burning hearth) through a separate nozzle(s). Grease will volatilize easily and introducing it too high in the furnace where the temperatures may be below 760°C (1400°F) will not provide effective burnout.

Advantages:

1. The retention or residence time in multiple hearth incinerators is usually higher for low volatile material than in other incinerator configuration.
2. Large quantities of water can be evaporated.
3. A wide variety of wastes with different chemical and physical properties can be handled.
4. Multiple hearth incinerators are able to utilize many fuels including natural gas, reformer gas, propane, butane, oil, coal dust, waste oils, and solvents.
5. Because of its multizone configuration, fuel efficiency is high and typically improves with the number of hearths used.
6. Fuel burners can be added to any of the hearths to maintain a desired temperature profile.
7. Multiple hearth incinerators are capable of a turndown ratio of 35%.
8. High fuel efficiency is allowed by the multizone configuration.

Disadvantages:

1. Due to the longer residence times of the waste materials, temperature response throughout the incinerator when the burners are adjusted is usually very slow.
2. It is difficult to control the firing of supplementary fuels as a result of this slow response.
3. Maintenance costs are high because of the moving parts (rabble arms, main shaft, etc.) subjected to combustion conditions.
4. Multiple hearth incinerators are susceptible to thermal shock resulting from frequent feed interruptions and excessive amounts of water in the feed. These conditions can lead to early refractory and hearth failures.
5. If used to dispose of hazardous wastes, a secondary combustion chamber probably will be necessary and different operating temperatures might be necessary.
6. Not well suited for wastes containing fusible ash, wastes which require extremely high temperature for destruction, or irregular bulky solids.

8.13.8 Liquid Injection Incineration:

Liquid Injection (LI) incinerators are one of the most widely used hazardous waste incineration systems in the United States. LI systems may be used to incinerate virtually any liquid hazardous waste, including most solvent hazardous wastes, due to their very basic design and high temperature and residence time capabilities. Liquid injection incinerators represent the most effective system available for most liquid hazardous waste solvents, from both a technical, i.e., destruction efficiency, and economic perspective.

Liquid injection incinerator systems typically employ a basic, fixed hearth combustion chamber. Pretreatment systems to blend wastes and fuels, remove solids and free water, and to lower viscosity through heating, are often used in conjunction with liquid injection incinerators. Ash recovery systems may not be required, at least on a continuous basis, because many liquid hazardous wastes fired in an LI system contain low volumes of ash or suspended solids.

The liquid waste feed system is the key element of the LI process. Liquid injection incinerators operate as "suspension burners," whose combustion efficiency (and hence destruction efficiency for constituents of hazardous wastes) is dependent upon the extent to which the feed mechanism can disperse the liquid waste within the combustion chamber and provide sufficient area for contacting waste with combustion air. There are two atomizer designs commonly employed in LI systems, denoted as fluid systems and mechanical systems.

Once liquid wastes enter into the liquid injection incinerator and are ignited at the burner, the turbulence imparted to the waste and good mixing with combustion air lead to efficient combustion. Combustion temperature capabilities of the systems can be very high, reaching over 3000°F in many cases. Residence times are generally within a 1 to 2 second range, depending upon liquid heat content.

Applicability of hazardous wastes to liquid injection incinerators is generally limited by the extent to which they may be atomized, and the physical effect they may have on the incinerator equipment (mostly notably, the atomizer). The primary restrictive waste characteristics of interest are the liquid viscosity, solids content, and corrosivity. Wastes with low heat value may also be restricted from burning in a liquid injection incinerator. A typical limiting value (below which waste must be mixed with a fuel, or a high heat value material), given by one incinerator operator, is 5,000 Btu/lb.

Liquid viscosity is regarded as the primary limiting waste characteristic for liquid injection systems since viscosity determines whether or not the liquid is pumpable and atomizable. Pretreatment may be needed to reduce the viscosity of wastes to a level where high combustion efficiencies may be achieved. The two most common viscosity reducing pretreatment operations are heating and dilution. In some cases, high energy mixing is done to produce a one– or two–phase emulsion of liquid waste in the carrier media. Energy for preheating is often supplied by heat recovery systems.

Suspended solids are another restrictive waste characteristic for liquid injection incinerators. Undissolved solids can impact negatively through abrasion or plugging. The best available technology to reduce the solids content is filtration or sedimentation. Filtered solids may be collected, washed of any retained liquids by an appropriate solvent, and disposed of separately. Wash solutions can be incinerated.

Highly corrosive wastes provide a potential limitation to effective performance of liquid injection systems. However, no pH limits for liquid injection incineration were described in the available literature other than those dealing with chloride content, and no detail was provided on chemical pretreatment of corrosive wastes. It can be assumed that such techniques are viable for LI systems, however, depending on the characteristics of wastes handled and process design.

In some cases, the applicability of an LI incinerator may be extended by the use of multiple injection systems. In this way, an injector may be fitted to more specific ranges in waste characteristics allowing a broader range of overall usage without requiring pretreatment. Certain atomization device designs are better suited to more viscous or higher suspended solids containing wastes than others. In addition, the use of multiple injection points may allow for concineration of incompatible wastes.

The Cyclin Cyclone Incinerator is an advanced liquid injection incinerator which provides high combustion–intensity and combustion–efficiency with low excess air. According to the developers, the Cyclin Cyclone Incinerator is capable of incinerating wastes of low (3,000 Btu/lb) and medium (3,000 to 6,000 Btu/lb) heating value.

The Cyclin Cyclone Incinerator features a compact incineration chamber with an ash/slag receiver and is capable of operating either in dry or molten slagging modes. The unit typically consists of a waste handling subsystem and is equipped with a heat recovery boiler, an air heater, and a baghouse.

The form of feed materials to the incinerator includes gaseous wastes, pumpable liquid wastes or slag. The incinerator can, however, be used to treat solid wastes if they are first brought to a molten state by pre–treatment process. The incinerator operates at temperatures between 1500° to 3000°F with gas phase residence time ranging from 0.1 to 5 seconds depending on the type of waste incinerated. The inside diameter of the incinerator is a variable depending on the flow rate and quantity of waste processed.

Advantages: The cyclonic incinerator, as described by the developers, allows better mixing and temperature uniformity, higher destruction efficiency, a wider range of operating parameters, greater flexibility to variation in waste properties, and more efficient heat recovery at reduced capital and operating costs. Advantages in the design parameters include centrifugal separation of ash and slag, which is conducive to clean exhaust gases, longer refractory life due to wall cooling and high combustion–intensity and combustion–efficiency with low excess air.

Limitations: Lower concentration of solids results in significantly higher natural gas consumption in the cyclonic liquid incinerator.

8.13.9 Mass Burn Combustion

Most large municipal waste incinerators in the United States are mass burn facilities. Refuse is burned in the same form as it is delivered with the exception that some large metal items are removed from the waste stream. This technology has been used since the 1970s and has experienced the greatest technical and financial operating success. Typical unit size is in the range of 400 to 1,000 tons per day (TPD) with some facilities as large as 3,000 TPD.

Typically, waste is loaded into a feed chute using an overhead crane. Rams or moving gate sections are then used to move the waste through the combustor and promote complete burning by agitating the fuel bed. The number of gate designs is large, but they all serve the same basic purpose. As the refuse enters the combustor, the first section of the gate conditions the refuse by driving off moisture and raising its heating value. The next section of the gate is the primary combustion zone, and this is followed by a section for a clinker burn–out. Underfire air is provided to support combustion in the bed, and overfire air is added to mix and combust volatile gases evolved from the bed. Variations in waste characteristics is handled by controlling the feed rate, grate speeds, and the amount and distribution of the combustion air.

Two variations in the traditional mass burn unit are the waterwall and refractory wall designs. In refractory wall units, combustion zone temperatures are somewhat hotter, and gases exiting the combustion zone are cooled below 1800°F with excess air levels of 100 to 200%. Heat recovery is generally not practiced, so additional cooling prior to entering the control device is usually accomplished with water sprays.

In the newer waterwall design, combustion gas temperatures are moderated by water tubes located in the furnace walls, and additional gas cooling is accomplished in a boiler located at the exit of the furnace. Heat recovery from the combustion process is used in the production of steam and/or electric power. Because of heat recovery, excess air levels are reduced, typically averaging about 80%.

Another mass burn combustor is the rotary furnace, which may be either refractory or waterwall. The Volund system uses a refractory lined rotary furnace in conjunction with upstream drying and ignition grates. The O'Connor system uses a waterwall rotary combustor followed by a traditional boiler. Of particular interest here is the unique design of the rotary cylinder, which consists of alternating perforated steel plates and watertubes. As the cylinder slowly rotates (10 to 20 rph), the refuse rides up with the wall and then breaks over and tumbles toward the bottom. Preheated combustion air enters through the perforated plates from six separately controlled zones arranged in side-by-side rows of three zones each along the length of the combustor. The row located along the rising wall provides underfire air, while the row near the bottom of the cylinder provides the overfire air. The ash falls out the lower end of the cylinder into a removal pit.

Facilities that utilize the mass burn technology can be classified according to the nature of construction, i.e., field erected or modular shop fabricated. Field-erected systems are usually medium- to large-scale (200 to 3,000 TPD) waterwall or refractory-lined furnaces that combust MSW under excess air conditions while modular systems are usually small-scale (up to 300 TPD) systems comprised of predesigned modules that are manufactured at a factory and assembled onsite. Modular systems also feature separate primary and secondary combustion chambers and separate heat recovery boilers.

The distinction between field-erected and shop-fabricated systems has become less clear in recent years as shop-fabricated installations have become larger. Many shop-fabricated systems have adopted features, such as moving grates and pit and crane systems, that were once limited to field-erected systems. In addition, large shop-fabricated system installations may require more on-site assembly and more substantial foundations and buildings due to the large size of their component modules. At the same time, modular construction techniques are beginning to be used to reduce the costs of smaller field-erected systems. For waterwall systems, the modularization of components can reduce the amount of field construction and thus reduce or slow the escalation of facility costs. This would enable waterwall systems to be more cost competitive with shop-fabricated incinerator systems.

Key advantages of mass burn facilities relate to their well established and proven technology, demonstrated long-term reliability, good thermal efficiency and minimal refuse processing requirements. Disadvantages relate to the long lead times required to design and build plants and their significant capital construction cost.

8.13.10 Molten Salt and Molten Metal Techniques

Molten salt incinerators involve the combustion of waste materials in a bed of molten salt. Using the molten salt incineration process, organic wastes may be burned while, at the same time scrubbing in situ any objectionable by-products of that burning, and thus preventing their emission in the effluent gas stream. Molten salt incinerators were developed by Rockwell International, specifically to burn hazardous organic wastes and, as designed, are applicable to both liquid and solid waste streams. However, wastes with high ash content, or a high percentage of water or noncombustible material, are not good candidates for molten salt destruction.

The molten salt incinerator can be used for destruction of hazardous liquids and solids. In this method wastes undergo catalytic destruction when they contact hot molten salt maintained at a temperature between 1382° and 1832°F. Hot gases rise through the molten salt bath, pass through a secondary reaction zone, and through an off-gas cleanup system before discharging to the atmosphere. Supplemental fuel may be required when wastes are not sufficiently combustible to maintain temperatures.

Liquid, free-flowing powders, sludges, and shredded solid wastes can be fed directly into the incinerator. The technology has been demonstrated to be highly effective for chlorinated hydrocarbons including PCBs, chlorinated solvents, and malathion. However, the process appears to be sensitive to materials containing high ash content or high chlorine content which must ultimately be removed in the purge system.

A variety of salts can be used, but the most recent studies have used sodium carbonate (Na_2CO_3) and potassium carbonate (K_2CO_3) in the 1450° to 2200°F (790° to 1200°Cd) temperature range.

The waste is fed to the bottom of a vessel containing the liquid salt along with air or oxygen-enriched air. The molten salt is maintained at temperatures of 800° to 1000°C. The high rate of heat transfer to the waste causes rapid destruction. Hydrocarbons are oxidized to carbon dioxide and water. Constituents of the feed such as phosphorous, sulfur, arsenic, and the halogens react with the salt, i.e., sodium carbonate, to form inorganic salts, which are retained in the melt. The operating temperatures are low enough to prevent NO_x emissions.

Eventually, the build-up of inorganic salts must be removed from the molten bed to maintain its ability to absorb acidic gases. Additionally, ash introduced by the waste must be removed to maintain the fluidity of the bed. Ash concentrations in the melt must be below 20% to preserve fluidity.

Melt removal can be performed continuously or in a batch mode. Continuous removal is generally used if the ash feed rates are high. The melt can be quenched in water and the ash can be separated by filtration while the salt remains in solution. The salt can then be recovered and recycled. Salt losses, necessary recycle rates, and recycling process design are strongly dependent on the waste feed characteristics.

Molten salt destruction (MSD) systems are limited in their applicability to various hazardous wastes. Although the system is capable of handling hazardous wastes in both the liquid and solid state, MSD is in practice limited to the incineration of hazardous organic wastes which have a relatively low percentage of solids or inorganics. Slurried wastes and most "dry" solid wastes (e.g., contaminated soils) are not good candidates for incineration by MSD. When ash accumulates in the bed, it tends to form a waste matrix, which eventually affects bed fluidity, the overall transfer of heat and will eventually limit waste by-product neutralization within the molten mass. Thus, 20% was determined to be the limit to which the system could effectively operate.

Wastes with high water content may pose a problem to the effectiveness of the molten salt destruction process. As moisture content increases, the waste will require more fuel and combustion air, to the point where the reactor volume is limited. Thus, many wastes must be dewatered by pretreatment to ensure that they are effectively destroyed in the MSD.

Molten Salt oxidation (MSO) has a high treatment potential for radioactive and hazardous forms of high–heating liquids (organic solvents, waste oils), low–heating value liquids (high–halogen content organic liquids), other wastes (pesticides, herbicides, PCBs, chemical warfare agents, explosives, propellants, infectious wastes), and extraction gases (volatile organic compounds and radionuclides, acids. By virtue of the latter, MSO could replace conventional wet scrubbers as a superior dry–scrubber system for use with incinerators. The typical residence time is two seconds for the treatment of wastes by the MSO process.

Aqueous sludges containing heavy metals are converted to oxides and retained in the melt. Organics in addition to combustible solids are destroyed but MSO is not suitable for treatment of inert solids, such as soils. The process also successfully destroys carbon in coal gasification demonstrations.

Advantages:

1. Achievement of high destruction efficiencies for many wastes, including highly toxic and highly halogenated wastes;
2. Low NO_x and heavy metal emissions;
3. Retention of halogens and metals in a manageable salt matrix;
4. Compact size, and the process has few moving parts, and acts as its own, highly efficient scrubber for acid combustion gases;
5. Especially well–suited to wastes whose combustion results in liberation of acids;
6. Improved reliability due to simple design;
7. Increased waste throughput possible.

Limitations:

1. Generally restricted to certain types of organic hazardous wastes;
2. Sensitive to high (20%) ash content in wastes;
3. Molten salt is corrosive to all but specific engineering alloys. Material and construction costs will therefore be high, and management of spent salt beds will be difficult.

Molten Metal Technology Inc. of Waltham, Massachusetts has developed a system for injecting toxic materials into a pool of **molten iron** at 3000°F. Such treatment causes almost all compounds to break down into their constituent elements because of the intense heat and the catalytic effect of the metal. Hydrocarbons are rendered into hydrogen, which bubbles off the top of the bath, and carbon, which similarly boils off as carbon monoxide or carbon dioxide gas if oxygen is provided. Using the elements produced and added reactants, other products can be produced such as synthesis gas. Recoverable inorganics, which float to the top of the bath, include silica and alumina, as well as calcium chloride from the chlorine content. Nonferrous metals such as nickel, chromium, and vanadium are reduced to the elements and dissolve in the bath (also see Section 8.13.1).

8.13.11 Oxygen Incineration

Oxygen incineration uses pure oxygen instead of air to achieve a higher destruction rate. In the past, use of oxygen injection was plagued by problems associated with improper mixing of fuel and oxygen, and excessive temperatures. Union Carbide designed the Linde "A" Burner System, an oxygen injection burner that solves both of these problems. Advantages of this technique include greater throughput for retrofitted incinerators; more effective and less costly gas cleaning (oxygen produces less flue gas, which means that existing scrubber systems would be oversized); lower fuel consumption due to less loss of heat in the flue gas; and higher levels of destruction efficiency. These advantages are offset by the costs to purchase, store, and handle liquid oxygen on-site. Oxygen incineration is claimed to be cost competitive with conventional air incineration.

8.13.12 Plasma Systems

Plasmas have been referred to as the fourth state of matter since they do not always behave as a solid, liquid or gas. A plasma may be defined as a conductive gas flow consisting of charged and neutral particles, having an overall charge of approximately zero, and all exhibiting collective behavior. The plasma, when applied to waste disposal, can best be understood by thinking of it as an energy conversion and energy transfer device. The electrical energy input is transformed into a plasma with a very high temperature at the centerline of the reactor. As the activated components of the plasma decay, their energy is transferred to waste materials exposed to the plasma. The wastes are then broken into atoms, ionized, pyrolyzed and finally destroyed as they interact with the decaying plasma species. The heart of this technology is that the breakdown of the wastes into atoms occurs virtually instantaneously and no large molecular intermediary compounds are produced during the kinetic recombination.

Since the process is pyrolytic, the scale of the equipment is small, especially considering the high throughput rates. This characteristic makes it potentially attractive for use as a mobile unit. Gaseous emissions (mostly H_2, CO), acid gases in the scrubber and ash components in scrubber water are the residuals. The system's advantages are that it can destroy refractory compounds and typically the process has a very short on/off cycle.

Direct heating involves the direct injection of liquid waste into the plasma plume. Indirect heating involves using the plasma to create a bath of molten solid material which is used to heat and decontaminate solid hazardous waste.

This section discusses plasma systems that handle liquids or finely divided, fluidizable sludges. Plasma systems that handle sludges and result in a bottoms product of a vitrified unleachable material are discussed in Chapter 4. Non-thermal plasmas are discussed in Chapter 7.

Plasma arc incineration utilizes a plasma generator to pyrolize hazardous waste. By passing an intense electrical current through air at low pressure, thermal plasma is created with temperatures ranging from 10,000° to 20,000°C. Upon being injected into the plasma, the waste molecules quickly disintegrate into individual atoms. After leaving the unit and cooling, these atoms recombine to form hydrogen, carbon monoxide, nitrogen, hydrogen chloride, and particulate carbon. The exhaust is scrubbed with caustic to remove hydrogen chloride and particulate carbon, and is flared to convert the hydrogen and carbon monoxide to water and carbon dioxide.

In a typical system, the plasma device is horizontally mounted in a refractory-lined pyrolysis chamber. Liquid wastes are injected through the colinear electrodes of the plasma device where the waste molecules dissociate into their atomic elements. These elements then enter the pyrolysis chamber which serves as a mixing zone where the atoms recombine to form hydrogen, carbon monoxide, hydrogen chloride and particulate carbon. The approximate residence times in the atomization zone and the recombination zone are 500 microseconds and 1 second, respectively. The temperature in the recombination zone is normally maintained at 900° to 1200°C.

After the pyrolysis chamber, the product gases are scrubbed with water and caustic soda to remove hydrochloric acid and particulate matter. The remaining gases, a high percentage of which are combustible, are drawn by an induction fan to the flare stack where they are electrically ignited. In the event of a power failure, the product gases are vectored through an activated carbon filter to remove any undestroyed toxic material.

Two residual streams are generated by this process. These are the exhaust gases released up the stack as a flare, and the scrubber water stream. Since the product gas (after scrubbing) is mainly hydrogen, carbon monoxide, and nitrogen, it burns with a clean flame after being ignited. Analysis of the flare exhaust gases, indicates virtually complete destruction of toxic constituents.

The scrubber water stream is composed mainly of salt water from neutralization of HCl and particulates, primarily carbon. Analyses of the scrubber water for the waste constituent of concern [e.g., carbon tetrachloride (CCl_4) and PCB in the feed material] have shown that the constituents were present at low ppb concentrations. The quality of scrubber water generated would depend on the water feed rate and corresponding product gas and scrubber waste flowrates.

Los Alamos has improved upon the basic rf plasma tube design using the concept of a transformer. The unique feature of the Los Alamos tube is a segmented, cooled, internal radiation shield. Several heavy-walled, water-cooled copper fingers are inserted inside the quartz mantle. These fingers couple power from the surrounding rf coil to the contained plasma, while their chevron cross section is such that they overlap each other and prevent ultraviolet radiation from impinging upon the surrounding quartz. This allows the Los Alamos torch to be operated at temperatures as high as 15,000 K. The shield also eliminates the arcing between the quartz mantle and the rf coil that typically occurs due to the ionization of the surrounding air from ultraviolet radiation. The Los Alamos shielded plasma torch routinely achieves temperatures exceeding 10,000 K and electron densities of $10^{16}/cm^3$ when operated continuously at one atm of argon. These highly energetic conditions are sufficient to dissociate most chemical compounds into their constituent atoms. Based upon these characteristics, Los Alamos is currently investigating the application of the shielded plasma torch technology to the destruction of organic and mixed hazardous wastes, as well as the direct production of actinide metals from the halides and oxides, without the cogeneration of contaminated wastes.

Argonne National Laboratories has been investigating the use of a microwave discharge plasma reactor.

Processes under development include:

1. Pyroplasma™ Process—Pyrolysis System, Inc. and Westinghouse.
2. PCB destruction system—ARC Technologies Company.
3. Plasma incinerator—Applied Energetics, Inc.
4. Oil/water interface—Al-Chem Fuels.
5. Microwave Plasma Generator—Cottrell and Efthimion.

6. Plasma Centrifugal Reactor—Retech, Inc.

Advantages:

1. Since radiative heat transfer proceeds as a function of the fourth power of temperature, a plasma system has very intense radiative power and therefore is capable of transferring its heat much faster than a conventional flame.
2. Organic chlorides are known to dehydrogenate when excited by the ultraviolet radiation which is abundant from thermal plasmas.
3. Because the plasma arc for waste destruction is a pyrolytic process, it virtually does not need oxygen at all. Compared to conventional incinerators which normally require about 150% excess air to ensure proper combustion, the plasma arc will save the energy required to heat the excess air to the combustion temperatures and will thereby produce significantly less oxygenated by-products that would otherwise need to be treated in downstream equipment.
4. The process has a very short on/off cycle.
5. Because of its compactness, a plasma arc system has potential for use in a mobile trailer for movement of the system from site to site.

Disadvantages:

1. Because the temperatures are so high (about 18,000°F at the arc's centerline), the durability of the arc and the refractory materials could be a potential problem.
2. Because the arc is very sensitive to many factors such as sudden drops in voltage, the operation of the system requires highly-trained professionals.

8.13.13 Pulse Combustion

The pulse combustor excites pulsations in the kiln and increases the completeness of combustion by promoting better mixing within the system. The addition of turbulence due to high amplitude acoustic pulsations has a strong tendency to reduce the amount of soot and/or semivolatile and non-volatile hydrocarbons. Since it is an acoustic process, it is more fully described in Chapter 7.

8.13.14 Pyrolysis

Pyrolysis is formally defined as chemical decomposition induced in organic materials by heat in the absence of oxygen. In practice, it is not possible to achieve a completely oxygen-free atmosphere; actual pyrolytic systems are operated with less than stoichiometric quantities of oxygen. Because some oxygen will be present in any pyrolytic system, nominal oxidation will occur. If volatile or semivolatile materials are present in the waste, thermal desorption will also occur.

Pyrolysis is a thermal process that transforms hazardous organic materials into gaseous components and a solid residue (coke) containing fixed carbon and ash. Upon cooling, the gaseous components condense, leaving an oil/tar residue. Pyrolysis typically occurs at operating temperatures above 800°F. Pyrolysis is applicable to a wide range of organic wastes and is generally not used in treating wastes consisting primarily of inorganics and metals. Pyrolysis of scrapped tires will be an important process.

Pyrolysis systems may be applicable to a number of organic materials that "crack" or undergo a chemical decomposition in the presence of heat. Pyrolysis has shown promise in treating organic contaminants in soils and oily sludges. Chemical contaminants for which treatment data exist include polychlorinated biphenyls (PCBs), dioxins, polycyclic aromatic hydrocarbons, and many other organics.

Pyrolysis is not effective in either destroying or physically separating inorganics from the contaminated medium. Volatile metals may be desorbed as a result of the higher temperatures associated with the process but are similarly not destroyed.

Although wastes with a wide range of chemical characteristics may be treated in a pyrolytic incinerator, certain wastes are clearly better candidates than others. Pyrolysis systems work best for wastes which fall into the following categories:

1. Too viscous to atomize in liquid incinerators, yet too fluid for spreader–stoker incinerators.
2. Low melting point materials that foul heat exchangers, spall refractories, and complicate residue discharge.
3. High residue materials (ash), with easily entrained solids, that would generally require substantial stack gas cleanup.
4. Material containing priority pollutants with excessive vapor pressure at incineration temperatures.
5. Any material, drummed or loose bulk, where controlled thermal treatment is desired to make clean gases for heat recovery or for discharge to the atmosphere.

The temperature in the pyrolysis chamber is typically between 800° and 2100°F, and the quantity of the oxygen present is not sufficient for the complete oxidation of all contaminants. In pyrolysis, organic materials are transformed into coke and gaseous components. Gas treatment options include: (1) condensation plus gas cleaning, and (2) incineration plus gas cleaning.

Pyrolysis forms new compounds whose presence could impact the design of the off–gas management system. For example, compounds such as hydrogen halides and sulfur–containing compounds may be formed. These must be accounted for within the design of the Air Pollution Control (APC) system.

The rate at which pyrolysis occurs increases with temperature. At low temperatures and in the presence of oxygen, the rates are typically negligible. In addition, the final percent weight loss for the treated material is directly proportional to the operating temperature. Similarly, the hydrogen fraction in the treated material is inversely proportional to the temperature.

The primary cleanup mechanisms in pyrolytic systems are destruction and removal. Destruction occurs when organics are broken down into lower molecular weight compounds. Removal occurs when pollutants are desorbed from the contaminated material and leave the pyrolysis portion of the system without being destroyed.

Pyrolysis systems typically generate solid, liquid, and gaseous products. Solid products include the treated (and dried) medium and the carbon residue (coke) formed from hydrocarbon decomposition. Various gases are produced during pyrolysis, and certain low–boiling compounds may volatilize rather than decompose. This is not typically a problem. Gases may be condensed, treated, incinerated in an afterburner, flared, or a combination of the above. Depending on the specific components, organic condensate may be reusable. Other liquid streams will include process water used throughout the system.

The system may involve the use of two combustive chambers. In the primary chamber the wastes are heated, separating the volatile components (combustible gases, water vapor, etc.) from the nonvolatile char and ash (metals and salts). In the secondary chamber (afterburner or fume incinerator) volatile components are burned under the proper air, temperature, time and turbulence to destroy any remaining hazardous components. Temperature in the pyrolysis section is controlled by the addition of auxiliary fuel. There are two ways to heat the material; the first is by direct heating where the material comes in contact with hot combustion gases from a burner or incinerator. The resulting off–gas is a combination of the combustion gases and the volatiles from the waste. The second method is by indirect heating by an electric resistance heating element or an external burner with its exhaust gases directly vented to the atmosphere. This approach allows product recovery, rather than incineration, from the gaseous stream leaving the primary chamber without contamination or dilution by the burner flue gases. Indirect heating is more complex and expensive than direct heating.

Advantages of pyrolysis include: (1) reduced amount of waste produced, (2) since wastes are broken down in the absence of oxygen, less hazardous by–products are produced, and (3) metals and energy may be recovered.

The primary technical factors affecting pyrolytic performance are the temperature, residence time, and heat transfer rate to the material. There are also several practical limitations which should be considered.

As the medium is heated and passes through a pyrolytic system, energy is consumed in heating moisture contained in the contaminated medium. A very high moisture content would result in lower throughput. High moisture content, therefore, causes increased treatment costs. For some wastes, dewatering prior to pyrolysis may be desirable.

The treated medium will typically contain less than 1% moisture. Dust can easily form in the transfer of the treated medium from the treatment unit, but this problem can be mitigated by water sprays.

A very high pH (greater than 11) or very low pH (less than 5) may corrode the system components. The pyrolysis of halogenated organics will yield hydrogen halides; the pyrolysis of sulfur–containing organics will yield various sulfur compounds including hydrogen sulfide (H_2S). Because hydrogen halides and hydrogen sulfide are corrosive chemicals, corrosion control measures should be taken for any pyrolytic system which will be processing wastes with high concentrations of halogenated or sulfur–containing organics.

A number of problems have been found in the application of pyrolysis systems to the destruction of solid waste. The temperatures developed in the reactor are sufficiently high to keep the ash and other residue components molten. It has been difficult in many of these systems to control the solidification of the molten materials as it leaves the reactor. Excessive slagging has occurred at the residue outlet and this slag, which is a result of uneven cooling of the molten ash, clogs the reactor, preventing ash discharge. The reactor must be shut down until the slag formation has been removed.

Another problem that has been found with pyrolysis systems is the generation of small particulate matter and organics in the off–gas. The nature of the pyrolysis reaction is that it does not burn the waste; it thermally decomposes the waste, generating a fairly dirty exhaust stream rich in organics and containing significant amounts of small carbonaceous solid particles. An afterburner must be employed to burn out organics in the off–gas.

Most pyrolysis systems include an emergency exhaust stack that is used to bypass critical equipment when operating problems occur such as a loss of cooling water in downstream equipment or the failure of the induced draft fan. An emergency exhaust stack must never be placed upstream of the afterburner. This stack discharges to the atmosphere and the hot gases, which are rich in organics, will burn upon contact with a source of air (or oxygen). Placing an exhaust stack between the pyrolysis reactor and the afterburner will result in, at best, uncontrolled burning out the stack, and at worst, the discharge of organics which may pose a threat to human health. There is also a high probability that such a discharge will result in an explosion as the hot organic–laden gases react with the surrounding air.

A number of pyrolysis products have been developed by various manufacturers.

8.13.15 RDF—Fired Combustion

Municipal waste that has been pre–processed, regardless of the degree, is termed refuse derived fuel or RDF. The degree of pre–processing can vary from bulky–item removal and shredding to removal of materials, glass and other inorganic materials. Additionally, the combustible fraction may be powdered or compressed into pellets or briquettes. Finally, the processed waste may be burned alone or in combination with coal.

The waste is injected into the furnace through an air–driven distributor. Partial burning takes place while the waste is in suspension, with larger material falling onto the grate and burning out on the fuel bed. Both underfire and overfire air are provided, typically at lower excess rates than for mass burn systems because of better waste uniformity. The underfire air is either uniform across the grate or disturbed on the basis of bed depth, depending on the design of the fuel distribution and grate systems. Heat release rates are comparable to mass burn combustors, but temperatures are often high because of smaller furnace volume and other factors.

Semi–suspension burning is suitable for the destruction of RDF as well as other materials requiring agitation, or turbulence, for effective combustion, e.g. powdered carbon, sawdust, etc. It also permits the reclamation of recyclable materials.

Although semi–suspension incineration systems are commercially available, they have not gained the widespread acceptance that mass–burning systems currently enjoy, largely due to the additional complexity and cost associated with waste preparation. They do, however, allow the reclamation of salvageable materials from the waste stream.

The basic guidelines for minimizing emissions of trace organics that apply to mass burn systems also apply to RDF systems. These guidelines require that:

1. Stable stoichiometries be maintained through proper distribution of fuel and combustion air;
2. Good mixing be achieved at a sufficiently high temperature to adequately destroy trace organic species; and
3. The design and operational performance of the system be verified through monitoring or performance tests.

These design, operation/control, and verification practices are expected to minimize trace organic emissions. Research has indicated that CDD/CDF formation may also occur at lower temperatures in downstream portions of the system through catalytic reactions. Therefore, another guideline should be included which addresses this phenomenon. This guideline is to minimize the retention time of flue gases in the temperature window where CDD/CDF formation occurs.

Good combustion guidelines for minimizing trace organic emissions from RDF–fired MWC's are as follows:

Design:

1. Temperature at fully mixed height
2. Underfire air control
3. Overfire air capacity
4. Overfire air injector design
5. Furnace exit gas temperature

Operation/Control:

1. Excess air
2. Turndown restrictions
3. Start–up procedures
4. Use of auxiliary fuel

Verification:

1. Oxygen in flue gas
2. CO in flue gas
3. Furnace temperature at fully mixed height
4. Temperature at APCD inlet
5. Adequate air distribution

RDF can be produced to a range of specifications, classified on the basis of particle size, density, and production process. The American Society for Testing and Materials (ASTM) through its E–38.01 Energy Subcommittee on Resource Recovery (currently part of D34–13) established classifications defining the different types of RDF. These characterizations are provided below:

Type of RDF	Description
RDF–1	Municipal solid waste used as a fuel in as–discarded form
RDF–2	MSW processed to coarse particle size, with or without ferrous metal separation, such that 95% by weight passes through a 6–inch square mesh screen
RDF–3	Shredded fuel derived from MSW and processed for the re–moval of metal, glass, and other entrained inorganics. The particle size of this material is such that 95% by weight passes through a 2–inch square mesh screen. Also called "fluff" RDF.
RDF–4	The combustible fraction processed into powdered form, 95% by weight passing through a 10–mesh (0.035–inch square) screen.
RDF–5	The combustible fraction densified into the form of pellets, slugs, cubettes, briquettes, or some similar form.
RDF–6	The combustible fraction processed into a liquid fuel (no standards have been developed).

RDF–7	The combustible fraction processed into a gaseous fuel (no standards have been developed).

8.13.16 Retort or Batch Incineration

Retort incinerators are the simplest type of incinerator. The operator preheats the waste burning chamber and places the waste inside. Preheating is not always conducted, but is recommended. Retort incinerators which are filled full of waste (i.e., "stuff–and–burn") cannot be preheated; but they may be designed to allow preheating of the after burner chamber. Fuel and air are introduced through burners. The incinerator operates until all the waste is burned; after a cool–down period, it is opened and the ash is quenched and removed. These incinerators could also be equipped with heat recovery, but in the past have rarely been controlled by add–on air pollution control equipment.

8.13.17 Rotary Kiln Incineration

Rotary kiln incinerators have gained widespread commercial acceptance in the hazardous waste management industry, despite being one of the more costly available alternatives. This acceptance is due primarily to the versatility of rotary kilns. There are many facilities, for example, which currently employ a rotary kiln as their sole means of disposing of both hazardous and nonhazardous wastes. In general, it is believed that as more emphasis is placed on utilizing available alternatives to land disposal of hazardous wastes, "multipurpose" technologies such as rotary kiln incinerators will gain more acceptance. In addition, the utilization of rotary kiln technology may increase significantly if high temperature industrial kiln processes are utilized as a means of hazardous waste disposal. Among those technologies currently being studied are cement and lime rotary kiln systems. A great reduction in cost may be realized by using existing industrial systems for hazardous waste disposal.

Rotary kilns are capable of burning a greater variety of wastes (gases, liquids, solids, and sludges) than liquid injection incinerators. Rotary kilns are often used when the size or the nature of the wastes precludes the use of other types of incinerators. Uniform combustion is achieved by rotation of a long, cylindrical rotating furnace lined with firebrick or other refractory material and mounted at a slight incline. Conventional rotary kilns can be designed for batch processing, but continuous feeding is used if reasonably homogeneous wastes are being treated.

There are two basic types of conventional rotary kiln incinerators. The first consists of a rotary kiln and a secondary combustion chamber. The second type handles large volumes of solid wastes, including entrained liquid. In the second type of rotary kiln, water and volatile organics are first vaporized in the drying section, and the vapors, by–passing the rotary kiln, are fed into a secondary combustion chamber. This type of rotary kiln, with drying section, is required when burning wastes containing large quantities of volatile and combustible materials.

Rotary kiln systems are considered the most versatile of the established incinerator technologies. Liquid, solid, and slurried hazardous wastes may all be burned in rotary kilns, without extensive adaptation of the design for specific waste types.

Rotary kiln systems employ a fairly basic design concept. The typical rotary kiln system involves two–stage combustion of waste materials with primary combustion occurring in the rotary kiln followed by secondary combustion of gaseous by-products. Heat recovery, ash recovery, and air pollution control devices are usually included in the overall system. Combustion by-products are most often scrubbed for both particulate matter and acidic by-products, e.g., HCl. Heat recovery is employed in the majority (70%, according to recent estimates) of cases.

Pretreatment of hazardous wastes is not often required for incineration in a rotary kiln, because of the great versatility of the system. The most common preparatory operations conducted at rotary kiln incinerators include size reduction, mixing of liquid wastes with solid wastes, and chemical neutralization. Wastes with an average heating value of 4,500 Btu/lb are reported adequate to sustain combustion at kiln temperatures between 1600° and 1800°F. In those cases where auxiliary fuel is required, No. 2 fuel oil is used most often. Size reduction of solid wastes, via crushing and grinding operations, is a common preparatory operation. This is often done both to preserve the life span of the kiln refractory lining and to increase the combustion efficiency of the system. Mixing of liquid wastes with solid wastes helps to increase the liquid waste residence time and thus enhance destruction efficiency. Highly corrosive wastes are often neutralized by chemical treatment before being fed to the rotary kiln. This helps preserve the working life of the kiln refractory.

Waste materials, following pretreatment, are fed to the elevated end of the rotary kiln. Waste feed mechanisms employed are typically simple hoppers which feed a regulated amount of material to the kiln. Vendors generally recommend continuous operation of a rotary kiln, although they may be operated intermittently. Waste materials flow through the rotary kiln as a consequence of the rotation and the angle of incineration. The kiln is often designed with baffles, which serve to regulate the flow rate through the unit, generally resulting in increased residence times. The rotation of the kiln serves to enhance the mixing of waste with combustion air and provides continuously renewed contact between waste material and the hot walls of the kiln. Combustion air is fed either concurrently or countercurrently. One feature of a rotary kiln is that it may be operated under substoichiometric (oxygen deficient) conditions to pyrolyze certain wastes.

As combustion of the waste progresses, ash flows to the bottom of the unit and is conveyed to the ash recovery system. Gaseous combustion products are exhausted to the secondary combustion unit.

Secondary combustion generally takes place in a fixed hearth type unit, where gaseous products of combustion, including incompletely combusted waste components, combustible waste products, and fly ash are fired. The gaseous products from the secondary combustion chamber are normally then passed through heat recovery and air pollution control systems, while ash is collected and transported to the ash recovery facility.

Most rotary kiln systems are equipped with a multistage scrubber system to control particulate matter, acid by-products of combustion, and oxides of sulfur and nitrogen. Heat recovery systems are often used not only for the conservation of energy, but also to reduce temperatures to allowable levels prior to introduction to the scrubbers.

Rotary kilns are generally large systems, and thus require a large capital expenditure. Due to their energy requirements, the operating costs associated with rotary kiln systems may also be higher than other incinerator systems. Their versatility may lead, however, to benefits measurable in overall reduced costs for hazardous waste management.

Rotary kilns can also be operated in a slagging mode if the furnace is operated properly, and this could result in an environmentally stable slag, which could be of interest for heavy metal containing wastes.

There are a number of rotary kiln manufacturers, some offering variations of the basic concept. The O'Connor rotary waterwall combustion system is one of the more unique designs in the existing population of municipal waste combustors. Another variant is the rotary reactor which is discussed in the fluidized bed section. A rocking kiln incinerator is similar to a rotary incinerator, however, in this case the waste is moved by rocking the kiln through an arch, instead of a full 360°.

Characteristics Affecting Feasibility:

1. Oversized debris such as large rocks, tree roots, and steel drums are difficult to handle and feed; may cause refractory loss through abrasion. Size reduction equipment such as shredders must be provided to reduce solid particle size.
2. Volatile metals (Hg, Pb, Cd, Zn, Ag, Sn) may result in high metals concentration in flue gas, thus requiring further treatment.
3. Alkali metal salts, particularly sodium and potassium sulfate ($NaSO_2$, KSO_4) can cause refractory attack and slagging at high temperatures. Slagging can impede solids removal from the kiln.
4. Fine particle size of soil feeds such as clay, silts can result in high particulate loading in flue gases due to the turbulence in the rotary kiln.
5. Spherical or cylindrical wastes may roll through the kiln before complete combustion can occur.
6. Operation of the kiln at or near the waste ash fusion temperature can cause melting and agglomeration of inorganic salts.
7. Auxiliary fuel is normally required to incinerate wastes with a heating value of less than 8,000 Btu.

Advantages:

1. Will incinerate a wide variety of liquids, slurries, sludges, tars, or solid wastes, either separately or in combination.
2. Adaptable to a wide variety of feed mechanism designs, including those for containerized wastes.
3. Characterized by high turbulence, thus provides good mixing of waste with combustion air, and good dispersion of waste to increase heat transfer surface area.
4. Can operate at temperatures up to or exceeding 2500°F.
5. Can control residence time by adjusting rotational speed. Thus, slow burning materials may be retained for a very long period of time.
6. Can achieve a turndown ratio (maximum to minimum feed rate) of approximately 2:1.
7. There are no moving parts within the kiln.
8. Continuous ash removal does not interfere with oxidation of wastes.
9. Requires minimal preparation of wastes.
10. Adaptable for use with wet gas scrubbing system.

Limitations:

1. High capital costs for installed system, particularly if secondary

combustion and heat recovery are included.
2. Maintenance costs are high due to refractory lining maintenance and replacement, and repair and maintenance of various rotating parts.
3. High energy costs required due to relatively low thermal efficiency.
4. Limited use for highly corrosive materials which could potentially damage refractory lining.
5. Air in-leakage from end seals is a common operational problem.
6. Some fusable material may collect in kiln.
7. May generate high levels of airborne particulates due to increase turbulence. Requires additional control for particulate matter.

8.13.18 Starved Air (Modular) Combustion

Starved air modular incinerators are used for combusting municipal waste, and are relatively small units ranging from 5 to 100 tons per day capacity. Multiple units can be installed when greater capacity is required.

In this type of system, wastes are pushed through the primary combustion chamber in new facilities by compressor rams, and loaded manually in older facilities. Air is blown up through the waste from below. The devices are known as starved air incinerators because controlled (substoichiometric) quantities of air are introduced into the primary chamber to partially burn the organic material. The partially burned organic compounds are discharged (leave) the primary chamber and enter into the secondary combustion chamber.

The exhaust gases flow from the primary chamber into the secondary chamber where additional air is added. Often auxiliary fuel is added to aid in the complete destruction of all of the unburned material in the gas. As in a rotary kiln, the gases leaving a modular starved air incinerator may pass through heat recovery equipment and/or air cleaning equipment, or may be emitted directly into the atmosphere. This type of incinerator has uncontrolled particulate emission levels of about 0.1 gram per day, standard cubic feet, if the unit is well-designed, well-operated, and well-controlled.

Primary air supplied to the first stage is approximately 40% of that required to theoretically combust the waste, causing it to function essentially as a gasifier. Air velocity through the chamber is low, minimizing the amount of fly ash that is entrained and carried into the second chamber.

Starved air units require close control of air injection into the furnace, particularly into the primary chamber. Any uncontrolled source of air (such as through the charging door or around the ash discharge ram) will increase burning and will reduce the heating value of the gas stream entering the secondary chamber. Normally, the secondary chamber is provided with burners sized to burn out flue gas with a significant organic content. It will not have sufficient heat release to effectively burn out gas without significant organic content. The more air introduced into the primary chamber, the poorer the waste destruction in the primary chamber and the more difficult it is to destroy organics in the secondary chamber.

Advantages:

1. Potential for by-product recovery.
2. Reduction of sludge volume without large amounts of supplementary fuel.
3. Thermal efficiency is higher than for normal incineration due to the lower

quantity of air required for this process.
4. Reduced air emissions are sometimes possible.
5. Converts carbonaceous solids into a gas which is more easily combustible.
6. Allows for the suppression of particulate emissions.
7. Allows for some treatment of the hot fuel gas stream prior to combustion to suppress the formation of acid gases.
8. Fast construction time.
9. Relatively low cost.
10. Flexibility.

Disadvantages:

1. Potential source of carcinogenic decomposition product formation.
2. Not capable of functioning very well on sludgy or caking material alone unless cake-breaking capabilities are included in the design.
3. Limited size.
4. Lower thermal efficiency.
5. Higher maintenance costs.
6. Shorter equipment life.

8.13.19 Steam Cracking

In the Neostar process (France), steam for cracking PCBs is produced at 2900°C by reacting hydrogen and oxygen over a burner in a refractory-lined furnace. The steam is fed to the reactor which operates under atmospheric pressure at 1500°C. Preheated liquid PCBs are injected into the reactor. The process breaks down PCBs to by-products of chlorine and a mixed stream of methane, ethane, and other substances that can be disposed of easily. The process does not form dioxins and furans because the PCB molecules are broken up using high-temperature, high-pressure steam without introducing oxygen into the reaction chamber. In a newly constructed pilot plant, the cracked products are separated, with the chlorine neutralized in caustic soda and the hydrocarbon stream recycled to feed the burner. In a commercial unit, the chlorine could be used to produce hydrochloric acid.

8.13.20 Submerged Quench Combustion

This equipment is manufactured by T-Thermal, Inc., in which the incinerator chamber of the SQl is a vertical cylinder instead of horizontal as is common for most other incinerator designs. The burner and waste injectors are located at the top of the chamber and are downfired. This orientation allows the salts, which are molten liquids at typical incineration temperature, to flow down the chamber walls carrying other inorganic metals with them.

The outlet of the incinerator chamber is the submerged quench system. The submerged quench is a unique design that not only cools the gases, but also provides for excellent mass transfer, thereby lowering the demands on the downstream pollution control system to neutralize acids and remove particulates. The hot corrosive gases and molten salts enter the quench via a "downcomer"—a metal tube extending into the quench water bath.

The bottom of the downcomer is open, allowing the salts to drop into the quench tank solution and redissolve. The quench solution is a concentrated salt solution to which caustic is added to react with the acid gases. The gases exit the downcomer through holes in its side.

8.13.21 Supercritical Water Oxidation

The supercritical water oxidation process is basically a high–temperature, high–pressure wet air oxidation. The unique properties of water in the supercritical region causes it to act as an excellent nonpolar solvent for nearly all organic materials. Aqueous solutions or slurries (organic content greater than 5%) are mixed with high pressure oxygen (3,200 to 3,600 psi or greater than 218 atm), to chemically oxidize waste in less than one minute at greater than 99.99% efficiency.

Two processing approaches have been evaluated, an above ground pressure vessel reactor (Modar) and the use of an 8,000 to 10,000 ft deep well as a reactor vessel (Vertox). The supercritical water process is best suited for large volume (200 to 1,000 gpm) dilute (in the range of 1 to 10,000 mg/ℓ COD) aqueous wastes that are of a volatile nature and that have a sufficiently high heat content to sustain the process. In many applications, high Btu, nonhazardous waste can be mixed with low Btu hazardous waste to provide the heat energy needed to make the process self sustaining. Emissions/residues include gaseous effluent (nitrogen and carbon dioxide), precipitate of inorganic salts and the liquid containing only soluble inorganic acids and salts. The advantages are rapid oxidation rates, complete oxidation of organics, efficient removal of inorganics and no off–gas processing is required.

The supercritical water (SCW) oxidation process utilizes the properties of water at pressures greater than 218 atmospheres combined with temperatures above 374°C to effect oxidation of organics. Above these temperatures and pressures, water is in its supercritical state and exhibits solubility characteristics which are the inverse of normal liquid water properties. Thus, organics become almost completely soluble and inorganic salts become only sparingly soluble and tend to precipitate.

Initially in the process developed by Modar, the waste (in the form of an aqueous solution or slurry) is pressurized and heated to supercritical conditions by mixing it with recycled reactor effluent. Compressed air is also mixed with the feed to serve as source of oxygen for the reactions. Oxygen and air are miscible with water under supercritical conditions, thereby enabling the homogeneous operation of the process. The homogenized mixture is then pumped to the oxidizer where organics are rapidly (residence times average 1 minute) oxidized. Oxidation is achieved under homogeneous conditions single–phase supercritical fluid) and therefore higher effective oxygen concentrations and destruction efficiencies can be achieved with shorter residence times than with other similar processes, i.e., the wet oxidation process.

The release of combustion heat from the oxidation reactions causes temperatures in the oxidizer reactor to rise to 1112° to 1202°F. The reactor effluent then enters a cyclone (solids separator) where inorganic salts are precipitated out (at temperatures above 450°C). The fluid effluent of the solid separator consists of superheated, supercritical water, nitrogen, and carbon dioxide. A portion of the superheated, supercritical water is directed to an eductor so that it can be recycled to heat the incoming waste feed (initial step in the process). Modar, Inc. suggests that the remaining effluent, which consists of a high–temperature, high–pressure fluid, can be cooled to subcritical temperatures in a heat exchanger and the resulting steam can be used with turbines to generate energy. However, the cost–effectiveness of the turbine power generation system is limited to certain cases.

The supercritical oxidation process results in conversion of carbon and hydrogen compounds from the organic compound to CO_2 and H_2O. Chlorine atoms are converted to chloride ions and can be precipitated as sodium chloride with the addition of basic materials to the feed. Gaseous emissions consist primarily of carbon dioxide with smaller amounts of oxygen and nitrogen gas, which do not require auxiliary treatment for off–gases. Solid emissions consist of precipitated inorganic salts (chlorine produces chloride salts, nitro compounds precipitate as nitrates, sulfur compounds as sulfates, and phosphorous compounds as phosphates). The liquid effluent consists of a purified water stream, which can be used for process water.

Certain restrictions exist concerning the types of waste that can be treated using the supercritical water oxidation system. These restrictions are:

1. Organic concentrations need to be less than 20% by weight in order for the process to be cost–effective; higher concentrations can be diluted by mixing with dilute wastewater or with pure water.
2. The waste needs to be in the form of an aqueous solution or slurry. Solids can be mixed with water to form a slurry.
3. Costs are higher if the waste has a fuel value of greater than 1,750 Btu/lb, a value equivalent to that exhibited by a waste consisting of 10% by weight of benzene or its equivalent. This is the optimal heat for achieving a reactor exit temperature of 600° to 650°C. Wastes with greater than a 10% benzene equivalent should be diluted, and fuel should be added to wastes with less than a 10% benzene–equivalent.

Oxidyne has proposed to conduct wet air oxidation and supercritical water oxidation in reactors which are placed underground in deep, well–like cavities. The process has been referred to as "downhole" oxidation.

Advantages:

1. Very high destruction of organic compounds including chemically–stable materials (such as PCBs).
2. The oxidation can be energy self–sufficient with as little as 2% organic concentration.
3. The process operates at conditions below which oxides of nitrogen are formed.
4. The process is self–scrubbing.
5. Organic contaminants such as salts and metals are separated from contaminated waste streams thus substantially reducing waste volumes.
6. The process provides for beneficial recovery of valuable components of the waste material, i.e., water for reuse, energy, and in some cases inorganic salts.
7. The system is closed loop and any process upset shuts the system down and the effluents are contained, not released to the environment.

Limitations:

1. Applicable to only organics in liquids or slurried solids.
2. The high temperatures and pressures require sophisticated equipment and operational techniques.

3. Corrosion requires the use of exotic metals.
4. Other problems include heavy metal speciation, ash leachability, charring, and encrustation.

8.13.22 Thermal Gas-Phase Reduction

Thermal gas-phase reduction of organic hazardous waste is an alternative to incineration suitable for processing aqueous waste such as harbor sediment, lagoon sludges, and landfill leachate. The reaction is conducted in a hydrogen-rich reducing atmosphere at approximately 900°C and atmospheric pressure. The products of the reaction depend on the waste constituents but usually include HCl from the reduction of chlorinated organics such as polychlorinated biphenyls (PCBs) and methane and ethylene from reduction of straight-chain and aromatic hydrocarbons. The absence of free oxygen in the reactor prevents the formation of dioxin compounds. The process is being developed by EcoLogic International.

The process is based on the gas-phase thermo-chemical reaction of hydrogen with organic and chlorinated organic compounds at elevated temperatures. AT 850°C or higher, hydrogen reacts with organic compounds in a process known as reduction to produce smaller, lighter hydrocarbons. In the case of chlorinated organic compounds, such as polychlorinated biphenyls (PCBs), the products of the reaction include hydrogen chloride, methane and ethylene. This reaction is enhanced by the presence of water, which can also act as a reducing agent. Bench-scale testing with trichlorobenzene (half of a PCB molecule) has shown that the reduction reaction will achieve 99.9999% destruction efficiency or better. The first reaction is the dechlorination and dismantling of a PCB molecule to produce hydrogen chloride and benzene. The second is the reduction of benzene to produce ethylene. The third is the reduction of straight-chain hydrocarbons to produce methane, and the fourth is the reduction of a polyaromatic hydrocarbon (PAH) compound, phenanthrene, to produce ethylene.

Because the process is not an incinerator, the reactor does not require a large volume for the addition of combustion air. The small reactor size and the capability to recirculate product gases from the reaction make the process equipment small enough to be mobile. As well, the smaller size reduces the capital cost of the process equipment. The main processing costs are for hydrogen, electricity, and personnel.

The thermo-chemical reaction takes place within a specially designed reactor. In the process, a mixture of preheated waste and hydrogen is injected through nozzles mounted tangentially near the top of the reactor. The mixture swirls around a central ceramic tube past glo-bar heaters. By the time the mixture passes through the ports at the bottom of the ceramic tube, it has been heated to 850°C. Particulate matter up to 5 millimeters in diameter not entrained in the gas stream will impact the hot refractory walls of the reactor.

Organic matter associated with the particulate is volatilized, and the particulate exits out of the reactor bottom to a quench tank, while finer particulate entrained in the gas stream flows up the ceramic tube into an exit elbow and through a retention zone. The reduction reaction takes place from the bottom of the ceramic tube onwards, and takes less than one second to complete. Gases enter a scrubber where hydrogen chloride fine particulates are removed. The gases that exit the scrubber consist only of excess hydrogen, methane, and a small amount of water vapor. Approximately 95% of this gas is recirculated back into the reactor. The remaining 5% is fed to a boiler where it is used as supplementary fuel to preheat the waste.

8.13.23 Thermocatalytic Conversion

The efficient utilization of waste produced during food processing operations is a topic of growing importance to the industry. While incineration is an attractive option for wastes with relatively low ash and moisture contents (i.e., under about 50 wt % moisture), it is not suitable for wastes with high moisture contents. Cheese whey, brewer's spent grain, and fruit pomace are examples of food processing wastes that are generally too wet to burn efficiently and cleanly. Pacific Northwest Laboratory (PNL) is developing a thermocataytic conversion process that can convert high–moisture wastes (up to 98 wt % moisture) to a medium–Btu fuel gas consisting primarily of methane and carbon dioxide. At the same time, the COD of these waste streams is reduced by 90 to 99%. Organic wastes are converted by thermocatalytic treatment at 350° to 400°C; and 3,000 to 4,000 psig. The process offers a relatively simple solution to waste treatment while providing net energy production from wastes containing as little as 2 wt % organic solids (this is equivalent to a COD of approximately 25,000 mg/ℓ).

8.13.24 Vortex/Rotary Hearth

Vortex: This is a relatively simple unit for liquids and occupies little area for the throughput achieved. A high–velocity air jet atomizes the feed and causes a spiral–flame effect. This flame insures a long residence time which generally assures fairly high combustion efficiency. Most operating problems are caused by refractory failure due to inadequate temperature control. This is caused by variations in feed quality or slagging. Erosion through impingement of materials on the internals is also a major problem. Incineration of halogenated materials requires efficient gas treatment.

Rotary Hearth: This unit is a slowly rotating, refractory–lined chamber. The design is similar to a vortex incinerator except that it is horizontal and rotates. The advantage over a vortex incinerator is that the rotary hearth can handle solids. Problems with linings similar to those of vortex incinerators can occur.

8.13.25 Wet Air Oxidation

Wet oxidation refers to processes for oxidizing suspended and dissolved organics in aqueous waste streams. The process operates on the principle that the rate of oxidation of organic compounds is significantly increased at high pressures. The oxidation reaction usually occurs at 350° to 625°F. Water moderates the reaction by removing excess heat by evaporation. Required oxygen is provided by an oxygen–containing gas, usually air, that passes through the liquid. Wet oxidation processes are very effective in detoxifying aqueous waste streams that are too dilute to incinerate economically, yet too toxic to treat biologically.

Three different wet oxidation processes have been described. They are (1) wet air oxidation; (2) catalyzed wet oxidation; and (3) supercritical water oxidation (described earlier).

Zimpro/Passavant, Inc. has developed a wet air oxidation process for the treatment of hazardous wastes. This process breaks down hazardous compounds to carbon dioxide and innocuous end products.

The wet air oxidation process features a high pressure pump for wastewater feed, an air compressor to supply air required for oxidation, a process heat exchanger with or without auxiliary heater to preheat wastewater, a reactor chamber for oxidation of wastewater, a cooler, a gas separator for depressurization and removal of off-gas, and an off-gas treatment system.

The wastewater liquid or slurry is brought to system pressure (300 to 3,000 psig) using the high pressure pump. Air from the compressor is added to the pressurized waste stream after the high pressure pump.

Preheat is necessary in order to raise the temperature of the wastewater-air mixture such that when the mixture enters the reactor vessel, the exothermic heat of reaction will raise the mixture temperature to the desired maximum. Preheat can be accomplished using an external source of heat or using the reactor effluent. Start-up energy is provided by the external heat source to the preheater or the auxiliary heater.

Oxidation is brought about by combining the wastewater with a gaseous source of oxygen (usually air) at temperatures in the range of 350° to 620°F. The enhanced solubility of oxygen in aqueous solutions at elevated temperatures provides a strong driving force for oxidation. The reactor serves to provide residence time for oxidation reaction. The temperature of the wastewater-air mixture rises as the reaction occurs.

The reactor effluent is cooled against cooling water or against the wastewater-air mixture. Cooling is usually to about 95° to 135°F.

The pressure of the oxidized liquor-spent air mixture is reduced through a control valve. The gas phase is disengaged from the liquid phase in a separator vessel.

The off-gas from a wet oxidation system is usually treated to reduce the concentration of hydrocarbons. Water scrubbing, which is commonly used to cool the gas stream, results in some reduction of hydrocarbons. Absorption columns using activated carbon provide more organics emissions reduction. Afterburning provides the most complete reduction in organic emissions.

A wide variety of hazardous wastes are acceptable feed materials for treatment in the wet air oxidation process. The feed materials include most EPA designated hazardous waste containing oxidizable materials, either of an organic or inorganic nature, which are soluble, colloidal, or suspended in an aqueous medium. Most waste streams can be treated by wet air oxidation process without any pretreatment as long as it is pumpable. Dilution may be necessary to ensure that excessive evaporation does not occur. Oxidation is brought about at temperatures in the range from about 350° to 620°F. Typical pressure ranges are from 0 to 3,000 psig. Residence time in the reactor for complete oxidation varies from 15 to 120 minutes. Usually the system is self-sustaining, except the start-up energy is provided by an external preheater.

Advantages: The developer states the following advantages:

1. The process is thermally self-sustaining when the amount of oxygen uptake is in the 15 to 20 g/ℓ range;
2. Condensed phase processing requires less equipment volume than gas phase processing, and the products of wet air oxidation stay in the liquid phase;
3. Off-gas from this process are free of nitrogen oxides, sulfur dioxide and

particulates;

4. This process can detoxify most of the EPA priority pollutants; and
5. It is well suited for wastes that are too dilute to incinerate economically, yet too toxic to treat biologically.

Limitations: Treatment of hazardous wastes by wet air oxidation is limited to waste streams containing oxidizable organic and inorganic compounds. The desired oxidation or destruction efficiency should be obtained within the temperature range of 325° to 620°F. Reduced destruction levels of oxidation resistant compounds, e.g., PCBs and chlorinated aromatics in the wet air oxidation process may limit the application of this technology. However, waste streams which are resistant to conventional wet air oxidation may be amenable to treatment by this technology using catalyst or by wet air oxidation at higher temperatures, e.g., supercritical wet air oxidation.

Catalyzed Wet Oxidation: Conventional (uncatalyzed) wet air oxidation achieves oxidation at very high temperatures and pressures. The advanced wet air oxidation process uses catalysts which can result in complete destruction of organic compounds at less severe temperatures and pressures.

International Technology Corporation (IT Corp.), Knoxville, Tennessee has developed a proprietary catalyzed wet oxidation process for destruction of both originally contaminated aqueous waste and organic residues. This process is based on the U.S. patent, originally assigned to the DOW Chemical Company and assigned to IT Enviroscience for development and commercialization. Research on the catalyzed wet air oxidation technology was supported by the U.S. EPA, Cincinnati, Ohio.

In the simplest form aqueous wastes or organic residues are pumped into a continuously stirred tank reactor (CSTR) containing a catalyst solution of bromide, nitrate and manganese. Air is pumped into the reactor. Organics are oxidized with the heat of the reaction driving off water. The only materials to leave the reactor are carbon dioxide, nitrogen, water vapor, excess air, any volatile organics formed, and any solids formed. Water is condensed and returned to the reactor, if necessary, as are condensable organics. Any inorganic salts or acids formed are removed by treatment of a closed stream of catalyst solution. Such treatment is individually designed, utilizing conventional technologies such as filtration or distillation. The vent gases from the reactor are treated by techniques such as absorption or scrubbing. The most important features of this process concept are that nonvolatile organics remain in the reactor until destroyed, and that there is no aqueous bottoms product. Therefore, very high destruction efficiencies and low reactor effluent concentrations are not required in reactor design. As long as the organics remain in the reactor, they will be destroyed ultimately.

The process design for destroying both aqueous waste and organic residues centers on utilizing the homogeneous catalysts in the continuously stirred reactor tank. The two variations of the basic reactor concept, one for the aqueous waste and one for organic residues, are different in the amount of water that is used. For dilute aqueous wastes, there is insufficient energy released by the oxidation of the organics to remove all of the incoming and formed water, and so it is necessary to recovery the catalyst solution for reuse. The basic recovery concept has the catalyst recovered by evaporation/concentration. The evaporation recovery process requires that all the water entering the system be vaporized. This water vapor is condensed and discharged separately from the off–gas. Auxiliary heat must be supplied from streams containing less than about 4% organics depending on the heat of combustion for the specific organics being oxidized. Evaporation must normally be used to supply sufficient heat transfer area. For wastes containing high levels of organics, no evaporator would be required.

The continuous process concept for treating nonaqueous organic wastes is substantially different from the process concept described above since the quantity of water which must be removed from the process is very low. In many cases, only the water formed as by-product from the oxidation reaction plus any amount of water entering with the compressed air must be considered. The only stream normally leaving the process is the off-gas containing principally nitrogen, unused oxygen, carbon dioxide, low levels of water vapor, traces of catalysts, and traces of volatile inorganic HCl and organic species which could be present in the reactor mixture. The heat generated from the oxidation must be removed by condensing and refluxing the water vapor leaving the reactor. This heat could be recovered by operating the reflux condenser as a steam generator. The catalyst is contained in the reactor and evaporation for catalyst recovery is not required.

Advantages: In comparison to straight wet oxidation, the catalyzed process achieves high levels of destruction with a variety of organic chemicals at significantly lower temperatures and pressures. It also produces no aqueous bottom products; all nonvolatile organics stay in the reactor system until oxidized. The unique homogeneous catalyst system enables it to treat water-insoluble compounds. The catalyzed wet oxidation process can also oxidize substances such as PCBs and TCDD which are not generally oxidized by conventional wet oxidation processes.

In comparison to incineration of hazardous wastes or aqueous wastes, the catalyzed wet oxidation process has several advantages. Little or no added energy is required; no auxiliary fuel is consumed. The catalyzed wet oxidation process operates at low temperatures and pressures. The vent gas volume and vent gas scrubber effluent are low in relation to an incinerator, and are readily adaptable for treatment if required for control of trace toxic releases.

Limitations: The process is best suited only for a select type of waste, i.e., moderate strength aqueous waste having high toxicity. Addition of catalysts increases costs of process.

Oxidyne has proposed to conduct wet air oxidation and supercritical water oxidation in reactors which are placed underground in deep, well-like cavities. The process has been referred to as "downhole" oxidation. The world's first commercial below ground vertical tube wet oxidation vessel has recently been completed in the Netherlands.

A low-temperature wet oxidation process (PETOX) for treating hazardous and mixed wastes is under development by Delphi Research, Inc. (Albuquerque, New Mexico). The patented oxidation process reacts ferric iron (Fe^{3+}) with organic material in the presence of a metal catalyst to produce ferrous iron (Fe^{2+}) and carbon dioxide (CO_2). The ferrous iron is subsequently oxidized back to Fe^{3+} in the presence of another metal catalyst in an acid solution. Both of these reactions occur within the same vessel. Relatively low oxidation temperatures of 300° to 500°F are attributable to the catalysts.

8.13.26 Others

(a) Barrel Grate Incinerators: A barrel grate incinerator uses a series of rotating barrels to provide movement of the waste materials through the combustion zone.

(b) Batch Feed Incinerators: As the name implies, the batch feed incinerator is non–continuous. Waste materials are fed into the incinerator periodically to allow ample time for combustion. The ash from these systems is removed in batches as well.

(c) Raker Arm Dryer/Incinerator: A raker arm dryer uses several chambers that "rake" waste from one chamber to another vertically from top to bottom.

(d) Ram Feed Incinerators: The ram feed incinerator is similar to the batch incinerator. The differences are in the waste feed and ash removal. In the ram feed incinerator, a ram (usually hydraulic) is used to move waste materials onto the "hearth" or burner grate and subsequent levels of combustion are taking place by "ram feeding," or indexing through the combustion zone. Automatic ash removal is typically used in these systems.

(e) Reciprocating Grate Incinerators: Material in this type of incinerator is moved from the hopper while the grate is stationary. The movement of waste through the unit comes from the reciprocating motion of the stoker bars.

(f) Traveling Grate Incinerators: The traveling grate incinerator uses a continuously moving feeder grate and one or more burner grates. The charge hopper is directly above the feeder grate to dry the solids before combustion.

8.14 VAPOR PHASE DESTRUCTION PROCESSES

8.14.1 Adiabatic Radiant Combustor

Alzeta Corporation developed the adiabatic radiant combustor under sponsorship of the Gas Research Institute (Chicago, Illinois), for the destruction of volatile organic compounds (VOCs).

It avoids the difficulty associated with some other VOC destruction technologies that include relatively long residence time, high temperatures, and turbulence, needed to ensure complete combustion and prevent formation of carbon monoxide (CO). Unfortunately, increasing residence time, temperature, and/or turbulence generally increases formation of NO_x. Furthermore, increases in these parameters affects costs, energy, and space requirements for the combustion process.

The process mixes VOC–laden gases with fuel, then forces the mixture through a porous ceramic burner. Because the individual molecules of the gaseous mixture are forced to pass through the combustion zone as they exit the burner, lower temperatures and residence times are needed for complete combustion. Furthermore, turbulence is essentially nonexistent with this technology, minimizing NO_x emissions. In addition, the burner can be heated to operating temperature in a matter of seconds, a useful feature for batch operations.

Modifications necessary for applying the technology to VOC concentrations high enough not to require additional fuel are also being studied. To further reduce supplemental fuel requirements, Alzeta has also developed a catalytic unit that can be used separately or in conjunction with the adiabatic radiant combustor.

Advantages:

1. Greater than 99.9% destruction and removal efficiency (DRE) of both chlorinated and non–chlorinated hydrocarbons;
2. NO_x and CO levels below 10 ppm (referenced to 3% oxygen);

3. Heating to acceptable operating temperature from a "cold" start in approximately 2 seconds, making the technology particularly well suited for batch operations;
4. Relatively small unit size;
5. Low fuel requirements; and
6. Automated operation.

8.14.2 Adsorption/Incineration Process

Equipment has been developed that combines activated–carbon adsorption with incineration. The adsorber concentrates the organic–laden air before treatment by incineration. This approach is particularly useful for organic streams with low concentrations and high volumes concentrations less than 100 ppm and flowrates over 20,000 cfm), such as paint spray booths. Combination systems provide the inherent advantages of the individual techniques—the high destruction efficiency and no generation of liquid or solid waste of incineration, and the low fuel consumption and good control efficiency of adsorption—without many of the disadvantages of each system.

The Catalytica, Inc. system which uses catalytic incineration in the second stage, mentions that the process is especially suitable for small companies. In laboratory tests, 96% of VOCs has been oxidized. Catalytica expects to achieve 99% destruction with a prototype unit that it is building under a grant from the U.S. EPA. It will handle gas streams of up to 10,000 ft^3/min and up to 2,500 ppm of VOCs.

8.14.3 Afterburners

For incinerators, afterburners are employed to destroy gaseous hydrocarbons not destroyed in the incinerator. Three types of afterburners utilized are: (1) direct flame, (2) thermal, and (3) catalytic. Direct flame and thermal afterburners are similar, but they destroy organic vapors by different methods. A high percentage of the vapors pass directly through the flame in a direct flame unit. In a thermal unit the vapors remain in a high temperature oxidizing atmosphere long enough for oxidation reactions to take place. Catalytic devices incorporate a catalytic surface to accelerate the oxidation reactions.

Thermal afterburners are usually an integral part of rotary kilns used in hazardous waste incineration. Thermal afterburners are also used with: liquid injection incinerators in a few instances; pyrolysis units when chemicals are not being recycled; and co–incineration units where the incinerator used normally requires an afterburner. Catalytic afterburners are a proven technology for nonhazardous gaseous material.

Thermal afterburners are suitable for any gaseous material that is also suitable for incineration or which has been produced by auxiliary equipment, i.e., a rotary kiln. Catalytic afterburners are applicable to the destruction of combustible materials in low concentrations (they are not applicable to chlorinated hydrocarbons due to the HCl formation).

From a chemical viewpoint, two main types of reactions occur in afterburner systems: oxidation and pyrolysis reactions. In general, the detailed mechanisms for the oxidation and pyrolysis of even the simplest organic compounds are not completely understood, but it is well established that the reactions occur in many complicated sequential and concurrent steps involving a multitude of chemical intermediates.

An auxiliary fuel is fired to supply the heat to warm the gases in a temperature that will promote oxidation of the organic vapors. Usually a portion of the gas stream supplies the oxygen necessary for organic vapor oxidation. Both gaseous and liquid fuels are used to fire afterburners. Gaseous fuels have the advantage of permitting firing in multiple jet (or distributed) burners. Oil firing has the disadvantage of producing sulfur oxides (from sulfur in the oil) and normally produces higher nitrogen oxides emissions.

Catalytic afterburners are applied to gaseous wastes containing low concentrations of combustible materials and air. Usually noble metals such as platinum and palladium are the catalytic agents. A catalyst is defined as a material which promotes a chemical reaction without taking a part in it. The catalyst does not change nor is it used up. However, it may become contaminated and lose its effectiveness.

Generally, catalytic afterburners are considered for operation with waste containing hydrocarbon levels that are less than 25% of the lower explosive limit. When the waste gas contains sufficient heating value to cause concern about catalyst burnout, the gas may be diluted by atmospheric air to ensure operating temperatures within the operating limits of the catalyst. Burned gases are discharged through a stack to the atmosphere if they are not sent to a waste heat recovery unit.

Advantages:

Thermal or Direct Flame

1. Destroys those pollutants that were not destroyed in the primary incineration.
2. Allows more flexibility in incinerator operation.

Catalytic

1. Carries out combustion at relatively low temperatures (more economical to operate than other afterburners).
2. Clean heated gas produced is well suited for waste heat recovery units.

Disadvantages:

Thermal or Direct Flame

1. Auxiliary fuel requirements.
2. Afterburner costs.

Catalytic

1. Burnout of the catalyst occurs at temperatures exceeding 1500°F.
2. Catalyst systems are susceptible to poisoning agents, activity suppressants, and fouling agents.
3. Occasional cleaning and eventual replacement of catalyst is required.
4. Maintenance costs are high.

8.14.4 Catalytic Vapor Incineration

Catalytic incineration is an air pollution control technique whereby VOCs in an emission stream are oxidized with the help of a catalyst. A catalyst is a substance that accelerates the rate of a reaction at a given temperature without being appreciably changed during the reaction. Catalysts typically used for VOC incineration include platinum and palladium; other formulations are also used, including metal oxides for emission streams containing chlorinated compounds. The catalyst bed (or matrix) in the incinerator is generally a metal mesh–mat, ceramic honeycomb, or other ceramic matrix structure designed to maximize catalyst surface area. The catalysts may also be in the form of spheres or pellets. Before passing through the catalyst bed, the emission stream is preheated, if necessary, in a natural gas–fired preheater and/or via heat exchange with the flue gas.

Recent advances in catalysts have broadened the applicability of catalytic incineration. Catalysts now exist that are relatively tolerant of compounds containing sulfur or chlorine. These new catalysts are often single or mixed metal oxides and are supported by a mechanically strong carrier. A significant amount of effort has been directed towards the oxidation of chlorine–containing VOCs. These compounds are widely used as solvents and degreasers, and are often encountered in emission streams. Catalysts such as chrome/alumina, cobalt oxide, and copper oxide/manganese oxide have been demonstrated to control emission streams containing chlorinated compounds. Platinum–based catalysts are often employed for control of sulfur containing VOCs but are sensitive to chlorine poisoning.

Despite catalyst advances, some compounds simply do not lend themselves well to catalytic oxidation. These include compounds containing atoms such as lead, arsenic, and phosphorus. Unless the concentration of such compounds is sufficiently low, or a removal system is employed upstream, catalytic oxidation should not be considered in these cases.

The performance of a catalytic incinerator is affected by several factors including: (a) operating temperature, (b) space velocity (reciprocal of residence time), (c) VOC composition and concentration, (d) catalyst properties, and (e) presence of poisons/inhibitors in the emission stream. In catalytic incinerator design, the important variables are the operating temperature at the catalyst bed inlet, the temperature rise across the catalyst bed, and the space velocity assuming adequate oxygen is present. The operating temperature for a particular destruction efficiency is dependent on the concentration and composition of the VOC in the emission stream and the type of catalyst used.

In a catalytic incinerator, the vent gas is introduced into a mixing chamber where it is heated to approximately 320°C (~600°F) by the hot combustion products of the auxiliary burners. The heated mixture then passes through the catalyst bed. Oxygen and organics diffuse onto the catalyst surface and are adsorbed in the pores of the catalyst. The oxidation reaction takes place at these active sites. Reaction products are desorbed from the active sites and diffuse back into the gas. The combusted gas can then be routed through a waste heat recovery device before exhausting into the atmosphere.

Combustion catalysts usually operate over a temperature range of 320° to 650°C (600° to 1200°F). Lower temperatures can slow down or stop the oxidation reaction. Higher temperatures can shorten the life of the catalyst or evaporate the catalyst from the inert substrate. Vent gas streams with high organic concentrations can result in temperatures high enough to cause catalyst failure. In such cases, dilution air may be required. Accumulations of particulate matter, condensed organics, or polymerized hydrocarbons on the catalyst can block the active sites and reduce efficiency. Catalysts can also be deactivated by compounds containing sulfur, bismuth, phosphorus, arsenic, antimony, mercury, lead, zinc, tin, or halogens. If these compounds deactivate the catalytic unit, organics will pass through unreacted or be partially oxidized to form compounds (aldehydes, ketones, and organic acids) that are highly reactive atmospheric pollutants that can corrode plant equipment. As a result, gases containing compounds with chlorine, sulfur, and other atoms that may deactivate the supported noble metal catalysts often used for VOC control were not suitably controlled by catalytic oxidation systems. Therefore, the use of catalytic oxidation for control of gaseous pollutants has generally been restricted to organic compounds containing only carbon, hydrogen, and oxygen.

The sensitivity of catalytic oxidizers to organic inlet stream flow conditions, their inability to handle high organic concentration off–gas streams, the sensitivity of the catalyst to deactivating compounds, and their higher cost for destruction efficiencies comparable to thermal oxidizers may limit the application of catalytic units for control of organics from process vent streams.

However, newer developments are helping to overcome these problems, since catalytic incineration is a more energy efficient process than thermal incineration whose temperatures of about 1800°F are required.

The major advantages of catalytic oxidation are:

1. Lower fuel consumption
2. Lower NO_x emissions
3. Lower CO emissions
4. Lower CO_2 emissions
5. Potential for lower costs as compared to carbon adsorption, and incineration.

8.14.5 Flares

Open flames used for disposing of waste gases during normal operations and emergencies are called flares. They are typically applied when the heating value of the waste gases cannot be recovered economically because of intermittent or uncertain flow, or when the value of the recovered product is low. In some cases, flares are operated in conjunction with baseload gas recovery systems, e.g., condensers. Flares handle process upset and emergency gas releases that the baseload system is not designed to recover.

Several types of flares exist, the most common of which are steam–assisted, air–assisted, and pressure head flares. Typical flare operations can be classified as "smokeless," "nonsmokeless," and "fired" or "endothermic." For smokeless operation, flares use outside momentum sources (usually steam or air) to provide efficient gas/air mixing and turbulence for complete combustion. Smokeless flaring is required for destruction of organics heavier than methane. Nonsmokeless operation is used for organic or other vapor streams which burn readily and do not produce smoke. Fired, or endothermic, flaring requires additional energy in order to ensure complete oxidation of the waste streams such as for sulfur tail gas and ammonia waste streams.

In general, flare performance depends on such factors as flare gas exit velocity, emission stream heating value, residence time in the combustion zone, waste gas/oxygen mixing, and flame temperature.

A steam–assisted flare, an air–assisted flare, and a flare with no assist are considered to be capable of achieving 98 wt % emission reduction.

If conditions in the flame zone are optimum (oxygen availability, adequate residence time, etc.), the VOC in the emission stream may be completely burned (~100% efficiency). In some cases, it may be necessary to add supplementary fuel (natural gas) to the emission stream to achieve destruction efficiencies of 98% and greater if the net heating value of the emission stream is less than 300 Btu/scf.

Flares can also be classified as: (1) elevated, and (2) ground-level.

Process off-gases are sent to the flare through the collection header. The off-gases entering the header can vary widely in volumetric flow rate; moisture content, organic concentration, and heat value. The knock-out drum removes water or hydrocarbon droplets that could create problems in the flare combustion zone. Off-gases are usually passed through a water seal before going to the flare. This prevents possible flame flashbacks, caused when the off-gas flow to the flare is too low and the flame front pulls down into the stack.

Purge gas (nitrogen, carbon dioxide, or natural gas) also helps to prevent flashback in the flare stack caused by low off-gas flow. The total volumetric flow to the flame must be controlled carefully to prevent low-flow flashback problems and to avoid a detached flame (a space between the stack and flame with incomplete combustion) caused by an excessively high flow rate. A gas barrier or a stack seal is sometimes used just below the flame head to impede the flow of air into the flare gas network.

The organic vapor stream enters at the base of the flame where it is heated by already burning fuel and pilot burners at the flare tip. Fuel flows into the combustion zone where the exterior of the microscopic gas pockets is oxidized. The rate of reaction is limited by the mixing of the fuel and oxygen from the air. If the gas pocket has sufficient oxygen and residence time in the flame zone, it can be burned completely. A diffusion flame receives its combustion oxygen by diffusion of air into the flame from the surrounding atmosphere. The high volume of fuel flow in a flare requires more combustion air at a faster rate than simple gas diffusion can supply, so flare designers add steam injection nozzles to increase gas turbulence in the flame boundary zones, thus drawing in more combustion air and improving combustion efficiency. This steam injection promotes smokeless flare operation by minimizing the cracking reactions that form carbon. Significant disadvantages of steam usage are the increased noise and cost. The steam requirement depends on the composition of the gas flared, the steam velocity from the injection nozzle, and the tip diameter. Although some gases can be flared smokelessly without any steam, typically 0.15 to 0.5 kg of steam per kilogram of flare gas is required.

Steam injection is usually controlled manually with the operator observing the flare (either directly or on a television monitor) and adding steam as required to maintain smokeless operation. Several flare manufacturers offer devices that sense a flare's flame characteristics and adjust the steam flow rate automatically to maintain smokeless operation.

Some elevated flares use forced air instead of steam to provide the combustion air and the mixing required for smokeless operation. These flares consist of two coaxial flow channels. The combustion gases flow in the center channel, and the combustion air (provided by a fan in the bottom of the flare stack) flows in the annulus. The principal advantage of air-assisted flares is that expensive steam is not required. Air assistance is rarely used on large flares because airflow is difficult to control when the gas flow is intermittent. About 597 W (0.8 hp) of blower capacity is required for each 45 kg/hr (100 lb/hr) of gas flared.

Ground flares are usually enclosed and have multiple burner heads that are staged to operate based on the quantity of gas released to the flare. The energy of the gas itself (because of the high nozzle pressure drop) is usually adequate to provide the mixing necessary for smokeless operation, and air or steam assist is not required. A fence or other enclosure reduces noise and light from the flare and provides some wind protection.

The flare is a useful emission control device and can be used for most nonhalogenated organic streams. It can handle fluctuations in organic concentration, flow rate, and inerts content very easily. However, the low volumetric flows typically associated with waste distillation–unit process vents and the low organic concentrations in process vent streams from air strippers are conditions that do not favor the use of flares. Flares are best suited and generally designed to control normal operating vents or emergency upsets that release large volumes of gases; and in the case of dilute gas streams, supplemental fuel costs can eliminate flares as a viable control alternative. On the other hand, it is possible (as is done in refineries) to combine a number of process vents in a common gas line, which can be sent to a flare.

8.14.6 Fume Incinerators

These systems are used to burn off vapors prior to emission to the atmosphere where recovery is not desired. Incineration facilities work best on air streams which contain solvents at 25% of their lower explosive limit (LEL). The minimum acceptable concentration is 15% of LEL. (For a typical solvent stream with a 1% LEL, this amounts to a 1,500 to 2,500 ppm concentration.) At 25% LEL, such equipment can provide a high calorific credit for the solvent burned. Incinerators are not efficient at low concentration effluents from, for example, a spray booth. Facilities such as paint or coating bake ovens, where solvent vapor concentrations are high, could profitably utilize fume incinerators.

8.14.7 Internal Combustion Engines

Internal combustion engines (ICEs) have been used for years to destroy landfill gas. The application of this method to hydrocarbon destruction is recent, with the first operational unit having been installed in 1986. Currently, over one hundred of these units are operating in southern California and providing good destruction and removal efficiencies.

The internal combustion engine used for this technique is simply an industrial or automotive engine with its carburetor modified to accept vapors rather than liquid fuel. Virtually any make of engine can be used: Volkswagen, Audi, Ford, Chevrolet and others have all been reported as having been used. The size of the engine (expressed in cubic inches) reportedly greatly affects the flow rate of air through the engine, with larger capacity engines able to handle larger flow volumes.

A second required modification to the engines is the addition of a supplemental fuel input valve. When the intake hydrocarbon concentration is too low to sustain complete combustion, a supplemental fuel source must be added to ensure complete combustion. Propane is the fuel used almost universally, although one vendor reported that tests with natural gas showed greatly reduced (by 50 to 75%) energy costs.

A catalytic converter is an integral component of the system, providing an important polishing step to reach the low discharge levels required by many regulatory agencies. A standard automobile catalytic converter, using a platinum–based catalyst, is normally used.

8.14.8 Silent Discharge Plasma

Researchers at the Los Alamos National Laboratory (Los Alamos, New Mexico) are developing a new, low-temperature plasma process that can be used to treat gaseous incinerator or thermal treatment unit emissions. The technology, called a silent discharge plasma (SDP) device, utilizes a cold plasma process in which gaseous emissions are combusted at low temperatures (from ambient temperatures to about 932°F or 500°C).

As an alternative to incineration, the Los Alamos scientists are proposing to use the SDP technology in conjunction with a high-temperature packed-bed reactor. The packed-bed reactor would destroy liquid organic and/or mixed wastes, and the SDP system would remove any remaining organic compounds from the reactor's off-gas. The system would be classified as a "thermal treatment unit" instead of an "incinerator" for regulatory purposes.

The related methods of corona processing are presently the focus of work at other institutions, particularly for flow gas processing. Both SDP and corona processes are characterized by the production of large quantities of highly reactive free radicals, especially atomic oxygen O(3P) and the hydroxyl OH, in the gaseous medium and their subsequent reaction with contaminants. Corona processing is discussed in Chapter 7.

8.14.9 Thermal Vapor Incineration

Thermal vapor incineration is a controlled oxidation process that occurs in an enclosed chamber. One type of thermal vapor incinerator consists of a refractory-lined chamber containing one or more discrete burners that premix the organic vapor gas stream with the combustion air and any required supplemental fuel. A second type of incinerator uses a plate-type burner firing natural gas to produce a flame zone through which the organic vapor gas stream passes. Packaged thermal vapor incinerators are commercially available in sizes capable of handling gas stream flow rates ranging from approximately 8 to 1400 m^3/min.

The two main types of thermal incinerators employed are thermal recuperative and thermal regenerative. The thermal recuperative type is the most common and nearly always employs a heat exchanger to preheat a gaseous stream prior to incineration. Regenerative type incinerators are newer and employ ceramics to obtain a more complete transfer of heat energy. Boilers and process heaters are also utilized.

Organic vapor destruction efficiency for a thermal vapor incinerator is a function of the organic vapor composition and concentration, combustion zone temperature, the period of time the organics remain in the combustion zone (referred to as "residence time"), and the degree of turbulent mixing in the combustion zone. Test results and combustion kinetics analyses indicated that thermal vapor incineration destroys at least 98% of non-halogenated organic compounds in the vapor stream at a temperature of 870°C and achieves a residence time of 0.75 second. If the vapor stream contains halogenated compounds, a temperature of 1100°C (2000°F) and a residence time of one second is needed to achieve a 98% destruction efficiency.

Incinerator performance is affected by the heating value and moisture content of the organic vapor stream, and the amount of excess combustion air. Combustion of organic vapor streams with a heating value less than 1.9 MJ/m^3 (or vapor concentrations below 12,000 ppm) usually requires the addition of supplemental fuel (also referred to as auxiliary fuel) to maintain the desired combustion temperature. Above this heating value, supplemental fuel may be used to maintain flame stability. Although either natural gas or fuel oil can be used as supplemental fuel, natural gas is preferred. Supplemental fuel requirements can be decreased if the combustion air or organic vapor stream is preheated.

A thermal incinerator handling vent gas streams with varying heating values and moisture content requires careful adjustment to maintain the proper chamber temperatures and operating efficiency. Water requires a great deal of heat to vaporize, so entrained water droplets in a vent gas stream can substantially increase auxiliary fuel requirements because of the additional energy needed to vaporize the water and raise it to the combustion chamber temperature. Combustion devices are always operated with some quantity of excess air to ensure a sufficient supply of oxygen. The amount of excess air used varies with the fuel and burner type, but it should be kept as low as possible. Using too much excess air wastes fuel because the additional air must be heated to the combustion chamber temperature. A large amount of excess air also increases flue gas volume and may increase the size and cost of the system.

The organic destruction efficiency of a thermal oxidizer can be affected by variations in chamber temperature, residence time, inlet organic concentration, compound type, and flow regime (mixing).

8.14.10 Flameless Techniques

Thermatrix Inc. has developed a Flameless Thermal Oxidizer. Its hot ceramic matrix destroys volatile organic compounds (VOCs), such as benzene and carbon tetrachloride, under controlled conditions.

Before the vapor stream is introduced to the TMX reactor, the inert aluminosilicate matrix is preheated to 1600°F (or 1800°F for chlorinated organics), using a gas–fired or electric heater. The heater is then turned off.

The incoming stream is thoroughly mixed (and supplemental fuel or dilution air is added, as needed) at ambient temperature in the inlet region of the vesse.

Exothermic energy released during oxidation maintains the temperature of the ceramic bed. In general, gas streams with a heat content ≥30 Btu/scf are self–sustaining. Leaner streams may require supplemental fuel and heat during operation, while richer streams may call for dilution air for optimum operation.

REFERENCES

1. Arienti, M., et al, *Dioxin–Containing Wastes, Treatment Technologies,* Noyes Data, 1988.
2. Barron, T., *Pyrolysis Struggles to Move Off Back Burner,* Env. Today, 1–2/92.
3. Bonner, T., et al, *Hazardous Waste Incineration Engineering,* Noyes Data, 1981.
4. Breton, M., et al, *Treatment Technologies for Solvent Containing Wastes,* Noyes Data, 1988.
5. Burton, D., et al, *Treatment of Hazardous Petrochemical and Petroleum Wastes,* Noyes Data, 1989.
6. Castaldini, C., et al, *Disposal of Hazardous Wastes in Industrial Boilers and Furnaces,* Noyes Data, 1986.
7. Chementator, *Detonation for CFCs,* Chem. Engr., 8/93.
8. Freeman, H., *Innovative Thermal Hazardous Organic Waste Treatment Processes,* Noyes Data, 1985.
9. Holden, T., et al, *How to Select Hazardous Waste Treatment Technologies for Soils and Sludges,* Noyes Data, 1989.
10. Jackson, A., et al, *Hazardous Waste Treatment Technologies,* Noyes Data, 1991.
11. Lee, C., et al, *Update of Innovative Thermal Destruction Technologies,* AIChE Meeting, 8/88.

12. McGowan, T., et al, *Hazardous Waste Incineration is Going mobile,* Chem. Eng. 10/91.
13. Moreno, F., *Adiabatic Radiant Combustor Offers Advantages for VOC Control,* Air Poll. Cons., 11–12/92.
14. EPA, *Alternative Control Technology Document—Organic Waste Process Vents,* EPA–450/3–91–007, 12/90.
15. EPA, *Control Technologies for Hazardous Air Pollutants,* EPA/625/6–91/014, 6/91.
16. EPA, *A Compendium of Technologies Used in the Treatment of Hazardous Wastes,* EPA/625/8–87/014, 9/87.
17. EPA, *Corrective Action: Technologies and Applications,* EPA/625/4–89/020, 9/89.
18. EPA, *Experience in Incineration Applicable to Superfund Situ Remediation,* EPA/625/9–88/008, 12/88.
19. EPA, *Forum on Innovative Hazardous Waste Treatment Technologies: Domestic and International,* EPA/540/2–89/056.
20. EPA, *Forum on Innovative Hazardous Waste Treatment Technologies: Domestic and International (Third),* EPA/540/2–91/015, 9/91.
21. EPA, *Innovative Treatment Technologies,* EPA/540/9–91/002, 10/91.
22. EPA, et al, *Medical Waste Management and Disposal,* Noyes Data, 1991.
23. EPA, *Municipal Waste Combustor,* EPA/450/3–89–27e, 8/89.
24. EPA, *Pyrolysis Treatment (Engineering Bulletin),* EPA/540/S–92/010, 10/92.
25. EPA, *Reactor Processes in Synthetic Organic Chemical Manufacturing Industry,* EPA/450/2–90/016a, 6/90.
26. EPA, *Remedial Action, Treatment and Disposal of Hazardous Waste, 15th,* EPA/600/9–90/006, 2/90.
27. EPA, *Remedial Action, Treatment and Disposal of Hazardous WAste, 16th,* EPA/600/9–90/037, 9/90.
28. EPA, *Remedial Action, Treatment and Disposal of Hazardous Waste, 17th,* EPA/600/9–91/002, 4/91.
29. EPA, *Risk Reduction Engineering Laboratory Research Symposium,* 18th Annual, EPA/600/R–92/028, 4/92.
30. EPA, et al, *Solvent Waste Reduction,* Noyes Data, 1990.
31. EPA, *Summary of Treatment Technology Effectiveness for Contaminated Soil,* EPA 9355.4–06, 6/90.
32. EPA, *Superfund Innovative Technology Evaluation Program (Fourth),* EPA/540/5–91/008, 11/91.
33. EPA, *Superfund Innovative Technology Evaluation Program (Fifth),* EPA/540/R–92/076, 10/92.
34. Gupta, B., et al, *Data Summary of Municipal Solid Waste Management Alternatives,* NREL, DOE, DE 92016433, 8/92.
35. Hazardous Waste Cleanup Project, *The Limits of Technology in Dealing with Hazardous Waste Site Cleanups,* HWCP, 6/93.
36. Johnson, N., *Transportable Incineration,* Haz. Mat. Control, 3–4/93.
37. Martin, R., *Selecting the Most Appropriate HAP Emission Control Technology,* Air Poll. Cons., 3–4/93.
38. Noyes, R., *Handbook of Pollution Control Processes,* Noyes Data, 1991.
39. Pedersen, T., et al, *Soil Vapor Extraction Technology,* Noyes Data, 1991.
40. Poll. Engr., 4/1/93.
41. Rood, M., *Technological and Economic Evaluation of Municipal Solid Waste Incineration,* OTT–2, Univ. of Illinois, 9/88.
42. Rosengrant, L., et al, *Treatment Technology Background Document,* OSW, EPA, 1/91.
43. Rosocha, L., *Cold Plasma Technology Used to Treat Incinerator Emissions,* Haz. Waste Cons. 9–10/92.
44. Schofield, B., et al, *Use of Chemical Additives to Reduce the Impact of Slag Formation in Hazardous Waste Incineration,* Haz. Mat. Cont., 9–10/92.
45. Tischler, J., et al, *Selecting State–of–the–Art Incinerators for Complex Aqueous Wastes,* Haz. Mat. Cont., 9–10/91.
46. Van Wyk, D., *Thermal Treatment and Heat Recovery Options,* Nat. Env. Jnl., 9–10/93.

INDEX

Other Noyes Publications

POLLUTION PREVENTION TECHNOLOGY HANDBOOK

Edited by

Robert Noyes

This book presents technical information relating to current and potential pollution prevention and waste minimization techniques in 36 industries. Many of these industries have similar problems, and there are many opportunities for **cross-fertilization** in adopting pollution prevention techniques across industry boundaries.

In general each chapter provides for each industry: (1) description of manufacturing processes, (2) types of waste generated, and (3) specific pollution prevention and waste minimization opportunities.

There are a number of benefits involved in adopting pollution prevention techniques, the most important of which is economic. When wastes are reduced or eliminated, substantial cost savings can be realized by reduced expenditures for pollution control equipment, and lower treatment and disposal costs. In some firms, substantial source reduction activities have been implemented with minor capital expenditures, with resultant payback within six months.

Other considerations include lessened liability problems, and improved public image.

The thousands of items of technological advice in this book make it a valuable source of current and potential pollution prevention technology.

CONTENTS

ISBN 0-8155-1311-9 (1993) **7" x 10"** **683 pages**

Other Noyes Publications

HANDBOOK OF LEAK, SPILL AND ACCIDENTAL RELEASE PREVENTION TECHNIQUES

Edited by

Robert Noyes

Leaks, spills, and accidental releases are a major source of release of toxic and hazardous substances to the environment. For example, equipment leaks alone account for 35% of all VOC emissions in the chemical industry.

Leaks, spills, and accidental releases can take many forms including (1) gaseous emissions; (2) liquid releases that allow VOCs to enter the atmosphere, as well as the liquid portion entering the ground; (3) heavier liquid releases that do not vaporize, or have very low volatility; and (4) solid waste emissions mainly in the form of dust.

Fugitive emissions are also distinguished from process (point-source) emissions. The term *fugitive emissions* includes the loss of chemicals through sealing mechanisms separating process fluid from the atmosphere. Examples of fugitive emissions are equipment leaks that come from the hundreds or thousands of valves, pumps, compressors, pressure relief devices, open-ended valves or lines, sampling connection systems, and flanges and other connectors within a processing plant.

The techniques used to control fugitive VOC emissions are quite different from those used to control process emissions, due in large part to the fact that process emissions are generally vented from a definable point or stack, while fugitive emission sources are more diffuse.

Emission sources in the chemical industry can be divided into six source types:

1) Process vent emissions
2) Storage tank emissions and leaks
3) Equipment and piping leak emissions
4) Transfer emissions, leaks and spills
5) Wastewater collection and treatment emissions
6) Waste storage piles

This book is designed to provide technical guidance to prevent leaks, spills or other accidental releases of hazardous substances from fixed facilities that produce hazardous substances, store them, or transfer them to and from transportation terminals. The audience addressed includes managerial and supervisory personnel as well as "hands on" personnel associated with both large and small manufacturers. As an aid to plant engineers and managers, federal workers, fire marshalls, and fire and casualty insurance inspectors, this document is offered as a guide to prevention of leaks, spills, and accidental releases. A **condensed contents** is given below.

ISBN 0-8155-1296-1 (1992) **6" x 9"** **487 pages**

Other Noyes Publications

HANDBOOK OF POLLUTION CONTROL PROCESSES

Edited by

Robert Noyes

This handbook presents a comprehensive and thorough overview of state-of-the-art technology for pollution control processes. It will be of interest to those engineers, consultants, educators, architects, planners, government officials, industry executives, attorneys, students and others concerned with solving environmental problems.

The pollution control processes are organized into chapters by broad **problem areas**; and appropriate technology for decontamination, destruction, isolation, etc. for each problem area is presented. Since many of these technologies are useful for more than one problem area, a specific technology may be included in more than one chapter, modified to suit the specific considerations involved.

The pollution control processes described are those that are actively in use today, as well as those innovative and emerging processes that have good future potential. An important feature of the book is that **advantages** and **disadvantages** of many processes are cited. Also, in many cases, **regulatory-driven trends** are discussed, which will impact the technology used in the future.

Where pertinent, regulations are discussed that relate to the technology under consideration. Regulations are continually evolving, frequently requiring modified or new treatment technologies. This should be borne in mind by those pursuing solutions to environmental problems.

Innovative and emerging technologies are also discussed; it is important to consider these new processes carefully, due to increasingly tighter regulatory restrictions, and possibly lower costs. For some pollutants specific treatment methods may be required; however for other pollutants, specific treatment levels must be obtained.

CONTENTS

In summary, a vast number of pollution control processes and process systems are discussed.

ISBN 0-8155-1290-2 (1991) **7" x 10"** **758 pages**